Progress in Sleep Research

The book series provides an overview of the physiological processes involved in sleep regulation and the essential role of sleep in various vital physiological processes. It also summarizes the different stages of sleep and the clinical features of common sleep disorders. The individual books discuss the neurological and neuropsychological implications of sleep and highlight the relation between sleep and brain structure-functions. It explores the impact of sleep on circadian biology, neuroanatomy, neurophysiology, neuropharmacology, neuroendocrinology, neuroimmunology, neuropathology, basic neurology, biological psychiatry, and the behavioral sciences. It reviews the genetic factors associated with normal sleep and sleep disorders. It also describes the recent advances in research aiming to elucidate the neurochemical mechanisms regulating sleep and wakefulness.

Ahmed S. BaHammam • S. R. Pandi-Perumal •
Haitham Jahrami
Editors

COVID-19 and Sleep: A Global Outlook

Springer

Editors
Ahmed S. BaHammam
The University Sleep Disorders Center
Department of Medicine
College of Medicine, King Saud University
Riyadh, Saudi Arabia

S. R. Pandi-Perumal
Saveetha Institute of Medical and Technical Sciences
Saveetha University
Chennai, Tamil Nadu, India

Haitham Jahrami
College of Medicine and Medical Sciences
Arabian Gulf University
Manama, Kingdom of Bahrain, Bahrain

ISSN 2730-5678 ISSN 2730-5686 (electronic)
Progress in Sleep Research
ISBN 978-981-99-0239-2 ISBN 978-981-99-0240-8 (eBook)
https://doi.org/10.1007/978-981-99-0240-8

This Springer imprint is published by the registered company Springer Nature Singapore Pte Ltd.
The registered company address is: 152 Beach Road, #21-01/04 Gateway East, Singapore 189721, Singapore

This volume is dedicated to our respective families.

Foreword

The COVID-19 pandemic had and continues to have profound effects on the world's health and economy. COVID-related morbidity and mortality, long COVID, COVID variants, and threat of new outbreaks mark the virus' impact on world health, while lockdowns, work stoppages, workdays lost, mobilization of healthcare systems, vaccine development efforts, and supply-chain breakdowns each add to COVID's burden to the world's economy.

The editors of COVID-19 and Sleep: A Global Outlook have chosen to examine the impact of the COVID-19 pandemic through its impact on and interactions with the essential physiological processes of sleep and circadian rhythms. They have assembled an international corps of experts in sleep and circadian rhythm research and medicine to explore sleep/COVID relationships. COVID impacts sleep directly in those who became infected and indirectly both in those who treat COVID patients and those who suffer the impact of the public health measures of lockdown and social distancing. Very often, these sleep disruptions were accompanied by cognitive and functional impairment as well as mental health disturbances such as increased depression or anxiety. Further, regular sufficient sleep promotes immune function and correspondingly disturbed sleep leads to immune function compromise.

COVID-19 and Sleep: A Global Outlook offers a comprehensive and authoritative exploration of the interactions of sleep and circadian rhythms and COVID-19. These explorations address a wide range of important topics, including, among others, sleep and circadian rhythm and immune system interactions, impact of COVID-19 and resultant lockdowns on sleep in various populations, interactions of COVID-19 with various sleep disorders, long-term impact of COVID-19 on sleep and circadian rhythms, interventions to improve sleep during COVID-19, and impact of the pandemic on the practice of sleep medicine.

The occurrence of another COVID pandemic is a potential threat to world health for which we must be prepared. Much has been learned from the world response to

COVID-19 that will allow for such a future pandemic to be better faced. COVID-19 and Sleep: A Global Outlook provides an outstanding compendium of the lessons learned, knowledge gained, and directions for future research concerning the role of sleep during such a world health crisis.

Psychiatry and Behavioral Sciences,
University of Washington, Seattle, WA,
USA

Michael V. Vitiello

Preface

The editors are pleased to present the first edition of COVID-19 and Sleep: A Global Outlook, which has been included in the prestigious Progress in Sleep Research series. As editors, we are pleased about this decision as our volume fits perfectly in this landmark biomedicine and clinical sciences series from Springer.

All animal species require sleep, which shows that sleep is essential for maintaining essential functions. However, the time people spend sleeping at night is decreasing, and this trend is becoming more pronounced in modern 24-h societies. According to epidemiological and experimental studies, sleep deprivation has been linked to deleterious effects on physical and mental health, well-being, longevity, and general health-related quality of life.

Immune interactions related to sleep are a common occurrence in real life and conventional wisdom. Being sick makes us sleepy and want to sleep more, which is why getting a good night's sleep is sometimes referred to as "nature's best medicine" for infections. It is believed that improving sleep during an infection will act as a feedback loop for the immune system to support host defense. Therefore, sleep promotes host resistance against infection and inflammatory stimuli by acting as an immune-supportive mechanism. Indeed, sleep influences several immune parameters, is linked to a lower risk of illness, and can enhance the course of infection and the effectiveness of vaccinations.

Numerous research conducted over the past few decades have explicitly demonstrated that sleep and immunity have mutually beneficial interactions. Notably, sleep and the immune system seem to work in tandem, particularly when the body is fending off illness. Immune system operations are strongly regulated by sleep and the circadian (~24-h periodicity) rhythm, which affects how well our body defends itself innately and adaptively. Along these lines, chronic sleep deprivation is believed to impair our body's immune system, making us more susceptible to infections such as the common cold. The immune system is stimulated by microbial challenges, which results in an inflammatory response. This reaction can either increase sleep duration and intensity or interrupt sleep depending on its severity and time course. A chronic inflammatory state and an increased risk for infectious/ inflammatory pathologies, such as cardiometabolic, neoplastic, autoimmune, and neurodegenerative dysfunctions, have been linked to sleep deprivation and altered innate and adaptive immunological parameters.

To enhance physical and mental well-being across animal species, including humans, sleep quality and adaptation to individual and environmental conditions are crucial. The connection between sleep and the immune system and how sleep disruptions may upset the delicate balance with serious consequences for health outcomes have been the subject of growing evidence in recent years.

Sleep has drawn particular attention in recent years due to its possible impact on the immune system. Several studies have shown that lack of sleep alters several immune system components, including the percentages of cell subpopulations and cytokine levels, making people less resistant to infections, particularly viral infections [1]. Additionally, the immune response to infections affects sleep patterns, indicating a possible bidirectional relationship between sleep and immune responses [2].

In this specialized volume, the editors have come together to address the current coronavirus disease (COVID-19) pandemic crisis. COVID-19 transformed into a horrible global public health problem that impacted the general people's and healthcare professionals' sleep and mental health. This stressful situation significantly impacted everyone's everyday activities and sleep patterns, affecting hundreds of millions worldwide. Moreover, systematic reviews have reported a high prevalence of sleep problems globally, with the most affected groups being patients with COVID-19 and healthcare workers, in addition to the general public [3–5]. During the pandemic and lockdown, insomnia symptoms, circadian rhythm disruption, dream recall frequency, and nightmares were common. Additionally, subsequent studies suggested that some sleep disorders could be linked to COVID-19. For example, evidence suggests that OSA patients may be more likely to experience severe COVID-19 [6].

The term "long COVID" has recently appeared, reflecting the persistence of some symptoms, after recovery from the acute illness, for weeks to months following a relapsing and remitting course. It has been shown that sleep problems may persist months after COVID-19 infection [7]. A recent international survey of 14,000 participants from 16 nations reported that long-lasting sleep problems, such as insomnia and excessive daytime sleepiness in COVID-19 patients, were linked to the disease's severity [8].

Moreover, the quarantine, curfew, and infection control measures have changed the medical practice during the pandemic, affecting sleep medicine practice and the performance of diagnostic and therapeutic sleep medicine procedures. Therefore, to meet the pandemic's demands and unforeseen circumstances, COVID-19 required a significant adjustment in healthcare policies and practices, such as the complete shift to home sleep apnea testing and telemedicine [9, 10].

To address current pandemics and upcoming ones, it is therefore vital to compile all of this accumulating data for academics and practitioners in one volume. This volume combines novel basic research and clinical aspects and is the first of its kind to take a critical look at the rapidly expanding areas linking COVID-19 with sleep science, sleep disorders, and sleep medicine practice. The practical importance and the availability of new data were taken into consideration while selecting the topics for this volume, though the majority of authors emphasize the need for more research. The volume brings together leading international scientists and clinical researchers to present research at the forefront of this field and explore the ramifications of advances in these cutting-edge areas, as well as their implications for an improved treatment paradigm. Additionally, each contributor represents a particular area of expertise in the study of sleep and COVID-19 and hails from a different world region. We expect this volume to provide an authoritative reference resource for researchers, practitioners, trainees, and students.

This specialty volume covers a wide range of important topics related to COVID-19 and sleep, including the link between sleep and circadian rhythm with the immune system, sleep disturbances and circadian rhythm disorders in various populations during the COVID-19 pandemic, impact of COVID-19 infection and pandemic on common sleep disorders, management of sleep disorders during the COVID-19 pandemic, role of telemedicine in the management of sleep disorders during pandemics, running of sleep medicine services during pandemics, and positive airway pressure management and the risk of aerosol during pandemics. This volume will be a great addition to the growing data on the impact of COVID-19 on sleep, sleep disorders, and sleep medicine.

We consider ourselves fortunate to have compiled this volume. We learned a lot while editing this important volume as part of our assignment. We sincerely hope that readers will find this volume extremely useful as a research and clinical resource. We sincerely hope that this volume will be useful to researchers and clinicians in the field. However, as the topic is evolving, this book aims as well to spark further discussion and stimulate more research ideas.

Riyadh, Saudi Arabia — Ahmed S. BaHammam
Manama, Kingdom of Bahrain, Bahrain — Haitham Jahrami
Chennai, Tamil Nadu, India — S. R. Pandi-Perumal

References

1. Ragnoli B, Pochetti P, Pignatti P, Barbieri M, Mondini L, Ruggero L, et al. Sleep deprivation, immune suppression and SARS-CoV-2 infection. Int J Environ Res Public Health. 2022;19(2).
2. Besedovsky L, Lange T, Haack M. The sleep-immune crosstalk in health and disease. Physiol Rev. 2019;99(3):1325-80.
3. AlRasheed MM, Fekih-Romdhane F, Jahrami H, Pires GN, Saif Z, Alenezi AF, et al. The prevalence and severity of insomnia symptoms during COVID-19: a global systematic review and individual participant data meta-analysis. Sleep Med. 2022;100:7-23.

4. Jahrami H, BaHammam AS, Bragazzi NL, Saif Z, Faris M, Vitiello MV. Sleep problems during the COVID-19 pandemic by population: a systematic review and meta-analysis. J Clin Sleep Med. 2021;17(2):299-313.
5. Jahrami HA, Alhaj OA, Humood AM, Alenezi AF, Fekih-Romdhane F, AlRasheed MM, et al. Sleep disturbances during the COVID-19 pandemic: a systematic review, meta-analysis, and meta-regression. Sleep Med Rev. 2022;62:101591.
6. Miller MA, Cappuccio FP. A systematic review of COVID-19 and obstructive sleep apnoea. Sleep Med Rev. 2021;55:101382.
7. Jahrami H, Haji EA, Saif ZQ, Aljeeran NO, Aljawder AI, Shehabdin FN, et al. Sleep quality worsens while perceived stress improves in healthcare workers over two years during the COVID-19 pandemic: results of a longitudinal study. Healthcare (Basel). 2022;10(8):1588. doi: https://doi.org/10.3390/healthcare10081588.
8. Merikanto I, Dauvilliers Y, Chung F, Wing YK, de Gennaro L, Holzinger B, et al. Sleep symptoms are essential features of long-COVID - Comparing healthy controls with COVID-19 cases of different severity in the international COVID sleep study (ICOSS-II). J Sleep Res. 2022:e13754.
9. Pirzada A, Awadh AA, Aleissi SA, Almeneessier AS, BaHammam AS. reopening sleep medicine services in the conundrum of an ongoing COVID-19 pandemic: a global view. Sleep Vigil. 2020;4(2):73-80.
10. Pirzada AR, Aleissi SA, Almeneessier AS, BaHammam AS. Management of aerosol during noninvasive ventilation for patients with sleep-disordered breathing: important messages during the COVID-19 pandemic. Sleep Vigil. 2020;4 (2):89-94.

Acknowledgements

COVID-19 and Sleep: A Global Outlook provides scientific and medical information to all healthcare workers interested in basic, translational, and clinical medicine. We are pleased to acknowledge the contributions of those who were instrumental in producing this book.

First and foremost, Prof. Michael Vitiello of Psychiatry and Behavioral Sciences at the University of Washington, who agreed to write the foreword for this volume, deserves our heartfelt gratitude. We would like to thank him for his contribution.

We want to express our deep appreciation to all the contributors for their scholarly contributions that facilitated the development of this volume. These authors have produced authoritative chapters that synthesize vast amounts of scientific and clinical data to create informative chapters. The expertise of contributors to COVID-19 reflects the broad diversity and knowledge concerning current research in this field, which has continued to grow over the last few years. These authors represent the cutting edge of basic and applied research and provide the most recent information regarding how such knowledge can be utilized in clinical settings. Their informed opinions and insights have significantly contributed to our scientific understanding of the interaction between COVID-19 and sleep and have provided essential interpretations regarding future research directions.

The highly talented people of Springer, USA, have made this project an incredibly pleasurable one. In addition, we were delighted to have the professional and highly enthusiastic support of Springer Nature's editorial and production team (specifically Dr. Bhavik Sawhney, Mahalakshmi Shankar and Ashok Kumar). Without their continuous and unstinting support, this volume would not have been possible.

It was a pleasure to work with the entire production team of Springer. Their guidance, technical expertise, and commitment to excellence were invaluable.

Finally, and most importantly, we want to thank our spouses and families for their support and understanding during the development of this book.

About This Volume

This volume combines novel basic research and clinical aspects and is the first of its kind to take a critical look at the rapidly expanding areas linking COVID-19 pandemic infection with sleep, circadian rhythm, and sleep disorders. The volume brings together leading international scientists and clinical researchers to present research at the forefront of this biomedical field and explore the ramifications of advances in these cutting-edge areas, as well as their implications for an improved treatment paradigm. Topics covered include, but are not limited to, the following themes:

- The link between sleep and circadian rhythm with the immune system
- Sleep disturbances in various populations during the COVID-19 pandemic
- Sleep during the COVID-19 pandemic: structural inequity and racial disparity
- The impact of COVID-19 infection and pandemic on common sleep disorders
- Management of sleep disorders during the COVID-19 pandemic
- Running of sleep medicine services during pandemics
- Positive airway pressure management and the risk of aerosol during pandemics
- The role of telemedicine in the management of sleep disorders during pandemics

Contents

Editors and Contributors

About the Editors

Ahmed S. BaHammam is a tenured Professor of Pulmonary and Sleep Medicine at King Saud University (KSU), Riyadh, Saudi Arabia. He is the Executive Director of Academic Affairs and the Director of Prince Naif Health Research Center at KSU, Medical City, and the Director of the University Sleep Disorders Center at KSU, Saudi Arabia. As a recipient of the Lifetime Achievement Award (2016) from KSU, he has published 320-plus scientific articles (H-index 59), more than 60 chapters, 3 books, and 30 papers about COVID-19 and its manifestations, including sleep and neuropsychiatric manifestations. He holds leadership positions in several high-profile journals. His research focuses on a wide range of sleep-related topics, including sleep and circadian rhythm changes during the COVID-19 pandemic, obesity hypoventilation syndrome (OHS), central hypersomnolence disorders, and impact of intermittent fasting and mealtime on sleep, circadian rhythm, and immunity. Further information about his work can be found at https://fac.ksu.edu.sa/ashammam/home.

S. R. Pandi-Perumal is a well-recognized sleep researcher both nationally and internationally and has authored many publications. His area of research interest includes sleep and biological rhythms. He has edited over 25 academic volumes related to sleep and biological rhythm research, and he serves on the editorial board of several leading journals. He was featured in the New York Times in 2004. His online biography can be viewed at http://pandi-perumal.blogspot.com.

Haitham Jahrami is the Director, Rehabilitation Services at Governmental Hospitals, Ministry of Health, Kingdom of Bahrain. He is also an Associate Professor, Department of Psychiatry, College of Medicine and Medical Sciences at Arabian Gulf University, Manama, Bahrain. He has published more than 250 peer-reviewed articles, with 50 papers on COVID-19 and central nervous sequelae. He is a peer reviewer and associate editor in several Q1 journals. His research uses diverse methodologies, including systematic reviews and meta-analyses, and emerging technologies such as artificial intelligence and machine learning.

Contributors

Hassen Al-Amin Department of Psychiatry, Weill Cornell Medicine, Ar-Rayyan, Qatar

Dana M. Alhasan Department of Health and Human Services, Epidemiology Branch, National Institute of Environmental Health Sciences, National Institutes of Health, Research Triangle Park, NC, USA

Maha H. Alhussain Department of Food Science and Nutrition, Chrononutrition, College of Food and Agriculture Sciences, King Saud University, Riyadh, Saudi Arabia

Mohammed Al Maqbali Department of Nursing Midwifery and Health, Northumbria University, Newcastle-Upon-Tyne, UK
Al Buraimi Hospital, Ministry of Health Oman, Al Buraimi, Oman

Sulaiman Alshakhs Mental Health Services, Hamad Medical Corporation, Doha, Qatar

Horacio Balam Álvarez-García Sleep Disorders Clinic, Research Division, School of Medicine, National Autonomous University of Mexico, UNAM, Ciudad de México, Mexico

Monica Levy Andersen Departamento de Psicobiologia, Universidade Federal de São Paulo (UNIFESP), São Paulo, Brasil

Ahmed S. BaHammam The University Sleep Disorders Center, Department of Medicine, College of Medicine, King Saud University, Riyadh, Saudi Arabia
Strategic Technologies Program of the National Plan for Sciences and Technology and Innovation in the Kingdom of Saudi Arabia, King Saud University, Riyadh, Saudi Arabia

Deirdre Barrett Department of Psychiatry, Harvard Medical School, Cambridge, MA, USA

Silvia Beatriz Boscardin Departamento de Parasitologia, Universidade de São Paulo, São Paulo, Brasil

Omar Boukhris SIESTA Research Group, School of Allied Health, Human Services and Sport, La Trobe University, Melbourne, Australia
Sport, Performance, and Nutrition Research Group, School of Allied Health, Human Services and Sport, La Trobe University, Melbourne, Australia

Nicola Luigi Bragazzi Department of Mathematics and Statistics, Laboratory for Industrial and Applied Mathematics (LIAM), York University, Toronto, ON, Canada

Gregory M. Brown Centre for Addiction and Mental Health, Department of Psychiatry, University of Toronto, Toronto, ON, Canada

Daniel Cardinali Faculty of Medical Sciences, Pontificia Universidad Católica Argentina, Buenos Aires, Argentina

Michael W. L. Chee Centre for Sleep and Cognition, Human Potential Program, Yong Loo Lin School of Medicine, National University of Singapore, Singapore, Singapore

Ulises Jiménez Correa Research Division, National Autonomous University of Mexico, Sleep Disorders Clinic, Medicine Faculty, Mexico City, Mexico

Milca Abda de Morais Faculdade de Ciências Aplicadas, Universidade Estadual de Campinas, Limeira, SP, Brazil
Faculdade de Medicina de Açailândia, FAMEAC/IDOMED, Açailândia, MA, Brazil

Andrea Maculano Esteves Universidade Estadual de Campinas, Faculdade de Ciências Aplicadas—Limeira, Sao Paulo, Brazil

Mo'ez Al-Islam E. Faris Department of Clinical Nutrition and Dietetics, College of Health Sciences/Research Institute of Medical and Health Sciences (RIMHS), University of Sharjah, Sharjah, UAE

Michele Ferrara Department of Biotechnological and Applied Clinical Sciences, University of L'Aquila, L'Aquila, Italy

Christian Franceschini Department of Medicine and Surgery, University of Parma, Parama, Italy

Dennys Franco-Avecilla Core Internal Medicine, The Ottawa Hospital, Ottawa, ON, USA

Beatriz Franco Faculdade de Educação Física, Universidade Estadual de Campinas, Campinas, SP, Brazil

Sergio Garbarino Department of Neuroscience, Rehabilitation, Ophthalmology, Genetics and Maternal/Child Sciences, Polyclinic Hospital San Martino Istituto di Ricovero e Cura a Carattere Scientifico (IRCCS), University of Genova, Genova, Italy

Symielle A. Gaston Department of Health and Human Services, Epidemiology Branch, National Institute of Environmental Health Sciences, National Institutes of Health, Research Triangle Park, NC, USA

Kamila Gautam Centre of Excellence & Advanced Research for Childhood Neurodevelopmental Disorders, Child Neurology Division, Department of Paediatrics, All India Institute of Medical Sciences, Delhi, India

Suhaila Ghuloum Department of Psychiatry, Hamad Medical Corporation, Doha, Qatar
Psychiatry Department, Weill Cornell Medicine, Doha, Qatar

Cara M. Gill Department of Respiratory Medicine, St James' Hospital, Dublin, Ireland

Jordan M. Glenn Department of Health, Exercise Science Research Center Human Performance and Recreation, University of Arkansas, Fayetteville, AR, USA

Max Hirshkowitz Baylor College of Medicine, Houston, TX, USA
Division of Public Mental Health and Population Sciences, Stanford University School of Medicine, Stanford, CA, USA

Chandra L. Jackson Epidemiology Branch, National Institute of Environmental Health Sciences, Department of Health and Human Services, National Institutes of Health, Research Triangle Park, NC, USA
Intramural Program, National Institute on Minority Health and Health Disparities, Department of Health and Human Services, National Institutes of Health, Bethesda, MD, USA

Haitham Jahrami Department of Psychiatry, Ministry of Health, Manama, Bahrain
Department of Psychiatry, College of Medicine and Medical Sciences, Arabian Gulf University, Manama, Bahrain

Tetyana Kendzerska Clinical Epidemiology Program, The Ottawa Hospital Research Institute, Ottawa, ON, Canada
Division of Respirology, Department of Medicine, The University of Ottawa, Ottawa, ON, Canada
The Ottawa Hospital Sleep Center, The Ottawa Hospital, Ottawa, ON, Canada

Brian D. Kent Department of Respiratory Medicine, St James' Hospital, Dublin, Ireland

Habibollah Khazaie Sleep Disorders and Research Center, Kermanshah University of Medical Sciences, Kermanshah, Iran

Laura Kortesoja Centre for Educational Assessment, University of Helsinki, Helsinki, Finland
Faculty of Education and Culture, Tampere University, Tampere, Finland

Isaias Vicente Lugo-Gonzalez Interdisciplinary Research Unit in Health Sciences and Education (IRUHSE), Faculty of Higher Studies Iztacala, National Autonomous University of Mexico, UNAM, Ciudad de México, Mexico

National Autonomous University of Mexico, Sleep Disorders Clinic, Medicine Faculty, Research Division, Mexico City, Mexico

Reaz Mahmud Department of Neurology, Dhaka Medical College, Dhaka, Bangladesh

S. A. A. Massar Sleep and Cognition Laboratory, Centre for Sleep and Cognition, Yong Loo Lin School of Medicine, National University of Singapore, Singapore, Singapore

Ilona Merikanto Department of Psychology and Logopedics and SleepWell Research Program, Faculty of Medicine, University of Helsinki, Helsinki, Finland
Finnish Institute for Health and Welfare, Helsinki, Finland
Orton Orthopaedics Hospital, Helsinki, Finland

J. L. Ong Sleep and Cognition Laboratory, Centre for Sleep and Cognition, Yong Loo Lin School of Medicine, National University of Singapore, Singapore, Singapore

S. R. Pandi-Perumal Somnogen Canada Inc., Toronto, ON, Canada
Saveetha Medical College and Hospitals, Saveetha Institute of Medical and Technical Sciences, Saveetha University, Chennai, Tamil Nadu, India

Sofia Pappa Department of Psychiatry, Imperial College London, London, UK
West London NHS Trust, London, UK

Laura M. Piggott Department of Respiratory Medicine, St James' Hospital, Dublin, Ireland

Gabriel Natan Pires Departamento de Psicobiologia, Universidade Federal de São Paulo (UNIFESP), São Paulo, Brasil

Rouf Pirzada North Cumbria Integrated Care (NCIC), NHS, Carlisle, UK

Giuseppe Plazzi IRCCS Istituto delle Scienze Neurologiche di Bologna, Bologna, Italy
Department of Biomedical, Metabolic and Neural Sciences, University of Modena and Reggio-Emilia, Bologna, Italy

Daniela Santoro Rosa Departamento de Microbiologia, Imunologia e Parasitologia, Universidade Federal de São Paulo (UNIFESP), São Paulo, Brasil

Nikolaos Sakkas Warwick Medical School, University of Warwick, Coventry, UK

Federico Salfi Department of Biotechnological and Applied Clinical Sciences, University of L'Aquila, L'Aquila, Italy

Mouaz Saymeh The Ottawa Hospital, Ottawa, ON, Canada

Roy Chowdhury Sayoni Centre of Excellence & Advanced Research for Childhood Neurodevelopmental Disorders, Child Neurology Division, Department of Paediatrics, All India Institute of Medical Sciences, Delhi, India

Egeria Scoditti National Research Council (CNR), Institute of Clinical Physiology (IFC), Lecce, Italy

Amir Sharafkhaneh Baylor College of Medicine, Full Tenured Professor of Medicine, Houston, TX, USA

Gulati Sheffali Centre of Excellence & Advanced Research for Childhood Neurodevelopmental Disorders, Child Neurology Division, Department of Paediatrics, All India Institute of Medical Sciences, Delhi, India

Jennifer Y. So Division of Pulmonary, Critical Care and Sleep Medicine, University of Maryland Sleep Laboratory, University of Maryland School of Medicine, Baltimore, MD, USA

Shanon Sullivan Division of Pediatric Pulmonary, Asthma, and Sleep Medicine, Department of Pediatrics, Stanford University School of Medicine, Palo Alto, CA, USA
Division of Sleep Medicine, Department of Psychiatry, Stanford University School of Medicine, Palo Alto, CA, USA

Khaled Trabelsi Higher Institute of Sport and Physical Education of Sfax, University of Sfax, Sfax, Tunisia
Research Laboratory: Education, Motricity, Sport and Health, EM2S, LR19JS01, University of Sfax, Sfax, Tunisia

Sergio Tufik Departamento de Psicobiologia, Universidade Federal de São Paulo (UNIFESP), São Paulo, Brasil

Giorgia Varallo Department of Medicine and Surgery, University of Parma, Parma, Italy

Chapter 1
The Effect of Sleep Disruption and Circadian Misalignment on the Immune System

Sergio Garbarino, Nicola Luigi Bragazzi, and Egeria Scoditti

Abstract Sleep disruption and circadian misalignment are associated with an increased risk for infectious and inflammatory pathologies, including cardiometabolic, neoplastic, autoimmune, and neurodegenerative diseases. Sleep and circadian rhythms are closely involved in the regulation of the immune system. Impairments of sleep quantity, quality, and timing, as well as circadian misalignment, result in derangements of innate and adaptive immune responses leading to a chronic inflammatory state and a decrease of the immune defense and reaction to threats (infection or injury). The immune system potentially plays an important mechanistic role in the relation between sleep disruption and circadian misalignment, and adverse health effects. By regulating the immune system, sleep- and circadian-centered intervention may beneficially impact overall health and on the prevention—and treatment—of infections and chronic diseases, especially in the modern lifestyles characterized by a multiplicity of social and environmental pressures on sleep and circadian rhythms, and in times of infectious disease outbreaks, such as COVID-19, where an effective immunity is of utmost importance.

Keywords Sleep · Circadian rhythm · Immune system · Infection · Chronic disease

1.1 Introduction

There is a time for many words, and there is also a time for sleep. Homer. The Odyssey.

S. Garbarino (✉)
Department of Neuroscience, Rehabilitation, Ophthalmology, Genetics and Maternal/Child Sciences, University of Genoa, Genoa, Italy

N. L. Bragazzi
Department of Mathematics and Statistics, Laboratory for Industrial and Applied Mathematics (LIAM), York University, Toronto, ON, Canada

E. Scoditti
National Research Council (CNR), Institute of Clinical Physiology (IFC), Lecce, Italy
e-mail: egeria.scoditti@ifc.cnr.it

A. S. BaHammam et al. (eds.), *COVID-19 and Sleep: A Global Outlook*, Progress in Sleep Research, https://doi.org/10.1007/978-981-99-0240-8_1

Since antiquity, sleep has been recognized as essential for overall health. Sleep is an active physiological process necessary for life, being involved in the regulation of physical, mental, and emotional health.

Key components in the regulation and maintenance of the sleep/wake cycle are the endogenous 24-h body clock/circadian system that drives wakefulness throughout the day and sleep during the night; and the wake-dependent homeostatic drive that generates a sleep pressure during the day, which is dissipated during sleep. However, an exogenous drive that is the product of our societal temporal structure (e.g., school and work times) forces an exogenous sleep/wake pattern influencing sleep opportunities and thereby sleep duration, quality, and efficiency, potentially resulting in sleep and circadian rhythm disruption (Zielinski et al. 2016).

On this background, patterns of sleep quantity, quality, and timing are influenced by health status, cultural, social, psychological, behavioral, and environmental factors (Grandner 2017). Expert consensus recommendations suggest that adults should sleep a minimum of 7 h per night on a regular night to promote optimal health (Consensus Conference et al. 2015a, b). Apart from sleep/circadian disorders however many factors, including jet lag syndrome, social jet lag, work schedule, growing frequency of technology use, higher rates of obesity and diabetes, inadequate nutrition, and other lifestyle factor changes occurring in the twenty-first century as well as the COVID-19 pandemic are increasingly and often negatively impacting on sleep quantity, quality, and timing, so that sleep problems have constantly arisen in the general population and special groups of vulnerable people, including children/adolescents or workers. Population-based studies reported that sleep duration among adults has changed over the past few decades and, in particular, the prevalence of adults sleeping less than 6 h per night has stably increased over a long period (Gilmour et al. 2013; Ford et al. 2015a, b; Zomers et al. 2017). This phenomenon is also occurring among children and adolescents (Matricciani et al. 2012), so 58% and 73% of middle and high school-aged adolescents, respectively, do not meet the American Academy of Sleep Medicine sleep duration recommendations (Wheaton et al. 2018). A parallel increasing prevalence of insomnia has been documented in several countries (Ford et al. 2015a, b).

In addition to excessive daytime sleepiness, fatigue, tiredness, depressed mood, poor daytime functioning, and impaired cognitive and safety-related performance, inadequate sleep is associated with an increased risk of adverse health outcomes, including weight gain, obesity, type 2 diabetes mellitus, hypertension, cardiovascular (CV) and neurodegenerative diseases, mental diseases, cancer as well as all-cause mortality (Vgontzas et al. 2009, 2013; Irwin 2015; Smagula et al. 2016; Cappuccio and Miller 2017).

Another integral part of human physiology and behavior tightly regulating sleep is the circadian rhythms, which are orchestrated by a central "master" clock, i.e., the suprachiasmatic nucleus (SCN) located in the anterior hypothalamus, that coordinates alignment between circadian clocks in other brain regions and peripheral tissues with external synchronizing agents (the environment and behaviors). Indeed, circadian rhythms are modulated by endogenous (genetic, physiological) as well as environmental (light) and behavioral (activity, feeding) factors. Various behaviors and physiological functions of the body show circadian rhythms, such as the sleep–wake cycle and food intake, blood pressure, blood lipids, coagulation/fibrinolysis system, heart rate, body temperature, locomotor activity, hormone levels, cell metabolism and proliferation (Sulli et al. 2018). The advantages of internal clocks are to enable individuals to anticipate, rather than react to, daily recurring events and align their physiology and behavior to the changing environment.

The "molecular circadian clock" refers to genes that maintain autoregulatory feedback loops in which oscillating outputs regulate their expression (circadian locomotor output cycles kaput [CLOCK], brain and muscle ARNT-like [BMAL], period [PER], REV-ERB/nuclear receptor subfamily 1, group D [NR1D], and cryptochrome [CRY]). The formation, trafficking, and degradation of different clock protein complexes throughout this transcriptional cycle establish the intrinsic 24 h period of the cellular clock. Furthermore, at the cellular level, molecular clock components generate circadian fluctuation in basic cellular functions, e.g., gene expression, protein translation, and intracellular signaling, which are all involved in fundamental processes, including cell cycle regulation, nutrient sensing/utilization, metabolism, stress response, redox regulation, detoxification, and cell defense (immunity and inflammation) in a tissue-specific manner (Mure et al. 2018). Circadian disruption may occur as a result of a misalignment between external factors (such as the natural light/dark cycle, social and work requirements, and behaviors such as sleep and meal timing) with the master circadian clock as well as with endogenous circadian clocks in other tissues. Misalignments can occur among two (or more) rhythms which may be both internal (central vs. peripheral rhythms), or one may be internal and the other external (e.g., central vs. light/dark or peripheral vs. feeding/fasting). Social jet lag, jet lag syndrome, shift work, and inappropriately timed light exposure (evening or night) are common causes of circadian (external-internal) misalignments in modern society.

Circadian disruption has important public health implications due to its prevalence in modern society as well as its association with serious safety- and performance-related issues (Mitler et al. 1988) and adverse health outcomes (Boivin et al. 2022). Subjects with a nondaytime working schedule (shift work), exposure to light pollution, social jet lag, or evening chronotype are at increased risk for circadian disruption. Circadian disruption in humans is associated with broad and significant consequences for mental and physical health, increasing the risk for the development of cancer (Sulli et al. 2019), neurodegenerative, psychiatric (Abbott et al. 2020), cardiometabolic (Scheer et al. 2009), and immune disorders (Fishbein et al. 2021). Importantly, changes in circadian function are often accompanied by sleep–wake disturbances. Therefore, increasing scientific efforts have been devoted to understanding the health consequences of sleep disruption and circadian misalignment, and to translating this science to directly impact human health.

The immune system functions to preserve the integrity of the body by sensing physiological disturbances (microbes or tissue injury) and reinstating homeostasis via both inflammatory immune responses and processes of tissue repair and physical barrier regulation. In many chronic diseases, a deregulated and/or exacerbated immune response shifts from repair/regulation towards immune-driven unresolved inflammatory responses (Hand et al. 2016). Several anatomic and molecular mechanisms, including neurons, glial cells, leukocytes, nerve fibers, soluble mediators, cellular receptors, the blood–brain barrier (BBB), the neuroendocrine hypothalamus-pituitary-adrenal (HPA) axis, and the autonomic nervous system, participate in orchestrating the bidirectional brain-immune crosstalk that fine tunes the immune response and its relationship with sleep and the circadian systems (Dantzer 2018). In this contest, the circadian system and sleep have emerged as important intertwined regulators of the immune system. It follows that sleep disruption and circadian misalignment may result in deregulated immune responses and pro-inflammatory responses, that contribute to an increase in the risk for the onset and/or worsening of infections as well as inflammation-related chronic diseases, including cancer, cardiometabolic and neurodegenerative diseases (Labrecque and Cermakian 2015; Scheiermann et al. 2018; Garbarino et al. 2021).

This Chapter focuses on the regulation of the immune system by sleep and the circadian rhythm, and the consequences for the immune function of sleep disruption and circadian misalignment.

1.2 Immune Regulation by Sleep

The sleep–immune interaction hypothesis was first suggested by pioneering studies attempting to identify substances involved in sleep regulation. The hypothesis of humoral regulation of sleep dates back to the early 1900s and posited that substances accumulated during waking could trigger subsequent sleep. Accordingly, experimental studies found that injection of cerebrospinal fluid from sleep-deprived dogs into rested recipient dogs caused the recipient dogs to fall into a narcosis-like sleep state (Ishimori 1909; Legendre and Piéron 1913). Subsequent animal studies replicated this result, pointing to the existence of hypnotoxins mediating sleep induction (endogenous factor(s) S, where S stands for sleep), and led to the identification of several sleep-promoting factors.

Among these hypnotoxins, muramyl peptide, a component of bacterial cell wall able of activating the immune system and inducing the release of immune regulators, such as cytokines, was recognized as the first molecular link between the immune system and sleep (Krueger et al. 1982). Other microbial-derived factors such as the endotoxin lipopolysaccharide (LPS), as well as mediators of inflammation, such as cytokines (interleukin [IL]-1, tumor necrosis factor [TNF]-α), prostaglandins (PG), uridine, growth factors, were found to regulate sleep (Zielinski and Krueger 2011). These factors, through the BBB or via afferent nerve fibers, establish a signaling network with other brain factors involved in sleep regulation, such as neurotransmitters (acetylcholine, dopamine, serotonin, norepinephrine, histamine), neuropeptides (orexin), nucleosides (adenosine), the hormone melatonin as well as the hypothalamus-pituitary (HPA) axis.

Animal studies have consistently reported a role for the cytokines IL-1 and TNF-α and the prostaglandin PGD2 in the physiologic, homeostatic nonrapid eye movement (NREM) sleep regulation in a dose- and time-of-day-dependent manner, so that the inhibition of the biological action of these substances resulted in a decrease of spontaneous NREM sleep, whereas administration of these substances enhanced NREM sleep amount and intensity and suppressed REM sleep (Opp 2005; Urade and Hayaishi 2011). Anti-inflammatory cytokines, including IL-4, IL-10, and IL-13, have been found to inhibit NREM sleep in animal models (Kubota et al. 2000). In humans, the circulating levels of IL-1, IL-6, TNF-α, and PGD2 are highest during sleep and, the available evidence, though indirect, converge to suggest the involvement of these immune mediators in the physiologic regulation of sleep (Besedovsky et al. 2019).

Accordingly, infection or inflammatory diseases induce immune activation and associated altered cytokine concentrations and profiles, which are transmitted to the central nervous system inducing adaptive and energy-saving responses, including sleep (Besedovsky et al. 2019). Acute mild immune activation enhances NREM sleep and suppresses REM sleep, and the increase in NREM sleep was a favorable prognostic factor for rabbits during infectious diseases (Toth et al. 1993). Contrarily, severe immune response with an upsurge of cytokine levels causes sleep disturbance with suppression of both NREM and REM sleep (Toth et al. 1993; Mullington et al. 2000; Sharpley et al. 2016). Chronic infectious, such as HIV infection (Chaponda et al. 2018), and inflammatory diseases, such as inflammatory bowel disease or rheumatoid arthritis (Ranjbaran et al. 2007; Uemura et al. 2016), are associated with sleep disturbances.

Evidence of the immune-supportive effect of sleep that may favor host defense is provided by vaccination studies. In humans, compared with nocturnal wakefulness, sleep after vaccination boosted both the memory phase and the effector phase of the immune response, underscoring the adjuvant-like effect of sleep on the immunological function (Lange et al. 2011). Similarly, habitual (and hence chronic) short sleep duration (less than 6 h) compared with longer sleep duration decreased the long-term clinical protection after vaccination against hepatitis B (Prather et al. 2012).

In exerting these effects, sleep may benefit the immune response through different mechanisms. Sleep influences the T helper (Th) phenotype and the cytokine balance between Th1 and Th2 cells thus determining the types of the effector mechanisms of the immune response. Th1 polarization state is typical of immune response to intracellular viral and bacterial challenges and is characterized by increased release of IFN-γ, IL-2, and TNF-α. It supports various cell-mediated responses, including macrophage activation, phagocytosis, the killing of intracellular microbes, and antigen presentation (Zhu and Zhu 2020). Th2 immunity is characterized by the expression of IL-4, IL-5, IL-10, and IL-13, and mediates humoral defense by stimulating mast cells, eosinophils, and B cells (with the production of IgG2,4 and IgE) against extracellular pathogens (Zhu and Zhu 2020).

The balance of Th1/Th2 immunity is critically involved in antimicrobial and antitumor immune responses. Th2 overactivity is found in some forms of allergic responses, and increases the susceptibility to infection (Moser et al. 2018), as well as to tumor development and progression, by limiting cytotoxic T lymphocytes proliferation and modulating other inflammatory cell types (Disis 2010). In contrast,

Th1 immunity supports cytotoxic lymphocytes with the potential of elimination or control of tumor cell growth; indeed, a Th1 adaptive immune response may be associated with improved survival or prognosis (Disis 2010; Lee et al. 2019).

Furthermore, sleep is associated with a reduction in circulating immune cells that most likely accumulate into lymphatic tissues thus increasing the probability to encounter antigens (immunological synapse) and trigger the immune response. An effective adaptive immune response to an immunological challenge may be facilitated by specific immune-active hormones associated with slow wave sleep (SWS)-rich early sleep, which is characterized by minimum concentrations of immunosuppressive hormones, such as cortisol, and high levels of immune-stimulating hormones such as growth hormone (GH), prolactin, and aldosterone, which support pro-inflammatory cytokine production and Th1 cell-mediated immunity (Besedovsky et al. 2019). Therefore, through pro-inflammatory hormones and cytokines night-time sleep facilitates the onset of adaptive immune responses, while during daytime activity, anti-inflammatory signals, hormones, and cytokines support immediate reactions to biological and other environmental challenges.

1.3 Sleep Disruption and Immune Consequences

In agreement with the sleep–immunity relationship, several lines of evidence from experimental and epidemiological studies converge on the significant effects of sleep disruption on immune function and related clinical outcomes.

Early animal studies found that sleep loss, besides being lethal after several weeks, was associated with dysfunction of host defense (Everson and Toth 2000; Everson et al. 2008, 2014) thus suggesting the importance of sleep for the immune system. More pertinently, the effect of sleep on immune function has emerged in studies in which immune parameters, including circulating levels of cytokines and cell adhesion molecules, leukocyte counts, and activity, were measured under the manipulation of sleep duration compared with undisturbed sleep.

Collectively most human and animal findings report on the supportive effect of sleep—and the detrimental effect of disturbed sleep—on several immune regulators. Indeed, compared with regular nocturnal sleep, acute and mostly sustained sleep loss has been found: (a) to alter circulating leukocyte counts with studies reporting increased numbers of total leukocytes and specific cell subsets mainly neutrophils, monocytes, B cells, decreased circulating natural killer (NK) cells, and changes in circulating CD4+ T cells (Born et al. 1997; Dimitrov et al. 2007; van Leeuwen et al. 2009; Lasselin et al. 2015; Said et al. 2019); (b) to alter the diurnal rhythm of circulating leukocytes, resulting in higher levels during the night and at awakening and a flattening of the rhythm (Born et al. 1997; Lasselin et al. 2015); (c) to increase the plasma levels of pro-inflammatory cytokines such as IL-1, IL-6, CRP, and, less consistently, TNF-α, MCP-1, and a homeostatic increase in endogenous inhibitors such as IL-1 receptor antagonist (IL-1ra) and TNF receptor I and II in an attempt to limit the increased cytokine levels and activity (Shearer et al. 2001; Hu et al. 2003;

Vgontzas et al. 2004; van Leeuwen et al. 2009); (d) to transiently decrease the cytotoxic activity of NK cells, the proliferation capacity of lymphocytes (Irwin et al. 1994), and the phagocytic activity of neutrophils, important against infection (Said et al. 2019); (e) to enhance circulating levels of endothelial adhesion molecules such as intercellular adhesion molecule (ICAM)-1 and E-selectin, suggesting endothelial activation and enhanced risk for vascular dysfunction (Sauvet et al. 2010); (f) to reduce the stimulated production of IL-2 and IL-12, which normally support the adaptive immune response (Dimitrov et al. 2007; Axelsson et al. 2013); (g) to reduce the levels of Mac-1 positive lymphocytes suggesting reduced migratory capacity of immune cells (Redwine et al. 2004). Compared to undisturbed sleep which is predominantly characterized by a Th1 response (mainly during early sleep), experimental sleep deprivation leads to a shift from a Th1 pattern towards a Th2 pattern in humans (Dimitrov et al. 2004; Axelsson et al. 2013). Elderly people (Ginaldi et al. 1999), alcoholic subjects (Redwine et al. 2003) as well as insomnia patients (Sakami et al. 2002), all characterized by disturbed sleep, show a cytokine shift towards Th2.

At the molecular level, findings demonstrate that a single night of partial sleep deprivation (4 h of sleep) (Irwin et al. 2006, 2015) or chronic partial sleep deprivation (4 h of sleep for five nights) (van Leeuwen et al. 2009) in healthy adults led to increased protein production and mRNA levels of inflammatory cytokines (IL-6, IL-1β, TNF-α, IL-17). Accordingly, prominent genome-wide gene expression changes have been found in response to acute (Irwin et al. 2006) or chronic (Aho et al. 2013; Moller-Levet et al. 2013) partial sleep deprivation in human circulating monocytes, so most of the genes and associated biological pathways upregulated after sleep loss compared with unrestricted sleep were related to immune and inflammatory processes (leukocyte activation and differentiation, cytokine positive regulation, innate and adaptive immunity, TLRs signaling), as well as to oxidative stress, response to stress, apoptosis, collectively indicating activation of the immune system. Interestingly, genes associated with B cell activation and Th2 cell differentiation were upregulated, whereas those associated with Th1 cell differentiation were downregulated (Aho et al. 2013), suggesting that the Th2 immune response driven by sleep deprivation, as observed in many studies, is regulated at the level of gene expression. In contrast, biological processes associated with genes downregulated following sleep deprivation compared with unrestricted sleep included chromatic organization and modification, gene expression, cellular macromolecule metabolism (Moller-Levet et al. 2013), cholesterol/lipid metabolism and transport, as well as NK cell function thus contributing to the reduced immune response against pathogens (Aho et al. 2013). The same expression profile of several genes identified in the experimental sleep deprivation was observed in a cohort of subjects with self-report of insufficient sleep (Aho et al. 2013), highlighting the physiological relevance of the experimental results at the population level in real-life conditions.

The pro-inflammatory transcriptomic response observed after sleep deprivation mainly involves the activation of the pro-inflammatory NF-κB family of transcription factors (Irwin et al. 2006, 2008; Aho et al. 2013). NF-κB mediates the expression of genes (e.g., cytokines, chemokines, growth factors, receptors/transporters, enzymes, adhesion molecules) involved in the activation of inflammation, adaptive and innate immunity, proliferation, and apoptosis, and is recognized as a promising therapeutic target in inflammatory diseases (Madonna and De Caterina 2012).

Sleep disruption is also associated with oxidative stress, which represents an imbalance in the production and/or detoxification of free radicals such as reactive oxygen species (ROS). Oxygen-derived free radicals are generated during oxidative metabolism and energy production processes and normally serve, at very low physiologic concentrations, as important second messengers in many intracellular signaling pathways for maintaining cell homeostasis and survival in response to stress (Liguori et al. 2018). At higher levels which are not counteracted by the antioxidant defenses of the cell, ROS can cause damage to cells and tissues resulting in cell senescence and injury, unbalanced local/systemic inflammation, metabolic dysfunction as well as immune derangements. Pertinently, ROS are essential, especially at low levels for a wide range of immune functions, including anti-viral, anti-bacterial, and anti-tumor responses through, for example, killing of pathogens, regulation of T cell activation, expansion and effector function, and induction of balanced inflammatory reaction (Sena and Chandel 2012). However, excessive uncompensated ROS production can lead to aberrant immune responses and unbalanced inflammatory reactions, including apoptosis and functional suppression of T cells with following reduced anti-tumor function and chronic activation of pro-inflammatory signaling pathways including NF-κB (Chen et al. 2016). Oxidative stress is now recognized to play a central role in the pathophysiology of many different disorders with immune components, mostly neurodegenerative, cardiovascular, and metabolic diseases as well as cancer (Liguori et al. 2018), which are also disease conditions triggered or exacerbated by sleep disturbance. Accordingly, most animal studies found an increase in oxidative stress markers and/or a decrease in endogenous antioxidants and antioxidant enzymes in brain regions and peripheral tissues (liver, heart, plasma, etc.) after sleep disturbance, while recovery sleep restored the antioxidant/oxidant balance (Villafuerte et al. 2015). Recent human findings agree with animal results, as shown in night workers with chronic sleep loss (Teixeira et al. 2019), and in young adults subjected to acute (overnight) sleep deprivation (Trivedi et al. 2017). Therefore, sleep shows an antioxidant function, responsible for eliminating ROS produced during wakefulness, and contrarily sleep curtailment may exert negative health effects by causing oxidative stress.

A breakdown of host defense against microorganisms has been found in animals subjected to insufficient sleep, as shown by the increased mortality after septic insult in sleep-deprived mice compared with control mice (Friese et al. 2009), or by systemic invasion by opportunistic microorganisms leading to increased morbidity and lethal septicemia in sleep deprived-rats (Everson and Toth 2000). Patients with sleep disorders exhibited a 1.23-fold greater risk of herpes zoster than did the comparison cohort, after adjustment for potential covariates (Chung et al. 2016). Accordingly, increased susceptibility to respiratory infections has been reported in sleep-deprived human subjects, as those with habitual short sleep ($\leq$5 h) compared with 7–8 h sleep, in cross-sectional and prospective studies (Patel et al. 2012; Prather and Leung 2016), and after an experimental viral challenge (Cohen et al. 2009; Prather et al. 2015). Similarly, compared with long sleep duration (around 7 h), short sleep duration (around 6 h) is associated with an increased risk of common illnesses, including cold, flu, gastroenteritis, and other common infectious diseases, in adolescents (Orzech et al. 2014).

In sum, most of these immune responses to sleep loss are suggestive of a systemic low-grade pro-inflammatory reaction. Following experimental findings, population-based studies in adult and younger individuals found that habitual short sleep duration (generally <5 or 6 h) is directly and independently associated with elevated circulating pro-inflammatory markers, such as acute phase proteins (CRP, IL-6), cytokines (TNF-α, IFN-γ, IL-1, etc.), adhesion molecules, white blood cell counts (Miller et al. 2009; Patel et al. 2009; Ferrie et al. 2013; Perez de Heredia et al. 2014; Bakour et al. 2017; Richardson and Churilla 2017). Furthermore, reduced NK cell activity (Fondell et al. 2011) and a decline in naive T cells (Carroll et al. 2017) were also found to be associated with habitual short sleep. A shortening of leukocyte telomere length, which is a marker of cellular senescence and inflammation damage, was also shown to be associated with shorter sleep duration in adults and children (Jackowska et al. 2012; James et al. 2017). Systemic low-grade inflammation has been shown to predict the risk of disease and mortality (Li et al. 2017); thus, it is suggested to mediate the increased risk of morbidity and mortality associated with sleep disruption (Hall et al. 2015; Smagula et al. 2016).

1.4 Circadian Clock Regulation of the Immune System

The circadian system encompasses a central clock representing a master circadian pacemaker located in the SCN of the hypothalamus, and peripheral circadian clocks are distributed in different cells or tissues outside the brain and functioning autonomously and flexibly (Patke et al. 2020). The central clock mainly responds to the light/dark cycle, so that the ambient light (one of the strongest zeitgebers, i.e., external time givers) is transmitted to the hypothalamus through the retinal ganglion cells leading to glutamate release at the nerve terminals and an increase in the SCN neuronal activity (i.e., wake pressure).

The central and peripheral clocks interact through neural, endocrine pathways and body temperature, to produce daily rhythms in sleep, physical activity, and nutrient metabolism via self-sustained near-24 h and alternating activation-repression cycles of core clock transcriptional and translational regulators (Clock, Bmal1, Npas2, Crys, Pers, Rors, and Rev-erbs) (Sulli et al. 2018). The peripheral clocks also respond to other synchronizers such as meal times and humoral factors, and body temperature. Several metabolites and proteins interact with the core clock components to influence their function and modulate specific outputs of the circadian system. Whole-genome transcriptomic studies have revealed circadian variation (i.e., with a ~24-h periodicity) in gene expression, and that almost every gene including those encoding drug targets show diurnal rhythmic expression in a tissue-specific manner and a bimodal distribution (with peaks predominantly occurring during the biological night and day) thus suggesting a marked temporal segmentation of biological processes and functions and contributing to the circadian rhythmicity of basic cellular functions, including metabolism, immune function, and tissue repair (Mure et al. 2018; Sulli et al. 2018; Christou et al. 2019).

Both the innate and adaptive arms of the immune response are under circadian control, which is instrumental to gate the immune functions, such as immune cell trafficking, production of cytokines, or host–pathogen interaction, and to increase organismal fitness, while minimizing metabolic costs of immune activation or collateral tissue damage due to uncontrolled immune response. Both circulating immune cell counts and inflammatory cytokine levels show variations during the sleep/wake cycle dependent on both sleep- and circadian rhythm-associated processes (Scheiermann et al. 2013): indeed, human natural killer (NK) cells and neutrophils peak at midday and show minimum levels during the night; while monocytes, B, and T cells peak during the first half of the night and have minimum levels during the day (Born et al. 1997). Similarly, blood levels of pro-inflammatory cytokines show a peak during early nocturnal sleep (Lange et al. 2010).

The rhythmic oscillation of immune function allows an appropriate magnitude of immune response and ensures the resolution of injury without progressing towards chronic inflammation. It follows that different aspects of innate immunity (e.g., production of cytokines and chemokines, expression of Toll-like receptors [TLRs], antimicrobial peptides, phagocytosis, secretion of complement and coagulation factors, barrier functions) are temporally gated (with nadir and peaks) to distinct phases of the circadian cycle preventing their synchronous activation and limiting the duration of the inflammatory response (Gibbs et al. 2012; Bellet et al. 2013). The exit of differentiated immune cells from the bone marrow and trafficking of innate and adaptive immune cells also display a circadian rhythm (Mendez-Ferrer et al. 2008; Druzd et al. 2017). The rhythmic cellular clock-based expression of TLRs as well as the temporal gating of T cell activation and proliferation can maximize the immune response at a specific time window, and contribute to the time-of-the-day dependence of immunization after vaccination (Silver et al. 2012). Accordingly, studies found that morning vaccination resulted in higher viral-specific antibody responses compared with afternoon vaccination (Long et al. 2016).

The rhythmic outputs of the immune function can be generated by both extrinsic (e.g., central clock) or intrinsic (cell-autonomous circadian clock) entrainment cues (Man et al. 2016). For example, studies found that the central clock-regulated rhythmic output of the sympathetic nervous system or glucocorticoids is the dominant entrainment cue for the recruitment of hematopoietic stem cells, innate immune cells, as well as adaptive immune cells, respectively. On the other hand, diurnal oscillations in the abundance of inflammatory monocytes are under the control of Bmal1 thus gating the host's vulnerability to infection and associated tissue damage (Man et al. 2016). Furthermore, the cellular circadian clock establishes the rhythmic oscillations in the expression of basal and inducible inflammatory genes such as cytokine and chemokines, which involves among others the regulation of the transcription factor NF-κB, the major transcriptional activator of inflammation. Bmal1 has been shown to modulate monocyte trafficking, immune response and NF-κB signaling (Man et al. 2016) as well as to regulate metabolic utilization in peripheral tissue (Peek et al. 2017) and to maintain the BBB function (Nakazato et al. 2017). Bmal1 knockout caused neuroinflammation, redox imbalance, and neurodegeneration (Musiek et al. 2013). Similarly, REV-ERB-α and REV-ERB-β, and CRY contribute to rhythmically repress inflammatory gene expression by acting on NF-κB signaling (Narasimamurthy et al. 2012; Man et al. 2016).

Diurnal oscillations have also been observed in the relative abundance and composition of gut microbial communities, which translate to rhythmic production of microbial metabolites, in turn, able to influence host circadian activity as well as immune cell functions (Thaiss et al. 2016). Genetic ablation of molecular clock components, circadian disruption due to jet lag, sleep disruption, or high-fat diet may lead to alterations of the gut microbiota ecosystem (dysbiosis), which subsequently impairs host metabolism and immune function (Murakami and Tognini 2019).

1.5 Immune Responses Under Circadian Misalignment

One potential common feature of the negative health consequences of circadian disruption may be the dysregulation of the immune system. The loss of immune regulation due to circadian disruption may increase the susceptibility to tissue damage in response to infection or other challenges. Several studies assessed the perturbation of circadian systems by environmental and/or genetic manipulations in animal models. Animal studies found that global, brain, or peripheral knockout of clock genes alters this circadian fluctuation and leads to an exacerbated inflammatory response to infection or other pathogenic stimuli, oxidative stress, and age-related phenotypes thus revealing a direct role for clock genes in suppressing chronic inflammation and ensuring its timely resolution (Nguyen et al. 2013; Scheiermann et al. 2013) Notably, cytokines, including TNF-α, IL-1β, and endotoxin, as observed in inflammation and infection, inhibit the expression of core clock genes and clock-controlled genes, resulting in reduced locomotor activity and prolonged rest time (Cavadini et al. 2007), and leading to the loss of basal oscillatory rhythm and to a reprogramming of the temporal relationships between gene expression, metabolites and leukocyte trafficking (Haspel et al. 2014). Modulation of clock molecules leads to increased replication of herpes, influenza, respiratory syncytial virus, parainfluenza type 3, and hepatitis C virus (Majumdar et al. 2017; Zhuang et al. 2019) pointing to an important role of the circadian clock in virus infection. Herpes viruses target molecular clock components of the host, which in turn affects the viral replication rate (Edgar et al. 2016).

A study in mice investigating the environmental perturbation of circadian rhythms shows that experimentally induced circadian disruption schedule (four consecutive weekly 6 h phase-advances of the light-dark) increased endotoxemic shock and mortality compared to unshifted control mice in response to the immune challenge LPS (Castanon-Cervantes et al. 2010). This result was associated with a heightened inflammation in shifted animals as exemplified by higher levels of pro-inflammatory cytokines and activation of peritoneal macrophages in response to LPS treatment. Rhythms in the expression of clock genes in the central clock, liver, thymus, and peritoneal macrophages were also altered and/or inhibited after chronic jet lag, while no sleep loss or stress was documented (Castanon-Cervantes et al. 2010). Furthermore, another study investigating the chronic effects of circadian misalignment in mice revealed that the long-term nonadjustive shifted condition of the light-dark cycle, simulating the chronic jet lag, induced chronic inflammation and accelerated immune senescence in association with a reduced survival rate (Inokawa et al. 2020). Therefore, the mouse model system of exposure to long-term nonadjustive shifted light conditions may mirror the pathophysiology of chronic circadian rhythm disruption in humans.

In humans, the most common forms—and causes—of circadian misalignment due to human behaviors are chronic jet lag, when traveling across several time zones; social jet lag, which corresponds to the time difference between routine sleep cycles during the work/school week and free time on weekends, involving a discrepancy social clock and individual's circadian rhythm; and shift work, an atypical working schedule where individuals experience an unnatural routine of activity during the dark phase and sleep during the light phase of the day so that work and sleep occur at times that conflict with the circadian rhythm. These are conditions associated with sleep disruption and psychosocial stress, and adversely affect the immune system due to circadian misalignment. Human studies have found that centrally controlled rhythms as well as peripheral clocks are disturbed in such conditions (Koshy et al. 2019). In night shift workers, light exposure at night has a large impact on the resetting of peripheral circadian clocks (Cuesta et al. 2017).

Compared with the non-shift workers, the shift work schedule was associated with a decline in innate immune response, e.g., NK cell activity (Okamoto et al. 2008; Nagai et al. 2011), and an increase in the number of circulating leukocytes (Lu et al. 2016; Wirth et al. 2017), a lower CD4/CD8 ratio, cytokines and systemic endotoxemia (Atwater et al. 2021), besides to shifted and desynchronized cytokine release by immune cells (Cuesta et al. 2016). A 4-day simulated night shift work protocol in healthy subjects changed the circadian regulation of the human transcriptome and mostly affected biological processes related to the adverse health effects associated with night shift work, notably the natural killer cell-mediated immune response and inflammatory pathways (Kervezee et al. 2018). A recent cross-sectional study showed that shift work, particularly night work, was associated with a 1.85-fold increased risk of COVID-19 infection (Fatima et al. 2021). In a case-control study on healthcare workers, people with sleep problems had greater odds of COVID-19 (Kim et al. 2021). Individuals with social jet lag were also significantly associated with a higher risk of COVID-19 infection (2.07-fold) (Coelho et al. 2022). A recent study on the vaccination response in people with circadian disruption, i.e., shift workers, found that compared with day workers shift workers had altered sleep architecture, with a lower slow wave sleep and REM duration, higher levels of cytokines and a weaker specific leukocyte-mediated immune response to vaccination against meningococcal C meningitis (Ruiz et al. 2020). Research concerning the impact of sleep disruption and circadian misalignment in shift workers on the immune response to COVID-19 vaccination is currently ongoing (Lammers-van der Holst et al. 2022).

A relationship between sleep and the circadian system has emerged in an investigation carried out in humans where sleep deprivation (5.70 h compared with 8.50 h sleep per 24 h, for 1 week) led in the blood transcriptome to a significant reduction in rhythmic transcripts (from 8.8% to 6.9%) and affected many genes associated with sleep homeostasis, oxidative stress, metabolism, inflammatory and stress responses, chromatin organization and modification (Moller-Levet et al. 2013). Among the genes whose rhythmic expression was altered by sleep loss, there were several genes classified as circadian, including classic clock genes, such as Bmal1, PERs, CRY, neuronal PAS domain-containing protein (NPAS), REV-ERB-α, REV-ERB-β, with a significant reduction in the number of rhythmic transcripts, in the circadian amplitude and the width of the period of expression. On the other hand, another set of genes, including those associated with RNA metabolic processes, became circadian following sleep loss (Moller-Levet et al. 2013).

These data are following previous animal studies (Barclay et al. 2012) and provide potential molecular mechanisms whereby sleep loss can lead to circadian disruption, and cause negative health consequences. Furthermore, while many transcripts in the human blood transcriptome have a circadian expression profile when sleep occurred in phase with the central circadian rhythm (as indexed by the melatonin rhythm), when sleep and associated locomotor, feeding, and metabolic rhythms are phase-shifted compared with the circadian clock (mistimed sleep or forced desynchrony), the majority of circadian transcripts became arrhythmic (reduction in the number of circadian transcripts from 6.4% to 1% and changes in the overall time course of expression of 34% of transcripts) (Archer et al. 2014). In addition, results from a mathematical modeling analysis, which separated the relative contribution of sleep and circadian rhythmicity on the temporal gene expression profile, suggests that circadian-driven transcripts are mainly associated with cellular metabolic and homeostatic processes, whereas sleep-driven transcripts are linked with the regulation of transcription and protein synthesis (Archer et al. 2014).

Overall, these data underscore the important role of sleep and circadian rhythmicity in the regulation of tissue gene expression and functions (Maret et al. 2007; Mongrain et al. 2010; Pellegrino et al. 2012; Anafi et al. 2013; Perrin et al. 2018), and indicate that a common feature is a 24-h organization of molecular processes, including the immune function, in humans (Fig. 1.1). They also suggest that appropriate sleep duration, quality, and timing significantly contribute to the overall temporal organization of the human transcriptome, and desynchrony of sleep and centrally driven circadian rhythms, as occurs in shift work and jet lag, may lead to disruption of rhythmicity in physiology and endocrinology.

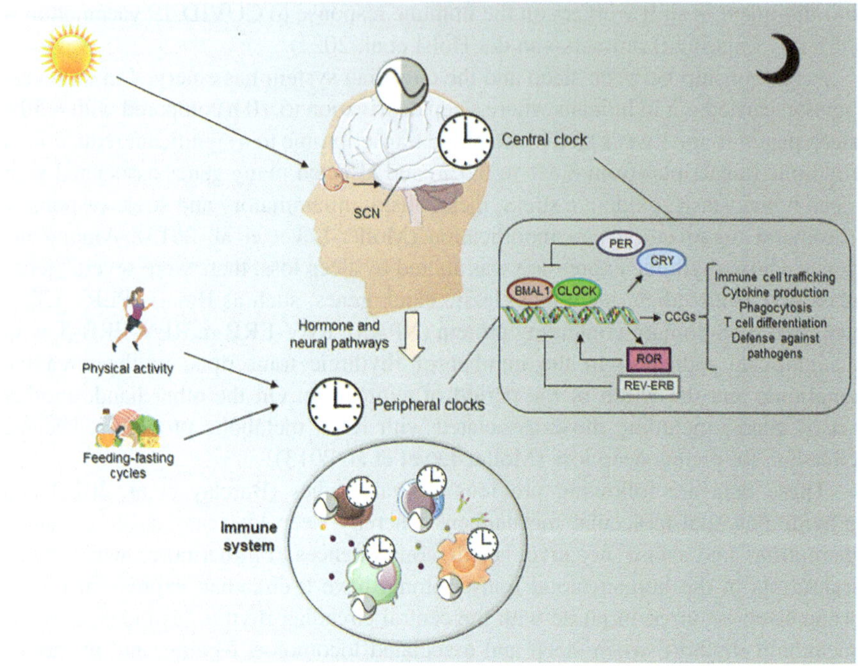

Fig. 1.1 Regulation of the immune system by sleep and the circadian system. Immune system functions, including cell proliferation, differentiation, trafficking, activity, and cytokine production, are regulated by both homeostatic and circadian drives of sleep. The circadian clocks, with the cell-autonomous transcriptional–translational feedback loop mechanism (depicted on the right inset), include the central clock, located in the suprachiasmatic nucleus (SCN) of the hypothalamus, which is entrained primarily by environmental cues from light. This SCN coordinates 24-h rhythms in physiology and behavior through neural and hormonal signals that synchronize peripheral clocks in peripheral organs or tissues, including the immune cells, through the expression of clock-controlled genes (CCGs). Peripheral clocks can also be synchronized by behaviors, including eating and physical activity. (Illustrations are adapted from Servier Medical Art (http://smart.servier.com/))

1.6 Conclusion and Perspectives

Epidemiological data suggest that increasing numbers of individuals are becoming sleep disturbed and circadian misaligned, and the accumulation of these disturbances over years may induce safety-related problems and have gradual cumulative adverse effects on health, increasing the risk of chronic diseases and the susceptibility to pathogens and pollution (Fig. 1.2). Sleep and the circadian rhythm exert immune-supportive and regulatory functions, and impairments of the immune inflammatory system are plausible mechanisms mediating the negative health effects of sleep and circadian disruption. Evidence-based molecular and clinical studies have provided a framework to understand the link between these disturbances and inflammatory immune diseases, positing a causal role for the immune derangements in the development and/or exacerbation of inflammatory and infectious diseases. Sleep

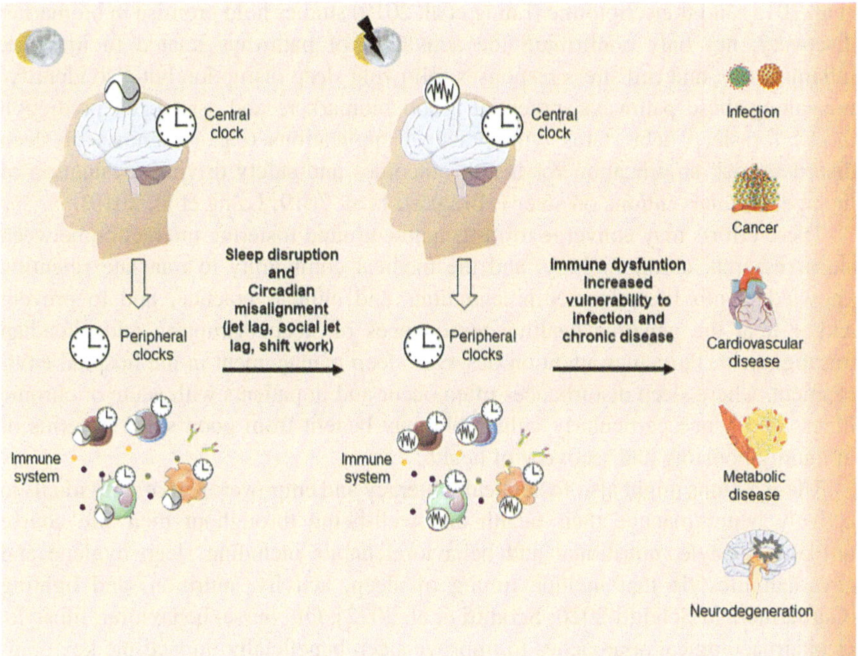

Fig. 1.2 Conditions of sleep disruption (e.g., sleep deprivation) and misaligned sleep–wake cycles alter immune processes and are associated with or increase the risk for many diseases (right). (Illustrations are adapted from Servier Medical Art (http://smart.servier.com/))

deprivation superimposed on preexisting sleep disorders, as occurred in abstinent alcohol-dependent subjects, more strongly increases susceptibility to inflammation compared with subjects without sleep problems (Irwin et al. 2004). Similarly, sleep deprivation can worsen the risk profile in individuals with hypertension, in which even half a night of sleep deprivation elevates blood pressure the next day (Lusardi et al. 1999).

A practical implication of these observations is that immune homeostasis may be a crucial target for intervention against diseases associated with sleep and circadian clock disruption. Currently, limited evidence exists in this context. Treatment of insomnia has been demonstrated to reverse humoral and cell inflammatory activation, as well as to reduce diabetic and cardiovascular risk, as assessed by a multisystem biological risk based on eight biomarkers (Carroll et al. 2015). Randomized controlled trials assessing the effect of the treatment of sleep and circadian disturbances on inflammatory immune dysfunction and/or health outcomes are needed to provide a cause–effect relationship. Knowledge of inflammatory and immunological signatures in response to sleep curtailment, mostly through omic-based approaches, may inform not only on the underlying molecular links but also contribute to refining risk profiles to be used for developing biomarkers of disturbed sleep and sleep disturbance-related health outcomes. Recent metabolomics (Weljie et al. 2015) and transcriptomic (Laing et al. 2019) studies hold promise in biomarker discovery, not only confirming the activation of pathways related to immune, inflammatory, and cell stress responses following sleep disruption but also identifying, along these pathways, potential blood biomarkers and associated prediction model for sleep debt status with practical applications (e.g., diagnosis of sleep disorders, risk stratification for health outcomes and safety driving, evaluation of therapeutic interventions on sleep) (Bragazzi et al. 2019; Laing et al. 2019).

These efforts may converge towards a new ground fostering interaction between sleep research, chronobiology, and the medical community to translate scientific knowledge into lifestyle recommendations and clinical practice, and to prevent and/or treat the negative health consequences of sleep disruption and circadian misalignment. Particular attention deserves sleep management in the hospital environment, where sleep disturbances often occur and in patients with acute or chronic illness and hence particularly vulnerable may benefit from good sleep in terms of immunity boosting and recovery of health.

These actions might also foster health literacy and empowerment of individuals to actively better manage their health and well-being throughout their life course utilizing lifestyle, nutritional and behavioral habits including sleep hygiene and circadian lifestyle that include timing of sleep, activity, nutrition, and lighting (Garbarino and Scoditti 2020; Scoditti et al. 2022). Of course, behavioral, lifestyle, or pharmacological approaches to improve sleep beneficially impact one key component of the circadian rhythm and may indirectly benefit other aspects of the daily rhythms thereby leading to a better quality of life. Further knowledge in the circadian system and its interaction with immunity may advance strategies to prevent or treat several chronic diseases, establishing how to maintain a regular circadian rhythm in sleep–wake, feeding–fasting, or light–dark cycles; how to optimize the timing of drug treatment; and how to directly target a circadian clock component (clock-targeting pharmacotherapy) for treating inflammatory disorders (Sulli et al. 2018).

Conclusively, in the perspective of staying healthy in this rapidly changing world, the sleep/circadian clock–immunity relationship raises relevant clinical implications for promoting health. During the COVID-19 pandemic, these issues have become particularly important because of sleep disturbances often reported in association with the heavy social, work, and lifestyle changes imposed to contain the virus spreading. Moreover, healthcare workers, which are subjected to night shifts, irregular sleep–wake schedules, changes in daily routine, and circadian rhythm problems, have been found to have more severe insomnia and a greater risk of coronavirus infection (Kim et al. 2021). Pending further research, sleep and circadian-based intervention may reduce the susceptibility to SARS-CoV-2 infection and the severity of COVID-19, and improve immune response to COVID-19 vaccination (Meira et al. 2020).

References

Abbott SM, Malkani RG, Zee PC (2020) Circadian disruption and human health: a bidirectional relationship. Eur J Neurosci 51:567–583. https://doi.org/10.1111/ejn.14298

Aho V, Ollila HM, Rantanen V, Kronholm E, Surakka I, Van Leeuwen WM, Lehto M, Matikainen S, Ripatti S, Harma M, Sallinen M, Salomaa V, Jauhiainen M, Alenius H, Paunio T, Porkka-Heiskanen T (2013) Partial sleep restriction activates immune response-related gene expression pathways: experimental and epidemiological studies in humans. PLoS One 8:e77184. https://doi.org/10.1371/journal.pone.0077184

Anafi RC, Pellegrino R, Shockley KR, Romer M, Tufik S, Pack AI (2013) Sleep is not just for the brain: transcriptional responses to sleep in peripheral tissues. BMC Genomics 14:362. https://doi.org/10.1186/1471-2164-14-362

Archer SN, Laing EE, Moller-Levet CS, Van Der Veen DR, Bucca G, Lazar AS, Santhi N, Slak A, Kabiljo R, Von Schantz M, Smith CP, Dijk DJ (2014) Mistimed sleep disrupts circadian regulation of the human transcriptome. Proc Natl Acad Sci U S A 111:E682–E691. https://doi.org/10.1073/pnas.1316335111

Atwater AQ, Immergluck LC, Davidson AJ, Castanon-Cervantes O (2021) Shift work predicts increases in lipopolysaccharide-binding protein, interleukin-10, and leukocyte counts in a cross-sectional study of healthy volunteers carrying low-grade systemic inflammation. Int J Environ Res Public Health 18(24):13158. https://doi.org/10.3390/ijerph182413158

Axelsson J, Rehman JU, Akerstedt T, Ekman R, Miller GE, Hoglund CO, Lekander M (2013) Effects of sustained sleep restriction on mitogen-stimulated cytokines, chemokines and T helper 1/T helper 2 balance in humans. PLoS One 8:e82291. https://doi.org/10.1371/journal.pone.0082291

Bakour C, Schwartz S, O'Rourke K, Wang W, Sappenfield W, Couluris M, Chen H (2017) Sleep duration trajectories and systemic inflammation in young adults: results from the National Longitudinal Study of Adolescent to Adult Health (Add Health). Sleep 40(11):zsx156. https://doi.org/10.1093/sleep/zsx156

Barclay JL, Husse J, Bode B, Naujokat N, Meyer-Kovac J, Schmid SM, Lehnert H, Oster H (2012) Circadian desynchrony promotes metabolic disruption in a mouse model of shiftwork. PLoS One 7:e37150. https://doi.org/10.1371/journal.pone.0037150

Bellet MM, Deriu E, Liu JZ, Grimaldi B, Blaschitz C, Zeller M, Edwards RA, Sahar S, Dandekar S, Baldi P, George MD, Raffatellu M, Sassone-Corsi P (2013) Circadian clock regulates the host response to Salmonella. Proc Natl Acad Sci U S A 110:9897–9902. https://doi.org/10.1073/pnas.1120636110

Besedovsky L, Lange T, Haack M (2019) The sleep-immune crosstalk in health and disease. Physiol Rev 99:1325–1380. https://doi.org/10.1152/physrev.00010.2018

Boivin DB, Boudreau P, Kosmadopoulos A (2022) Disturbance of the circadian system in shift work and its health impact. J Biol Rhythm 37:3–28. https://doi.org/10.1177/07487304211064218

Born J, Lange T, Hansen K, Molle M, Fehm HL (1997) Effects of sleep and circadian rhythm on human circulating immune cells. J Immunol 158:4454–4464

Bragazzi NL, Guglielmi O, Garbarino S (2019) SleepOMICS: how big data can revolutionize sleep science. Int J Environ Res Public Health 16(2):291. https://doi.org/10.3390/ijerph16020291

Cappuccio FP, Miller MA (2017) Sleep and cardio-metabolic disease. Curr Cardiol Rep 19:110. https://doi.org/10.1007/s11886-017-0916-0

Carroll JE, Seeman TE, Olmstead R, Melendez G, Sadakane R, Bootzin R, Nicassio P, Irwin MR (2015) Improved sleep quality in older adults with insomnia reduces biomarkers of disease risk: pilot results from a randomized controlled comparative efficacy trial. Psychoneuroendocrinology 55:184–192. https://doi.org/10.1016/j.psyneuen.2015.02.010

Carroll JE, Irwin MR, Levine M, Seeman TE, Absher D, Assimes T, Horvath S (2017) Epigenetic aging and immune senescence in women with insomnia symptoms: findings from the Women's

Health Initiative Study. Biol Psychiatry 81:136–144. https://doi.org/10.1016/j.biopsych.2016.07.008

Castanon-Cervantes O, Wu M, Ehlen JC, Paul K, Gamble KL, Johnson RL, Besing RC, Menaker M, Gewirtz AT, Davidson AJ (2010) Dysregulation of inflammatory responses by chronic circadian disruption. J Immunol 185:5796–5805. https://doi.org/10.4049/jimmunol.1001026

Cavadini G, Petrzilka S, Kohler P, Jud C, Tobler I, Birchler T, Fontana A (2007) TNF-alpha suppresses the expression of clock genes by interfering with E-box-mediated transcription. Proc Natl Acad Sci U S A 104:12843–12848. https://doi.org/10.1073/pnas.0701466104

Chaponda M, Aldhouse N, Kroes M, Wild L, Robinson C, Smith A (2018) Systematic review of the prevalence of psychiatric illness and sleep disturbance as co-morbidities of HIV infection in the UK. Int J STD AIDS 29:704–713. https://doi.org/10.1177/0956462417750708

Chen X, Song M, Zhang B, Zhang Y (2016) Reactive oxygen species regulate T cell immune response in the tumor microenvironment. Oxidative Med Cell Longev 2016:1580967. https://doi.org/10.1155/2016/1580967

Christou S, Wehrens SMT, Isherwood C, Moller-Levet CS, Wu H, Revell VL, Bucca G, Skene DJ, Laing EE, Archer SN, Johnston JD (2019) Circadian regulation in human white adipose tissue revealed by transcriptome and metabolic network analysis. Sci Rep 9:2641. https://doi.org/10.1038/s41598-019-39668-3

Chung WS, Lin HH, Cheng NC (2016) The incidence and risk of Herpes Zoster in patients with sleep disorders: a population-based cohort study. Medicine (Baltimore) 95:e2195. https://doi.org/10.1097/MD.0000000000002195

Coelho J, Micoulaud-Franchi JA, Wiet AS, Nguyen D, Taillard J, Philip P (2022) Circadian misalignment is associated with Covid-19 infection. Sleep Med 93:71–74. https://doi.org/10.1016/j.sleep.2022.03.015

Cohen S, Doyle WJ, Alper CM, Janicki-Deverts D, Turner RB (2009) Sleep habits and susceptibility to the common cold. Arch Intern Med 169:62–67. https://doi.org/10.1001/archinternmed.2008.505

Consensus Conference Panel, Watson NF, Badr MS, Belenky G, Bliwise DL, Buxton OM, Buysse D, Dinges DF, Gangwisch J, Grandner MA, Kushida C, Malhotra RK, Martin JL, Patel SR, Quan SF, Tasali E (2015a) Joint Consensus Statement of the American Academy of Sleep Medicine and Sleep Research Society on the recommended amount of sleep for a healthy adult: methodology and discussion. J Clin Sleep Med 11:931–952. https://doi.org/10.5664/jcsm.4950

Consensus Conference Panel, Watson NF, Badr MS, Belenky G, Bliwise DL, Buxton OM, Buysse D, Dinges DF, Gangwisch J, Grandner MA, Kushida C, Malhotra RK, Martin JL, Patel SR, Quan SF, Tasali E, Non-Participating Observers, Twery M, Croft JB, Maher E, American Academy of Sleep Medicine Staff, Barrett JA, Thomas SM, Heald JL (2015b) Recommended amount of sleep for a healthy adult: a Joint Consensus Statement of the American Academy of Sleep Medicine and Sleep Research Society. J Clin Sleep Med 11:591–592. https://doi.org/10.5664/jcsm.4758

Cuesta M, Boudreau P, Dubeau-Laramee G, Cermakian N, Boivin DB (2016) Simulated night shift disrupts circadian rhythms of immune functions in humans. J Immunol 196:2466–2475. https://doi.org/10.4049/jimmunol.1502422

Cuesta M, Boudreau P, Cermakian N, Boivin DB (2017) Rapid resetting of human peripheral clocks by phototherapy during simulated night shift work. Sci Rep 7:16310. https://doi.org/10.1038/s41598-017-16429-8

Dantzer R (2018) Neuroimmune interactions: from the brain to the immune system and vice versa. Physiol Rev 98:477–504. https://doi.org/10.1152/physrev.00039.2016

Dimitrov S, Lange T, Tieken S, Fehm HL, Born J (2004) Sleep associated regulation of T helper 1/T helper 2 cytokine balance in humans. Brain Behav Immun 18:341–348. https://doi.org/10.1016/j.bbi.2003.08.004

Dimitrov S, Lange T, Nohroudi K, Born J (2007) Number and function of circulating human antigen presenting cells regulated by sleep. Sleep 30:401–411. https://doi.org/10.1093/sleep/30.4.401

Disis ML (2010) Immune regulation of cancer. J Clin Oncol 28:4531–4538. https://doi.org/10.1200/JCO.2009.27.2146

Druzd D, Matveeva O, Ince L, Harrison U, He W, Schmal C, Herzel H, Tsang AH, Kawakami N, Leliavski A, Uhl O, Yao L, Sander LE, Chen CS, Kraus K, De Juan A, Hergenhan SM, Ehlers M, Koletzko B, Haas R, Solbach W, Oster H, Scheiermann C (2017) Lymphocyte circadian clocks control lymph node trafficking and adaptive immune responses. Immunity 46:120–132. https://doi.org/10.1016/j.immuni.2016.12.011

Edgar RS, Stangherlin A, Nagy AD, Nicoll MP, Efstathiou S, O'Neill JS, Reddy AB (2016) Cell autonomous regulation of herpes and influenza virus infection by the circadian clock. Proc Natl Acad Sci U S A 113:10085–10090. https://doi.org/10.1073/pnas.1601895113

Everson CA, Toth LA (2000) Systemic bacterial invasion induced by sleep deprivation. Am J Physiol Regul Integr Comp Physiol 278:R905–R916. https://doi.org/10.1152/ajpregu.2000.278.4.R905

Everson CA, Thalacker CD, Hogg N (2008) Phagocyte migration and cellular stress induced in liver, lung, and intestine during sleep loss and sleep recovery. Am J Physiol Regul Integr Comp Physiol 295:R2067–R2074. https://doi.org/10.1152/ajpregu.90623.2008

Everson CA, Henchen CJ, Szabo A, Hogg N (2014) Cell injury and repair resulting from sleep loss and sleep recovery in laboratory rats. Sleep 37:1929–1940. https://doi.org/10.5665/sleep.4244

Fatima Y, Bucks RS, Mamun AA, Skinner I, Rosenzweig I, Leschziner G, Skinner TC (2021) Shift work is associated with increased risk of COVID-19: findings from the UK Biobank cohort. J Sleep Res 30:e13326. https://doi.org/10.1111/jsr.13326

Ferrie JE, Kivimaki M, Akbaraly TN, Singh-Manoux A, Miller MA, Gimeno D, Kumari M, Davey Smith G, Shipley MJ (2013) Associations between change in sleep duration and inflammation: findings on C-reactive protein and interleukin 6 in the Whitehall II Study. Am J Epidemiol 178: 956–961. https://doi.org/10.1093/aje/kwt072

Fishbein AB, Knutson KL, Zee PC (2021) Circadian disruption and human health. J Clin Invest 131(19):e14828. https://doi.org/10.1172/JCI148286

Fondell E, Axelsson J, Franck K, Ploner A, Lekander M, Balter K, Gaines H (2011) Short natural sleep is associated with higher T cell and lower NK cell activities. Brain Behav Immun 25: 1367–1375. https://doi.org/10.1016/j.bbi.2011.04.004

Ford ES, Cunningham TJ, Croft JB (2015a) Trends in self-reported sleep duration among US adults from 1985 to 2012. Sleep 38:829–832. https://doi.org/10.5665/sleep.4684

Ford ES, Cunningham TJ, Giles WH, Croft JB (2015b) Trends in insomnia and excessive daytime sleepiness among U.S. adults from 2002 to 2012. Sleep Med 16:372–378. https://doi.org/10.1016/j.sleep.2014.12.008

Friese RS, Bruns B, Sinton CM (2009) Sleep deprivation after septic insult increases mortality independent of age. J Trauma 66:50–54. https://doi.org/10.1097/TA.0b013e318190c3a1

Garbarino S, Scoditti E (2020) On the role of sleep hygiene in health management during COVID-19 pandemic. Sleep Med 77:74. https://doi.org/10.1016/j.sleep.2020.11.036

Garbarino S, Lanteri P, Bragazzi NL, Magnavita N, Scoditti E (2021) Role of sleep deprivation in immune-related disease risk and outcomes. Commun Biol 4:1304. https://doi.org/10.1038/s42003-021-02825-4

Gibbs JE, Blaikley J, Beesley S, Matthews L, Simpson KD, Boyce SH, Farrow SN, Else KJ, Singh D, Ray DW, Loudon AS (2012) The nuclear receptor REV-ERBalpha mediates circadian regulation of innate immunity through selective regulation of inflammatory cytokines. Proc Natl Acad Sci U S A 109:582–587. https://doi.org/10.1073/pnas.1106750109

Gilmour H, Stranges S, Kaplan M, Feeny D, Mcfarland B, Huguet N, Bernier J (2013) Longitudinal trajectories of sleep duration in the general population. Health Rep 24:14–20

Ginaldi L, De Martinis M, D'Ostilio A, Marini L, Loreto MF, Corsi MP, Quaglino D (1999) The immune system in the elderly: I. Specific humoral immunity. Immunol Res 20:101–108. https://doi.org/10.1007/BF02786466
Grandner MA (2017) Sleep, health, and society. Sleep Med Clin 12:1–22. https://doi.org/10.1016/j.jsmc.2016.10.012
Hall MH, Smagula SF, Boudreau RM, Ayonayon HN, Goldman SE, Harris TB, Naydeck BL, Rubin SM, Samuelsson L, Satterfield S, Stone KL, Visser M, Newman AB (2015) Association between sleep duration and mortality is mediated by markers of inflammation and health in older adults: the Health, Aging and Body Composition Study. Sleep 38:189–195. https://doi.org/10.5665/sleep.4394
Hand TW, Vujkovic-Cvijin I, Ridaura VK, Belkaid Y (2016) Linking the microbiota, chronic disease, and the immune system. Trends Endocrinol Metab 27:831–843. https://doi.org/10.1016/j.tem.2016.08.003
Haspel JA, Chettimada S, Shaik RS, Chu JH, Raby BA, Cernadas M, Carey V, Process V, Hunninghake GM, Ifedigbo E, Lederer JA, Englert J, Pelton A, Coronata A, Fredenburgh LE, Choi AM (2014) Circadian rhythm reprogramming during lung inflammation. Nat Commun 5:4753. https://doi.org/10.1038/ncomms5753
Hu J, Chen Z, Gorczynski CP, Gorczynski LY, Kai Y, Lee L, Manuel J, Gorczynski RM (2003) Sleep-deprived mice show altered cytokine production manifest by perturbations in serum IL-1ra, TNFa, and IL-6 levels. Brain Behav Immun 17:498–504. https://doi.org/10.1016/j.bbi.2003.03.001
Inokawa H, Umemura Y, Shimba A, Kawakami E, Koike N, Tsuchiya Y, Ohashi M, Minami Y, Cui G, Asahi T, Ono R, Sasawaki Y, Konishi E, Yoo SH, Chen Z, Teramukai S, Ikuta K, Yagita K (2020) Chronic circadian misalignment accelerates immune senescence and abbreviates lifespan in mice. Sci Rep 10:2569. https://doi.org/10.1038/s41598-020-59541-y
Irwin MR (2015) Why sleep is important for health: a psychoneuroimmunology perspective. Annu Rev Psychol 66:143–172. https://doi.org/10.1146/annurev-psych-010213-115205
Irwin M, Mascovich A, Gillin JC, Willoughby R, Pike J, Smith TL (1994) Partial sleep deprivation reduces natural killer cell activity in humans. Psychosom Med 56:493–498. https://doi.org/10.1097/00006842-199411000-00004
Irwin M, Rinetti G, Redwine L, Motivala S, Dang J, Ehlers C (2004) Nocturnal proinflammatory cytokine-associated sleep disturbances in abstinent African American alcoholics. Brain Behav Immun 18:349–360. https://doi.org/10.1016/j.bbi.2004.02.001
Irwin MR, Wang M, Campomayor CO, Collado-Hidalgo A, Cole S (2006) Sleep deprivation and activation of morning levels of cellular and genomic markers of inflammation. Arch Intern Med 166:1756–1762. https://doi.org/10.1001/archinte.166.16.1756
Irwin MR, Wang M, Ribeiro D, Cho HJ, Olmstead R, Breen EC, Martinez-Maza O, Cole S (2008) Sleep loss activates cellular inflammatory signaling. Biol Psychiatry 64:538–540. https://doi.org/10.1016/j.biopsych.2008.05.004
Irwin MR, Witarama T, Caudill M, Olmstead R, Breen EC (2015) Sleep loss activates cellular inflammation and signal transducer and activator of transcription (STAT) family proteins in humans. Brain Behav Immun 47:86–92. https://doi.org/10.1016/j.bbi.2014.09.017
Ishimori K (1909) True causes of sleep: a hypnogenic substance as evidenced in the brain of sleep-deprived animals. Tokyo Igakki Zasshi 23:429–457
Jackowska M, Hamer M, Carvalho LA, Erusalimsky JD, Butcher L, Steptoe A (2012) Short sleep duration is associated with shorter telomere length in healthy men: findings from the Whitehall II cohort study. PLoS One 7:e47292. https://doi.org/10.1371/journal.pone.0047292
James S, Mclanahan S, Brooks-Gunn J, Mitchell C, Schneper L, Wagner B, Notterman DA (2017) Sleep duration and telomere length in children. J Pediatr 187:247–252.e241. https://doi.org/10.1016/j.jpeds.2017.05.014
Kervezee L, Cuesta M, Cermakian N, Boivin DB (2018) Simulated night shift work induces circadian misalignment of the human peripheral blood mononuclear cell transcriptome. Proc Natl Acad Sci U S A 115:5540–5545. https://doi.org/10.1073/pnas.1720719115

Kim H, Hegde S, Lafiura C, Raghavan M, Luong E, Cheng S, Rebholz CM, Seidelmann SB (2021) COVID-19 illness in relation to sleep and burnout. BMJ Nutr Prev Health 4:132–139. https://doi.org/10.1136/bmjnph-2021-000228

Koshy A, Cuesta M, Boudreau P, Cermakian N, Boivin DB (2019) Disruption of central and peripheral circadian clocks in police officers working at night. FASEB J 33:6789–6800. https://doi.org/10.1096/fj.201801889R

Krueger JM, Pappenheimer JR, Karnovsky ML (1982) The composition of sleep-promoting factor isolated from human urine. J Biol Chem 257:1664–1669

Kubota T, Fang J, Kushikata T, Krueger JM (2000) Interleukin-13 and transforming growth factor-beta1 inhibit spontaneous sleep in rabbits. Am J Physiol Regul Integr Comp Physiol 279:R786–R792. https://doi.org/10.1152/ajpregu.2000.279.3.R786

Labrecque N, Cermakian N (2015) Circadian clocks in the immune system. J Biol Rhythm 30:277–290. https://doi.org/10.1177/0748730415577723

Laing EE, Moller-Levet CS, Dijk DJ, Archer SN (2019) Identifying and validating blood mRNA biomarkers for acute and chronic insufficient sleep in humans: a machine learning approach. Sleep 42(1):zsy186. https://doi.org/10.1093/sleep/zsy186

Lammers-Van Der Holst HM, Lammers GJ, Van Der Horst GTJ, Chaves I, De Vries RD, Geurtsvankessel CH, Koch B, Van Der Kuy HM (2022) Understanding the association between sleep, shift work and COVID-19 vaccine immune response efficacy: protocol of the S-CORE study. J Sleep Res 31:e13496. https://doi.org/10.1111/jsr.13496

Lange T, Dimitrov S, Born J (2010) Effects of sleep and circadian rhythm on the human immune system. Ann N Y Acad Sci 1193:48–59. https://doi.org/10.1111/j.1749-6632.2009.05300.x

Lange T, Dimitrov S, Bollinger T, Diekelmann S, Born J (2011) Sleep after vaccination boosts immunological memory. J Immunol 187:283–290. https://doi.org/10.4049/jimmunol.1100015

Lasselin J, Rehman JU, Akerstedt T, Lekander M, Axelsson J (2015) Effect of long-term sleep restriction and subsequent recovery sleep on the diurnal rhythms of white blood cell subpopulations. Brain Behav Immun 47:93–99. https://doi.org/10.1016/j.bbi.2014.10.004

Lee HL, Jang JW, Lee SW, Yoo SH, Kwon JH, Nam SW, Bae SH, Choi JY, Han NI, Yoon SK (2019) Inflammatory cytokines and change of Th1/Th2 balance as prognostic indicators for hepatocellular carcinoma in patients treated with transarterial chemoembolization. Sci Rep 9:3260. https://doi.org/10.1038/s41598-019-40078-8

Legendre R, Piéron H (1913) Recherches sur le besoin de sommeil consécutif à une veille prolongée. Z Physiol 14:235–262

Li Y, Zhong X, Cheng G, Zhao C, Zhang L, Hong Y, Wan Q, He R, Wang Z (2017) Hs-CRP and all-cause, cardiovascular, and cancer mortality risk: a meta-analysis. Atherosclerosis 259:75–82. https://doi.org/10.1016/j.atherosclerosis.2017.02.003

Liguori I, Russo G, Curcio F, Bulli G, Aran L, Della-Morte D, Gargiulo G, Testa G, Cacciatore F, Bonaduce D, Abete P (2018) Oxidative stress, aging, and diseases. Clin Interv Aging 13:757–772. https://doi.org/10.2147/CIA.S158513

Long JE, Drayson MT, Taylor AE, Toellner KM, Lord JM, Phillips AC (2016) Morning vaccination enhances antibody response over afternoon vaccination: a cluster-randomised trial. Vaccine 34:2679–2685. https://doi.org/10.1016/j.vaccine.2016.04.032

Lu LF, Wang CP, Tsai IT, Hung WC, Yu TH, Wu CC, Hsu CC, Lu YC, Chung FM, Jean MC (2016) Relationship between shift work and peripheral total and differential leukocyte counts in Chinese steel workers. J Occup Health 58:81–88. https://doi.org/10.1539/joh.15-0137-OA

Lusardi P, Zoppi A, Preti P, Pesce RM, Piazza E, Fogari R (1999) Effects of insufficient sleep on blood pressure in hypertensive patients: a 24-h study. Am J Hypertens 12:63–68. https://doi.org/10.1016/s0895-7061(98)00200-3

Madonna R, De Caterina R (2012) Relevance of new drug discovery to reduce NF-kappaB activation in cardiovascular disease. Vasc Pharmacol 57:41–47. https://doi.org/10.1016/j.vph.2012.02.005

Majumdar T, Dhar J, Patel S, Kondratov R, Barik S (2017) Circadian transcription factor BMAL1 regulates innate immunity against select RNA viruses. Innate Immun 23:147–154. https://doi.org/10.1177/1753425916681075

Man K, Loudon A, Chawla A (2016) Immunity around the clock. Science 354:999–1003. https://doi.org/10.1126/science.aah4966

Maret S, Dorsaz S, Gurcel L, Pradervand S, Petit B, Pfister C, Hagenbuchle O, O'Hara BF, Franken P, Tafti M (2007) Homer1a is a core brain molecular correlate of sleep loss. Proc Natl Acad Sci U S A 104:20090–20095. https://doi.org/10.1073/pnas.0710131104

Matricciani L, Olds T, Petkov J (2012) In search of lost sleep: secular trends in the sleep time of school-aged children and adolescents. Sleep Med Rev 16:203–211. https://doi.org/10.1016/j.smrv.2011.03.005

Meira ECM, Miyazawa M, Gozal D (2020) Putative contributions of circadian clock and sleep in the context of SARS-CoV-2 infection. Eur Respir J 55(6):2001023. https://doi.org/10.1183/13993003.01023-2020

Mendez-Ferrer S, Lucas D, Battista M, Frenette PS (2008) Haematopoietic stem cell release is regulated by circadian oscillations. Nature 452:442–447. https://doi.org/10.1038/nature06685

Miller MA, Kandala NB, Kivimaki M, Kumari M, Brunner EJ, Lowe GD, Marmot MG, Cappuccio FP (2009) Gender differences in the cross-sectional relationships between sleep duration and markers of inflammation: Whitehall II study. Sleep 32:857–864

Mitler MM, Carskadon MA, Czeisler CA, Dement WC, Dinges DF, Graeber RC (1988) Catastrophes, sleep, and public policy: consensus report. Sleep 11:100–109. https://doi.org/10.1093/sleep/11.1.100

Moller-Levet CS, Archer SN, Bucca G, Laing EE, Slak A, Kabiljo R, Lo JC, Santhi N, Von Schantz M, Smith CP, Dijk DJ (2013) Effects of insufficient sleep on circadian rhythmicity and expression amplitude of the human blood transcriptome. Proc Natl Acad Sci U S A 110:E1132–E1141. https://doi.org/10.1073/pnas.1217154110

Mongrain V, Hernandez SA, Pradervand S, Dorsaz S, Curie T, Hagiwara G, Gip P, Heller HC, Franken P (2010) Separating the contribution of glucocorticoids and wakefulness to the molecular and electrophysiological correlates of sleep homeostasis. Sleep 33:1147–1157. https://doi.org/10.1093/sleep/33.9.1147

Moser EK, Field NS, Oliver PM (2018) Aberrant Th2 inflammation drives dysfunction of alveolar macrophages and susceptibility to bacterial pneumonia. Cell Mol Immunol 15:480–492. https://doi.org/10.1038/cmi.2016.69

Mullington J, Korth C, Hermann DM, Orth A, Galanos C, Holsboer F, Pollmacher T (2000) Dose-dependent effects of endotoxin on human sleep. Am J Physiol Regul Integr Comp Physiol 278: R947–R955. https://doi.org/10.1152/ajpregu.2000.278.4.R947

Murakami M, Tognini P (2019) The circadian clock as an essential molecular link between host physiology and microorganisms. Front Cell Infect Microbiol 9:469. https://doi.org/10.3389/fcimb.2019.00469

Mure LS, Le HD, Benegiamo G, Chang MW, Rios L, Jillani N, Ngotho M, Kariuki T, Dkhissi-Benyahya O, Cooper HM, Panda S (2018) Diurnal transcriptome atlas of a primate across major neural and peripheral tissues. Science 359(6381):eaao0318. https://doi.org/10.1126/science.aao0318

Musiek ES, Lim MM, Yang G, Bauer AQ, Qi L, Lee Y, Roh JH, Ortiz-Gonzalez X, Dearborn JT, Culver JP, Herzog ED, Hogenesch JB, Wozniak DF, Dikranian K, Giasson BI, Weaver DR, Holtzman DM, Fitzgerald GA (2013) Circadian clock proteins regulate neuronal redox homeostasis and neurodegeneration. J Clin Invest 123:5389–5400. https://doi.org/10.1172/JCI70317

Nagai M, Morikawa Y, Kitaoka K, Nakamura K, Sakurai M, Nishijo M, Hamazaki Y, Maruzeni S, Nakagawa H (2011) Effects of fatigue on immune function in nurses performing shift work. J Occup Health 53:312–319. https://doi.org/10.1539/joh.10-0072-oa

Nakazato R, Kawabe K, Yamada D, Ikeno S, Mieda M, Shimba S, Hinoi E, Yoneda Y, Takarada T (2017) Disruption of Bmal1 impairs blood-brain barrier integrity via pericyte dysfunction. J Neurosci 37:10052–10062. https://doi.org/10.1523/JNEUROSCI.3639-16.2017

Narasimamurthy R, Hatori M, Nayak SK, Liu F, Panda S, Verma IM (2012) Circadian clock protein cryptochrome regulates the expression of proinflammatory cytokines. Proc Natl Acad Sci U S A 109:12662–12667. https://doi.org/10.1073/pnas.1209965109

Nguyen KD, Fentress SJ, Qiu Y, Yun K, Cox JS, Chawla A (2013) Circadian gene Bmal1 regulates diurnal oscillations of Ly6C(hi) inflammatory monocytes. Science 341:1483–1488. https://doi.org/10.1126/science.1240636

Okamoto H, Tsunoda T, Teruya K, Takeda N, Uemura T, Matsui T, Fukazawa S, Ichikawa K, Takemae R, Tsuchida K, Takashima Y (2008) An occupational health study of emergency physicians in Japan: health assessment by immune variables (CD4, CD8, CD56, and NK cell activity) at the beginning of work. J Occup Health 50:136–146. https://doi.org/10.1539/joh.16084

Opp MR (2005) Cytokines and sleep. Sleep Med Rev 9:355–364. https://doi.org/10.1016/j.smrv.2005.01.002

Orzech KM, Acebo C, Seifer R, Barker D, Carskadon MA (2014) Sleep patterns are associated with common illness in adolescents. J Sleep Res 23:133–142. https://doi.org/10.1111/jsr.12096

Patel SR, Zhu X, Storfer-Isser A, Mehra R, Jenny NS, Tracy R, Redline S (2009) Sleep duration and biomarkers of inflammation. Sleep 32:200–204. https://doi.org/10.1093/sleep/32.2.200

Patel SR, Malhotra A, Gao X, Hu FB, Neuman MI, Fawzi WW (2012) A prospective study of sleep duration and pneumonia risk in women. Sleep 35:97–101. https://doi.org/10.5665/sleep.1594

Patke A, Young MW, Axelrod S (2020) Molecular mechanisms and physiological importance of circadian rhythms. Nat Rev Mol Cell Biol 21:67–84. https://doi.org/10.1038/s41580-019-0179-2

Peek CB, Levine DC, Cedernaes J, Taguchi A, Kobayashi Y, Tsai SJ, Bonar NA, Mcnulty MR, Ramsey KM, Bass J (2017) Circadian clock interaction with HIF1alpha mediates oxygenic metabolism and anaerobic glycolysis in skeletal muscle. Cell Metab 25:86–92. https://doi.org/10.1016/j.cmet.2016.09.010

Pellegrino R, Sunaga DY, Guindalini C, Martins RC, Mazzotti DR, Wei Z, Daye ZJ, Andersen ML, Tufik S (2012) Whole blood genome-wide gene expression profile in males after prolonged wakefulness and sleep recovery. Physiol Genomics 44:1003–1012. https://doi.org/10.1152/physiolgenomics.00058.2012

Perez de Heredia F, Garaulet M, Gomez-Martinez S, Diaz LE, Warnberg J, Androutsos O, Michels N, Breidenassel C, Cuenca-Garcia M, Huybrechts I, Gottrand F, Ferrari M, Santaliestra-Pasias AM, Kafatos A, Molnar D, Sjostrom M, Widhalm K, Moreno LA, Marcos A, HELENA Study Group (2014) Self-reported sleep duration, white blood cell counts and cytokine profiles in European adolescents: the HELENA study. Sleep Med 15:1251–1258. https://doi.org/10.1016/j.sleep.2014.04.010

Perrin L, Loizides-Mangold U, Chanon S, Gobet C, Hulo N, Isenegger L, Weger BD, Migliavacca E, Charpagne A, Betts JA, Walhin JP, Templeman I, Stokes K, Thompson D, Tsintzas K, Robert M, Howald C, Riezman H, Feige JN, Karagounis LG, Johnston JD, Dermitzakis ET, Gachon F, Lefai E, Dibner C (2018) Transcriptomic analyses reveal rhythmic and CLOCK-driven pathways in human skeletal muscle. Elife 7:e34114. https://doi.org/10.7554/eLife.34114

Prather AA, Leung CW (2016) Association of insufficient sleep with respiratory infection among adults in the United States. JAMA Intern Med 176:850–852. https://doi.org/10.1001/jamainternmed.2016.0787

Prather AA, Hall M, Fury JM, Ross DC, Muldoon MF, Cohen S, Marsland AL (2012) Sleep and antibody response to hepatitis B vaccination. Sleep 35:1063–1069. https://doi.org/10.5665/sleep.1990

Prather AA, Janicki-Deverts D, Hall MH, Cohen S (2015) Behaviorally assessed sleep and susceptibility to the common cold. Sleep 38:1353–1359. https://doi.org/10.5665/sleep.4968

Ranjbaran Z, Keefer L, Stepanski E, Farhadi A, Keshavarzian A (2007) The relevance of sleep abnormalities to chronic inflammatory conditions. Inflamm Res 56:51–57. https://doi.org/10.1007/s00011-006-6067-1

Redwine L, Dang J, Hall M, Irwin M (2003) Disordered sleep, nocturnal cytokines, and immunity in alcoholics. Psychosom Med 65:75–85. https://doi.org/10.1097/01.psy.0000038943.33335.d2

Redwine L, Dang J, Irwin M (2004) Cellular adhesion molecule expression, nocturnal sleep, and partial night sleep deprivation. Brain Behav Immun 18:333–340. https://doi.org/10.1016/j.bbi.2004.01.001

Richardson MR, Churilla JR (2017) Sleep duration and C-reactive protein in US adults. South Med J 110:314–317. https://doi.org/10.14423/SMJ.0000000000000632

Ruiz FS, Rosa DS, Zimberg IZ, Dos Santos Quaresma MV, Nunes JO, Apostolico JS, Weckx LY, Souza AR, Narciso FV, Fernandes-Junior SA, Goncalves B, Folkard S, Bittencourt L, Tufik S, Tulio De Mello M (2020) Night shift work and immune response to the meningococcal conjugate vaccine in healthy workers: a proof of concept study. Sleep Med 75:263–275. https://doi.org/10.1016/j.sleep.2020.05.032

Said EA, Al-Abri MA, Al-Saidi I, Al-Balushi MS, Al-Busaidi JZ, Al-Reesi I, Koh CY, Idris MA, Al-Jabri AA, Habbal O (2019) Sleep deprivation alters neutrophil functions and levels of Th1-related chemokines and CD4(+) T cells in the blood. Sleep Breath 23:1331–1339. https://doi.org/10.1007/s11325-019-01851-1

Sakami S, Ishikawa T, Kawakami N, Haratani T, Fukui A, Kobayashi F, Fujita O, Araki S, Kawamura N (2002) Coemergence of insomnia and a shift in the Th1/Th2 balance toward Th2 dominance. Neuroimmunomodulation 10:337–343. https://doi.org/10.1159/000071474

Sauvet F, Leftheriotis G, Gomez-Merino D, Langrume C, Drogou C, Van Beers P, Bourrilhon C, Florence G, Chennaoui M (2010) Effect of acute sleep deprivation on vascular function in healthy subjects. J Appl Physiol (1985) 108:68–75. https://doi.org/10.1152/japplphysiol.00851.2009

Scheer FA, Hilton MF, Mantzoros CS, Shea SA (2009) Adverse metabolic and cardiovascular consequences of circadian misalignment. Proc Natl Acad Sci U S A 106:4453–4458. https://doi.org/10.1073/pnas.0808180106

Scheiermann C, Kunisaki Y, Frenette PS (2013) Circadian control of the immune system. Nat Rev Immunol 13:190–198. https://doi.org/10.1038/nri3386

Scheiermann C, Gibbs J, Ince L, Loudon A (2018) Clocking in to immunity. Nat Rev Immunol 18: 423–437. https://doi.org/10.1038/s41577-018-0008-4

Scoditti E, Tumolo MR, Garbarino S (2022) Mediterranean diet on sleep: a health alliance. Nutrients 14(14):2998. https://doi.org/10.3390/nu14142998

Sena LA, Chandel NS (2012) Physiological roles of mitochondrial reactive oxygen species. Mol Cell 48:158–167. https://doi.org/10.1016/j.molcel.2012.09.025

Sharpley AL, Cooper CM, Williams C, Godlewska BR, Cowen PJ (2016) Effects of typhoid vaccine on inflammation and sleep in healthy participants: a double-blind, placebo-controlled, crossover study. Psychopharmacology 233:3429–3435. https://doi.org/10.1007/s00213-016-4381-z

Shearer WT, Reuben JM, Mullington JM, Price NJ, Lee BN, Smith EO, Szuba MP, Van Dongen HP, Dinges DF (2001) Soluble TNF-alpha receptor 1 and IL-6 plasma levels in humans subjected to the sleep deprivation model of spaceflight. J Allergy Clin Immunol 107:165–170. https://doi.org/10.1067/mai.2001.112270

Silver AC, Arjona A, Walker WE, Fikrig E (2012) The circadian clock controls toll-like receptor 9-mediated innate and adaptive immunity. Immunity 36:251–261. https://doi.org/10.1016/j.immuni.2011.12.017

Smagula SF, Stone KL, Redline S, Ancoli-Israel S, Barrett-Connor E, Lane NE, Orwoll ES, Cauley JA, Osteoporotic Fractures in Men Research Group (2016) Actigraphy- and polysomnography-measured sleep disturbances, inflammation, and mortality among older men. Psychosom Med 78:686–696. https://doi.org/10.1097/PSY.0000000000000312

Sulli G, Manoogian ENC, Taub PR, Panda S (2018) Training the circadian clock, clocking the drugs, and drugging the clock to prevent, manage, and treat chronic diseases. Trends Pharmacol Sci 39:812–827. https://doi.org/10.1016/j.tips.2018.07.003

Sulli G, Lam MTY, Panda S (2019) Interplay between circadian clock and cancer: new frontiers for cancer treatment. Trends Cancer 5:475–494. https://doi.org/10.1016/j.trecan.2019.07.002

Teixeira KRC, Dos Santos CP, De Medeiros LA, Mendes JA, Cunha TM, De Angelis K, Penha-Silva N, De Oliveira EP, Crispim CA (2019) Night workers have lower levels of antioxidant defenses and higher levels of oxidative stress damage when compared to day workers. Sci Rep 9:4455. https://doi.org/10.1038/s41598-019-40989-6

Thaiss CA, Levy M, Korem T, Dohnalova L, Shapiro H, Jaitin DA, David E, Winter DR, Gury-Benari M, Tatirovsky E, Tuganbaev T, Federici S, Zmora N, Zeevi D, Dori-Bachash M, Pevsner-Fischer M, Kartvelishvily E, Brandis A, Harmelin A, Shibolet O, Halpern Z, Honda K, Amit I, Segal E, Elinav E (2016) Microbiota diurnal rhythmicity programs host transcriptome oscillations. Cell 167:1495–1510.e1412. https://doi.org/10.1016/j.cell.2016.11.003

Toth LA, Tolley EA, Krueger JM (1993) Sleep as a prognostic indicator during infectious disease in rabbits. Proc Soc Exp Biol Med 203:179–192. https://doi.org/10.3181/00379727-203-43590

Trivedi MS, Holger D, Bui AT, Craddock TJA, Tartar JL (2017) Short-term sleep deprivation leads to decreased systemic redox metabolites and altered epigenetic status. PLoS One 12:e0181978. https://doi.org/10.1371/journal.pone.0181978

Uemura R, Fujiwara Y, Iwakura N, Shiba M, Watanabe K, Kamata N, Yamagami H, Tanigawa T, Watanabe T, Tominaga K, Arakawa T (2016) Sleep disturbances in Japanese patients with inflammatory bowel disease and their impact on disease flare. Springerplus 5:1792. https://doi.org/10.1186/s40064-016-3408-6

Urade Y, Hayaishi O (2011) Prostaglandin D2 and sleep/wake regulation. Sleep Med Rev 15:411–418. https://doi.org/10.1016/j.smrv.2011.08.003

Van Leeuwen WM, Lehto M, Karisola P, Lindholm H, Luukkonen R, Sallinen M, Harma M, Porkka-Heiskanen T, Alenius H (2009) Sleep restriction increases the risk of developing cardiovascular diseases by augmenting proinflammatory responses through IL-17 and CRP. PLoS One 4:e4589. https://doi.org/10.1371/journal.pone.0004589

Vgontzas AN, Zoumakis E, Bixler EO, Lin HM, Follett H, Kales A, Chrousos GP (2004) Adverse effects of modest sleep restriction on sleepiness, performance, and inflammatory cytokines. J Clin Endocrinol Metab 89:2119–2126. https://doi.org/10.1210/jc.2003-031562

Vgontzas AN, Liao D, Bixler EO, Chrousos GP, Vela-Bueno A (2009) Insomnia with objective short sleep duration is associated with a high risk for hypertension. Sleep 32:491–497. https://doi.org/10.1093/sleep/32.4.491

Vgontzas AN, Fernandez-Mendoza J, Liao D, Bixler EO (2013) Insomnia with objective short sleep duration: the most biologically severe phenotype of the disorder. Sleep Med Rev 17:241–254. https://doi.org/10.1016/j.smrv.2012.09.005

Villafuerte G, Miguel-Puga A, Rodriguez EM, Machado S, Manjarrez E, Arias-Carrion O (2015) Sleep deprivation and oxidative stress in animal models: a systematic review. Oxidative Med Cell Longev 2015:234952. https://doi.org/10.1155/2015/234952

Weljie AM, Meerlo P, Goel N, Sengupta A, Kayser MS, Abel T, Birnbaum MJ, Dinges DF, Sehgal A (2015) Oxalic acid and diacylglycerol 36:3 are cross-species markers of sleep debt. Proc Natl Acad Sci U S A 112:2569–2574. https://doi.org/10.1073/pnas.1417432112

Wheaton AG, Jones SE, Cooper AC, Croft JB (2018) Short sleep duration among middle school and high school students - United States, 2015. MMWR Morb Mortal Wkly Rep 67:85–90. https://doi.org/10.15585/mmwr.mm6703a1

Wirth MD, Andrew ME, Burchfiel CM, Burch JB, Fekedulegn D, Hartley TA, Charles LE, Violanti JM (2017) Association of shiftwork and immune cells among police officers from the Buffalo Cardio-Metabolic Occupational Police Stress study. Chronobiol Int 34:721–731. https://doi.org/10.1080/07420528.2017.1316732

Zhu X, Zhu J (2020) CD4 T helper cell subsets and related human immunological disorders. Int J Mol Sci 21(21):8011. https://doi.org/10.3390/ijms21218011

Zhuang X, Magri A, Hill M, Lai AG, Kumar A, Rambhatla SB, Donald CL, Lopez-Clavijo AF, Rudge S, Pinnick K, Chang WH, Wing PC, Brown R, Qin X, Simmonds P, Baumert TF, Ray D, Loudon A, Balfe P, Wakelam M, Butterworth S, Kohl A, Jopling CL, Zitzmann N, Mckeating JA (2019) The circadian clock components BMAL1 and REV-ERBalpha regulate flavivirus replication. Nat Commun 10:377. https://doi.org/10.1038/s41467-019-08299-7

Zielinski MR, Krueger JM (2011) Sleep and innate immunity. Front Biosci (Schol Ed) 3:632–642. https://doi.org/10.2741/s176

Zielinski MR, Mckenna JT, Mccarley RW (2016) Functions and mechanisms of sleep. AIMS Neurosci 3:67–104. https://doi.org/10.3934/Neuroscience.2016.1.67

Zomers ML, Hulsegge G, Van Oostrom SH, Proper KI, Verschuren WMM, Picavet HSJ (2017) Characterizing adult sleep behavior over 20 years-the population-based Doetinchem Cohort Study. Sleep 40(7):zsx085. https://doi.org/10.1093/sleep/zsx085

Chapter 2
Changes in the Sleep and Circadian Pattern of Sleep During the COVID-19 Pandemic

Laura Kortesoja and Ilona Merikanto

Abstract This chapter provides a broad perspective on the pandemic's effects on sleep and sleep–wake behavior in the general population. Studies on these topics are currently largely cross-sectional and national online survey studies or small patient-sample studies, except for a few multinational studies and large-scale studies by an international collaboration initiative ICOSS.

Changes in sleep–wake behavior during the COVID-19 pandemic included prominent delays in bedtimes and wake-up times, increased sleep duration, and an overall increase in sleep problems. Insomnia symptoms, dream recall frequency, dream-enactment behaviors, and nightmares were prevalent during the pandemic. Mental health symptoms and sleep problems were reflected as a decrease in sleep quality during the pandemic.

In this chapter, we consider the differences between circadian types on the pandemic effects on sleep. Of different circadian types, many small-scale national studies and one large-scale international study showed that sleep–wake schedules were delayed, especially among evening-types, who also reported an increase in sleep duration. Evening-types had, during lockdown, an opportunity to break away from the misalignment of their natural diurnal rhythms and to choose sleep schedules that were more aligned with their natural diurnal rhythms. However, since evening-types were at increased risk for sleep problems and mental health issues already before the pandemic, the flexibility in everyday life schedules did not override the negative effects of the pandemic. An international study with a large general

L. Kortesoja
Centre for Educational Assessment, University of Helsinki, Helsinki, Finland

Faculty of Education and Culture, Tampere University, Tampere, Finland
e-mail: laura.kortesoja@helsinki.fi; https://researchportal.helsinki.fi/en/persons/laura-kortesoja

I. Merikanto (✉)
SleepWell Research Program, Faculty of Medicine, University of Helsinki, Helsinki, Finland

Finnish Institute for Health and Welfare, Helsinki, Finland

Orton Orthopaedics Hospital, Helsinki, Finland
e-mail:
Ilona.merikanto@helsinki.fi; https://researchportal.helsinki.fi/fi/persons/ilona-merikanto

A. S. BaHammam et al. (eds.), *COVID-19 and Sleep: A Global Outlook*, Progress in Sleep Research, https://doi.org/10.1007/978-981-99-0240-8_2

population sample showed, that sleep disturbances and risk for poor mental health during the pandemic were emphasized among evening-types, moderated by financial suffering and time spent in confinement.

Keywords Chronotype · Coronavirus · Eveningness · Sleep quality · Sleep–wake behavior

2.1 Introduction

The Coronavirus disease 2019 (COVID-19) has left a globally noticeable impact on human lifestyle, including sleep habits and daily rhythms. There is a growing body of literature that recognizes the wide-ranging effects of the pandemic on everyday life. The pandemic not only had a dramatic effect on health and everyday life but also forced individuals to reorganize their daily lives with, for example, home, work, mobility, social interaction, and eating habits. This has resulted in redefining what a good life is during a pandemic. The disease has also brought undesirable changes in the form of various health symptoms and difficulties such as sleep problems. In this chapter, the pandemic effects are examined in the perspective of changes in sleep and in relation to individual circadian timing of physiological and behavioral functions (Czeisler and Gooley 2007; Merikanto et al. 2021).

2.2 Pandemic Effects on Sleep Quality and Sleep Problems

The pandemic has had a profound impact on sleep quality. The prevalence of sleep disturbances, such as insomnia symptoms and nightmares, have all become more common among the general population during the pandemic compared to the time before the pandemic (Bhat and Chokroverty 2022; Jahrami et al. 2021b; Merikanto et al. 2022a; Partinen et al. 2021). According to a large-scale international ICOSS I study on the general adult population in 14 different countries across four continents, the prevalence of poor sleep quality increased during the pandemic from 12.5% to 28.2% (Partinen et al. 2021). Concerning gender, a French survey study with mainly adult sample showed that poor sleep quality occurred more often in women than men during lockdown (Bertrand et al. 2022). In addition, sleep disturbances were more common among individuals who had chronic conditions in an Italian survey study (Gualano et al. 2020). Besides sleep disorders and problems, according to the large-scale international ICOSS I study sample, hypnotic use increased from 7.8% to 12.2% during the pandemic (Partinen et al. 2021). This result was in line with a Canadian survey study on adolescents and adults, which also found that the use of sleeping medications increased during the pandemic (Robillard et al. 2021). A systematic review of 44 papers with a total of 54,231 participants from 13 countries showed that especially in individuals who had covid-19 the prevalence of sleep problems, e.g., insomnia symptoms and nightmares, have become more common (Jahrami et al. 2021b).

2.2.1 Insomnia Symptoms During the Pandemic

Insomnia refers to a condition where individuals report problems with sleep (Roth 2007), such as difficulties falling asleep, staying asleep/disrupted sleep, or early morning awakenings (Ohayon 2002). A large-scale international ICOSS I sample-based study showed that the prevalence rate for clinical insomnia according to Insomnia Severity Index (ISI) score was approximately 36.7% during the pandemic around the globe and 17.4% met the criteria for a probable insomnia disorder (Morin et al. 2021). Based on this international study, the prevalence of insomnia symptoms was significantly higher during the first wave of the pandemic as compared to the time before the pandemic, and the risk for insomnia was emphasized among individuals who had COVID-19 disease, reported financial burden, had been isolated for 4–5 weeks or lived alone or with more than five people (Morin et al. 2021). In the same sample, confinement and physiological, socioeconomic, and psychological factors were also associated with the increase in sleep problems in general (Merikanto et al. 2022a). It is well known that there is a potential bidirectional relationship between mental health, e.g., stress, anxiety and depression, and insomnia symptoms, although most of the pandemic studies on insomnia and mental health can only indicate an association between insomnia and mental status, not a causal relationship. According to an online survey study of Greek adults, pandemic-related worrying, loneliness, and depressive symptoms were associated with insomnia (Voitsidis et al. 2020). Also, in the ICOSS I sample-based study, depression, and anxiety were associated with insomnia symptoms (Morin et al. 2021), and a multinational sample with 2724 participants from 67 countries showed that especially post-pandemic insomnia symptoms were associated with higher levels of anxiety, depression, and stress in contrast to pre-pandemic insomnia symptoms (Meaklim et al. 2021). The increase in insomnia symptoms during the pandemic has also been supported by findings on Chinese adults, the risk group for insomnia being especially healthcare workers in China (Li et al. 2020; Pappa et al. 2020). Yet, a review of 98 studies on different subpopulations indicated that there were no differences in the prevalence of insomnia symptoms between the early and late stages of the pandemic (Li et al. 2022). Furthermore, a large international ICOSS II study across 16 countries with a control group of healthy participants showed that long-lasting insomnia symptoms were more prevalent among COVID-infected cases and associated with higher disease severity (Merikanto et al. 2022b). All in all, insomnia symptoms seemed to increase in the general population as well as in covid-19 cases during the pandemic, and several pandemic-related risk factors have been discovered.

2.2.2 Dreams and Nightmares During the Pandemic

International ICOSS I study showed that high nightmare frequency increased from 13.24% to 22.35% during the pandemic and was associated with mental illness and sleep problems (Holzinger et al. 2022). Nightmares were also related to financial suffering in this same international sample (Partinen et al. 2021). In a case-control study on ICOSS I study sample, sleep duration, dream recall frequency, insomnia, the severity of COVID-19 symptoms as well as mental health problems, such as post-traumatic stress disorder and anxiety, associated with the frequency of experienced nightmares among those who reported having had COVID-19 disease (Scarpelli et al. 2022). In addition, increased dream recall frequency has been reported during the pandemic in a large Finnish crowdsourcing sample of responders between the ages of 10 and 99 years old (Pesonen et al. 2020). All in all, dreams and nightmares were more frequent during the pandemic and associated more with worsened rather than improved sleep quality and mental health as well as increased financial suffering.

2.2.3 Sleep Quality During the Pandemic

Sleep quality decreased in the general population during the pandemic. In a self-selected Chinese survey sample, sleep quality was impaired during the pandemic (Huang and Zhao 2020). A registry study by the Netherlands Sleep Registry showed that sleep quality decreased especially among individuals who had reported good sleep quality before the COVID-19 pandemic (Kocevska et al. 2020). Few studies have also reported especially healthcare workers being at high risk for poor sleep quality and moderate to severe stress during the pandemic (Huang and Zhao 2020; Jahrami et al. 2021a). Increased self-perceived burden had also a negative effect on the quality of sleep during the pandemic based on an online study with German, Austrian, and Swiss adults (Blume et al. 2020). In a large-scale international ICOSS I study sample, high dream recall frequency during the pandemic was related to poor sleep quality (Fränkl et al. 2021). In that study, 29.2% of individuals with high dream recall frequency had good sleep quality, whereas 44.9% of individuals with high dream recall frequency had poor sleep quality (Fränkl et al. 2021).

2.2.4 Parasomnias During the Pandemic

Parasomnias refer to nonrapid eye movement (NREM)-related parasomnias, rapid eye movement (REM)-related parasomnias, and other parasomnias (Sateia 2014). Parasomnia symptoms related to NREM include, e.g., sleepwalking, sleep terrors, and arousals, whereas symptoms associated with REM sleep include nightmares or

REM-sleep behavior (Plante and Winkelman 2006). There was only one study that looked at parasomnias in a comprehensive sample during a pandemic internationally (Liu et al. 2022). A large-scale international ICOSS I sample-based study showed that during the pandemic, dream-enactment behaviors increased during the pandemic in the general population compared to the time before the pandemic (Liu et al. 2022). High dream-enactment behavior was associated with the international study of olfactory impairment, mood, post-traumatic stress disorder, sleep apnea symptoms, and COVID-19 infection and severity (Liu et al. 2022). Therefore, as shown in the international study by Liu et al. (2022), post-traumatic stress during the pandemic could increase the prevalence of parasomnias indicated also prior to the pandemic (Elliott et al. 2020). Since alcohol disrupts sleep architecture and is associated with sleep disturbances including parasomnias (Roehrs and Roth 2001), the increased use of alcohol during the pandemic could also play a role in the increased prevalence of parasomnias (Pérez-Carbonell et al. 2020).

2.2.5 Sleep Apnea Symptoms During the Pandemic

Sleep apnea symptoms appear as sleep-disordered breathing, and can predispose to obstructive sleep apnea (OSA), which is a complex and heterogeneous disorder (Ayas et al. 2015). A large-scale international ICOSS I study showed that sleep apnea symptoms are also associated with the severity of COVID-19 symptoms. Individuals at high risk of obstructive sleep apnea had increased odds of having covid-19 disease and were two times more likely to be hospitalized or treated in an intensive care unit (Chung et al. 2021). In addition, a small Italian sample of sleep apnea patients showed that sleep apnea symptoms were associated with decreased sleep quality during the pandemic (Spicuzza et al. 2022). Individuals that suffered from obstructive sleep apnea reported an increase in sleep disturbance from 54% to 66%, whereas controls reported sleep disturbances from 29% to 40% (Spicuzza et al. 2022). There are many national and international studies, especially on the association between the risk of COVID-19 and obstructive sleep apnea and here we mention only a few. However, the long-term effects between sleep apnea symptoms and the pandemic need more investigation in the future.

2.2.6 Fatigue and Excessive Daytime Sleepiness During the Pandemic

In a large-scale international ICOSS I study, fatigue and excessive sleepiness increased globally during the first phase of the COVID-19 pandemic. Fatigue increased from 20.7% to 29.9%, and excessive sleepiness from 18.5% to 27.7% in the general population during the first phase of the COVID-19 pandemic and was associated with confinement and financial suffering (Partinen et al. 2021). In addition, those who reported having had COVID-19 were more likely to report daytime sleepiness (Partinen et al. 2021). A systematic review showed that around one in three individuals experienced fatigue 12 or more weeks after the COVID-19 disease (Ceban et al. 2022). A large international ICOSS II study showed that long-lasting excessive daytime sleepiness and fatigue were more prevalent among COVID-infected cases than among healthy controls and associated with higher disease severity (Merikanto et al. 2022b).

2.3 Pandemic Effects on Sleep–Wake Behavior and Sleep Duration in General Population

The circadian rhythm is a natural, internal process that regulates the sleep–wake cycle and repeats approximately every 24 h. The internal clock has a genetic basis reflected also as differences in genetic variants between circadian types (Jones et al. 2019; Merikanto et al. 2021). Circadian rhythms determine the ability to maintain alertness during the day and sleep at night by interacting with the homeostatic regulation of the sleep–wake cycle (Czeisler and Gooley 2007). The internal circadian rhythm produced by the biological clock is influenced by external time markers (zeitgeber), of which one of the most important are the environmental light conditions, and others include, e.g., social interaction (Duffy and Dijk 2002). Recurring lockdown protocols that COVID-19 brought along worldwide had a significant impact on the timing of daily behaviors. During the lockdown, the stay-at-home lifestyle brought more freedom to live more in line with the natural diurnal rhythms leading to changes in sleep–wake behavior that was more in line with the morningness-eveningness preference (Blume et al. 2020; Korman et al. 2020; Merikanto et al. 2022c; Roitblat et al. 2020). Lockdown protocols reduced the exposure to different zeitgebers, i.e., clues that regulate the circadian rhythms. It is hypothesized that individuals who spend most of the day indoors, expose themselves less to daylight, which is likely to delay and reduce the amplitude of nocturnal melatonin release, resulting in a delay in sleep onset (Kutana and Lau 2021). This hypothesis is supported by an online survey study in the French general population that showed a delay in sleep schedules, a decrease in exposure to morning and evening natural light during the pandemic, and also a significant increase in screen exposure time (Bertrand et al. 2022). Previous research has shown that social and economic daily life follows different behavioral timing compared to a situation, where an individual can choose the timing for their daily routines and sleep schedule according to their preference (Morin et al. 2020; Roenneberg et al. 2019). It is indicated in large-scale population-based studies that the increased flexibility in everyday life during the pandemic allowed adjustment of sleep and wake times more according to individual preference and circadian typology, which increased sleep duration and decreased social jetlag (Blume et al. 2020; Korman et al. 2020; Merikanto et al. 2022c).

2.3.1 Sleep Duration During the Pandemic

During confinement, different survey data have shown that there was an increase in overall sleep duration (Blume et al. 2020; Gupta et al. 2020; Pépin et al. 2021), especially among evening-types based on large international ICOSS I study sample (Merikanto et al. 2022c). However, concerning the changes reported in sleep duration during the pandemic, changes may have diminished along with the reduced restrictions. For example, a large multinational sample assessing sleep based on smartphone applications from around the world and large survey sample of US Fitbit users showed that time in bed as an indicator for sleep duration increased compared to time before the pandemic rather in the earlier months of the pandemic than later on during pandemic (Rezaei and Grandner 2021; Yuan et al. 2022). Taken together, sleep duration increased during the pandemic in the general population, especially among evening-types. More research is needed to investigate the changes in sleep duration at later stages of the pandemic.

2.3.2 Sleep Schedules During the Pandemic

The confinement brought also changes in sleep schedules (Merikanto et al. 2022c). A large international ICOSS I study on adults showed that sleep schedules were delayed especially during working days (Merikanto et al. 2022c). Delayed bedtimes during the pandemic have been reported also in a large-scale Argentinean sample (Leone et al. 2020) and delays both in bedtime and wake-up times in a small survey sample of high school students (Genta et al. 2021). There was also a decrease in the time difference between sleep–wake schedules on workdays and free days (social jetlag) during the early stages of the pandemic when lockdowns were common (Capodilupo and Miller 2021; Roitblat et al. 2020; Brandão et al. 2021; Bottary et al. 2022; Staller and Randler 2021). A large international ICOSS I survey study showed that while the majority (46%) exhibited a reduction in social jetlag, 20% showed an increase and 34% no change in the time difference between sleep–wake schedules between workdays and free days (Brandão et al. 2021). The confinement brought not only changes in sleep–wake schedules but also in other daily rhythms. In a small Swedish adult study, the first meal of the day was also scheduled later which was associated with a shift in sleep timing (Benedict et al. 2021).

2.4 Pandemic Effects on Sleep and Mental Well-Being Among Different Circadian Types

Sleep disturbances such as social jetlag, insomnia symptoms, and insufficient sleep are generally more pronounced among evening-types (Merikanto et al. 2012; Merikanto and Partonen 2020) and this was shown also during the pandemic in sleep and mental health problems such as insomnia, poor sleep quality, nightmares, fatigue, depression, anxiety, and stress being emphasized among evening-types in a large-scale international ICOSS I study (Merikanto et al. 2022c). For example, according to the insomnia severity index (ISI), moderately severe insomnia was highest (26.5%) among definite evening-types, 19.8% for moderate evening-types, 12.9% for intermediate-types, 14.7% for moderate morning-types, and 16.8% for definite morning-types amid pandemic (Merikanto et al. 2022c). Even though social jetlag generally decreased during the pandemic, it remained to be higher in definite evening-types compared to other circadian types in a small survey sample of Australian athletes (Facer-Childs et al. 2021). Overall, it seems that evening-types suffered from lockdowns more compared to other circadian types.

2.4.1 Sleep–Wake Behavior During the Pandemic by Circadian Type

A large international ICOSS I study showed that during the pandemic, sleep–wake schedules were delayed especially among evening-types, who also reported an increase in sleep duration (Merikanto et al. 2022c). Yet, definite evening-types suffered from insufficient sleep more often than definite morning-types in ICOSS I study (Merikanto et al. 2022c). Intermediate-types were less vulnerable to changes in sleep schedules compared to extreme circadian types (Merikanto et al. 2022c). Concerning social jetlag, both an increase and decrease were associated with a later sleep midpoint (Brandão et al. 2021). These findings might indicate a higher sensitivity to societal changes in evening-types.

2.4.2 Sleep Quality and Sleep Problems During the Pandemic by Circadian Type

Both large international and Italian general adult population samples showed that during the pandemic, sleep quality was significantly lower among evening-types compared to other circadian types (Bazzani et al. 2021; Merikanto et al. 2022c). In contrast, morning-types showed a lower predisposition to sleep problems based on large international and national samples (Merikanto et al. 2022c; Salfi et al. 2021). In a large international ICOSS I study sample, the association between circadian type and sleep difficulties was moderated by financial suffering and confinement during the pandemic (Merikanto et al. 2022c).

2.4.3 Mental Well-Being During the Pandemic by Circadian Type

In general, evening-types seem to be more susceptible to mental health issues as compared to other circadian types, mediated at least partly by insufficient sleep among evening-types (Merikanto and Partonen 2021). Also during the pandemic, especially evening-types reported more mental health symptoms, such as anxiety and depression, than other circadian types (Merikanto et al. 2022c). In a small sample of students and adults in the UK, evening-type reported more loneliness in contrast to other circadian types (Norbury 2021). A survey sample of Italian adults showed that during the pandemic, evening-types had also lower levels of resilience, a psychological capacity to cope in challenging situations, compared to non-evening types during lockdown (Bazzani et al. 2021). This might also reflect the abovementioned high prevalence of sleep and mental health problems among evening-types during the pandemic.

2.5 Conclusion

The COVID-19 pandemic seemed to have a variety of effects on individuals' sleep depending on the circadian type, time in confinement, health background and family, and socioeconomic situation. During the pandemic, there were novel opportunities to shift sleep and work schedules more closely aligned to individual biological rhythms. Nevertheless, in general, there was an increase in sleep and mental health problems during the pandemic. As our daily routines help our sleep and wake cycles stay synchronized with daylight, the pandemic-related changes in social schedules and reduced exposure to light were likely to disrupt sleep and circadian rhythms.

References

Ayas NT, Owens RL, Kheirandish-Gozal L (2015) Update in sleep medicine 2014. Am J Respir Crit Care Med 192(4):415–420. https://doi.org/10.1164/rccm.201503-0647UP

Bazzani A, Bruno S, Frumento P, Cruz-Sanabria F, Turchetti G, Faraguna U (2021) Sleep quality mediates the effect of chronotype on resilience in the time of COVID-19. Chronobiol Int 38(6): 883–892. https://doi.org/10.1080/07420528.2021.1895199

Benedict C, Brandão LEM, Merikanto I, Partinen M, Bjorvatn B, Cedernaes J (2021) Meal and sleep timing before and during the COVID-19 pandemic: a cross-sectional anonymous survey study from Sweden. Clocks Sleep 3(2):251–258. https://doi.org/10.3390/clockssleep3020015

Bertrand L, Schröder C, Bourgin P, Maruani J, Atoui Y, d'Ortho M, Lejoyeux M, Geoffroy PA (2022) Sleep and circadian rhythm characteristics in individuals from the general population during the French COVID-19 full lockdown. J Sleep Res 31(2):e13480. https://doi.org/10.1111/jsr.13480

Bhat S, Chokroverty S (2022) Sleep disorders and COVID-19. Sleep Med 91:253–261. https://doi.org/10.1016/j.sleep.2021.07.021

Blume C, Schmidt MH, Cajochen C (2020) Effects of the COVID-19 lockdown on human sleep and rest-activity rhythms. Curr Biol 30(14):R795–R797. https://doi.org/10.1016/j.cub.2020.06.021

Bottary R, Fields EC, Kensinger EA, Cunningham TJ (2022) Age and chronotype influenced sleep timing changes during the first wave of the COVID-19 pandemic. J Sleep Res 31(2):e13495. https://doi.org/10.1111/jsr.13495

Brandão LEM, Martikainen T, Merikanto I, Holzinger B, Morin CM, Espie CA, Bolstad C, Leger D, Chung F, Plazzi G, Dauvilliers Y, Matsui K, De Gennaro L, Sieminski M, Nadorff MR, Chan NY, Wing YK, Mota-Rolim SA, Inoue Y et al (2021) Social jetlag changes during the COVID-19 pandemic as a predictor of insomnia – a multi-national survey study. Nat Sci Sleep 13:1711–1722. https://doi.org/10.2147/NSS.S327365

Capodilupo ER, Miller DJ (2021) Changes in health promoting behavior during COVID-19 physical distancing: utilizing wearable technology to examine trends in sleep, activity, and cardiovascular indicators of health. PLoS ONE 16(8):e0256063. https://doi.org/10.1371/journal.pone.0256063

Ceban F, Ling S, Lui LMW, Lee Y, Gill H, Teopiz KM, Rodrigues NB, Subramaniapillai M, Di Vincenzo JD, Cao B, Lin K, Mansur RB, Ho RC, Rosenblat JD, Miskowiak KW, Vinberg M, Maletic V, McIntyre RS (2022) Fatigue and cognitive impairment in post-COVID-19 syndrome: a systematic review and meta-analysis. Brain Behav Immun 101:93–135. https://doi.org/10.1016/j.bbi.2021.12.020

Chung F, Waseem R, Pham C, Penzel T, Han F, Bjorvatn B, Morin CM, Holzinger B, Espie CA, Benedict C, Cedernaes J, Saaresranta T, Wing YK, Nadorff MR, Dauvilliers Y, De Gennaro L, Plazzi G, Merikanto I, Matsui K, Leger D, Sieminski M, Mota-Rolim S, Inoue Y, Partinen M, For the International COVID Sleep Study (ICOSS) Group (2021) The association between high risk of sleep apnea, comorbidities, and risk of COVID-19: a population-based international harmonized study. Sleep Breath 25(2):849–860. https://doi.org/10.1007/s11325-021-02373-5

Czeisler CA, Gooley JJ (2007) Sleep and circadian rhythms in humans. Cold Spring Harb Symp Quant Biol 72(1):579–597. https://doi.org/10.1101/sqb.2007.72.064

Duffy JF, Dijk D-J (2002) Getting through to circadian oscillators: why use constant routines? J Biol Rhythm 17(1):4–13. https://doi.org/10.1177/074873002129002294

Elliott JE, Opel RA, Pleshakov D, Rachakonda T, Chau AQ, Weymann KB, Lim MM (2020) Posttraumatic stress disorder increases the odds of REM sleep behavior disorder and other parasomnias in veterans with and without comorbid traumatic brain injury. Sleep 43(3):zsz237

Facer-Childs ER, Hoffman D, Tran JN, Drummond SP, Rajaratnam SM (2021) Sleep and mental health in athletes during COVID-19 lockdown. Sleep 44(5):zsaa261

Fränkl E, Scarpelli S, Nadorff MR, Bjorvatn B, Bolstad CJ, Chan NY, Chung F, Dauvilliers Y, Espie CA, Inoue Y, Leger D, Macêdo T, Matsui K, Merikanto I, Morin CM, Mota-Rolim S, Partinen M, Penzel T, Plazzi G et al (2021) How our dreams changed during the COVID-19 pandemic: effects and correlates of dream recall frequency - a multinational study on 19,355 adults. Nat Sci Sleep 13:1573–1591. https://doi.org/10.2147/NSS.S324142

Genta FD, Rodrigues Neto GB, Sunfeld JPV, Porto JF, Xavier AD, Moreno CR et al (2021) COVID-19 pandemic impact on sleep habits, chronotype, and health-related quality of life among high school students: a longitudinal study. J Clin Sleep Med 17(7):1371–1377

Gualano MR, Lo Moro G, Voglino G, Bert F, Siliquini R (2020) Effects of Covid-19 lockdown on mental health and sleep disturbances in Italy. Int J Environ Res Public Health 17(13):4779. https://doi.org/10.3390/ijerph17134779

Gupta R, Grover S, Basu A, Krishnan V, Tripathi A, Subramanyam A, Nischal A, Hussain A, Mehra A, Ambekar A, Saha G, Mishra K, Bathla M, Jagiwala M, Manjunatha N, Nebhinani N, Gaur N, Kumar N, Dalal P et al (2020) Changes in sleep pattern and sleep quality during COVID-19 lockdown. Indian J Psychiatry 62(4):370. https://doi.org/10.4103/psychiatry.IndianJPsychiatry_523_20

Holzinger B, Nierwetberg F, Chung F, Bolstad CJ, Bjorvatn B, Chan NY, Dauvilliers Y, Espie CA, Han F, Inoue Y, Leger D, Macêdo T, Matsui K, Merikanto I, Morin CM, Mota-Rolim SA, Partinen M, Plazzi G, Penzel T et al (2022) Has the COVID-19 pandemic traumatized us collectively? The impact of the COVID-19 pandemic on mental health and sleep factors via traumatization: a multinational survey. Nat Sci Sleep 14:1469–1483. https://doi.org/10.2147/NSS.S368147

Huang Y, Zhao N (2020) Generalized anxiety disorder, depressive symptoms and sleep quality during COVID-19 outbreak in China: a web-based cross-sectional survey. Psychiatry Res 288: 112954. https://doi.org/10.1016/j.psychres.2020.112954

Jahrami H, BaHammam AS, AlGahtani H, Ebrahim A, Faris M, AlEid K, Saif Z, Haji E, Dhahi A, Marzooq H, Hubail S, Hasan Z (2021a) The examination of sleep quality for frontline healthcare workers during the outbreak of COVID-19. Sleep Breath 25(1):503–511. https://doi.org/10.1007/s11325-020-02135-9

Jahrami H, BaHammam AS, Bragazzi NL, Saif Z, Faris M, Vitiello MV (2021b) Sleep problems during the COVID-19 pandemic by population: a systematic review and meta-analysis. J Clin Sleep Med 17(2):299–313. https://doi.org/10.5664/jcsm.8930

Jones SE, Lane JM, Wood AR, van Hees VT, Tyrrell J, Beaumont RN, Jeffries AR, Dashti HS, Hillsdon M, Ruth KS, Tuke MA, Yaghootkar H, Sharp SA, Jie Y, Thompson WD, Harrison JW, Dawes A, Byrne EM, Tiemeier H, Allebrandt KV, Bowden J, Ray DW, Freathy RM, Murray A, Mazzotti DR, Gehrman PR, Lawlor DA, Frayling TM, Rutter MK, Hinds DA, Saxena R, Weedon MN (2019) Genome-wide association analyses of chronotype in 697,828 individuals provides insights into circadian rhythms. Nat Commun 10(1):343. https://doi.org/10.1038/s41467-018-08259-7

Kocevska D, Blanken TF, Van Someren EJW, Rösler L (2020) Sleep quality during the COVID-19 pandemic: not one size fits all. Sleep Med 76:86–88. https://doi.org/10.1016/j.sleep.2020.09.029

Korman M, Tkachev V, Reis C, Komada Y, Kitamura S, Gubin D, Kumar V, Roenneberg T (2020) COVID-19-mandated social restrictions unveil the impact of social time pressure on sleep and body clock. Sci Rep 10(1):22225. https://doi.org/10.1038/s41598-020-79299-7

Kutana S, Lau PH (2021) The impact of the 2019 coronavirus disease (COVID-19) pandemic on sleep health. Can Psychol 62(1):12–19. https://doi.org/10.1037/cap0000256

Leone MJ, Sigman M, Golombek DA (2020) Effects of lockdown on human sleep and chronotype during the COVID-19 pandemic. Curr Biol 30(16):R930–R931. https://doi.org/10.1016/j.cub.2020.07.015

Li Y, Qin Q, Sun Q, Sanford LD, Vgontzas AN, Tang X (2020) Insomnia and psychological reactions during the COVID-19 outbreak in China. J Clin Sleep Med 16(8):1417–1418. https://doi.org/10.5664/jcsm.8524

Li Y, Chen B, Hong Z, Sun Q, Dai Y, Basta M, Tang X, Qin Q (2022) Insomnia symptoms during the early and late stages of the COVID-19 pandemic in China: a systematic review and meta-analysis. Sleep Med 91:262–272. https://doi.org/10.1016/j.sleep.2021.09.014

Liu Y, Partinen E, Chan NY, Dauvilliers Y, Inoue Y, De Gennaro L, Plazzi G, Bolstad CJ, Nadorff MR, Merikanto I, Bjorvatn B, Han F, Zhang B, Cunha AS, Mota-Rolim S, Léger D, Matsui K, Espie CA, Chung F et al (2022) Dream-enactment behaviours during the COVID-19 pandemic: an international COVID-19 sleep study. J Sleep Res. https://doi.org/10.1111/jsr.13613

Meaklim H, Junge MF, Varma P, Finck WA, Jackson ML (2021) Pre-existing and post-pandemic insomnia symptoms are associated with high levels of stress, anxiety, and depression globally during the COVID-19 pandemic. J Clin Sleep Med 17(10):2085–2097. https://doi.org/10.5664/jcsm.9354

Merikanto I, Partonen T (2020) Increase in eveningness and insufficient sleep among adults in population-based cross-sections from 2007 to 2017. Sleep Med 75:368–379. https://doi.org/10.1016/j.sleep.2020.07.046

Merikanto I, Partonen T (2021) Eveningness increases risks for depressive and anxiety symptoms and hospital treatments mediated by insufficient sleep in a population-based study of 18,039 adults. Depress Anxiety 38(10):1066–1077. https://doi.org/10.1002/da.23189

Merikanto I, Kronholm E, Peltonen M, Laatikainen T, Lahti T, Partonen T (2012) Relation of chronotype to sleep complaints in the general Finnish population. Chronobiol Int 29(3): 311–317. https://doi.org/10.3109/07420528.2012.655870

Merikanto I, Kantojärvi K, Partonen T, Pesonen A-K, Paunio T (2021) Genetic variants for morningness in relation to habitual sleep-wake behavior and diurnal preference in a population-based sample of 17,243 adults. Sleep Med 80:322–332. https://doi.org/10.1016/j.sleep.2021.01.054

Merikanto I, Dauvilliers Y, Chung F, Holzinger B, De Gennaro L, Wing YK, Korman M, Partinen M II, ICOSS members, Benedict C, Bjelajac A, Bjorvatn B, Yin Chan N, Hrubos-Strøm H, Inoue Y, Kirov R, Landtblom A, Léger D, Matsui K et al (2022a) Disturbances in sleep, circadian rhythms and daytime functioning in relation to coronavirus infection and Long-COVID – a multinational ICOSS study. J Sleep Res 31(4). https://doi.org/10.1111/jsr.13542

Merikanto I, Dauvilliers Y, Chung F, Wing YK, De Gennaro L, Holzinger B, Bjorvatn B, Morin CM, Penzel T, Benedict C, Koscec Bjelajac A, Chan NY, Espie CA, Hrubos-Strøm H, Inoue Y, Korman M, Landtblom A, Léger D, Matsui K et al (2022b) Sleep symptoms are essential features of long-COVID – comparing healthy controls with COVID-19 cases of different severity in the international COVID sleep study (ICOSS-II). J Sleep Res. https://doi.org/10.1111/jsr.13754

Merikanto I, Kortesoja L, Benedict C, Chung F, Cedernaes J, Espie CA, Morin CM, Dauvilliers Y, Partinen M, De Gennaro L, Wing YK, Chan NY, Inoue Y, Matsui K, Holzinger B, Plazzi G, Mota-Rolim SA, Leger D, Penzel T, Bjorvatn B (2022c) Evening-types show highest increase of sleep and mental health problems during the COVID-19 pandemic—multinational study on 19 267 adults. Sleep 45(2):zsab216. https://doi.org/10.1093/sleep/zsab216

Morin CM, Carrier J, Bastien C, Godbout R, On behalf of the Canadian Sleep and Circadian Network (2020) Sleep and circadian rhythm in response to the COVID-19 pandemic. Can J Public Health 111(5):654–657. https://doi.org/10.17269/s41997-020-00382-7

Morin CM, Bjorvatn B, Chung F, Holzinger B, Partinen M, Penzel T, Ivers H, Wing YK, Chan NY, Merikanto I, Mota-Rolim S, Macêdo T, De Gennaro L, Léger D, Dauvilliers Y, Plazzi G, Nadorff MR, Bolstad CJ, Sieminski M et al (2021) Insomnia, anxiety, and depression during the COVID-19 pandemic: an international collaborative study. Sleep Med 87:38–45. https://doi.org/10.1016/j.sleep.2021.07.035

Norbury R (2021) Loneliness in the time of COVID. Chronobiol Int 38(6):817–819. https://doi.org/10.1080/07420528.2021.1895201

Ohayon MM (2002) Epidemiology of insomnia: what we know and what we still need to learn. Sleep Med Rev 6(2):97–111. https://doi.org/10.1053/smrv.2002.0186

Pérez-Carbonell L, Meurling IJ, Wassermann D, Gnoni V, Leschziner G, Weighall A, Ellis J, Durrant S, Hare A, Steier J, on behalf of the BSS Task Force on Early Detection Screening (2020) Impact of the novel coronavirus (COVID-19) pandemic on sleep. J Thorac Dis 12(S2): S163–S175. https://doi.org/10.21037/jtd-cus-2020-015

Pappa S, Ntella V, Giannakas T, Giannakoulis VG, Papoutsi E, Katsaounou P (2020) Prevalence of depression, anxiety, and insomnia among healthcare workers during the COVID-19 pandemic: a systematic review and meta-analysis. Brain Behav Immun 88:901–907. https://doi.org/10.1016/j.bbi.2020.05.026

Partinen M, Holzinger B, Morin CM, Espie C, Chung F, Penzel T, Benedict C, Bolstad CJ, Cedernaes J, Chan RNY, Dauvilliers Y, De Gennaro L, Han F, Inoue Y, Matsui K, Leger D, Cunha AS, Merikanto I, Mota-Rolim S et al (2021) Sleep and daytime problems during the COVID-19 pandemic and effects of coronavirus infection, confinement and financial suffering: a multinational survey using a harmonised questionnaire. BMJ Open 11(12):e050672. https://doi.org/10.1136/bmjopen-2021-050672

Pépin J-L, Bailly S, Mordret E, Gaucher J, Tamisier R, Ben Messaoud R, Arnal PJ, Mignot E (2021) Greatest changes in objective sleep architecture during COVID-19 lockdown in night owls with increased REM sleep. Sleep 44(9):zsab075. https://doi.org/10.1093/sleep/zsab075

Pesonen A-K, Lipsanen J, Halonen R, Elovainio M, Sandman N, Mäkelä J-M, Antila M, Béchard D, Ollila HM, Kuula L (2020) Pandemic dreams: network analysis of dream content during the COVID-19 lockdown. Front Psychol 11:573961. https://doi.org/10.3389/fpsyg.2020.573961

Plante DT, Winkelman JW (2006) Parasomnias. Psychiatr Clin N Am 29(4):969–987. https://doi.org/10.1016/j.psc.2006.08.006

Rezaei N, Grandner MA (2021) Changes in sleep duration, timing, and variability during the COVID-19 pandemic: large-scale Fitbit data from 6 major US cities. Sleep Health 7(3): 303–313. https://doi.org/10.1016/j.sleh.2021.02.008

Robillard R, Dion K, Pennestri MH, Solomonova E, Lee E, Saad M, Murkar A, Godbout R, Edwards JD, Quilty L, Daros AR, Bhatla R, Kendzerska T (2021) Profiles of sleep changes during the COVID-19 pandemic: demographic, behavioural and psychological factors. J Sleep Res 30(1):e13231

Roehrs T, Roth T (2001) Sleep, sleepiness, sleep disorders and alcohol use and abuse. Sleep Med Rev 5(4):287–297. https://doi.org/10.1053/smrv.2001.0162

Roenneberg T, Pilz LK, Zerbini G, Winnebeck EC (2019) Chronotype and social jetlag: a (self-) critical review. Biology 8(3):54. https://doi.org/10.3390/biology8030054

Roitblat Y, Burger J, Leit A, Nehuliaieva L, Umarova GS, Kaliberdenko V, Kulanthaivel S, Buchris N, Shterenshis M (2020) Stay-at-home circumstances do not produce sleep disorders: an international survey during the COVID-19 pandemic. J Psychosom Res 139:110282. https://doi.org/10.1016/j.jpsychores.2020.110282

Roth T (2007) Insomnia: definition, prevalence, etiology, and consequences. J Clin Sleep Med 3(5 Suppl):S7–S10. https://doi.org/10.5664/jcsm.26929

Salfi F, Lauriola M, D'Atri A, Amicucci G, Viselli L, Tempesta D, Ferrara M (2021) Demographic, psychological, chronobiological, and work-related predictors of sleep disturbances during the COVID-19 lockdown in Italy. Sci Rep 11(1):11416. https://doi.org/10.1038/s41598-021-90993-y

Sateia MJ (2014) International classification of sleep disorders-third edition. Chest 146(5): 1387–1394. https://doi.org/10.1378/chest.14-0970

Scarpelli S, Nadorff MR, Bjorvatn B, Chung F, Dauvilliers Y, Espie CA, Inoue Y, Matsui K, Merikanto I, Morin CM, Penzel T, Sieminski M, Fang H, Macêdo T, Mota-Rolim SA, Leger D, Plazzi G, Chan NY, Partinen M et al (2022) Nightmares in people with COVID-19: did coronavirus infect our dreams? Nat Sci Sleep 14:93–108. https://doi.org/10.2147/NSS.S344299

Staller N, Randler C (2021) Changes in sleep schedule and chronotype due to COVID-19 restrictions and home office. Somnologie 25(2):131–137. https://doi.org/10.1007/s11818-020-00277-2

Spicuzza L, Mancuso S, Campisi R, Vancheri C (2022) Sleep quality and mental health during the COVID-19 pandemic in patients with severe obstructive sleep apnea. J Patient-Reported Outcomes 6(1):46. https://doi.org/10.1186/s41687-022-00454-x

Voitsidis P, Gliatas I, Bairachtari V, Papadopoulou K, Papageorgiou G, Parlapani E, Syngelakis M, Holeva V, Diakogiannis I (2020) Insomnia during the COVID-19 pandemic in a Greek population. Psychiatry Res 289:113076

Yuan RK, Zitting K-M, Maskati L, Huang J (2022) Increased sleep duration and delayed sleep timing during the COVID-19 pandemic. Sci Rep 12(1):10937. https://doi.org/10.1038/s41598-022-14782-x

Chapter 3
Anxiety and Depression During the COVID-19 Pandemic and Their Impact on Sleep

Mohammed Al Maqbali

Abstract The effects of COVID-19 on mental health are severe. It is widely acknowledged that there is a link between anxiety, depression, and sleep disorders. Although we are aware that such disorders can be exacerbated during an active infection, the exact effects of the COVID-19 pandemic on the relationship between anxiety, depression, and sleep disorders are not yet clear. Sleep disorders that appear to be associated with high levels of mental disorders are typically disregarded and classified as nonspecific occurrences. However, they should still be expected to cause problems. This is because the link between mental health conditions and other problems is often shared. This chapter aims to provide a comprehensive overview of how anxiety and depression impacted sleep disorders during the COVID-19 pandemic. It also briefly explores the prevalence and treatment options available for these conditions. The overall prevalence of anxiety, depression, and sleep disorders was higher during the COVID-19 pandemic compared with before the pandemic. In addition, bidirectionally significant associations were found between sleep disorders, anxiety, and depression. This suggests that anxiety and depression predict sleep disorders and vice versa. Thus, we might recommend that treating sleep disorders can prevent anxiety and depression from developing.

Keywords Sleep disorders · Anxiety · Depression · COVID-19 · Insomnia

3.1 Introduction

Prior to 2020, sleep disorders were ubiquitous and associated with a wide range of psychopathologies, especially anxiety and depression. However, at the end of December 2019, the new severe acute respiratory syndrome coronavirus (COVID-19) was identified in Wuhan City, Hubei province, China, and has since spread worldwide (Li et al. 2020). Multiple studies have shown that psychological outcomes—mainly anxiety and depression—are major factors that significantly

M. Al Maqbali (✉)
Fatima College of Health Sciences, Al Ain, United Arab Emirates

A. S. BaHammam et al. (eds.), *COVID-19 and Sleep: A Global Outlook*, Progress in Sleep Research, https://doi.org/10.1007/978-981-99-0240-8_3

increase the poor sleep outcomes of individuals prior to the COVID-19 pandemic (Oh et al. 2019; Johansson et al. 2021).

The sudden onset of a threatening infectious disease can affect the mental health and sleep of the individual. COVID-19 significantly impacted the daily lifestyle activities of individuals as governments all over the world simply introduced various restrictions (Dai et al. 2021) such as lockdowns and avoiding social activities and self-isolation and quarantine at home for individuals infected by COVID-19. In addition, individuals might have received worrying information such as daily COVID-19 death statistics reported by news outlets and social media (Zhou et al. 2021b). Consequently, those in receipt of lifestyle changes and worrying information developed a great fear of infection, which might have increased the incidence of sleep disorders among individuals.

The etiology of the relationship between poor sleep and depression and anxiety remains unclear during the COVID-19 pandemic. Understanding this relationship could help to improve treatment, health management, and staff education for further assessment of sleep in the healthcare system. This chapter aims to provide a framework for understanding how anxiety and depression impacted sleep during the COVID-19 pandemic. To do this, we will initially review the existing literature on the prevalence of sleep disorders, anxiety, and depression. We will then assess the impact of anxiety and depression on sleep separately, before and during the pandemic. Lastly, we will assess the available evidence concerning treating sleep disorders in the context of anxiety and depression disorders.

3.2 Overview of Anxiety

Anxiety disorders are a widespread mental health problem worldwide, which has created new challenges for healthcare systems in terms of improving diagnoses and treatment. In 2019 anxiety disorders were ranked as the sixth leading cause of disability as they are responsible for about 28.6 million disability-adjusted life years according to the Global Burden of Diseases, Injuries, and Risk Factors Study 2019 (Murray et al. 2020). People suffering from anxiety disorders may experience excessive fear and worries that are out of proportion to the actual threat in the environment or are perceived by themselves (Craske et al. 2017).

Anxiety disorder is a complex and multidimensional condition that incorporates subjective and objective experiences involving physical, affective, cognitive, and behavioral factors (Barros et al. 2022). Anxiety disorders may include generalized anxiety disorder, specific phobias, panic disorder, social anxiety disorder, agoraphobia, and separation anxiety disorder (Association American Psychiatric 2013). A specific diagnostic criterion for anxiety disorders was outlined in two of most classification systems—the Diagnostic and Statistical Manual of Mental Disorders, Fifth Edition (DSM-5) (Association American Psychiatric 2013) and the International Classification of Diseases, Tenth Edition (ICD-10) (Organization World Health 1992).

According to the Global Burden of Diseases, Injuries, and Risk Factors Study 2019, anxiety and depression were the two most common mental disorders that caused disability worldwide in 2019. These disorders were ranked among the top 25 leading causes of burden globally (GBD 2019 Mental Disorders Collaborators 2022). The number of individuals suffering from anxiety disorders has increased significantly over the past 30 years. In 2019, around 45.8 million individuals were diagnosed with anxiety disorders globally. The prevalence of anxiety disorders has been estimated to be around 301 million people (Yang et al. 2021). However, globally, the lifetime prevalence estimate of individuals suffering from anxiety was 7.3% (Stein et al. 2017).

It is well-known that anxiety can affect various aspects of daily life, such as the ability to perform basic activities. Evidence indicates that anxiety is associated with impaired individuals' ability to perform an activity (Edwards et al. 2022). This condition can manifest itself in terms of various functional limitations. Furthermore, a study revealed that those individuals who displayed anxiety were more prone to experiencing physical disability (Behrens-Wittenberg and Wedegaertner 2020). In addition, several studies found a positive relationship between anxiety and the development of hypertensives (Johnson 2019), heart disease (De Hert et al. 2018), and diabetes (Smith et al. 2018). Furthermore, individuals with late-onset anxiety disorder are more prone to developing immune dysregulation, which could make those individuals more susceptible to disease (Gaspersz et al. 2017).

3.3 Relationship Between Sleep Disorders and Anxiety

Since Dement's Report (Dement 1960) in the 1960s about the possibility that the increase in anxiety was attributed to lack of sleep, this relationship has been widely studied. The cognitive model of sleep disorders states that excessive worry and anxiety can negatively affect sleep quality (Harvey 2002). It also suggests that people with high levels of worry are more prone to experiencing sleep problems. In addition, several studies have suggested that sleep disruption can trigger a physiological response that can help predict the development of anxiety (Kalmbach et al. 2019; Richards et al. 2020).

Sleep is acknowledged as a vital process that can help maintain and improve various physiological and neural systems. For instance, it can help clear metabolic waste from the brain and improve cognitive function (Kaur et al. 2021). However, losing sleep can also have detrimental effects. Some of these include the development of deficits in cognitive functioning and affect the regulation of various circadian processes (Hudson et al. 2020). Poor sleep can also negatively affect the emotional function of people. It can also lead to deficits in regulating emotion-related abilities (Vandekerckhove and Wang 2017). In addition, sleep disturbance is also prevalent in various psychiatric disorders (Freeman et al. 2020).

The rapid spread of COVID-19 has caused many people around the world to feel scared and anxious. Regular sleep is essential for human beings to maintain their daily functions. However, several reviews suggest that the pandemic has had an overwhelmingly negative effect on mental health (Salari et al. 2020; Santabárbara et al. 2021). For instance, a meta-analysis by Santabárbara et al. (2021) suggested a threefold increase in the prevalence of anxiety in various countries compared to before the pandemic. Therefore, it is important to understand the association between sleep disorder and anxiety before and during the pandemic.

3.4 Prior to the Pandemic

To understand the relationship between sleep disorders and anxiety during the pandemic, it is vital to know the association between sleep disorders and anxiety prior to the pandemic (Freeman et al. 2020). In psychiatry, sleep disorders such as insomnia, hypersomnia, and sleep-regulation dysregulation have been regarded as secondary to the condition of anxiety. This means that if anxiety worsens, sleep disturbances would also do so. A large body of evidence supports the idea that sleep disturbances can be in a bidirectional relationship with anxiety. A meta-analysis of 18 studies of 34 experiments revealed that those with sleep disorders were 3.9 times more likely to develop anxiety than those who did not experience sleep disorders (Pires et al. 2016). Similarly, another meta-analysis conducted by Hertenstein et al. (2019), involving 13 primary studies, found that insomnia is a significant predictor of anxiety. It suggests that there is a threefold increase in the prevalence of anxiety in individuals with insomnia. A previous meta-analysis indicated that a reduction in sleep duration is a key factor that can contribute to the development of anxiety (Baglioni et al. 2016; Cox and Olatunji 2020).

Garbarino et al. (2020) found a bidirectional relationship between obstructive sleep apnea and anxiety. In addition, a bidirectional association has been found between poor sleep quality and anxiety among different types of individuals, whether in the general population (Pires et al. 2016), college students (Friedrich and Schlarb 2018), healthcare professionals (Weaver et al. 2018), individuals with chronic diseases (Naranjo et al. 2019), the elderly (Press et al. 2018), and adolescents (Vancampfort et al. 2019).

Over the past two decades, two previous pathogenic pandemic outbreaks have impacted the global population—Severe Acute Respiratory Syndrome (SARS) in 2003 and Middle East Respiratory Syndrome (MERS). A longitudinal study in Taiwan found an association between anxiety and sleep disturbances during the SARS outbreak (Chen et al. 2006). In the case of the MERS outbreaks, a retrospective study conducted in Korea among MERS patients found that patients with insomnia reported moderate to severe anxiety (Kim et al. 2018). However, a limited number of studies have reported an association between sleep and anxiety during the MERS and SARS pandemics.

3.5 During the Pandemic

The COVID-19 pandemic is creating concerns about the mental health of individuals around the globe, in that it has introduced significant disruptions to their daily lives. As a result, there has been a call for research funding and for researchers to deploy resources to understand the implications of the psychological effects of the COVID-19 pandemic and the ensuing mental health crises.

Several researchers have shown that large-scale disasters, such as those that occurred during the pandemic, can increase the prevalence of mental health issues in the affected population (Makwana 2019; Morganstein and Ursano 2020). These correlations have also remained true throughout the pandemic. Multiple studies have indicated that the number of people with anxiety and associated sleep disorders has increased significantly since the onset of the pandemic (Vindegaard and Benros 2020; McCracken et al. 2021).

Furthermore, in response to the COVID-19 pandemic, several governments have enacted various measures such as lockdowns to limit the spread of the disease. Some studies longitudinally addressed the significant impact of anxiety on sleep disorders during the COVID-19 lockdown (Salfi et al. 2020; Mirolli et al. 2021; Falkingham et al. 2022). However, several longitudinal studies conducted in several countries have investigated the prevalence of anxiety during different waves of the COVID-19 pandemic; thus, specifically, they considered the situation when restrictions were eased (Czeisler et al. 2021; Niederkrotenthaler et al. 2022; Wetherall et al. 2022). These studies did not find significant differences concerning the prevalence of anxiety during the different waves of the pandemic.

In the context of the relationship between sleep disorders and anxiety during the restrictions and after they eased, Lam et al. (2021) conducted a follow-up study to assess the factors associated with sleep changes across two waves of the COVID-19 pandemic among adults in Hong Kong. He found that insomnia was negatively associated with anxiety across the two waves. Similarly, a study conducted in Ireland found a bidirectional sleep quality and anxiety relationship during two waves of the COVID-19 pandemic (Raman et al. 2022). These findings were in line with a longitudinal study conducted among Chinese adolescents during three waves of the pandemic, which found a bidirectional association between poor sleep quality and anxiety (Wang et al. 2022).

A recent systematic review and meta-analysis, which included 177 observational studies, attempted to understand the relationship between sleep disorders and psychological distress during the COVID-19 pandemic (Alimoradi et al. 2021). The study found that sleep disorders were moderately (effect size 0.55) correlated with anxiety. Moreover, sleep disorders were positively associated with anxiety among the general population, COVID-19 patients, and healthcare professionals, with effect sizes of 0.48, 0.49, and 0.55, respectively. In a longitudinal study conducted to examine the underlying factors in terms of mental health problems among college students in China, Zhang et al. (2020) noted the direct and indirect significant impact of anxiety (indirect effect $= 0.40$, $p < 0.001$) on sleep quality.

Regarding the association between sleep disorders and anxiety among different populations, several studies have found that there is a significant relationship between sleep disorders and anxiety in patients infected with COVID-19 (Fernández-de-las-Peñas et al. 2021), healthcare professionals (Th'ng et al. 2021), individuals with chronic diseases (Idrissi et al. 2020), university students (Gardani et al. 2022), children and adolescents (Zhao et al. 2022), and the general population (Alqahtani et al. 2022). Due to the varying methods used in the study, it is difficult to determine precisely how the pandemic has affected the health of individuals. However, consistently, across different countries, results show that significant proportions of people are affected by the various mental health and well-being issues that are assessed.

3.6 Management of Anxiety

Understanding the link between sleep disturbances and anxiety disorders is the next step in developing effective interventions. Although sleep problems are typically severe and complex in people with anxiety, much evidence has supported the idea of using Cognitive Behavioural Therapy (CBT) to treat these conditions. A meta-analysis suggested that there may be a bidirectional relationship between the treatment of anxiety and sleep disorders. An earlier meta-analysis included 25 studies that assessed the impact of CBT for anxiety on comorbid sleep disturbances and reported a moderate effect (effect size 0.53) of anxiety treatment on sleep disturbance (Belleville et al. 2010). Conversely, another meta-analysis of 43 randomized controlled trials (RCTs) involving 5945 trial participants, investigating the impact of CBT sleep intervention on anxiety showed that CBT sleep intervention has a moderate reduction in anxiety (Staines et al. 2022). Ho et al. (2015) and van Straten and Cuijpers (2009) conducted systematic reviews and meta-analyses on the effectiveness of self-help CBT for sleep on anxiety and found a small effect.

The results of individual cognitive therapy sessions are similar to those of group therapy, with individual sessions being more likely to produce significant effects in terms of both anxiety and insomnia (Freeman et al. 2020; Wallsten et al. 2021). In clinical trials, the link between sleep disturbances and anxiety has been established, suggesting that it can be a part of a treatment strategy.

3.7 Overview of Depression

Depression is a major cause of disability worldwide (World Health Organization 2017) with a high disease burden and societal cost (Walker et al. 2015; Liu et al. 2020). It is a known contributor to death by suicide (World Health Organization 2017). It is also associated with other chronic diseases such as cardiovascular and cancer. In addition, it can lead to poor treatment and increased mortality (Machado et al. 2018). Depression is a common mental illness that can affect an individual's mental and physical health. Its symptoms include a lack of interest in regular activities and insomnia, and the disease can cause poor quality of life (Fang et al. 2019). Depression is a chronic disease that can impair the functioning of people in most societies.

The word depression comes from the Latin word "depressio," which means "To press down, depress." Depression is a type of mental disorder that typically involves feelings of sadness, loss of pleasure, low self-worth, tiredness, and anxiety; it can be long-lasting or recurrent (Mancini 2021). According to the DSM-5 criteria of major depressive disorders, it is a constellation of symptoms that include depressed mood, anhedonia, loss of energy, motor retardation or agitation, insomnia, or hypersomnia; at least five or more symptoms should be present for at least 2 weeks (Association American Psychiatric 2013).

Around 322 million people globally are affected by depression, a common mental disorder affecting around 3.8% of the world's population (World Health Organization 2017). It is considered a leading cause of disability, and can severely affect a person's quality of life. The effects of depression can last for a long time, and it can also affect a person's work and personal life. The health burden of depression is estimated by the GBD study, which uses disability-adjusted life years (DALYs) to calculate the impact of the disorder. In 2019, the number of DALYs caused by depression was 37.3% and was ranked 13th in terms of the leading cause of DALYs (GBD 2019 Mental Disorders Collaborators 2022). Lim et al. (2018) studied 1,112,573 adults worldwide, and calculated the prevalence of depression over 1 year as 7.2%, while the lifetime prevalence was 10.8%.

One of the most common mental disorders globally, major depressive disorder is considered to be very burdensome. Depression is a leading cause of poor health and disability worldwide and is expected to rank first globally by 2030 (Malhi and Mann 2018). Furthermore, depression has been linked to suicide. A meta-analysis of 20 studies examined the association between suicide and depression and found that the presentation of depression increased the risk of suicide sevenfolds (Moitra et al. 2021).

It is widely believed that depression can play a role in the development of medical illnesses. For example, studies found that individuals with depression had a higher risk of developing diabetes (Sartorius 2018), hypertension (Herrera et al. 2021), cardiovascular diseases (Zhang et al. 2022b), stroke (Hu et al. 2021), asthma (Cooley et al. 2022), headaches (Baker et al. 2020), obesity (Blasco et al. 2020), dementia, including Alzheimer's disease (Hayley et al. 2021), and an increased risk of hospitalization and accidents (Gialluisi et al. 2021).

3.8 Relationship Between Sleep Disorders and Depression

Individuals suffering from depression are prone to experiencing sleep problems. One of the most important factors researchers are currently looking into when it comes to understanding depression is its relationship with sleep disorders. In many cases, the condition is revealed through sleep disorders. Various forms of sleep disorders can also be associated with depression. These include difficulty falling asleep, frequent awakening during the night, early morning awakening, and non-restorative sleep.

According to the DSM-5 and ICD-10, sleep disturbance is a core diagnostic criterion for major depressive disorder (World Health Organization 1991; Association American Psychiatric 2013). Due to the complexity of depression, it is commonly believed that it is a heterogeneous construct (Spellman and Liston 2020). Therefore, it is important to understand the biology of depression to develop effective treatment strategies. The National Institute of Mental Health's Research Domain Criteria initiative has provided a framework for the study of mental health disorders by focusing on the multiple measurable traits that can be found in each disorder (Morris et al. 2022). As with depression, it is also known that sleep disturbance is often associated with major depression. Thus, the discussion of sleep complaints in both clinical and research settings must not be superficial. While short-term measures can be used to assess sleep disturbance, they are not enough to meaningfully advance the understanding of the link between sleep and depression.

The concept of the relationship between sleep architecture and depression is supported by the alteration in the sleep architecture of individuals in depressive states. Depression is known to have significant effects on the sleep architecture of individuals. These include shortened rapid eye movement (REM) latency, increased REM sleep duration, and REM density. All these factors have been used as biological markers of depression, and predict the likelihood of relapse (Freeman et al. 2020). At the basic level, brain regions involved in regulating sleep include the thalamic nuclei, the forebrain, the hypothalamus, the brain stem, and the limbic mechanisms that affect arousal. All of these are associated with the pathophysiology of both sleep disturbances and depression disorders (Shen et al. 2022).

The rapid spread of COVID-19 has caused anxiety in people around the world to increase. Consequently, it might lead to the development of depressive disorders. Recent meta-analyses found that depression significantly increased during the COVID-19 pandemic compared with before the pandemic, with a standard mean change of 0.216 (Robinson et al. 2022).

3.9 Prior to the Pandemic

The standard psychiatric diagnosis for depression is that hypersomnia and insomnia are symptoms of the disorder. There have been numerous studies that suggest that sleep disturbance is a contributing factor to the high frequency of depression in patients. Studies have shown that patients with depression and comorbid sleep disorders are more likely to visit psychiatric clinics for treatment. Additionally, they are also more likely to be resistant to treatment (Bishop et al. 2020). In some cases, persistent sleep problems can be a contributing factor to the development of depression and suicide (Wang et al. 2019).

It is estimated that around 90% of patients with depression are prone to experiencing sleep complaints such as insomnia, sleep-disordered breathing, narcolepsy, and restless legs syndrome (RLS) (Wiersema et al. 2018). However, about two-thirds of patients who are experiencing a depressive episode will also experience insomnia (Li et al. 2016), whereas 40% of depression patients may report one or all of experiencing difficulties initiating sleep, maintaining sleep, and/or early morning awakenings (Vorvolakos et al. 2020). In addition, a meta-analysis involving 73 studies, showed that the prevalence of depression was 35% among patients with obstructive sleep apnea (OSA) (Garbarino et al. 2020).

Although depression was initially regarded as a risk factor for sleep disturbances, other studies have shown that this condition can also be an independent cause of depression. Poor sleep quality is known to be a prodromal manifestation of depression. Sleep disorders have been known to be a risk factor for major depression. A longitudinal study involving 5481 patients with mental disorders revealed that the number of patients experiencing depression was significantly higher in those who experience sleep disturbance (Schennach et al. 2019). Three meta-analyses found that nondepressed individuals with insomnia have more than a twofold risk of developing depression compared to individuals without sleep disorders (Baglioni et al. 2011; Li et al. 2016; Hertenstein et al. 2019).

Evidence suggests that a bidirectional relationship exists between sleep disorders and depression. A recent Mendelian randomization study showed a bidirectional causal association between insomnia and depression (Zhou et al. 2021a). Similarly, Nguyen et al. (2022) conducted a longitudinal study that included 3294 participants and examined the proposition of sleep disorders and predictions of depressive symptoms at an 18-year follow-up. The study concluded that sleep disorders significantly mediated depression symptoms and vice versa. A reciprocal relationship existed between sleep disorders and depression in diverse adolescents (Vazsonyi et al. 2022), individuals with chronic diseases (Zhang et al. 2022a), pregnant women (Chan et al. 2022), and community-dwelling adults (Sun et al. 2018).

Regarding the relationship between sleep disorders and depression during the previous MRSA and SARS pandemics, a longitudinal study during the SARS outbreak revealed that the link between depression and sleep disturbance was bidirectional (Chen et al. 2006). However, only a limited number of research studies have investigated the relationship between sleep and depression during the MRSA pandemic.

In individuals with depression, those with poor sleep quality are more prone to experiencing disorder symptoms. This is because the link between these conditions and the development of depression is known to be bidirectional. In other words, if you have insomnia and poor sleep quality, you are more likely to develop depression and vice versa.

3.10 During the Pandemic

The effects of COVID-19 on sleep can be expected to raise awareness about the importance of regular sleep. Sleep is vital to maintaining healthy and balanced physical and psychological processes. Specifically, it can affect the prognosis and quality of life of individuals with COVID-19.

During the lockdown, the main sleep–wake cycle synchronizations are disrupted, usually triggered by the disruption of the circadian rhythm, which is a natural part of our daily routine. Consequently, sleep disorders, depression, and other mental health problems might develop. A meta-analysis found that the prevalence of sleep disturbance was higher during lockdown (42.29%) compared to a non-lockdown period (37.97%) (Jahrami et al. 2022). Several longitudinal studies have assessed the relationship between sleep disorders during the lockdown and a non-lockdown period and found a bidirectional relationship between sleep disorders and depression (Viselli et al. 2021; Gorgoni et al. 2021; Bi and Chen 2022).

Many countries implemented strict social distancing measures, which included banning group meetings, closing schools, and restricting nonessential commercial activities. Implementing these measures can lead to changes in an individual's circadian rhythm (Salehinejad et al. 2022). For instance, it can reduce the pressure to wake up early in the morning on workdays, which can help people avoid experiencing social jet lag. However, studies have shown that circadian rhythm abnormalities in people are positively associated with depression (Reynaud et al. 2022; Felician et al. 2022).

It is widely acknowledged that there is a link between poor sleep quality and depression. A meta-analysis of 177 papers comprising 345,270 participants from 39 countries examined the relationship between sleep disorders and depression during the COVID-19 pandemic (Alimoradi et al. 2021). Sleep disorders were found to be moderately associated with depression (effect size = 0.54). Additionally, in subgroup analysis, an association between sleep disorders and depression was found among healthcare professionals (effect size = −0.28), the general population (effect size = −0.30), and COVID-19 patients (effect size = −0.36).

Insomnia is a common sleep disorder that can be triggered by depression, specifically during the COVID-19 pandemic. Several longitudinal studies have been conducted that examined the relationship between insomnia symptoms and depression during the pandemic. These studies support bidirectional associations between insomnia and depression (Tsaras et al. 2021; Pizzonia et al. 2021). Insomnia can affect the integrity of the sleep cycle, and it can also lead to a reduction in the rapid eye movement cycle, which is believed to be caused by the secretion of monoamines (Ye et al. 2022). Consequently, this reduction in the level of serum monoamines may trigger depressive symptoms (Filatova et al. 2021).

3.11 Management of Depression

The concept of sleep disorders as a transdiagnostic process has been supported by the accumulating evidence related to their relationship with depression. Furthermore, treatment for sleep disturbance may result in a reduction in depressive symptoms. The management of sleep disorders to reduce depression has been divided into non-pharmacological and pharmacological interventions.

3.12 Non-pharmacological Interventions

There are various non-pharmacologic intervention approaches available to treat sleep disorders. For example, there is a multicomponent CBT for insomnia, which may include sleep restriction, stimulus control therapy, sleep hygiene or psychoeducation, and relaxation. Additionally, there are other non-pharmacologic approaches, including light therapy, exercise acupuncture, acupressure, moxibustion, aromatherapy, foot reflexology, meditative movement therapies, homeopathy, music therapy, and yoga. A meta-analysis of 49 RCTs was conducted to determine if non-pharmacological interventions can help reduce depression symptoms (Gee et al. 2019). The author revealed that the effects of CBT for insomnia were associated with a significant reduction in depression symptoms to a great extent (effect size = −0.81).

Another meta-analysis included 65 RCTs involving 8608 participants. It was conducted to examine the effects of any sleep intervention composite on depression (Scott et al. 2021). Most interventions were multicomponent CBT for insomnia, including acupuncture, sleep hygiene alone, sleep restriction alone, Tai Chi, pharmacological treatments, CBT for nightmares, walking, and yoga. The researchers found that these sleep interventions had a significant albeit moderate effect on the symptoms of depression (effect size = −0.63). The results of the CBT interventions suggest that treating sleep disorders with the help of evidence-based treatments can reduce the risk of depression. This finding is consistent with the idea that sleep disturbance can contribute to the development of depression.

3.13 Pharmacological Interventions

Based on the pathophysiology of sleep, pharmaceutical interventions have been developed to treat sleep disorders. These drugs include benzodiazepines, antidepressants, antihistamines, melatonin, and phytotherapeutic substances. Several meta-analyses have attempted to examine the efficacy of phytotherapeutics in treating sleep disorders, and have concluded there is a small to moderate effect in terms of treating sleep disorders (Kuriyama et al. 2014; Winkler et al. 2014; Yeh et al. 2021). It should also be noted that depression was not evaluated in those meta-analyses.

Moon et al. (2022) conducted meta-analyses involving 30 RCTs to report the effect of melatonin agonists on sleep disorders in the case of healthy participants and patients with psychiatric disorders, including depression. Four trials reported patients with depression. However, due to the very limited data, the authors could not provide enough evidence of the benefits of melatonin for sleep disorders in terms of improving depression. Further large single-blinded RCTs are required to establish if the use of pharmaceutical interventions can help treat sleep disorders as well as depression.

3.14 Conclusion

The COVID-19 pandemic has been a global phenomenon, disrupting various aspects of life for most people. Understanding the relationship between anxiety, depression, and sleep disorders could help identify the factors that contribute to their persistence, and improve the quality of life for individuals. The prevalence of the symptoms of anxiety, depression, and sleep disorders was higher during the COVID-19 incident than in previous years, suggesting that the effects of the event on mental health were more severe than previously known. The prevalence of anxiety, depression, and sleep disorders is higher than in previous mass traumatic events. The effects of COVID-19 on the economy and society are also more apparent.

Generally, most individuals experience sleep difficulties at some point in their lives. Sleep disturbance is a common component of anxiety and depression. It is also known that this sleep disorder is often included as a symptom of anxiety and depression. In recent years, the use of longitudinal assessment methods has led to the development of solid evidence supporting the bidirectional associations between sleep and anxiety and depression disorders.

In a clinical setting, it should be routine to assess anxiety and depression to improve the quality of life of individuals with sleep disorders, as a means of helping identify the appropriate treatment and management. These problems should then be treated appropriately through CBT treatment.

Disclosure Statement

The authors do not have any disclosures to report.

Authors Contributorship MA completed the data collection, statistical support and provided analysis. All authors contributed to the conceptualization and design and the manuscript preparation.

References

Alimoradi Z, Broström A, Tsang HWH et al (2021) Sleep problems during COVID-19 pandemic and its' association to psychological distress: a systematic review and meta-analysis. EClinicalMedicine 36:100916. https://doi.org/10.1016/j.eclinm.2021.100916

Alqahtani JS, AlRabeeah SM, Aldhahir AM et al (2022) Sleep quality, insomnia, anxiety, fatigue, stress, memory and active coping during the COVID-19 pandemic. Int J Environ Res Public Health 19:4940. https://doi.org/10.3390/ijerph19094940

Association American Psychiatric (2013) Diagnostic and statistical manual of mental disorders, 5th Edition: DSM-5, 5th edn. American Psychiatric Publishing, Washington, DC

Baglioni C, Battagliese G, Feige B et al (2011) Insomnia as a predictor of depression: a meta-analytic evaluation of longitudinal epidemiological studies. J Affect Disord 135:10–19. https://doi.org/10.1016/j.jad.2011.01.011

Baglioni C, Nanovska S, Regen W et al (2016) Sleep and mental disorders: a meta-analysis of polysomnographic research. Psychol Bull 142:969–990. https://doi.org/10.1037/bul0000053

Baker VB, Sowers CB, Hack NK (2020) Lost productivity associated with headache and depression: a quality improvement project identifying a patient population at risk. J Headache Pain 21: 50. https://doi.org/10.1186/s10194-020-01107-4

Barros F, Figueiredo C, Brás S et al (2022) Multidimensional assessment of anxiety through the State-Trait Inventory for Cognitive and Somatic Anxiety (STICSA): from dimensionality to response prediction across emotional contexts. PLoS One 17:e0262960. https://doi.org/10.1371/journal.pone.0262960

Behrens-Wittenberg E, Wedegaertner F (2020) Identifying individuals at high risk for permanent disability from depression and anxiety. Front Psychiatry 11:740

Belleville G, Cousineau H, Levrier K et al (2010) The impact of cognitive-behavior therapy for anxiety disorders on concomitant sleep disturbances: a meta-analysis. J Anxiety Disord 24:379–386. https://doi.org/10.1016/j.janxdis.2010.02.010

Bi K, Chen S (2022) Sleep profiles as a longitudinal predictor for depression magnitude and variability following the onset of COVID-19. J Psychiatr Res 147:159–165. https://doi.org/10.1016/j.jpsychires.2022.01.024

Bishop TM, Walsh PG, Ashrafioun L et al (2020) Sleep, suicide behaviors, and the protective role of sleep medicine. Sleep Med 66:264–270. https://doi.org/10.1016/j.sleep.2019.07.016

Blasco BV, García-Jiménez J, Bodoano I, Gutiérrez-Rojas L (2020) Obesity and depression: its prevalence and influence as a prognostic factor: a systematic review. Psychiatry Investig 17: 715–724. https://doi.org/10.30773/pi.2020.0099

Chan C, Poon SH, Chua T-E et al (2022) A prospective study of the relationship between sleep quality and depression in pregnancy. Proc Singapore Healthc 31:20101058211068590. https://doi.org/10.1177/20101058211068591

Chen R, Chou K-R, Huang Y-J et al (2006) Effects of a SARS prevention programme in Taiwan on nursing staff's anxiety, depression and sleep quality: a longitudinal survey. Int J Nurs Stud 43: 215–225. https://doi.org/10.1016/j.ijnurstu.2005.03.006

Cooley C, Park Y, Ajilore O et al (2022) Impact of interventions targeting anxiety and depression in adults with asthma. J Asthma 59:273–287. https://doi.org/10.1080/02770903.2020.1847927

Cox RC, Olatunji BO (2020) Sleep in the anxiety-related disorders: a meta-analysis of subjective and objective research. Sleep Med Rev 51:101282. https://doi.org/10.1016/j.smrv.2020.101282

Craske MG, Stein MB, Eley TC et al (2017) Anxiety disorders. Nat Rev Dis Primers 3:1–19. https://doi.org/10.1038/nrdp.2017.24

Czeisler MÉ, Wiley JF, Facer-Childs ER et al (2021) Mental health, substance use, and suicidal ideation during a prolonged COVID-19-related lockdown in a region with low SARS-CoV-2 prevalence. J Psychiatr Res 140:533–544. https://doi.org/10.1016/j.jpsychires.2021.05.080

Dai J, Sang X, Menhas R et al (2021) The influence of COVID-19 pandemic on physical health–psychological health, physical activity, and overall well-being: the mediating role of emotional regulation. Front Psychol 12:667461

De Hert M, Detraux J, Vancampfort D (2018) The intriguing relationship between coronary heart disease and mental disorders. Dialog Clin Neurosci 20:31–40. https://doi.org/10.31887/DCNS.2018.20.1/mdehert
Dement W (1960) The effect of dream deprivation: the need for a certain amount of dreaming each night is suggested by recent experiments. Science 131:1705–1707. https://doi.org/10.1126/science.131.3415.1705
Edwards G, Jones C, Pearson E et al (2022) Prevalence of anxiety symptomatology and diagnosis in syndromic intellectual disability: a systematic review and meta-analysis. Neurosci Biobehav Rev 138:104719. https://doi.org/10.1016/j.neubiorev.2022.104719
Falkingham JC, Evandrou M, Qin M, Vlachantoni A (2022) Prospective longitudinal study of 'Sleepless in Lockdown': unpacking differences in sleep loss during the coronavirus pandemic in the UK. BMJ Open 12:e053094. https://doi.org/10.1136/bmjopen-2021-053094
Fang H, Tu S, Sheng J, Shao A (2019) Depression in sleep disturbance: a review on a bidirectional relationship, mechanisms and treatment. J Cell Mol Med 23:2324–2332. https://doi.org/10.1111/jcmm.14170
Felician J, Peter-Derex L, Nourredine M et al (2022) Association of delayed sleep/wake rhythm with depression during the first COVID-19 lockdown in France. Sleep Med 100:S43–S44. https://doi.org/10.1016/j.sleep.2022.05.130
Fernández-de-las-Peñas C, Gómez-Mayordomo V, de-la-Llave-Rincón AI et al (2021) Anxiety, depression and poor sleep quality as long-term post-COVID sequelae in previously hospitalized patients: a multicenter study. J Infect 83:496–522. https://doi.org/10.1016/j.jinf.2021.06.022
Filatova EV, Shadrina MI, Slominsky PA (2021) Major depression: one brain, one disease, one set of intertwined processes. Cells 10:1283. https://doi.org/10.3390/cells10061283
Freeman D, Sheaves B, Waite F et al (2020) Sleep disturbance and psychiatric disorders. Lancet Psychiatry 7:628–637. https://doi.org/10.1016/S2215-0366(20)30136-X
Friedrich A, Schlarb AA (2018) Let's talk about sleep: a systematic review of psychological interventions to improve sleep in college students. J Sleep Res 27:4–22. https://doi.org/10.1111/jsr.12568
Garbarino S, Bardwell WA, Guglielmi O et al (2020) Association of anxiety and depression in obstructive sleep apnea patients: a systematic review and meta-analysis. Behav Sleep Med 18:35–57. https://doi.org/10.1080/15402002.2018.1545649
Gardani M, Bradford DRR, Russell K et al (2022) A systematic review and meta-analysis of poor sleep, insomnia symptoms and stress in undergraduate students. Sleep Med Rev 61:101565. https://doi.org/10.1016/j.smrv.2021.101565
Gaspersz R, Lamers F, Wittenberg G et al (2017) The role of anxious distress in immune dysregulation in patients with major depressive disorder. Transl Psychiatry 7:1268. https://doi.org/10.1038/s41398-017-0016-3
GBD 2019 Mental Disorders Collaborators (2022) Global, regional, and national burden of 12 mental disorders in 204 countries and territories, 1990–2019: a systematic analysis for the Global Burden of Disease Study 2019. Lancet Psychiatry 9:137–150. https://doi.org/10.1016/S2215-0366(21)00395-3
Gee B, Orchard F, Clarke E et al (2019) The effect of non-pharmacological sleep interventions on depression symptoms: a meta-analysis of randomised controlled trials. Sleep Med Rev 43:118–128. https://doi.org/10.1016/j.smrv.2018.09.004
Gialluisi A, Costanzo S, Castelnuovo AD et al (2021) Combined influence of depression severity and low-grade inflammation on incident hospitalization and mortality risk in Italian adults. J Affect Disord 279:173–182. https://doi.org/10.1016/j.jad.2020.10.004
Gorgoni M, Scarpelli S, Mangiaruga A et al (2021) Persistence of the effects of the COVID-19 lockdown on sleep: a longitudinal study. Brain Sci 11:1520. https://doi.org/10.3390/brainsci11111520
Harvey AG (2002) A cognitive model of insomnia. Behav Res Ther 40:869–893. https://doi.org/10.1016/S0005-7967(01)00061-4

Hayley S, Hakim AM, Albert PR (2021) Depression, dementia and immune dysregulation. Brain 144:746–760. https://doi.org/10.1093/brain/awaa405

Herrera PA, Campos-Romero S, Szabo W et al (2021) Understanding the relationship between depression and chronic diseases such as diabetes and hypertension: a Grounded Theory Study. Int J Environ Res Public Health 18:12130. https://doi.org/10.3390/ijerph182212130

Hertenstein E, Feige B, Gmeiner T et al (2019) Insomnia as a predictor of mental disorders: a systematic review and meta-analysis. Sleep Med Rev 43:96–105. https://doi.org/10.1016/j.smrv.2018.10.006

Ho FY-Y, Chung K-F, Yeung W-F et al (2015) Self-help cognitive-behavioral therapy for insomnia: a meta-analysis of randomized controlled trials. Sleep Med Rev 19:17–28. https://doi.org/10.1016/j.smrv.2014.06.010

Hu J, Wang L, Fan K et al (2021) The association between systemic inflammatory markers and post-stroke depression: a prospective stroke cohort. Clin Interv Aging 16:1231–1239. https://doi.org/10.2147/CIA.S314131

Hudson AN, Van Dongen HPA, Honn KA (2020) Sleep deprivation, vigilant attention, and brain function: a review. Neuropsychopharmacology 45:21–30. https://doi.org/10.1038/s41386-019-0432-6

Idrissi A, Lamkaddem A, Benouajjit A et al (2020) Sleep quality and mental health in the context of COVID-19 pandemic and lockdown in Morocco. Sleep Med 74:248–253. https://doi.org/10.1016/j.sleep.2020.07.045

Jahrami HA, Alhaj OA, Humood AM et al (2022) Sleep disturbances during the COVID-19 pandemic: a systematic review, meta-analysis, and meta-regression. Sleep Med Rev 62: 101591. https://doi.org/10.1016/j.smrv.2022.101591

Johansson M, Jansson-Fröjmark M, Norell-Clarke A, Linton SJ (2021) Changes in insomnia as a risk factor for the incidence and persistence of anxiety and depression: a longitudinal community study. Sleep Sci Pract 5:1–9. https://doi.org/10.1186/s41606-020-00053-z

Johnson HM (2019) Anxiety and hypertension: is there a link? A literature review of the comorbidity relationship between anxiety and hypertension. Curr Hypertens Rep 21:66. https://doi.org/10.1007/s11906-019-0972-5

Kalmbach DA, Abelson JL, Arnedt JT et al (2019) Insomnia symptoms and short sleep predict anxiety and worry in response to stress exposure: a prospective cohort study of medical interns. Sleep Med 55:40–47. https://doi.org/10.1016/j.sleep.2018.12.001

Kaur J, Fahmy LM, Davoodi-Bojd E et al (2021) Waste clearance in the brain. Front Neuroanat 15: 665803

Kim H-C, Yoo S-Y, Lee B-H et al (2018) Psychiatric findings in suspected and confirmed Middle East respiratory syndrome patients quarantined in hospital: a retrospective chart analysis. Psychiatry Investig 15:355–360. https://doi.org/10.30773/pi.2017.10.25.1

Kuriyama A, Honda M, Hayashino Y (2014) Ramelteon for the treatment of insomnia in adults: a systematic review and meta-analysis. Sleep Med 15:385–392. https://doi.org/10.1016/j.sleep.2013.11.788

Lam CS, Yu BY-M, Cheung DST et al (2021) Sleep and mood disturbances during the COVID-19 outbreak in an urban Chinese population in Hong Kong: a longitudinal study of the second and third waves of the outbreak. Int J Environ Res Public Health 18:8444. https://doi.org/10.3390/ijerph18168444

Li L, Wu C, Gan Y et al (2016) Insomnia and the risk of depression: a meta-analysis of prospective cohort studies. BMC Psychiatry 16:1–16. https://doi.org/10.1186/s12888-016-1075-3

Li Q, Guan X, Wu P et al (2020) Early transmission dynamics in Wuhan, China, of novel coronavirus–infected pneumonia. N Engl J Med 382:1199–1207. https://doi.org/10.1056/NEJMoa2001316

Lim GY, Tam WW, Lu Y et al (2018) Prevalence of depression in the community from 30 countries between 1994 and 2014. Sci Rep 8:2861. https://doi.org/10.1038/s41598-018-21243-x

Liu Q, He H, Yang J et al (2020) Changes in the global burden of depression from 1990 to 2017: findings from the Global Burden of Disease study. J Psychiatr Res 126:134–140. https://doi.org/10.1016/j.jpsychires.2019.08.002

Machado MO, Veronese N, Sanches M et al (2018) The association of depression and all-cause and cause-specific mortality: an umbrella review of systematic reviews and meta-analyses. BMC Med 16:112. https://doi.org/10.1186/s12916-018-1101-z

Makwana N (2019) Disaster and its impact on mental health: a narrative review. J Family Med Prim Care 8:3090–3095. https://doi.org/10.4103/jfmpc.jfmpc_893_19

Malhi GS, Mann JJ (2018) Depression. Lancet 392:2299–2312. https://doi.org/10.1016/S0140-6736(18)31948-2

Mancini MA (2021) Screening and assessment for depression and anxiety disorders. In: Mancini MA (ed) Integrated behavioral health practice. Springer International Publishing, Cham, pp 237–267

McCracken LM, Badinlou F, Buhrman M, Brocki KC (2021) The role of psychological flexibility in the context of COVID-19: associations with depression, anxiety, and insomnia. J Contextual Behav Sci 19:28–35. https://doi.org/10.1016/j.jcbs.2020.11.003

Mirolli M, Simione L, Martoni M, Fabbri M (2021) Accept anxiety to improve sleep: the impact of the COVID-19 lockdown on the relationships between mindfulness, distress, and sleep quality. Int J Environ Res Public Health 18:13149. https://doi.org/10.3390/ijerph182413149

Moitra M, Santomauro D, Degenhardt L et al (2021) Estimating the risk of suicide associated with mental disorders: a systematic review and meta-regression analysis. J Psychiatr Res 137:242–249. https://doi.org/10.1016/j.jpsychires.2021.02.053

Moon E, Partonen T, Beaulieu S, Linnaranta O (2022) Melatonergic agents influence the sleep-wake and circadian rhythms in healthy and psychiatric participants: a systematic review and meta-analysis of randomized controlled trials. Neuropsychopharmacology 47:1523–1536. https://doi.org/10.1038/s41386-022-01278-5

Morganstein JC, Ursano RJ (2020) Ecological disasters and mental health: causes, consequences, and interventions. Front Psychiatry 11:1

Morris SE, Sanislow CA, Pacheco J et al (2022) Revisiting the seven pillars of RDoC. BMC Med 20:220. https://doi.org/10.1186/s12916-022-02414-0

Murray CJL, Aravkin AY, Zheng P et al (2020) Global burden of 87 risk factors in 204 countries and territories, 1990–2019: a systematic analysis for the Global Burden of Disease Study 2019. Lancet 396:1223–1249. https://doi.org/10.1016/S0140-6736(20)30752-2

Naranjo C, Del Reguero L, Moratalla G et al (2019) Anxiety, depression and sleep disorders in patients with diabetic neuropathic pain: a systematic review. Expert Rev Neurother 19:1201–1209. https://doi.org/10.1080/14737175.2019.1653760

Nguyen VV, Zainal NH, Newman MG (2022) Why sleep is key: poor sleep quality is a mechanism for the bidirectional relationship between major depressive disorder and generalized anxiety disorder across 18 years. J Anxiety Disord 90:102601. https://doi.org/10.1016/j.janxdis.2022.102601

Niederkrotenthaler T, Laido Z, Kirchner S et al (2022) Mental health over nine months during the SARS-CoV2 pandemic: representative cross-sectional survey in twelve waves between April and December 2020 in Austria. J Affect Disord 296:49–58. https://doi.org/10.1016/j.jad.2021.08.153

Oh C-M, Kim HY, Na HK et al (2019) The effect of anxiety and depression on sleep quality of individuals with high risk for insomnia: a population-based study. Front Neurol 10:849. https://doi.org/10.3389/fneur.2019.00849

Organization World Health (1992) The ICD-10 classification of mental and behavioural disorders: clinical descriptions and diagnostic guidelines, 1st edn. World Health Organization, Geneva

Pires GN, Bezerra AG, Tufik S, Andersen ML (2016) Effects of acute sleep deprivation on state anxiety levels: a systematic review and meta-analysis. Sleep Med 24:109–118. https://doi.org/10.1016/j.sleep.2016.07.019

Pizzonia KL, Koscinski B, Suhr JA et al (2021) Insomnia during the COVID-19 pandemic: the role of depression and COVID-19-related risk factors. Cogn Behav Ther 50:246–260. https://doi.org/10.1080/16506073.2021.1879241

Press Y, Punchik B, Freud T (2018) The association between subjectively impaired sleep and symptoms of depression and anxiety in a frail elderly population. Aging Clin Exp Res 30:755–765. https://doi.org/10.1007/s40520-017-0837-1

Raman S, Hyland P, Coogan AN (2022) Temporal associations between insomnia and depression symptoms in adults during the COVID-19 pandemic: a cross-lagged path modelling analysis. Psychiatry Res 312:114533. https://doi.org/10.1016/j.psychres.2022.114533

Reynaud E, Pottelette J, Rabot J et al (2022) Differential effects of COVID-related lockdown on sleep–wake rhythms in adults with autism spectrum disorder compared to the general population. Autism Res 15:945–956. https://doi.org/10.1002/aur.2692

Richards A, Kanady JC, Neylan TC (2020) Sleep disturbance in PTSD and other anxiety-related disorders: an updated review of clinical features, physiological characteristics, and psychological and neurobiological mechanisms. Neuropsychopharmacology 45:55–73. https://doi.org/10.1038/s41386-019-0486-5

Robinson E, Sutin AR, Daly M, Jones A (2022) A systematic review and meta-analysis of longitudinal cohort studies comparing mental health before versus during the COVID-19 pandemic in 2020. J Affect Disord 296:567–576. https://doi.org/10.1016/j.jad.2021.09.098

Salari N, Hosseinian-Far A, Jalali R et al (2020) Prevalence of stress, anxiety, depression among the general population during the COVID-19 pandemic: a systematic review and meta-analysis. Glob Health 16:57. https://doi.org/10.1186/s12992-020-00589-w

Salehinejad MA, Azarkolah A, Ghanavati E, Nitsche MA (2022) Circadian disturbances, sleep difficulties and the COVID-19 pandemic. Sleep Med 91:246–252. https://doi.org/10.1016/j.sleep.2021.07.011

Salfi F, Lauriola M, Amicucci G et al (2020) Gender-related time course of sleep disturbances and psychological symptoms during the COVID-19 lockdown: a longitudinal study on the Italian population. Neurobiol Stress 13:100259. https://doi.org/10.1016/j.ynstr.2020.100259

Santabárbara J, Lasheras I, Lipnicki DM et al (2021) Prevalence of anxiety in the COVID-19 pandemic: an updated meta-analysis of community-based studies. Prog Neuro-Psychopharmacol Biol Psychiatry 109:110207. https://doi.org/10.1016/j.pnpbp.2020.110207

Sartorius N (2018) Depression and diabetes. Dialog Clin Neurosci 20:47–52. https://doi.org/10.31887/DCNS.2018.20.1/nsartorius

Schennach R, Feige B, Riemann D et al (2019) Pre- to post-inpatient treatment of subjective sleep quality in 5,481 patients with mental disorders: a longitudinal analysis. J Sleep Res 28:e12842. https://doi.org/10.1111/jsr.12842

Scott AJ, Webb TL, Martyn-St James M et al (2021) Improving sleep quality leads to better mental health: a meta-analysis of randomised controlled trials. Sleep Med Rev 60:101556. https://doi.org/10.1016/j.smrv.2021.101556

Shen Y-C, Sun X, Li L et al (2022) Roles of neuropeptides in sleep-wake regulation. Int J Mol Sci 23:4599. https://doi.org/10.3390/ijms23094599

Smith KJ, Deschênes SS, Schmitz N (2018) Investigating the longitudinal association between diabetes and anxiety: a systematic review and meta-analysis. Diabet Med 35:677–693. https://doi.org/10.1111/dme.13606

Spellman T, Liston C (2020) Toward circuit mechanisms of pathophysiology in depression. Am J Psychiatry 177:381–390. https://doi.org/10.1176/appi.ajp.2020.20030280

Staines AC, Broomfield N, Pass L et al (2022) Do non-pharmacological sleep interventions affect anxiety symptoms? A meta-analysis. J Sleep Res 31:e13451. https://doi.org/10.1111/jsr.13451

Stein DJ, Scott KM, de Jonge P, Kessler RC (2017) Epidemiology of anxiety disorders: from surveys to nosology and back. Dialog Clin Neurosci 19:127–136. https://doi.org/10.31887/DCNS.2017.19.2/dstein

Sun Y, Shi L, Bao Y et al (2018) The bidirectional relationship between sleep duration and depression in community-dwelling middle-aged and elderly individuals: evidence from a longitudinal study. Sleep Med 52:221–229. https://doi.org/10.1016/j.sleep.2018.03.011

Th'ng F, Rao KA, Ge L et al (2021) A one-year longitudinal study: changes in depression and anxiety in frontline emergency department healthcare workers in the COVID-19 Pandemic. Int J Environ Res Public Health 18:11228. https://doi.org/10.3390/ijerph182111228

Tsaras K, Tsiantoula M, Papathanasiou IV et al (2021) Predictors of depression and insomnia in community-dwelling elderly people: a cross-sectional evidence of their bidirectional relationship. Cureus 13:e13965. https://doi.org/10.7759/cureus.13965

van Straten A, Cuijpers P (2009) Self-help therapy for insomnia: a meta-analysis. Sleep Med Rev 13:61–71. https://doi.org/10.1016/j.smrv.2008.04.006

Vancampfort D, Van Damme T, Stubbs B et al (2019) Sedentary behavior and anxiety-induced sleep disturbance among 181,093 adolescents from 67 countries: a global perspective. Sleep Med 58:19–26. https://doi.org/10.1016/j.sleep.2019.01.048

Vandekerckhove M, Wang Y (2017) Emotion, emotion regulation and sleep: an intimate relationship. AIMS Neurosci 5:1–17. https://doi.org/10.3934/Neuroscience.2018.1.1

Vazsonyi AT, Liu D, Blatny M (2022) Longitudinal bidirectional effects between sleep quality and internalizing problems. J Adolesc 94:448–461. https://doi.org/10.1002/jad.12039

Vindegaard N, Benros ME (2020) COVID-19 pandemic and mental health consequences: systematic review of the current evidence. Brain Behav Immun 89:531–542. https://doi.org/10.1016/j.bbi.2020.05.048

Viselli L, Salfi F, D'Atri A et al (2021) Sleep quality, insomnia symptoms, and depressive symptomatology among Italian University students before and during the Covid-19 lockdown. Int J Environ Res Public Health 18:13346. https://doi.org/10.3390/ijerph182413346

Vorvolakos T, Leontidou E, Tsiptsios D et al (2020) The association between sleep pathology and depression: a cross-sectional study among adults in Greece. Psychiatry Res 294:113502. https://doi.org/10.1016/j.psychres.2020.113502

Walker ER, McGee RE, Druss BG (2015) Mortality in mental disorders and global disease burden implications: a systematic review and meta-analysis. JAMA Psychiat 72:334–341. https://doi.org/10.1001/jamapsychiatry.2014.2502

Wallsten D, Norell-Clarke A, Alfonsson S et al (2021) Treating co-morbid insomnia and social anxiety disorder with sequential CBT protocols: a single-case experimental study. Behav Cogn Psychother 49:641–657. https://doi.org/10.1017/S1352465821000278

Wang X, Cheng S, Xu H (2019) Systematic review and meta-analysis of the relationship between sleep disorders and suicidal behaviour in patients with depression. BMC Psychiatry 19:303. https://doi.org/10.1186/s12888-019-2302-5

Wang W, Guo Y, Du X et al (2022) Associations between poor sleep quality, anxiety symptoms, and depressive symptoms among Chinese adolescents before and during COVID-19: a longitudinal study. Front Psychiatry 12:786640

Weaver MD, Vetter C, Rajaratnam SMW et al (2018) Sleep disorders, depression and anxiety are associated with adverse safety outcomes in healthcare workers: a prospective cohort study. J Sleep Res 27:e12722. https://doi.org/10.1111/jsr.12722

Wetherall K, Cleare S, McClelland H et al (2022) Mental health and well-being during the second wave of COVID-19: longitudinal analyses of the UK COVID-19 Mental Health and Wellbeing study (UK COVID-MH). BJPsych Open 8:e103. https://doi.org/10.1192/bjo.2022.58

Wiersema C, Zelst WV, Voshaar RO (2018) When a patient with depression is feeling sleepy, be aware of sleep apnoea. BMJ Case Rep 2018:bcr2018224873. https://doi.org/10.1136/bcr-2018-224873

Winkler A, Auer C, Doering BK, Rief W (2014) Drug treatment of primary insomnia: a meta-analysis of polysomnographic randomized controlled trials. CNS Drugs 28:799–816. https://doi.org/10.1007/s40263-014-0198-7

World Health Organization (1991) International classification of diseases (ICD-10). World Health Organization, Geneva

World Health Organization (2017) Depression and other common mental disorders: global health estimates. World Health Organization, Geneva

Yang X, Fang Y, Chen H et al (2021) Global, regional and national burden of anxiety disorders from 1990 to 2019: results from the Global Burden of Disease Study 2019. Epidemiol Psychiatr Sci 30:e36. https://doi.org/10.1017/S2045796021000275

Ye H, Ji M, Wang C et al (2022) Integrated functional neuroimaging, monoamine neurotransmitters, and behavioral score on depressive tendency in intensive care unit medical staffs induced by sleep deprivation after night shift work. Front Psychiatry 13:848709. https://doi.org/10.3389/fpsyt.2022.848709

Yeh W-C, Lu S-R, Wu M-N et al (2021) The impact of antiseizure medications on polysomnographic parameters: a systematic review and meta-analysis. Sleep Med 81:319–326. https://doi.org/10.1016/j.sleep.2021.02.056

Zhang Y, Zhang H, Ma X, Di Q (2020) Mental health problems during the COVID-19 pandemics and the mitigation effects of exercise: a longitudinal study of college students in China. Int J Environ Res Public Health 17:3722. https://doi.org/10.3390/ijerph17103722

Zhang C, Chang Y, Yun Q et al (2022a) The impact of chronic diseases on depressive symptoms among the older adults: the role of sleep quality and empty nest status. J Affect Disord 302:94–100. https://doi.org/10.1016/j.jad.2022.01.090

Zhang Y, Li X, Chan VKY et al (2022b) Depression duration and risk of incident cardiovascular disease: a population-based six-year cohort study. J Affect Disord 305:188–195. https://doi.org/10.1016/j.jad.2022.03.005

Zhao J, Xu J, He Y, Xiang M (2022) Children and adolescents' sleep patterns and their associations with mental health during the COVID-19 pandemic in Shanghai, China. J Affect Disord 301:337–344. https://doi.org/10.1016/j.jad.2021.12.123

Zhou F, Guo Y, Wang Z et al (2021a) Assessing the causal associations of insomnia with depressive symptoms and subjective well-being: a bidirectional Mendelian randomization study. Sleep Med 87:85–91. https://doi.org/10.1016/j.sleep.2021.08.025

Zhou Y, Draghici A, Abbas J et al (2021b) Social media efficacy in crisis management: effectiveness of non-pharmaceutical interventions to manage COVID-19 challenges. Front Psychiatry 12:626134. https://doi.org/10.3389/fpsyt.2021.626134

Chapter 4
Sleep Problems During the COVID-19 Pandemic in Children and Adolescents

Roy Chowdhury Sayoni, Kamila Gautam, and Gulati Sheffali

Abstract Adequate sleep is vital for the optimal functioning of diverse cognitive and psychological functions such as learning, memory integration, attention, socio-emotional regulation, and coping mechanisms. Although highly prevalent in the pediatric population, sleep problems remain poorly recognized and addressed. As the pandemic has unfolded, researchers have studied the negative consequences the pandemic had on sleep in children. Due to the lockdown and social restrictions brought on by the COVID-19 pandemic, children and adolescents were suddenly confronted with major lifestyle changes such as home confinement, online lessons, and altered daily routines. The prevalence of sleep disruptions has risen considerably amidst the pandemic, with delayed sleep–wake phase disorder being the most frequently reported sleep issue among children and youth. Management of these sleep problems entails a structured approach centered on optimizing sleep hygiene, treating specific COVID-19-related sleep issues, excluding confounders, and treating modifiable risk factors while implementing telemedicine services for the delivery of care.

Keywords Sleep disturbances · Circadian phase delay · Physical activity · Screen time · Teleconsultation

4.1 Importance of Sleep and Circadian Rhythm in the Maintenance of Well-Being and Its Effect on Cognition

Sleep is essential for the maintenance of physical health and mental well-being. It is defined as a reversible state of interruption of sensorimotor interactions with the external environment, typically characterized by immobilization and recumbency

---

R. C. Sayoni · K. Gautam · G. Sheffali (✉)
Centre of Excellence & Advanced Research for Childhood Neurodevelopmental Disorders, Child Neurology Division, Department of Paediatrics, All India Institute of Medical Sciences, Delhi, India

A. S. BaHammam et al. (eds.), *COVID-19 and Sleep: A Global Outlook*, Progress in Sleep Research, https://doi.org/10.1007/978-981-99-0240-8_4

(Brand and Kirov 2011). It is additionally thought of as a state of adaptive inactivity (Siegel 2009). Numerous physiological processes such as immunological, metabolic, thermoregulatory, and cardiorespiratory functions necessitate sleep for their effective functioning. Sleep being a complex neurobiological process is inextricably linked to circadian rhythm regulation. A complex interplay of these two processes bears a crucial role in neuroplasticity and cortical excitability by maintaining synaptic homeostasis and organizing neuronal restoration networks (Tononi and Cirelli 2006). Adequate sleep thus lays the foundation for the sound functioning of diverse cognitive and psychological abilities including learning, memory consolidation, attention, socio-emotional regulation, and coping mechanisms (Maquet 2001; van der Helm et al. 2010; Mikoteit et al. 2012). Reduced functional connectivity of the default mode network (DMN), seen in sleep-deprived adolescents and adults across studies, offers a plausible explanation for how poor sleep contributes to higher order cognitive deficits (Tashjian et al. 2018). Circadian rhythm however is dependent on the internal clock (biological rhythm) as well as extrinsic factors like social cues, auditory stimuli, exercise, and daily routines. These external influences interfere with social zeitgebers which, in turn, disrupt biological rhythms (Salehinejad et al. 2022).

4.2 Sleep Architecture and Development in the Pediatric Population

Non-rapid eye movement (NREM) and rapid eye movement (REM) are the two phases of sleep that the human body alternates between. NREM sleep is further categorized into three stages N1 to N3, exemplifying a continuum of relative depth. Each phase and stage of sleep is characterized by alterations in electrographic waves, muscle tone, and ocular movements. Loss of muscle tone is a defining feature of REM sleep, which is also when we experience most of our dreams. NREM sleep duration shortens as REM sleep duration lengthens as the night advances. For a healthy adult, the total sleep cycle lasts for 90–110 min with an average of 4–6 sleep cycles per night (Ophoff et al. 2018).

However, in children, sleep patterns tend to evolve with age with the most consistent trend being a decrease in sleep duration. Table 4.1 represents the genesis of normal sleep patterns from the neonatal period through adolescence. It might be challenging to provide specific recommendations regarding how much sleep a child needs and what time he should go to bed because every child has unique sleep demands. A child who wakes up easily and without prompting has likely had enough sleep.

Table 4.1 Sleep architecture and development in the pediatric population (Ophoff et al. 2018; Paruthi et al. 2016; MacLean et al. 2015)

Age group	Sleep architecture and development	
Neonates 	• Three sleep stages: active sleep (REM), quiet sleep (NREM), and indeterminate sleep • Sleep onset through REM • No differentiation between day and night	Sleep needs: 16–18 h/day
Infants 	• Two sleep phases: NREM, REM • Day/night differentiation from 1 month • By 3 months, sleep onset through NREM and the diurnal pattern is established (cycling of melatonin and cortisol in a circadian rhythm)	Sleep needs: 12–16 h/day for 4–-12 months old infants, two daytime naps
Toddlers and pre-schoolers 	• REM sleep duration decreases while NREM increases	Sleep needs: 11–14 h/day: 1–2 years old (one daytime nap) 10–13 h/day: 3–5 years old (no daytime nap)
Schoolers 	• N3 duration peaks due to prolonged REM latency • By 6 years, circadian rhythm preferences develop: night owl or an early riser	Sleep needs: 9–12 h/day
Adolescents 	• N3 duration dips • Mid-puberty: daytime sleepiness increases	Sleep needs: 8–10 h/day

4.3 Common Sleep Issues in Children and Adolescents During the Pre-pandemic Era

4.3.1 Typically Developing Children and Adolescents

Although highly prevalent in the pediatric population, sleep problems remain poorly recognized and addressed. An estimated 20–30% of children struggle with sleep issues (Owens 2008; Trosman and Ivanenko 2021). Age-dependent variability in sleep patterns as well as the spectrum of sleep disturbances is a well-known fact. With advancing age, bedtimes tend to get delayed with declining sleep duration. While night-time awakenings are more common in infancy and early childhood days, sleep problems caused by poor sleep hygiene or circadian rhythm disturbances tend to predominate in adolescence. Considering gender differences, boys during childhood years and adolescent girls are prone to sleep-related difficulties. Lower socioeconomic status was also more often associated with sleep issues in the adolescent age group than in children (Lewien et al. 2021). A schematic representation of sleep difficulties in the pediatric population as per age is depicted in Fig. 4.1.

4.3.2 Children with Neurodevelopmental Disorders (NDDs)

Up to 85% of children with NDDs suffer from sleep difficulties as compared to 25% of healthy children (Horwood et al. 2019).

Attention deficit hyperactivity disorder (ADHD): The behavioral dysregulation deficits that comprise ADHD often interfere with maintaining age-appropriate

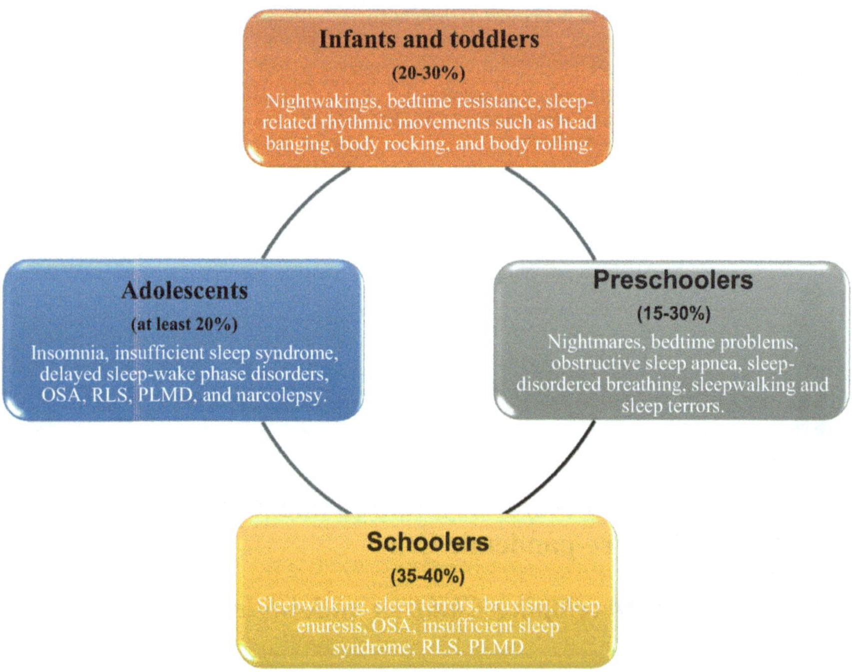

Fig. 4.1 Common sleep problems in children and adolescents in the pre-pandemic era (Ophoff et al. 2018). *OSA* obstructive sleep apnea, *RLS* restless leg syndrome, *PLMD* periodic limb movement disorder

healthy sleeping habits. About 25–50% of children and teenagers with ADHD are at risk for a wide spectrum of sleep disruptions, ranging from delayed onset, short sleep duration, and night-awakenings to developing secondary sleep disorders. While insomnia is more common in the hyperactive-impulsive ADHD group, the inattentive ADHD type typically has late bedtimes. Additionally, these children are also prone to certain sleep disorders such as circadian rhythm abnormalities, sleep-disordered breathing, and restless leg syndrome. Due to the overlapping symptoms of ADHD and sleep deprivation, these sleep disorders frequently go undiagnosed, necessitating a high index of clinical suspicion (Sleep Foundation 2021).

Autism spectrum disorder (ASD): An estimated 40–80% of children with ASD suffer from sleep issues such as delayed sleep onset, bedtime resistance, frequent awakenings, daytime sleepiness, and parasomnias. All these can be explained by the complex interplay of factors such as disturbances in melatonin homeostasis, impaired GABAergic neuronal maturation, and sensory impairment in ASD individuals. Coexistent medical and behavioral comorbidities can also trigger sleep disruption, often leading to daytime challenging behaviors and worsening of core ASD symptoms (Hyman et al. 2020).

Cerebral palsy: Children with cerebral palsy frequently report sleep disturbances in up to 25% of cases. Difficulty in falling and maintaining sleep were the most common sleep problems encountered (Horwood et al. 2019).

4.4 The Spectrum of Sleep Problems in Children and Adolescents During the COVID-19 Pandemic

Two and half years have passed since the first case of the 2019 novel coronavirus (2019-nCoV/SARS-CoV-2) was reported in December 2019 from Wuhan city of China. The World Health Organization (WHO) declared COVID-19 a Public Health Emergency of International Concern on January 30, 2020, and a pandemic was subsequently declared on March 11, 2020. Globally, as of October 13, 2022, there have been 620,301,709 confirmed cases of COVID-19, including 6,540,487 deaths (WHO Coronavirus (COVID-19) Dashboard 2022). In the absence of definitive treatments and vaccines, measures implemented to control the sequential waves of the COVID-19 pandemic primarily focused on mask-wearing, social distancing, hand hygiene, lockdown, and quarantine. All these measures posed a considerable challenge to the healthcare system, government revenue, and educational institutions.

The contagion effect of SARS-CoV-2 has readily percolated through every tier of society. While COVID-19 continues to affect millions worldwide and society struggles to adapt to a new normal, the significance of sleep issues has flown under the radar. Healthy sleeping habits are key for sustaining good physical, emotional, and mental health. As the pandemic has unfolded, researchers have studied the negative consequences the pandemic had on sleep. The psychosocial

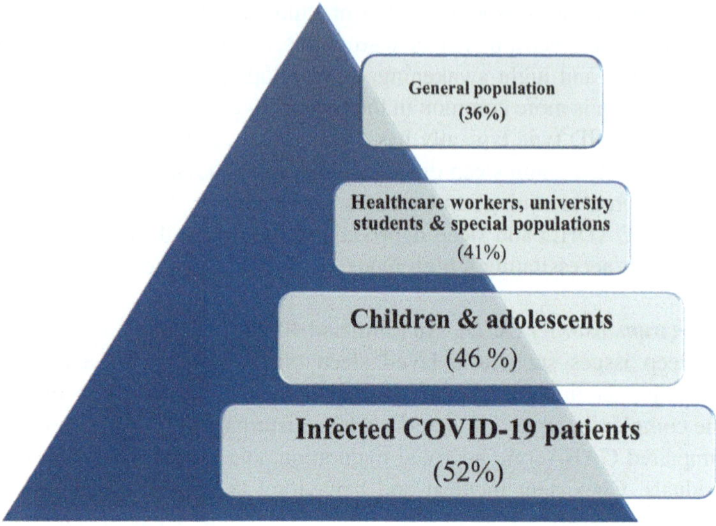

Fig. 4.2 Prevalence of sleep disturbances brought on by the COVID-19 pandemic (Jahrami et al. 2022)

burden imposed by the ongoing global health crisis has been a major contributor to sleep problems irrespective of age, sex, and socioeconomic background. COVID-19 patients, children and adolescents, healthcare workers, and university students suffered major hits. The prevalence of sleep difficulties, independent of any covariate, was 40% (Jahrami et al. 2022). Following actively infected COVID-19 patients, children and adolescents were the second most affected group. The prevalence of any sleep disturbance in children and adolescents ranged from 46% to 59%, nearly doubling from the pre-pandemic times. Approximately, half of the children did not meet any sleep recommendations (Jahrami et al. 2022; Sharma et al. 2021; Wearick-Silva et al. 2022). Remote learning due to school closures coupled with social restrictions due to the lockdown has led to a major lifestyle change in the pediatric population typified by less physical activity, increased screentime exposure, and irregular sleep patterns. The frequency and spectrum of sleep disturbances have however been observed to differ between toddlers, school-goers, and teenagers (Fig. 4.2).

4.4.1 Typically Developing Younger Children (Toddlers, Preschoolers)

4.4.1.1 Sleep–Wake Schedule, Sleep Quality, and Duration

An altered sleep–wake cycle with late bedtime and rise time was most frequently reported by parents of toddlers and preschoolers. This phase shift during lockdown was also associated with shorter and reduced daytime naps. Approximately, one-third of children experienced a change in sleep duration and poor sleep quality. These observations showed a positive correlation with maternal factors such as COVID-19 anxiety, worse sleep quality, insomnia symptoms, and harsh parenting style (Liu et al. 2021; Zreik et al. 2021; Carroll et al. 2020). Caregiver's use of mindfulness strategies and the presence of siblings and pets in the house were the few variables that favored a child's sound night sleep. Even though bedtime and wake-up time have been delayed by around an hour, overall sleep duration in preschoolers remained inconsistent: it either stayed the same, increased, or decreased from the pre-pandemic period. Infants and toddlers, on the otherhand, more frequently reported shorter sleep duration, later bedtimes, and longer sleep latency than preschoolers. The sleep trajectory, both quality, and duration however followed a quadratic pattern with an initial decline followed by a stabilization phase as the pandemic unfolded (Di Giorgio et al. 2021; Markovic et al. 2021; Dellagiulia et al. 2020).

Interestingly, age-specific sleep disorders like bedtime resistance, sleep onset anxiety, night-awakenings, parasomnias, and sleep-disordered breathing improved during the early confinement days. Relative to pre-pandemic times, data is controversial regarding the overall prevalence of sleep disturbances in younger children. Some studies reported a fairly constant or a downhill trend of sleep issues. This might be attributed to the fact that good parenting practices and nutritious dietary habits are more effective with the additional time available during the lockdown, especially for younger kids for whom increased family time offered the greatest benefit (Liu et al. 2021; Di Giorgio et al. 2021). On the contrary, a single study observed an astonishing rise in the prevalence of altered sleep habits from 20% in the pre-pandemic era to 58.6% during the pandemic in the youngest ones (0–3 years) (Wearick-Silva et al. 2022).

As the pandemic advanced with subsequent waves, a gradual shift towards a worsening sleep quality with trouble falling asleep, bedtime anxiety, night-awakenings, nightmares, and sleep terrors was observed in children up to 5 years (Bruni et al. 2022a).

4.4.2 Healthy School-Goers and Adolescents

School-going children and adolescents behaved differently concerning sleep throughout the pandemic as compared to preschoolers. They may be more prone to the adverse effects of social isolation owing to the lack of peer ties and may be faced with distinct parent-teen conflicts. School closures, online classes, lack of outdoor activities, increased screentime, COVID-19-related fear, psychological issues, and uncertainty about the future also contributed largely to inadequate sleep in them. The majority experienced delayed bedtime and rise time, a worsened sleep quality, and longer sleep duration during 24 h including naps (Wearick-Silva et al. 2022; Bruni et al. 2022a; Genta et al. 2021; Zhou et al. 2020; Fidanci et al. 2020; Windiani et al. 2021).

4.4.2.1 Sleep–Wake Schedule

Nearly, one-third of schoolers and two-thirds of adolescents fell asleep after 11 p.m. during the lockdown, compared to 0.9% and 12%, respectively before the pandemic. This late sleep–wake phase also persisted during the weekends, although less pronounced. The average reported bedtime delay was over an hour, which may be related to protracted sleep latency (Bruni et al. 2022a). A minimum sleep latency of 30 min was evident among a significant proportion of children of all ages, which has nearly doubled during this unprecedented health crisis. Teenagers who spent more leisure screen time at night became more alert, had less melanin production induced by the blue light of the screen, and experienced a phase delay in their circadian rhythm, all of which affected sleep latency and efficiency (Baby et al. 2021). On the contrary, flexible sleep routines due to school closings and late bedtimes were the main predictors of a later wake-up time. The majority of them woke up after 8 am on weekdays and after 9 am on weekends. This sleep–wake pattern, in the pre-pandemic times, is typically observed on weekends and holidays. The circadian dysregulation, which was most obvious during the initial days of home confinement, lasted throughout the successive COVID-19 waves.

4.4.2.2 Sleep Duration and Quality

The use of electronic gadgets, female sex, mental health issues, and having an acquaintance suffering from COVID-19 were all found to be strongly associated with poor sleep quality in adolescents. Nutritive dietary choices and regular physical exercise on the flipside correlated with good sleep quality in them (Zhai and Du 2020). An increase in sleep duration by approximately 1 h was noted among high school-goers. This can be well ascribed to the late start of daily routines, including virtual classroom sessions and adjustable timetables. Surprisingly, the pandemic witnessed a significant proportion of adolescents to be certainly gaining the recommended amount of sleep with the elimination of social jet lag (Tyack et al. 2022). It proved to be beneficial for some youth who suffered from chronic sleep deprivation due to early school hours. This pandemic indeed evoked a circadian shift, in alignment with the endogenous biological clock of evening chronotypes and night owls, due to additional flexibility offered by home learning and variations in work/school timings (Becker and Gregory 2020). Shorter sleep duration and higher bedtime arousal before the pandemic were however predictive of an increased perceived stress level among teenagers during the pandemic, which in turn negatively impacted their sleeping patterns. With the evolution of the pandemic, sleep quality improved with reduced somnolence, a narrower gap in sleep duration between weekdays and weekends, and fewer midday naps (Gruber et al. 2021; AMHSI Research Team et al. 2020).

4.4.2.3 Insomnia

The prevalence of insomnia was 21–38% among school-going adolescents. Early worries and sufferings associated with insomnia can evolve into primed cognitive and somatic hyperarousal, which causes persistent difficulties in falling and staying asleep as these folks simply lie awake in bed. Aside from the primary stressors brought on by the pandemic (social isolation, fear of contagion, juggling new roles), a few of the several risk factors for insomnia included female gender, living in cities, anxiety, depression, gaming disorder, and senior high school attendies. On the contrary, lower rates of sleep irregularities were observed in those who had optimistic COVID-19 predictions, stronger social support networks, and fewer negative consequences of the pandemic on their studies and daily lives (Zhou et al. 2020; Lu et al. 2020; Chi et al. 2021).

Overall, adolescents had lesser sleep difficulties than younger children with better sleep quality, longer sleep duration, and later bedtimes between 10 p.m. and 12 a.m. Prolonged sleep latency was the most noticeable sleep pattern among teenagers (Sharma et al. 2021; Bruni et al. 2022a; Tyack et al. 2022; Richter et al. 2022) (Fig. 4.3).

4.4.3 Children and Adolescents with Neurodevelopmental Disorders

NDDs perse pose an independent risk for sleep disturbances. These children are more vulnerable to stressful situations due to a lack of effective coping mechanisms. The intricacy of the pandemic situation with its social constraints surpasses the general understanding of these children, which in turn incites psychological distress and exacerbates sleep issues.

4.4.3.1 Attention Deficit Hyperactivity Disorder (ADHD)

Disruptions of daily activity due to the COVID-19 pandemic, with the remote or hybrid modes of learning, limitations on social interactions with friends, and changes to parental work routines often exacerbated the impairments associated with ADHD. About 70% of caregivers reported disturbed sleep in children and adolescents with ADHD during the lockdown. This included trouble initiating sleep (85%), irregular sleep–wake schedule (60–70%), arousal disorders (50%), altered sleep duration (40–50%), sleep-disordered breathing (40%), and somnolence (40%) (Bruni et al.

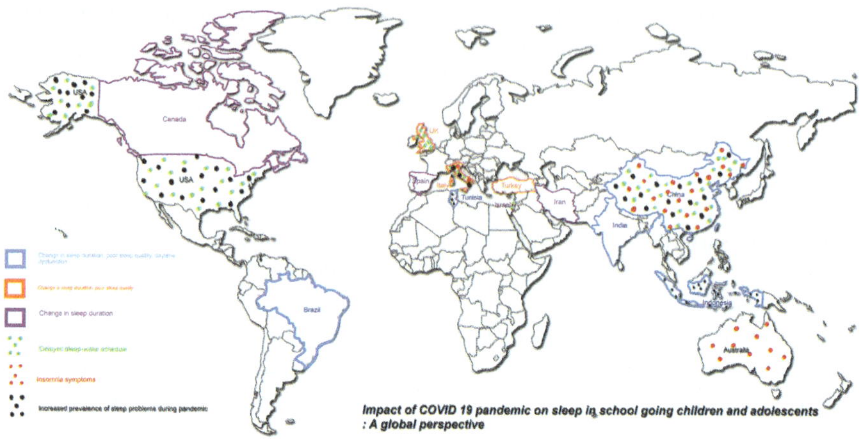

Fig. 4.3 Global impact of COVID-19 pandemic on sleep in school-going children and adolescents (Jahrami et al. 2022; Sharma et al. 2021; Wearick-Silva et al. 2022; Zreik et al. 2021; Di Giorgio et al. 2021; Dellagiulia et al. 2020; Bruni et al. 2022a; Genta et al. 2021; Zhou et al. 2020; Fidanci et al. 2020; AMHSI Research Team et al. 2020; Lu et al. 2020; Chi et al. 2021; Mitra et al. 2020; Medrano et al. 2021; Baptista et al. 2021; Zhao et al. 2022; Lopez-Leon et al. 2022; Abid et al. 2021; Cellini et al. 2021; Bacaro et al. 2021; Dondi et al. 2021; Lavigne-Cerván et al. 2021; Üstündağ et al. 2022; Dutta et al. 2022; Ranjbar et al. 2021; Albrecht et al. 2022; Knowland et al. 2022; Murata et al. 2021; Chin et al. 2022; Li et al. 2022)

2021). Like typically developing children, later bedtime and shorter sleep duration in ADHD children correlated with increased screen use. Among children with developmental disabilties, ADHD group suffered from a more disrupted and unstable sleep–wake cycle than children with ASD. The variable and maladaptive nocturnal sleep pattern, a core characteristic of ADHD, provides a compelling explanation for this (Gruber et al. 2000).

4.4.3.2 Autism Spectrum Disorders (ASD)

Parents reported that children with ASD experienced a significant increase in sleep problems during the lockdown, ranging from a delayed sleep–wake schedule (55–70%), difficulty falling asleep (35%), bedtime anxiety, (23%), night-awakenings, daytime sleepiness (14%) to parasomnias like sleep terrors (5%) (Bruni et al. 2022b). An intolerability to adapt to the abrupt change in lifestyle measures, social restrictions, and a lack of access to behavioral intervention therapies were the main reasons behind this. Sleep issues demonstrated a direct correlation with ASD severity score during home confinement: the more sleep disturbances, the more severe the core symptoms of ASD (Mutluer et al. 2020; Türkoğlu et al. 2020).

4.4.3.3 Cerebral Palsy (CP)

Sleep difficulties were reported by 20–25% of parents of children with CP, roughly comparable to the pre-pandemic period. An older age, a severe CP phenotype, and the existence of additional comorbidities like epilepsy, visual impairment, hearing deficit, and intellectual disability are common risk factors for sleep problems, both before as well as during the pandemic.

4.4.4 Metrics for Sleep Evaluation in Children and Adolescents

The measures used to assess sleep parameters varied across studies from validated online available sleep questionnaires/scales and their modified versions to self-made questionnaires developed by the authors, resulting in a quite heterogenous set of data. *Sleep Disturbance Scale for Children, SDSC* (Bruni et al. 2022a; Wearick-Silva et al. 2022; Cellini et al. 2021; Dondi et al. 2021); *Children's Sleep Habit Questionnaire, CSHQ* (Liu et al. 2021) and *Brief Infant Sleep Questionnaire, BISQ* (Markovic et al. 2021; Wearick-Silva et al. 2022) were the most commonly used measures for evaluating sleep characteristics of infants and young children. On the other hand, the sleep profile of adolescents was frequently analyzed using *Pittsburgh Sleep Quality Index, PQSI* (Wearick-Silva et al. 2022; Windiani et al. 2021; Genta et al. 2021; Zhou et al. 2020) followed by *SDSC* (Bruni et al. 2021; Fidanci et al. 2020) and *Epworth Sleepiness Scale, ESS* (Dutta et al. 2022; Genta et al. 2021; Wearick-Silva et al. 2022). While some authors used the total scores as a barometer of interrupted sleep, others only looked at specific items of scales or questionnaires. Irrespective of the scales used, the majority of the studies on sleep research pointed towards a rising prevalence of sleep problems with worsening sleep quality, variable sleep duration, and delayed sleep–wake patterns across all ages during the pandemic than before.

4.5 Prevalence of Primary Sleep Disorders in Children and Adolescents Amid the COVID-19 Pandemic

Due to the COVID-19 pandemic, children and adolescents were suddenly confronted with major lifestyle changes such as home confinement, online lessons, and altered daily routines. Besides the circadian rhythm abnormalities (as discussed in previous sections), the prevalence of primary sleep disorders also showed an uphill trend, as per some researchers. Contrary to the adolescents, who faced the most significant delayed sleep–wake phase schedule amid the outbreak, younger

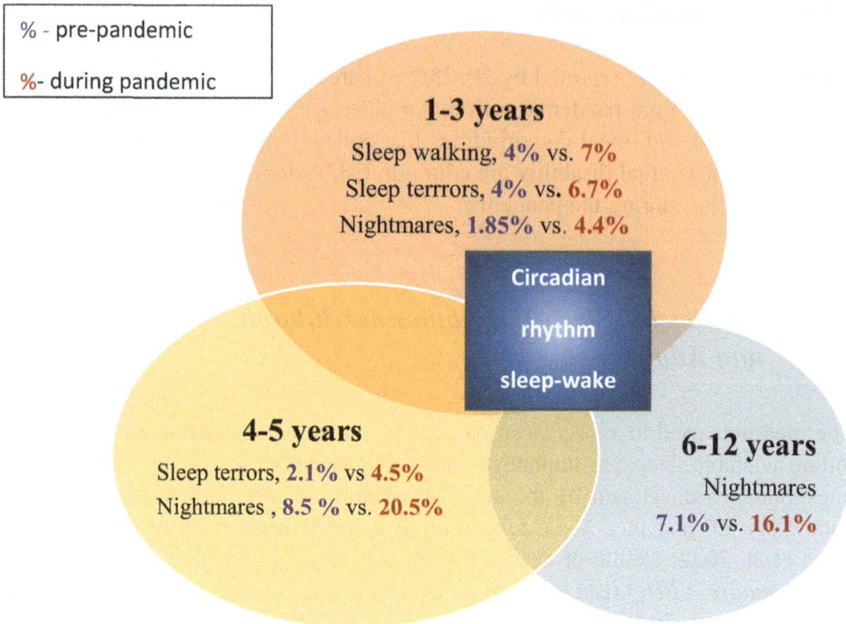

Fig. 4.4 A pictorial representation depicting the prevalence of primary sleep disorders in younger children, pre vs. during the COVID-19 pandemic (Bruni et al. 2022a)

ones suffered from the worse hits of nightmares, sleepwalking, and sleep terrors (Bruni et al. 2022a; Dondi et al. 2021).

Children and adolescents with preexisting narcolepsy type 1 took more frequent naps and slept more during day hours, although the duration of daytime naps did not differ from the pre-lockdown period. The majority of children experiencing difficulty in initiating/maintaining sleep (34–70%) and sleep–wake transition disorders (18%) are aged between 4 and 12 years (Wearick-Silva et al. 2022).

Parents also perceived poor oral hygiene during home confinement to be a risk factor for sleep-related breathing disorders, sleep–wake transition disorders, and excessive sleepiness (Baptista et al. 2021). The fact that dental problems might interfere with sleep patterns can help to explain this novel observation partly, but further research is necessary to validate this.

The escalating prevalence of obesity, driven by sedentary lifestyles and altered eating habits due to the pandemic, heightened the risk for obstructive sleep apnoea too (Fig. 4.4).

4.6 Impact of Acute SARS-CoV-2 Infection and Long-COVID on Sleep in Children and Adolescents

The prevalence of sleep disturbances was highest (52%) among acute SARS-CoV-2 virus-infected patients, including both children and adults. The core symptoms of the viral illness, such as cough, fever, respiratory distress, myalgia, and side effects of medications, which all contribute to disrupted sleep, could provide a compelling explanation for this (Jahrami et al. 2022).

Although severe COVID-19 disease is less frequent in children, long-COVID is a known but poorly characterized long-term complication of it. Long-COVID, as defined by the National Institute for Health and Care Excellence (NICE) guideline (2022), encompasses both ongoing symptomatic COVID-19 (from 1 month to 3 months) and post-COVID-19 syndrome (3 months onwards) (Clements et al. 2021). The prevalence of long-COVID in children and adolescents was estimated to be around 25%. Sleep difficulties were noted in approximately 8% of them, ranking third among clinical manifestations after mood changes (16%) and fatigue (10%). Poor sleep quality, insomnia, and hypersomnia were the most frequent sleep issues reported by caregivers of those children (Lopez-Leon et al. 2022). However, there is considerable overlap between long-COVID and pandemic-driven symptoms. The majority of children who did even not contract the SARS-CoV-2 virus also reported fatigue, headache, disturbed sleep, and concentration difficulties amid the pandemic, making it a challenging task for researchers to determine the true estimate of long COVID-19 (Zimmermann et al. 2022).

4.7 Epidemiological Characteristics of Sleep Difficulties in Children and Adolescents

- *Age*: The spectrum of sleep disruptions varied widely between preschoolers and school-going children/adolescents due to different risk factors. Maladaptive sleep patterns with a delayed sleep–wake schedule and poor sleep quality prevailed across all age groups. Additionally, parasomnias, trouble falling asleep and staying asleep, and bedtime anxiety were more common in younger kids (Bruni et al. 2022a).
- *Sex*: Female sex carried a higher risk for insomnia in the adolescent age group. Considerable gender differences were noted concerning overall sleep disturbances, sleep quality, and global PQSI scores, with girls being more affected than boys in a regional population. Although these can be partly supported by hormonal influences, higher anxiety and screen use in girls, it would be imprudent at this point to extrapolate this finding to an entire pediatric population. In contrast to this, no age or gender difference moderated the prevalence of sleep issues among adults (Jahrami et al. 2022; Abid et al. 2021).

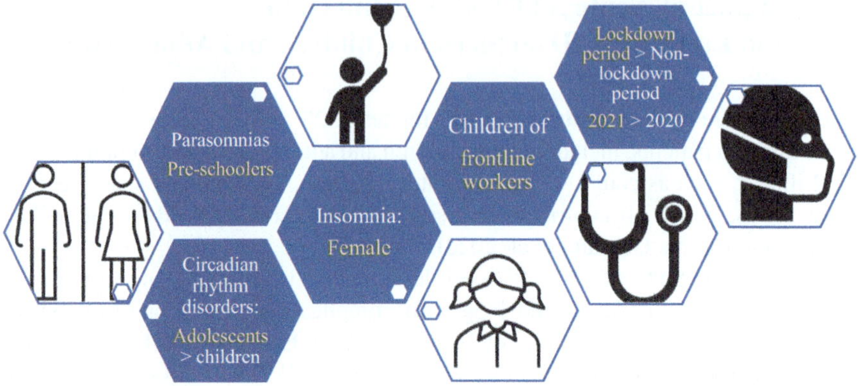

Fig. 4.5 Epidemiology of sleep issues in children and adolescents during the COVID-19 pandemic (Bruni et al. 2022a; Jahrami et al. 2022; Abid et al. 2021; Bucak et al. 2021)

- *Country-wise variation*: A significant difference was observed across the globe, with the magnitude of sleep disturbances varying from 30% in China and India to 60% in Spain (Jahrami et al. 2022). Community viral transmission, local containment measures, type of media usage, and prevailing mental health symptoms could be contributing to this cross-country discrepancy.
- *Lockdown vs. non-lockdown period*: Lockdown period witnessed a higher frequency of sleep problems in children and adults, relative to the non-lockdown period (43% vs. 38%) (Jahrami et al. 2022).
- *Children of frontline workers*: School-age children of parents, who work in healthcare sectors, were more adversely affected by disrupted sleep patterns than children of non-healthcare staff. The fear of their family members actively contracting the deadly virus owing to heightened risk could be the cause of this (Bucak et al. 2021).
- *Other factors*: Researchers found a 10% rise in the frequency of sleep disturbances in children and adults in studies published in 2021 vs. 2020 (Jahrami et al. 2022) (Fig. 4.5).

4.8 Risk Factors and Predictors for Sleep Issues Due to the COVID-19 Pandemic and Its Psychopathological Basis in Children and Adolescents

A multitude of variables such as host (child) and paternal and environmental factors adversely modulated the sleeping habits of children and youth during these unprecedented times. Refer to Fig. 4.6 for various risk factors contributing to poor sleep due to the COVID-19 pandemic, few of them are discussed in the text in detail.

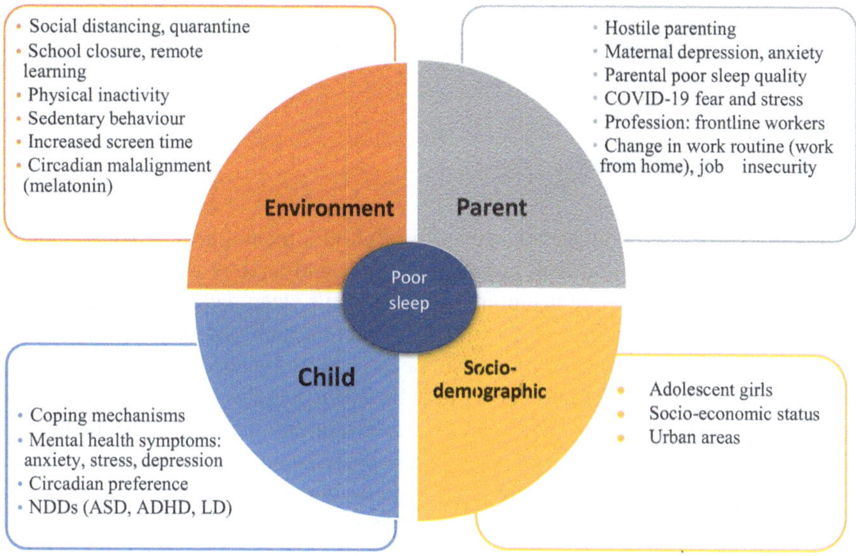

Fig. 4.6 Risk factors and predictors for sleep issues in children and adolescents during the COVID-19 pandemic

4.8.1 Environmental Factors

Social limitations and lockdown measures implemented to contain the viral transmission reduced 24-h movement behaviors and provided ample opportunities for a sedentary lifestyle pattern, both of which independently and negatively impacted sleep hygiene to a large extent. School-going children and youth (5–17 years) barely met the recommended guideline of 1 h of moderately intense physical activity and nearly half spent more than 2 h/day of recommended recreational screen time, even before lockdown (Roman-Viñas et al. 2016; Fakhouri et al. 2013). A further three- to fourfold reduction in physical activity (12.7–3.6%) coupled with a substantial rise of 20–66% in leisure screen time was observed since the onset of the pandemic (Moore et al. 2020; Rhodes et al. 2019). Children aged 5–13 years were engaged five times more in sitting related to recreational activities (>8 h/day) than they did for online classes. Preschoolers with enough space for play at home and living in rural regions however faced an attenuated effect of pandemic sequelae (Aguilar-Farias et al. 2021).

Home confinement (>65 days), lifestyle change, circadian misalignment with evening-ness chronotype, diminished sunlight exposure and COVID-19-related stress and anxiety were found to be bidirectionally linked to sleep disturbances during the ongoing COVID-19 health crisis (Salehinejad et al. 2022). The COVID-19 pandemic indeed disrupted the entire socio-ecological model related to behavior change at every level (individual, inter-individual, and environment), leading to negative health consequences (Bates et al. 2020).

4.8.2 Parent Factors

Parental well-being modulated the severity of child's mental health issues, largely through sleep. Pandemic-driven cumulative stress and anxiety in caregivers led to harsh parenting attitudes, contributing to maladaptive sleep patterns and behavioral issues in children (Wang et al. 2021). Maternal factors such as depression (moderate or severe), anxiety, and insomnia also negatively influenced sleep quality and duration in infants and preschoolers (Liu et al. 2021; Schultz et al. 2020). Children of frontline workers suffered from greater interruptions in sleep than kids of the general public (Bucak et al. 2021). Healthcare professionals experienced significant psychological distress (anxiety, depression, stress) due to the inevitable health crisis, which in turn affected their own and their children's sleep quality as well (Alimoradi et al. 2021). Parents who worked from home during the pandemic had the greatest gains in sleep duration with a later wake-up time. Children's sleep duration, especially those of preschoolers and primary school students, mirrored that of parents (Aishworiya et al. 2021).

4.8.3 Child Factors and Psychopathological Basis of Sleep Issues

Mood disturbances (loneliness, sadness, worry) in children and youth showed a positive correlation with difficulty falling and staying asleep at night (Dondi et al. 2021). Teenagers with preexisting psychiatric disorders (depression, anxiety), being more prone to COVID-19-related stress and fear, were particularly vulnerable to sleep issues during the pandemic (Becker and Gregory 2020). Irregular sleeping patterns confounded or even aggravated psychological distress symptoms in them and vice-versa. Pre-pandemic shorter sleep duration posed a higher risk for depressive symptoms among adolescents during the pandemic, with the reverse also being true (Liao et al. 2021; Panda et al. 2021).

Children with later chronotypes demonstrated a higher likelihood of mood disruptions, both before and during the pandemic (Wirz-Justice 2003). Sleep issues too worsened in children with developmental disabilities (ASD, ADHD) during stressful situations due to a complex interplay of various risk factors: circadian misalignment, medication use, and comorbidities (Dondi et al. 2021). Over the course of the pandemic, children displayed more negative coping emotional expressions (crying, screaming, blaming others) than before. The reciprocal interdependency between sleep patterns and coping mechanisms (adaptive behavior) however persisted in the midst of the pandemic (Lokhandwala et al. 2021).

Even in otherwise healthy children, chronic sleep issues led to poor attention span and oppositional defiant traits, that resembled ADHD symptoms.

Female gender, urban residency, university students, social media usage, presence of comorbidities, and daytime naps were independent predictors of insomnia during the outbreak (Al Mamun et al. 2022).

4.9 Diagnostic Approach and Recommendations for Pediatric Sleep Issues During the COVID-19 Pandemic

As sleep traverses across all disciplines, all general pediatricians and any superspecialist should be well-versed in common pediatric sleep issues. Aiding children in attaining quality sleep with the restoration of child and family well-being should be the universal goal in the wake of this pandemic. A meticulous clinical history followed by a comprehensive physical examination, while simultaneously assessing the risk factors should be the key strategy for dealing with any sleep concern. Nevertheless, it should be noted that any child who requires a psychologist or psychiatric consult needs a formal sleep assessment too (Fig. 4.7).

The underpinning for managing sleep disorders entails a structured approach centered on optimizing sleep hygiene, treating specific COVID-19-related sleep issues, excluding confounders, and treating modifiable risk factors while implementing telemedicine services for the delivery of care (Fig. 4.8).

Sleep hygiene, encompassing both bedroom environment and daily routines, lays the foundation for regular, uninterrupted sleep, regardless of a pandemic crisis. The fundamentals of ideal sleep hygiene commence with a relaxing bedtime routine, stable sleep–wake schedule, conducive bedroom environment, and healthy day-to-day activities. A consistent bedtime and wake-up time, morning sunshine exposure, daytime physical activity, and scheduled meal times are crucial for entraining the circadian rhythm through the presence of social time cues. These practices can be tailored to every child's need, enabling them to develop productive habits for pleasant overnight sleep and waking up feeling refreshed (Ophoff et al. 2018; Crew et al. 2020) (Table 4.2).

The primary goals for addressing *COVID-1-related sleep concerns* should be directed towards *eliminating the risk factors causing sleep disruption* (screen time at night, sedentary habits, naps), *promoting activities that build sleep needs* (daily exercise, outdoor recreational activities), while adopting regular sleep–wake cycle at the forefront. Owing to school closings, altered routine activities, and working from home, families might have to renegotiate their sleep routines. While setting therapeutic targets, one has to be more pragmatic and flexible in light of this pandemic. *Nightmares* can be sensibly managed by addressing the triggers, utilizing relaxation

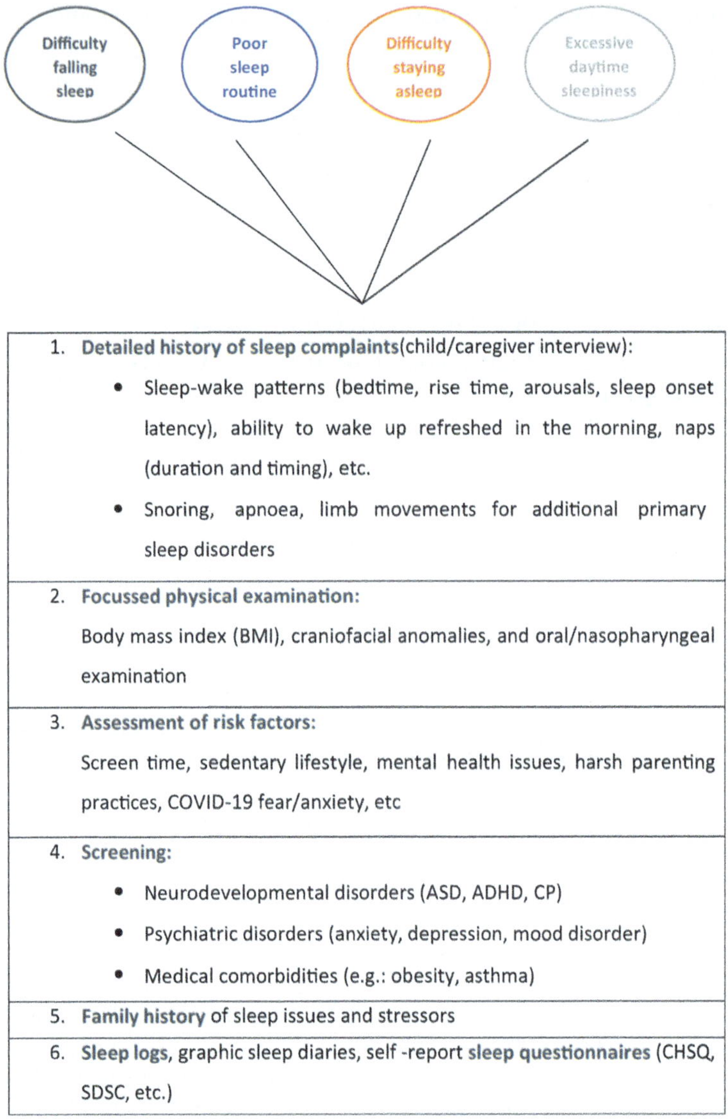

Fig. 4.7 Diagnostic approach to COVID-19-related pediatric sleep complaints in children and adolescents (Moturi and Avis 2010)

techniques (deep breathing, mindfulness meditation, progressive muscle relaxation), and modifying the sleeping environment. Concerns regarding *excessive daytime sleepiness* can be tackled by regularizing sleep routines and implementing scheduled naps. Pediatricians, child neurologists, and psychologists should assist children as they confront their unpleasant thoughts (anxiety, fear of contagion, and death) and

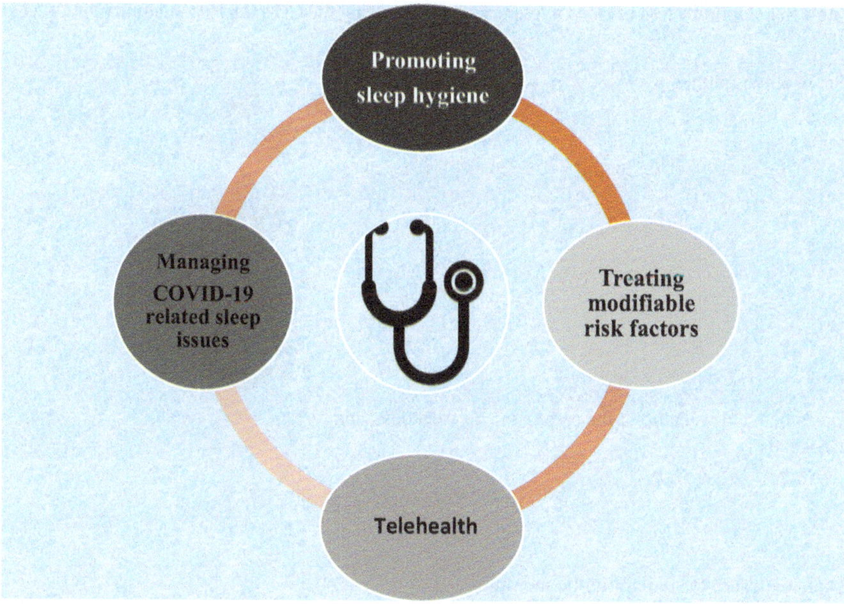

Fig. 4.8 Essential elements of management strategy for COVID-19-related sleep issues in children and adolescents

deviate from the ought's of sleep while focusing on principal behaviors which they can control (Crew et al. 2020; DelRosso et al. 2021).

The recommended first-line treatment approach for *pandemic-triggered insomnia* symptoms should focus on adopting conventional cognitive/behavioral interventions and environmental control measures rather than using sleeping pills. Cognitive behavioral therapy for adolescents, delivered either through physical or teleconsults, aims towards lowering arousal and bedtime anxiety by applying cognitive strategies (alleviating bedtime negative thoughts, enhancing positive thinking, maintaining worry journal), encouraging behavioral changes (sleep restriction, stimulus control), and relaxation exercises. However, in younger children, parent-guided behavioral modifications through the use of extinction, positive routines, faded bedtime, and scheduled awakenings should be the primary treatment modality. To date, no FDA-approved medications are available for insomnia in the paediatric population. Recent years have seen an upsurge in melatonin usage for off-label indications, notably in children with NDDs. It reduces sleep onset latency and prolongs total sleep duration with minimal addictive potential. Whether short-term use of melatonin, in adjunct to behavioral approaches, can temporarily ameliorate the sleep difficulties imposed by the ongoing health crisis is difficult to discern at this juncture. A consultation with a sleep specialist in advance to explore the risk-benefit ratio and dosing information is advisable in this context (Moturi and Avis 2010; Crew et al. 2020; DelRosso et al. 2021).

Table 4.2 Optimization of sleep hygiene and habits during the COVID-19 pandemic (Ophoff et al. 2018)

Set bedtime routine	
Go to bed and wake up at the same time on weekdays and weekends	
Daily exposure to sunlight in the morning	
Daily exercise	
Bedroom conducive to sleep	

(continued)

Table 4.2 (continued)

Bed reserved for sleeping only	
Food hygiene	

Interventions should also focus on strengthening the parent–child bond as well as helping parents establish sustainable and healthy sleep schedules, which may have an impact on their children's sleep quality (Crew et al. 2020). While establishing healthy standards for balancing sleep alongside recreation and social activities, parents should be compassionate towards teenagers.

Overall, managing sleep difficulties during the pandemic should adhere to the pre-pandemic guidelines with a greater emphasis on mitigating the environmental and psychosocial risk factors that contribute to insufficient sleep.

4.10 Role of Telehealth in Sleep Medicine Practice

Social restrictions imposed by the pandemic jeopardized sleep medicine practices by hindering outpatient physical visits and in-patient sleep studies. During this critical transition, telemedicine opened a newer horizon for the delivery of care, bridging the treatment gap. Guidelines for incorporating teleconsultations into sleep medicine have been proposed by various national and international sleep research societies. There is a slight variation in the recommendations for initial and follow-up teleconsults, primarily regarding the mode of teleconsultation and prescribing medications. Teleconsultation can be delivered either synchronously, wherein the patient consults the physician in real time via audio–video platforms (Skype, WhatsApp, Zoom), or asynchronously, wherein the patient communicates information through email, SMS, and the physician responds at his convenience. A family member or caregiver's presence is essential during a teleconsultation as children suffering from parasomnias or apnoea may not be aware of their sleep-related difficulties. Most

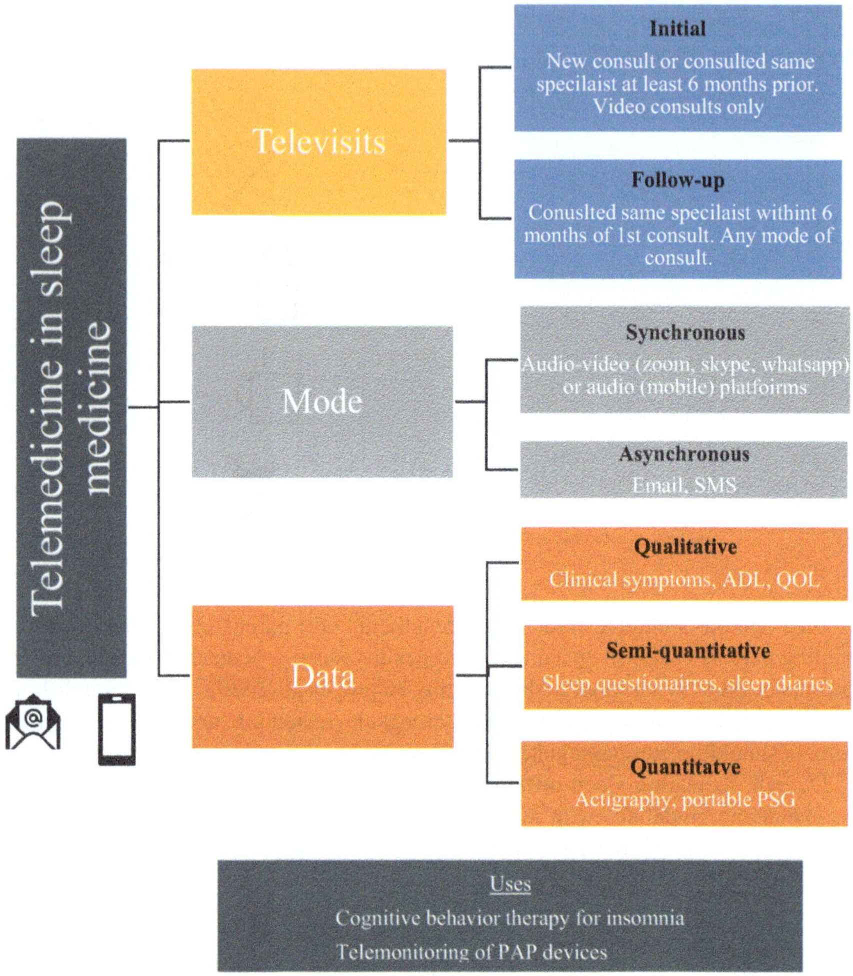

Fig. 4.9 Role of teleservices in sleep medicine practice (Paruthi 2020; Gupta et al. 2020). ADL-activities of daily, QOL-quality of life, PSG- polysomnography

sleep issues pertaining to the COVID-19 pandemic, in particular circadian rhythm disorders and insomnia, can be initially assessed and effectively managed using virtual platforms as the physical examination may add little extra benefit, and sleep studies in these clinical situations are not warranted. Continuity of care can also be provided to some extent to children with chronic sleep disorders requiring frequent visits, such as those on prescription medications or positive airway pressure (PAP) for OSA. Results of teleservices have been reported to be quite promising and at par with face-to-face consults in the field of cognitive behavior therapy for insomnia and telemonitoring of PAP devices. However, quite a few downsides to telemedicine exist, especially when it comes to performing sleep studies and initiating treatment for OSA (Paruthi 2020; Gupta et al. 2020) (Fig. 4.9).

4.11 Consequences of Altered Sleep Patterns Due to the COVID-19 Pandemic on Neurodevelopmental Trajectory

Poor sleep often bears a detrimental impact on a child's physical, emotional, and cognitive development. This holds true even for the sleep disruptions spurred by the COVID-19 pandemic. Neurocognitive dysfunction and psychological impairments are most often the repercussions of this. A growing body of evidence supports a significant reciprocal relationship between the perplexing rise in mental health concerns and pandemic-driven sleep disturbances. Children frequently exhibit maladaptive behavioral, cognitive, socio-emotional generation, and regulatory processes leading to inattention, hyperactivity, mood swings, anxiety, excessive daytime sleepiness, and learning difficulties. Additionally, as a coping mechanism, adolescents frequently engage in risky behavior such as substance abuse (drugs, alcohol, nicotine), which places them even at a higher risk for road traffic accidents. Therefore, timely identification of sleep problems is imperative for maintaining a healthy mind and body (Ophoff et al. 2018; Spruyt 2019) (Fig. 4.10).

4.12 Transition to the Post-pandemic Era

Since the COVID-19 pandemic was declared, two and a half years have passed. While adjusting to a new normal life brought on by the lockdown and social restrictions, a sizeable proportion of children and adolescents struggled with sleep issues, which in turn jeopardized their daily functioning and also evoked behavioral/cognitive concerns. As the pandemic continues to evolve and normalcy sets in with the curtailment of social distancing measures, implementation of vaccination, and reopening of schools and offices; the long-term consequences of this pandemic on

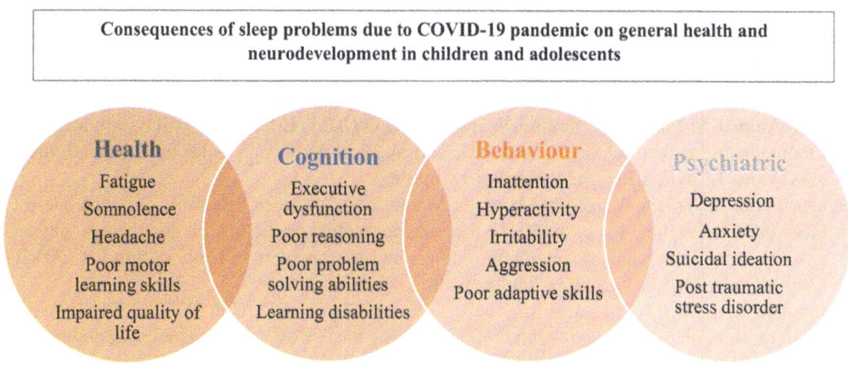

Fig. 4.10 Consequences of sleep problems due to the COVID-19 pandemic on general health and neurodevelopment in children and adolescents (Ophoff et al. 2018; Spruyt 2019)

the sleep patterns and habits of children are yet to be determined. This bears pivotal implications for research and practice in the future. In this context, addressing the research question "Will a subset of children and youth still continue to endure long-term sleep disturbances emerging out of the pandemic?" is a pressing need of the moment. Despite great uncertainty about the near future, our main goal should be to help kids and teenagers attain good sleep health by fostering their innate resilience and helping them acclimatize to the post-pandemic world.

Salient Points

- The COVID-19 pandemic witnessed an alarming rise in sleep disturbances among children and adolescents, nearly doubling from pre-pandemic times.
- The majority experienced a circadian phase delay (late bedtime and wake-up times), poor sleep quality, variable sleep duration, and insomnia symptoms. The pandemic-driven circadian shift, on the contrary, helped some teenagers overcome their social jet lag.
- Social distancing and lockdown-induced remote learning, lack of recreational activities, sedentary habits, and mental health issues were the major contributors to sleep disruptions.
- Against the backdrop of this pandemic, children and adolescents seeking psychologists for their behavioral and psychiatric concerns must go through a formal sleep assessment as part of their comprehensive evaluation.
- Management should primarily focus on optimizing sleep hygiene and addressing the environmental and psychosocial risk factors contributing to insufficient sleep.

Author Contributions **Sayoni Roy Chowdhury**: Literature search, writing the initial draft, and reviewing the final draft.

Gautam Kamila: Literature search, writing the initial draft, and reviewing the final draft.

Sheffali Gulati: Conceptualization, conceived and designed the flow of the draft, writing the initial draft, and reviewing the final draft.

References

Abid R, Ammar A, Maaloul R, Souissi N, Hammouda O (2021) Effect of COVID-19-related home confinement on sleep quality, screen time and physical activity in Tunisian boys and girls: a survey. Int J Environ Res Public Health 18(6):3065

Aguilar-Farias N, Toledo-Vargas M, Miranda-Marquez S, Cortinez-O'Ryan A, Cristi-Montero C, Rodriguez-Rodriguez F et al (2021) Sociodemographic predictors of changes in physical activity, screen time, and sleep among toddlers and preschoolers in Chile during the COVID-19 pandemic. Int J Environ Res Public Health 18(1):176

Aishworiya R, Lim MTC, Ramamurthy MB, Tran AP, Rajgor DD, Goh DYT (2021) Impact of work routines on parents' and children's sleep during the COVID-19 pandemic lockdown. Sleep Med 88:61–67

Al Mamun F, Gozal D, Hosen I, Misti JM, Mamun MA (2022) Predictive factors of insomnia during the COVID-19 pandemic in Bangladesh: a GIS-based nationwide distribution. Sleep Med 91:219–225

Albrecht JN, Werner H, Rieger N, Jenni OG, Huber R (2022) Sleep and health-related characteristics among adolescents during COVID-19: an update. Int J Environ Res Public Health 19(9): 5078

Alimoradi Z, Broström A, Tsang HWH, Griffiths MD, Haghayegh S, Ohayon MM et al (2021) Sleep problems during COVID-19 pandemic and its' association to psychological distress: a systematic review and meta-analysis. EClinicalMedicine 36:100916

AMHSI Research Team, Milken Research Team, Roitblat Y, Burger J, Leit A, Nehuliaieva L et al (2020) Stay-at-home circumstances do not produce sleep disorders: an international survey during the COVID-19 pandemic. J Psychosom Res 139:110282

Baby RS, Issac A, Vasudev A, Sabu DK, Gopika TS, Pal M et al (2021) Impact of screen time on sleep quality. Indian J Psychiatr Nurs 18(1):29

Bacaro V, Chiabudini M, Buonanno C, De Bartolo P, Riemann D, Mancini F, Baglioni C (2021) Sleep characteristics in Italian children during home confinement due to Covid-19 outbreak. Clin Neuropsychiatry 18(1):13–27

Baptista AS, Prado IM, Perazzo MF, Pinho T, Paiva SM, Pordeus IA et al (2021) Can children's oral hygiene and sleep routines be compromised during the COVID-19 pandemic? Int J Paediatr Dent 31(1):12–19

Bates LC, Zieff G, Stanford K, Moore JB, Kerr ZY, Hanson ED et al (2020) COVID-19 impact on behaviors across the 24-hour day in children and adolescents: physical activity, sedentary behavior, and sleep. Children 7(9):138

Becker SP, Gregory AM (2020) Editorial Perspective: Perils and promise for child and adolescent sleep and associated psychopathology during the COVID-19 pandemic. J Child Psychol Psychiatry 61(7):757–759

Brand S, Kirov R (2011) Sleep and its importance in adolescence and in common adolescent somatic and psychiatric conditions. Int J Gen Med 4:425–442

Bruni O, Giallonardo M, Sacco R, Ferri R, Melegari MG (2021) The impact of lockdown on sleep patterns of children and adolescents with ADHD. J Clin Sleep Med 17(9):1759–1765

Bruni O, Malorgio E, Doria M, Finotti E, Spruyt K, Melegari MG et al (2022a) Changes in sleep patterns and disturbances in children and adolescents in Italy during the Covid-19 outbreak. Sleep Med 91:166–174

Bruni O, Melegari MG, Breda M, Cedrone A, Finotti E, Malorgio E et al (2022b) Impact of COVID-19 lockdown on sleep in children with autism spectrum disorders. J Clin Sleep Med 18(1):137–143

Bucak IH, Almis H, Tasar SO, Uygun H, Turgut M (2021) Have the sleep habits in children of health workers been more affected during the COVID-19 pandemic? Sleep Med 83:235–240

Carroll N, Sadowski A, Laila A, Hruska V, Nixon M, Ma DWL et al (2020) The impact of COVID-19 on health behavior, stress, financial and food security among middle to high income Canadian families with young children. Nutrients 12(8):2352

Cellini N, Di Giorgio E, Mioni G, Di Riso D (2021) Sleep and psychological difficulties in Italian school-age children during COVID-19 lockdown. J Pediatr Psychol 46(2):153–167

Chi X, Liang K, Chen ST, Huang Q, Huang L, Yu Q et al (2021) Mental health problems among Chinese adolescents during the COVID-19: the importance of nutrition and physical activity. Int J Clin Health Psychol 21(3):100218

Chin W-C, Yao T-C, Tang I, Lee P-Y, Huang Y-S (2022) The impact of COVID-19 lockdown on sleep patterns, emotions, and behaviors of children and adolescents in Taiwan. Front Psychiatry 13:975399

Clements W, Joseph T, Koukounaras J (2021) UK NICE guidelines for EVAR: cost implications for post-COVID Australian public health. Cardiovasc Intervent Radiol 44(8):1286–1288

Crew EC, Baron KG, Grandner MA, Ievers-Landis CE, McCrae CS, Nadorff MR et al (2020) The Society of Behavioral Sleep Medicine (SBSM) COVID-19 Task Force: Objectives and

summary recommendations for managing sleep during a pandemic. Behav Sleep Med 18(4): 570–572
Dellagiulia A, Lionetti F, Fasolo M, Verderame C, Sperati A, Alessandri G (2020) Early impact of COVID-19 lockdown on children's sleep: a 4-week longitudinal study. J Clin Sleep Med 16(9): 1639–1640
DelRosso LM, Mogavero MP, Ferri R, Bruni O, Chen ML (2021) Update and progress in pediatric sleep disorders. J Pediatr 239:16–23
Di Giorgio E, Di Riso D, Mioni G, Cellini N (2021) The interplay between mothers' and children behavioral and psychological factors during COVID-19: an Italian study. Eur Child Adolesc Psychiatry 30(9):1401–1412
Dondi A, Fetta A, Lenzi J, Morigi F, Candela E, Rocca A et al (2021) Sleep disorders reveal distress among children and adolescents during the Covid-19 first wave: results of a large web-based Italian survey. Ital J Pediatr 47(1):130
Dutta K, Mukherjee R, Sen D, Sahu S (2022) Effect of COVID-19 lockdown on sleep behavior and screen exposure time: an observational study among Indian school children. Biol Rhythm Res 53(4):628–639
Fakhouri THI, Hughes JP, Brody DJ, Kit BK, Ogden CL (2013) Physical activity and screen-time viewing among elementary school-aged children in the United States from 2009 to 2010. JAMA Pediatr 167(3):223–229
Fidanci I, Güleryüz OD, Fidanci I (2020) An analysis on sleep quality of the healthcare professionals during the COVID-19 pandemic. Acta Medica Mediterr 36:3797–3800
Genta FD, Rodrigues NGB, Sunfeld JPV, Porto JF, Xavier AD, Moreno CRC et al (2021) COVID-19 pandemic impact on sleep habits, chronotype, and health-related quality of life among high school students: a longitudinal study. J Clin Sleep Med 17(7):1371–1377
Gruber R, Sadeh A, Raviv A (2000) Instability of sleep patterns in children with attention-deficit/hyperactivity disorder. J Am Acad Child Adolesc Psychiatry 39(4):495–501
Gruber R, Gauthier-Gagne G, Voutou D, Somerville G, Saha S, Boursier J (2021) Pre-pandemic sleep behavior and adolescents' stress during Covid-19: a prospective longitudinal study. Child Adolesc Psychiatry Ment Health 15(1):43
Gupta R, Kumar VM, Tripathi M, Datta K, Narayana M, Ranjan Sarmah K, Bhatia M, Devnani P, Das S, Shrivastava D, Gourineni RD, Singh TD, Jindal A, Mallick HN (2020) Guidelines of the Indian Society for Sleep Research (ISSR) for practice of sleep medicine during COVID-19. Sleep Vigil 4(2):61–72
Horwood L, Li P, Mok E, Shevell M, Constantin E (2019) A systematic review and meta-analysis of the prevalence of sleep problems in children with cerebral palsy: how do children with cerebral palsy differ from each other and from typically developing children? Sleep Health 5(6):555–571
Hyman SL, Levy SE, Myers SM, COUNCIL ON CHILDREN WITH DISABILITIES, SECTION ON DEVELOPMENTAL AND BEHAVIORAL PEDIATRICS, Kuo DZ, Apkon S et al (2020) Identification, evaluation, and management of children with autism spectrum disorder. Pediatrics 145(1):e20193447
Jahrami HA, Alhaj OA, Humood AM, Alenezi AF, Fekih-Romdhane F, AlRasheed MM et al (2022) Sleep disturbances during the COVID-19 pandemic: a systematic review, meta-analysis, and meta-regression. Sleep Med Rev 62:101591
Knowland VCP, van Rijn E, Gaskell MG, Henderson L (2022) UK children's sleep and anxiety during the COVID-19 pandemic. BMC Psychol 10(1):76
Lavigne-Cerván R, Costa-López B, Juárez-Ruiz de Mier R, Real-Fernández M, Sánchez-Muñoz de León M, Navarro-Soria I (2021) Consequences of COVID-19 confinement on anxiety, sleep and executive functions of children and adolescents in Spain. Front Psychol 12:565516
Lewien C, Genuneit J, Meigen C, Kiess W, Poulain T (2021) Sleep-related difficulties in healthy children and adolescents. BMC Pediatr 21(1):82
Li SH, Beames JR, Newby JM, Maston K, Christensen H, Werner-Seidler A (2022) The impact of COVID-19 on the lives and mental health of Australian adolescents. Eur Child Adolesc Psychiatry 31(9):1465–1477

Liao S, Luo B, Liu H, Zhao L, Shi W, Lei Y et al (2021) Bilateral associations between sleep duration and depressive symptoms among Chinese adolescents before and during the COVID-19 pandemic. Sleep Med 84:289–293

Liu Z, Tang H, Jin Q, Wang G, Yang Z, Chen H et al (2021) Sleep of preschoolers during the coronavirus disease 2019 (COVID-19) outbreak. J Sleep Res 30(1):e13142

Lokhandwala S, Holmes JF, Mason GM, St. Laurent CW, Delvey C, Hanron O et al (2021) Sleep and coping in early childhood during the COVID-19 pandemic. Front Pediatr 9:716608

Lopez-Leon S, Wegman-Ostrosky T, Ayuzo del Valle NC, Perelman C, Sepulveda R, Rebolledo PA et al (2022) Long-COVID in children and adolescents: a systematic review and meta-analyses. Sci Rep 12(1):9950

Lu C, Chi X, Liang K, Chen ST, Huang L, Guo T et al (2020) Moving more and sitting less as healthy lifestyle behaviors are protective factors for insomnia, depression, and anxiety among adolescents during the COVID-19 pandemic. Psychol Res Behav Manag 13:1223–1233

MacLean JE, Fitzgerald DA, Waters KA (2015) Developmental changes in sleep and breathing across infancy and childhood. Paediatr Respir Rev 16(4):276–284

Maquet P (2001) The role of sleep in learning and memory. Science 294(5544):1048–1052

Markovic A, Mühlematter C, Beaugrand M, Camos V, Kurth S (2021) Severe effects of the COVID-19 confinement on young children's sleep: a longitudinal study identifying risk and protective factors. J Sleep Res 30(5):e13314

Medrano M, Cadenas-Sanchez C, Oses M, Arenaza L, Amasene M, Labayen I (2021) Changes in lifestyle behaviours during the COVID-19 confinement in Spanish children: a longitudinal analysis from the MUGI project. Pediatr Obes 16(4):e12731

Mikoteit T, Brand S, Beck J, Perren S, von Wyl A, von Klitzing K et al (2012) Visually detected NREM Stage 2 sleep spindles in kindergarten children are associated with stress challenge and coping strategies. World J Biol Psychiatry 13(4):259–268

Mitra R, Moore SA, Gillespie M, Faulkner G, Vanderloo LM, Chulak-Bozzer T et al (2020) Healthy movement behaviours in children and youth during the COVID-19 pandemic: exploring the role of the neighbourhood environment. Health Place 65:102418

Moore SA, Faulkner G, Rhodes RE, Brussoni M, Chulak-Bozzer T, Ferguson LJ et al (2020) Impact of the COVID-19 virus outbreak on movement and play behaviours of Canadian children and youth: a national survey. Int J Behav Nutr Phys Act 17(1):85

Moturi S, Avis K (2010) Assessment and treatment of common pediatric sleep disorders. Psychiatry Edgmont 7(6):24–37

Murata S, Rezeppa T, Thoma B, Marengo L, Krancevich K, Chiyka E et al (2021) The psychiatric sequelae of the COVID-19 pandemic in adolescents, adults, and health care workers. Depress Anxiety 38(2):233–246

Mutluer T, Doenyas C, Aslan Genc H (2020) Behavioral implications of the Covid-19 process for autism spectrum disorder, and individuals' comprehension of and reactions to the pandemic conditions. Front Psychiatry 11:561882

Ophoff D, Slaats MA, Boudewyns A, Glazemakers I, Van Hoorenbeeck K, Verhulst SL (2018) Sleep disorders during childhood: a practical review. Eur J Pediatr 177(5):641–648

Owens J (2008) Classification and epidemiology of childhood sleep disorders. Prim Care 35(3): 533–46, vii

Panda PK, Gupta J, Chowdhury SR, Kumar R, Meena AK, Madaan P, Sharawat IK, Gulati S (2021) Psychological and behavioral impact of lockdown and quarantine measures for COVID-19 pandemic on children, adolescents and caregivers: a systematic review and meta-analysis. J Trop Pediatr 67(1):fmaa122

Paruthi S (2020) Telemedicine in pediatric sleep. Sleep Med Clin 15(3S):e1–e7

Paruthi S, Brooks LJ, D'Ambrosio C, Hall WA, Kotagal S, Lloyd RM et al (2016) Recommended amount of sleep for pediatric populations: a consensus statement of the American Academy of Sleep Medicine. J Clin Sleep Med 12(6):785–786

Ranjbar K, Hosseinpour H, Shahriarirad R, Ghaem H, Jafari K, Rahimi T et al (2021) Students' attitude and sleep pattern during school closure following COVID-19 pandemic quarantine: a web-based survey in south of Iran. Environ Health Prev Med 26(1):33
Rhodes RE, Spence JC, Berry T, Faulkner G, Latimer-Cheung AE, O'Reilly N et al (2019) Parental support of the Canadian 24-hour movement guidelines for children and youth: prevalence and correlates. BMC Public Health 19(1):1385
Richter SA, Ferraz-Rodrigues C, Schilling LB, Camargo NF, Nunes ML (2022) Effects of the COVID-19 pandemic on sleep quality in children and adolescents: a systematic review and meta-analysis. J Sleep Res 2022:e13720
Roman-Viñas B, Chaput JP, Katzmarzyk PT, Fogelholm M, Lambert EV, Maher C et al (2016) Proportion of children meeting recommendations for 24-hour movement guidelines and associations with adiposity in a 12-country study. Int J Behav Nutr Phys Act 13(1):123
Salehinejad MA, Azarkolah A, Ghanavati E, Nitsche MA (2022) Circadian disturbances, sleep difficulties and the COVID-19 pandemic. Sleep Med 91:246–252
Schultz LF, Kroll C, Constantino B, Trombelli MCMC, El Rafihi-Ferreira R, Mastroeni MF (2020) Association of maternal depression and anxiety symptoms with sleep duration in children at preschool age. Matern Child Health J 24(1):62–72
Sharma M, Aggarwal S, Madaan P, Saini L, Bhutani M (2021) Impact of COVID-19 pandemic on sleep in children and adolescents: a systematic review and meta-analysis. Sleep Med 84:259–267
Siegel JM (2009) Sleep viewed as a state of adaptive inactivity. Nat Rev Neurosci 10:747–753
Sleep Foundation. ADHD and sleep problems: how are they related? [Internet]. 2021 [cited 2022 Oct 14]. https://www.sleepfoundation.org/mental-health/adhd-and-sleep
Spruyt K (2019) A review of developmental consequences of poor sleep in childhood. Sleep Med 60:3–12
Tashjian SM, Goldenberg D, Monti MM, Galván A (2018) Sleep quality and adolescent default mode network connectivity. Soc Cogn Affect Neurosci 13(3):290–299
Tononi G, Cirelli C (2006) Sleep function and synaptic homeostasis. Sleep Med Rev 10(1):49–62
Trosman I, Ivanenko A (2021) Classification and epidemiology of sleep disorders in children and adolescents. Child Adolesc Psychiatr Clin N Am 30(1):47–64
Türkoğlu S, Uçar HN, Çetin FH, Güler HA, Tezcan ME (2020) The relationship between chronotype, sleep, and autism symptom severity in children with ASD in COVID-19 home confinement period. Chronobiol Int 37(8):1207–1213
Tyack C, Unadkat S, Voisnyte J (2022) Adolescent sleep - lessons from COVID-19. Clin Child Psychol Psychiatry 27(1):6–17
Üstündağ G, Özyurt G, Kara Aksay A, Şahin A, EkemenKeleş Y, Öztürk Y et al (2022) Evaluation of depression, anxiety, and sleep quality in children diagnosed with COVID-19. J Nerv Ment Dis 210(8):629–632
van der Helm E, Gujar N, Walker MP (2010) Sleep deprivation impairs the accurate recognition of human emotions. Sleep 33(3):335–342
Wang P, Sun X, Li W, Wang Z, He S, Zhai F et al (2021) Mental health of parents and preschool-aged children during the COVID-19 pandemic: the mediating role of harsh parenting and child sleep disturbances. Front Psychiatry 12:746330
Wearick-Silva LE, Richter SA, Viola TW, Nunes ML, COVID-19 Sleep Research Group (2022) Sleep quality among parents and their children during COVID-19 pandemic. J Pediatr 98(3): 248–255
WHO Coronavirus (COVID-19) Dashboard [Internet] [cited 2022 Oct 14]. https://covid19.who.int
Windiani IGAT, Noviyani NMR, Adnyana IGANS, Murti NLSP, Soetjiningsih S (2021) Prevalence of sleep disorders in adolescents and its relation with screen time during the COVID-19 pandemic era. Open Access Maced J Med Sci 9(B):297–300
Wirz-Justice A (2003) Chronobiology and mood disorders. Dialogues Clin Neurosci 5(4):315–325
Zhai Y, Du X (2020) Mental health care for international Chinese students affected by the COVID-19 outbreak. Lancet Psychiatry 7(4):e22

Zhao J, Xu J, He Y, Xiang M (2022) Children and adolescents' sleep patterns and their associations with mental health during the COVID-19 pandemic in Shanghai, China. J Affect Disord 301: 337–344

Zhou SJ, Wang LL, Yang R, Yang XJ, Zhang LG, Guo ZC et al (2020) Sleep problems among Chinese adolescents and young adults during the coronavirus-2019 pandemic. Sleep Med 74: 39–47

Zimmermann P, Pittet LF, Curtis N (2022) Long covid in children and adolescents. BMJ 376:o143

Zreik G, Asraf K, Haimov I, Tikotzky L (2021) Maternal perceptions of sleep problems among children and mothers during the coronavirus disease 2019 (COVID-19) pandemic in Israel. J Sleep Res 30(1):e13201

Chapter 5
Sleep Disturbances Among Healthcare Workers During the COVID-19 Pandemic

Nikolaos Sakkas and Sofia Pappa

Abstract The COVID-19 pandemic has had a significant impact on the sleep of healthcare workers (HCWs), with an estimated 40% experiencing some form of sleep disturbance during this time. A growing number of studies on the topic implies a need to synthesize the best available evidence and gain a better understanding of the epidemiology, etiology, impact, and management of sleep problems in this population. Female gender, frontline status, COVID-19 infection, lockdown, poor mental health, low level of social support, stigmatization, high workload, and a high-risk work environment have been associated with an elevated rate of sleep dysfunction during COVID-19 and identified as common risk factors among HCWs. Potential mechanisms underlying this phenomenon have also been suggested, and these include a disproportionate risk of SaRS-COV-2 transmission and infection, a change in sleep habits, and a substantial rate of comorbid psychological distress. Finally, the consequences of sleeplessness in this group have been felt on a personal, socioeconomic, and institutional level. As such, there is a need to mitigate risk and prevent harm among HCWs using effective measures, such as extensive screening and personal and professional support.

Keywords Healthcare workers · COVID-19 · Pandemic · Sleep · Insomnia

5.1 Introduction

In December 2019, the world witnessed the emergence of a novel SARS-CoV-2 virus, in Wuhan, China. Its implications, societal and economic, were swift and onerous. Placed at the core of the outbreak, healthcare workers (HCWs) found

N. Sakkas
Warwick Medical School, University of Warwick, Coventry, UK

S. Pappa (✉)
Department of Brain Sciences, Imperial College London, London, UK

West London NHS Trust, London, UK
e-mail: sofia.pappa@westlondon.nhs.uk; s.pappa@imperial.ac.uk

A. S. BaHammam et al. (eds.), *COVID-19 and Sleep: A Global Outlook*, Progress in Sleep Research, https://doi.org/10.1007/978-981-99-0240-8_5

themselves in a particularly vulnerable position. Unforeseen and unprecedented work pressures and conditions meant that they were exposed to an increased risk of psychological distress and sleep dysfunction (Pappa et al. 2022a).

In this chapter, we aim to summarize the extent and impact of sleep disturbances in HCWs during the COVID-19 pandemic. We also explore potential risk factors and mechanisms involved in the development of poor sleep in HCWs, as well as ways to prevent and manage its onset.

HCWs have been broadly defined as anyone working in primary, secondary, or tertiary care, including healthcare students undertaking a clinical placement during the COVID-19 pandemic. Sleep dysfunction pertains to a change in the quantity and quality of sleep. It typically manifests in one of three ways: insufficient quantity/ quality of sleep (sleep deprivation), sleep fragmentation, and events that occur during sleep (e.g., sleep apnoea) (Medic et al. 2017). This includes insomnia, which is defined as "experiencing persistent trouble falling asleep, staying asleep, or having quality, restful sleep despite having the appropriate circumstances and opportunity to do so along with a noticeable level of detriment to one's life" (Roth 2007).

Poor sleep quality and insomnia account for a large majority of sleep dysfunction in HCWs, both in the literature and in practice; hence, the terms insomnia, sleep dysfunction, sleep problem, and sleep disturbance will be used interchangeably in this text unless otherwise specified. That said, an attempt has been made to also include other, less common sleep conditions.

5.2 Epidemiology

5.2.1 Prevalence During the COVID-19 Pandemic

It is well known that HCWs are susceptible to sleep disturbances more than the general population. The nature of their job predisposes them to a range of harmful factors at sociodemographic and organizational levels (Pappa et al. 2022a). In normal times, an estimated prevalence between 20% and 30% has been suggested (Booker et al. 2018), though figures as high as 40% have been recorded in the literature (Weaver et al. 2018).

Additional stressors experienced during an outbreak place them at an even higher risk of developing sleep-related problems, and indeed, during the COVID-19 pandemic, the rate of insomnia and other sleep disorders among HCWs increased significantly. Several studies and systematic reviews have reported on its prevalence, and, despite a degree of variation, reports of disturbed sleep were consistently high: in five different systematic reviews with meta-analysis, pooled prevalence rates for sleep problems in HCWs ranged from 36% (Dutta et al. 2021), 39% (Pappa et al. 2020), 42% (Jahrami et al. 2022), 44% (de Pablo et al. 2020), and 45% (Mahmud et al. 2021). Meta-epidemiological data support these findings; a quality umbrella review of ten meta-analyses estimated a pooled insomnia prevalence rate of 36.6% in

HCWs during COVID-19 (Sahebi et al. 2021). Another recent review of 26 meta-analyses, and arguably the most comprehensive evidence to date, reported a rate of 39.4% (Dragioti et al. 2022). This included a range of 23.1%–64.3%, and a 95% CI of 36.9–42.0% across all studies, suggesting that around four in ten HCWs experienced a degree of sleep dysfunction during the pandemic.

Evidence from previous pandemics and epidemics also demonstrated high rates of sleep dysfunction and psychological distress in HCWs; the aggregate prevalence rate for sleeping difficulties and insomnia across Swine Flu, MERS, SARS, H1N1, Ebola, and SARS-Cov-2 was reported at approximately 40%, which is comparable to SARS-Cov-2 alone (Busch et al. 2021).

Sleep-related difficulties experienced by HCWs during COVID-19 include difficulty falling asleep at night, waking up in the early hours, and using sedative medication regularly, among others (Pappa et al. 2020). There is limited insight into the prevalence of specific sleep conditions in this population, with most studies using established outcome measures to define and quantify sleep disturbance. In this vein, the most frequently reported sleep concern was diminished sleep quality and quantity, with many HCWs meeting clinical diagnostic scores for insomnia (Jahrami et al. 2022).

5.2.2 *Prevalence After the COVID-19 Pandemic*

It is unclear if this surge in sleep dysfunction has been sustained post-COVID. While point prevalence rates may have dropped, a complete return to pre-COVID levels seems unlikely. In fact, given the prognosis and trajectory of sleep disorders, it is reasonable to assume that a percentage of HCWs are still affected in the aftermath of the pandemic, and there is some evidence to support this. The reported prevalence of sleep disturbances in publications in 2021 (47.1%) appeared to be higher than in 2020 (36.2%), which may suggest that the pandemic is having an ongoing negative impact on sleep (Jahrami et al. 2022). The emergence and spread of new COVID variants have led to "pandemic fatigue," while successive lockdowns and a rising infection load increase the risk of developing long-lasting sleep dysfunction (Hassan and Mahmoud 2021; Reicher and Drury 2021). Longitudinal studies have demonstrated a worsening of sleep outcomes over observational waves (Mahmud et al. 2021).

What is more, HCWs have reported higher levels of long-COVID than the general population, with approximately a third of those who contracted COVID-19 during the first wave still suffering from persistent symptoms 3–4 months after (Gaber et al. 2021). Given that sleep disturbance is a common feature both postinfection and as part of the long-COVID syndrome (Pappa et al. 2022b), this may partly explain the persistent or worsening levels of sleep dysfunction in HCWs following the peak of the COVID-19 pandemic (Crook et al. 2021; Katsarou et al. 2022).

5.2.3 Prevalence by Population

Comparison to other populations reveals that HCWs are likely the third most affected group. Patients infected with COVID-19 were the most affected group, at a rate of 52.4% (Jahrami et al. 2022). Children and adolescents ranked second and HCWs third, while special populations with healthcare needs and university students demonstrated similar values. The general population was the least affected by sleep disturbances during the pandemic, according to a comprehensive review with meta-analysis. This included 250 papers and a total of 493,475 participants from 49 countries, though significant variation within the literature was noted (Jahrami et al. 2022).

5.2.4 Prevalence by Country

There is significant heterogeneity in the rate of sleep dysfunction by country, and a lack of focused studies, alongside overall poor or variable study quality, makes valid comparisons difficult. There appear to be significant differences among the general population of different countries, with figures ranging from 27.3% (India) to 58.6% (Spain) (Jahrami et al. 2022), though whether this applies to HCWs in a similar pattern is unclear. A general trend observed among HCWs is that prevalence rates from Chinese and Southeast Asian studies tend to be lower than other countries, although this too remains somewhat speculative (Jahrami et al. 2022; Pappa et al. 2022c).

5.2.5 Prevalence by Mental Health Condition

Sleep problems in this group are more prevalent than other mental health concerns. One review suggested that the prevalence of anxiety and depression symptoms among hospital staff was 29.9% for anxiety and 28.4% for depression (Dragioti et al. 2022), while another reported values of 24.9% and 24.8%, respectively (Sahebi et al. 2021). The prevalence of insomnia and sleep problems was much higher, with values of 39.5% and 36.4% (Sahebi et al. 2021; Dragioti et al. 2022). Dragioti also reported a prevalence of 18.8% for post-traumatic stress disorder and 6.2% for suicidal ideation (Dragioti et al. 2022). Acute stress appeared to be the most common complaint during the pandemic, at 44.3%. This order is observed across a large body of studies (Dutta et al. 2021; Pappa et al. 2020; Hu et al. 2022; Marvaldi et al. 2021; Chirico et al. 2021; Zhang et al. 2021).

5.2.6 *Prevalence by Outcome Measure Used*

How sleep disturbance is measured may influence the rate observed (Pappa et al. 2021a). Outcome measures most commonly used to assess the presence and severity of insomnia in HCWs include the Pittsburgh Sleep Quality Index (PSQI), Insomnia Severity Index (ISI), Athens Insomnia Scale (AIS), and researcher-developed measures (Pappa et al. 2022a). Subgroup analysis demonstrates significant heterogeneity in the pooled prevalence rate indicated by each measure (Dragioti et al. 2022). During COVID-19, the highest prevalence of sleep problems was obtained by studies that used the PSQI (51.9%), followed by the AIS (47.2%) and the ISI (31.0%). Other measures, including researcher-developed ones, yielded a rate of 35.7% (Jahrami et al. 2022). Summary estimates of sensitivity and specificity did not seem to differ significantly among the three measures, but they did in researcher-developed measures (Chiu et al. 2016). In practice, a sleep problem is only labeled as such when measured with a valid and reliable psychometric test and cut-off points have been established (Fabbri et al. 2021). One meta-analysis found that the use of lower cut-off scores was associated with a higher prevalence of sleep disturbance in HCWs during the pandemic (Qiu et al. 2020).

5.2.7 *Prevalence by HCW Characteristics*

Certain characteristics seem to affect the prevalence rate of sleep disturbance in this group. These factors can be personal, societal, or institutional and can play a protective or offensive role. They are discussed further in Sect. 5.3.

5.3 Etiology (Risk Factors)

5.3.1 *Importance of Risk Factors*

Although some factors are unmodifiable, others can be identified and then dealt with to mitigate the risk that they entail. A basic understanding of risk and protective factors is therefore important; it helps gauge vulnerability and enables us to stratify mental health needs, permitting effective prevention and management.

5.3.2 *General Risk Factors for Sleep Disturbance*

Generally, sleep dysfunction has several known risk factors, some of which include female gender, low socioeconomic status, suffering from physical or psychological illnesses, and being single (LeBlanc et al. 2007). Other risk factors contributing to sleep deprivation and disruption include:

- Lifestyle—Excessive caffeine, alcohol and drug abuse, shift work, attending university, and jet lag.
- Environmental—Excessive noise and light.
- Psychosocial—Presence of anxiety or stress, parenting, and caregiving responsibilities.
- Sleep disorders—Presence of insomnia, obstructive sleep apnoea, narcolepsy, and circadian rhythm disorders.
- Medical conditions—Medication side effects, comorbid lung disease, kidney disease, diabetes, and psychiatric disorders (Medic et al. 2017).

5.3.3 Risk Factors Identified During COVID-19

In previous pandemics, being an HCW has proven to be a leading risk factor for the occurrence of diminished sleep, and SARS, H1N1, Ebola, and MERS have all induced worse mental health outcomes among frontline healthcare workers (Busch et al. 2021). Likewise, emerging research during COVID-19 has identified various individual factors with the potential to influence the onset of sleep problems in this population. These can be sorted into three different levels: personal, societal, and organizational (Pappa et al. 2022a; Centers for Disease Control and Prevention 2015). There is less insight into the latter two themes, with most empirical data focusing on personal factors.

5.3.4 Strength of Risk Factors

Despite the publication of numerous trials, the quality of evidence remains variable. There is conflicting evidence for some factors, and lacking evidence for others. In larger studies, no single risk factor managed to reach statistical significance when meta-regression was performed, and study heterogeneity acts as a confounding factor in many pooled estimates (Sahebi et al. 2021; Dragioti et al. 2022; Chirico et al. 2021). Thus, while associations have been reported, causality cannot be assumed, and the link between sleep dysfunction and various demographic parameters remains ambiguous.

However, the combined effect of multiple weak risk factors could be clinically significant, and interactions between individual factors could further increase risk. Furthermore, there is varying consensus on the significance of each factor, and some appear to be stronger than others. The weight of each factor has been marked accordingly, and factors that are stronger (but not necessarily statistically significant) have been signposted with an **(S)**.

5.3.5 Personal (Sociodemographic) Factors

Personal factors attributed to differences in the rate of sleep dysfunction in HCWs during COVID-19 include:

- Profession **(S)**—Nurses experienced higher rates of sleep dysfunction than doctors, and this may be due to differing responsibilities, health care tasks, and working hours between the two groups (Dragioti et al. 2022).
- Sex **(S)**—Female hospital staff experienced higher rates of sleep dysfunction than male hospital staff (Dragioti et al. 2022; Al Maqbali et al. 2021). Potential reasons for this difference include the fact that women are more likely to report psychological problems, as well as underlying brain structural differences across sexes, and a clear gender dimorphism in conditions such as insomnia (Bangasser and Valentino 2014; Power et al. 2022).
- Frontline Status **(S)**—Frontline hospital staff experienced higher rates than second-line hospital staff or nonclinical staff (Dragioti et al. 2022; Zhou et al. 2020). Potential reasons include increased workplace stress, extended work hours, and higher infection rates in those caring for COVID-19 patients directly (Power et al. 2022).
- Lockdown Status **(S)**—The prevalence of sleep disorders was greater in lockdown than in no lockdown for all populations, including HCWs (Jahrami et al. 2022). This may be explained partly via reports of loneliness, social isolation, and changes in sleep habits during the lockdown, all of which are known to affect mental health and sleep function.
- Infection Status **(S)**—HCWs with a current or past COVID-19 infection were more likely to report worse sleep (Jahrami et al. 2021). Sleep dysfunction is a feature of COVID-19 infection. HCWs had increased contact with COVID-19-positive patients (Nguyen et al. 2020) and were at an increased risk of infection transmission during the pandemic, particularly on the frontline (Shah et al. 2020).
- Mental Health Status **(S)**—HCWs with a preexisting or comorbid mental health condition experienced a greater rate of sleep dysfunction than those without (Power et al. 2022; Alimoradi et al. 2021; Pappa et al. 2021b).
- Age—Older HCWs experienced a higher magnitude of sleep disturbances than young HCWs (Jahrami et al. 2022).
- Geographical Distribution—HCWs residing in areas where COVID-19 was more severe (e.g., large urban areas) may be at increased risk of developing sleep problems (Jahrami et al. 2021). An increased rate of viral transmission here may precipitate this.
- BMI—HCWs with a high BMI experienced shorter sleep time and lower sleep efficiency during normal times (Brum et al. 2020). As such, obesity/overweight may be a risk factor for insomnia development during COVID-19, although evidence is lacking.
- Ethnicity—BAME HCWs were disproportionately affected by COVID-19 and are shown to experience higher insomnia rates and severity compared to their non-BAME counterparts in normal life (Ertel et al. 2011). Given the relationship

between infection status, psychological well-being, and sleep quality, further insight into the role of race and ethnicity in sleep dysfunction during COVID-19 would be valuable (Pappa et al. 2022a).

- Lifestyle Changes—Changes in physical activity, smoking, alcohol consumption, and substance use may play a role in mediating sleep among HCWs during the COVID-19 pandemic (Pappa et al. 2020).
- Education Level—HCWs with a lower level of education experienced increased sleep dysfunction during the pandemic (Zhang et al. 2021; Serrano-Ripoll et al. 2021).

5.3.6 Societal (Social) Factors

Social factors attributed to differences in the rate of sleep dysfunction in HCWs during COVID-19 include:

- Social Support **(S)**—Reduced family and social support may increase the risk of sleep dysfunction in HCWs (Xiao et al. 2020; Que et al. 2020).
- Stigma—Greater social stigma, e.g., regarding frontline status during COVID-19, may be linked with an increased onset of sleep dysfunction (Xiao et al. 2020; Taylor et al. 2020).
- Caregiving—Living with a chronically ill person during the COVID-19 pandemic may increase the risk of insomnia development in this group (Dosil et al. 2020).
- Media—Lack of social support in mass media and social networks may increase the risk of sleep disturbance among HCWs (Zhang et al. 2021).
- Change—A significant change to a normal routine may be linked to negative sleep effects in HCWs during the pandemic (Pappa et al. 2022a).
- Financial—Economic uncertainty was associated with reduced sleep duration and quality in the general population during the pandemic, and this may apply to HCWs too (Partinen et al. 2021).

5.3.7 Organizational (Occupational) Factors

Organizational factors attributed to differences in the rate of sleep dysfunction in HCWs during COVID-19 include:

- Work environment **(S)**—Working in a high-risk environment has been linked to an increased rate of sleep dysfunction during COVID-19 (Serrano-Ripoll et al. 2021). Underlying reasons include high pressure and workload, excessive fatigue, direct contact with positive patients (frontline), and work in isolated environments (i.e., isolation units) (Papoutsi et al. 2020).

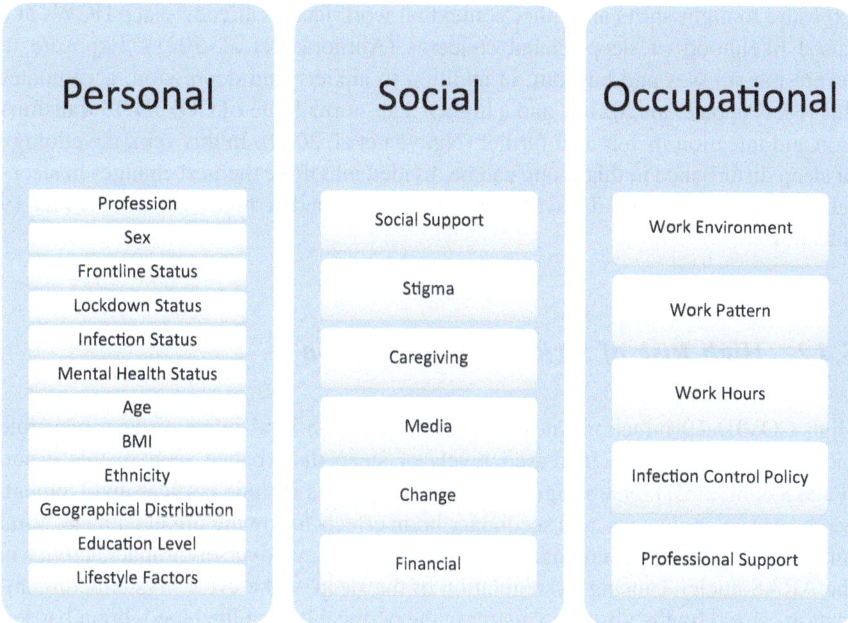

Fig. 5.1 Risk factors for sleep dysfunction in HCWs during the COVID-19 pandemic

- Work pattern **(S)**—HCWs are exposed to more pronounced patterns of shift work, which has been shown to increase the risk of insomnia development during the pandemic (Brum et al. 2020).
- Work hours **(S)**—Extended work hours have been shown to correlate positively with reduced sleep quantity and quality in HCWs during COVID-19 (Power et al. 2022; Serrano-Ripoll et al. 2021).
- Infection Control Policy—Reduced satisfaction with hospital infection control policy, fear of contracting the virus, and reduced availability of PPE may be related to worse sleep outcomes in HCWs during the pandemic (Pappa et al. 2022a).
- Professional Support—Uncertainty about the effective management of COVID-19, lack of availability of support services, and occupational health concerns may prevent healthy sleep in HCWs (Pappa et al. 2022a; Moll 2014) (Fig. 5.1).

5.4 Etiology (Mechanisms Involved)

5.4.1 Why Sleep Disturbance Occurs

No one mechanism can be attributed to it as the sole cause of sleep disturbance in HCWs, which is often multifactorial and complex. Unusual work schedules, exposure to night shifts and other contextual work factors already place HCWs at a raised likelihood of sleep-related concerns (Alimoradi et al. 2021). Exposure to severe job stresses and burnout, in addition to anxiety and depression, accentuates this risk (Pappa et al. 2021c), and a higher-than-normal rate of COVID-19 transmission and infection increases it further (Nguyen et al. 2020). In this vein, the etiology of sleep disturbance in this group can be divided into three themes: changes in sleep-related factors, comorbid psychological distress, and a high risk of COVID-19 infection.

5.4.2 High Risk of SARS-CoV-2 Infection

High COVID-19 infection rates among HCWs open viral infection as a plausible biological basis for the increased levels of sleep disturbance seen in this group (Pataka et al. 2021). Neurological effects of the virus include cortical involvement, cerebrovascular changes, and secondary brain effects following organ damage. Viral infiltration of the hypothalamic/brainstem regions may adversely impact activity in the ARAS nuclei, causing dysregulation of the sleep–wake cycle. Systemic inflammation caused by the virus may increase the permeability of the blood–brain barrier, with the entry of inflammatory markers in the brain causing further circadian dysfunction. In the long term, post-viral neurodegeneration could play a role in making sleep disturbances chronic. Respiratory complications may also preempt the development of sleep-related breathing disorders in those infected with COVID-19 (Gupta and Pandi-Perumal 2021). The symptomatology of the disease is characterized by cough, fever, and difficulty breathing, while pain levels, medication side effects, and disease stress factors could also contribute to a lack of quality sleep (Shi et al. 2020). Finally, it is worth noting that COVID-19 infection and sleep dysfunction demonstrate a bidirectional relationship, and a lack of sufficient sleep can increase susceptibility to viral transmission and infection in HCWs (Nguyen et al. 2020).

5.4.3 Changes in Sleep-Related Factors

HCWs experienced significant changes in sleep-related factors during the pandemic. This includes changes in sleep–wake habits with delayed bedtime, lights off time, and sleep onset time (Salehinejad et al. 2022), as well as home quarantine/confinement, extensive and isolated shift work, stress/anxiety, and changes to the level of activity during the daytime. An abrupt change to any component can precede circadian malalignment, leading to abnormalities in the balance of neurotransmitter release in the CNS and disrupting many physiologic, biologic, and behavioral parameters that are responsible for normal REM and NREM sleep (Altevogt and Colten 2006). Dysregulation of sleep-wake homeostasis is thought to affect multiple bodily pathways and aspects of health, including our body's immune response (Borrmann et al. 2021). Immune parameters follow a cyclical pattern regulated by our circadian rhythm, and disruption of the immune cycle can increase susceptibility to viral respiratory diseases, such as SARS-Cov-2 (Salehinejad et al. 2022). In short, changes in sleep-related factors disrupt circadian rhythm, adversely impacting sleep and increasing the risk of COVID-19 infection in HCWs.

5.4.4 Comorbid Psychological Distress

Sleep in this population has been a concern long before the emergence of COVID-19, with workplace distress and physician burnout having been described as a public health crisis (Altevogt and Colten 2006). There is a moderate association between the two phenomena, and, despite their resilience, HCWs are vulnerable to psychological affliction and harm (Dragioti et al. 2022; Pappa et al. 2021c). Stress responses, faced in everyday life, create an imbalance by suppressing REM sleep and increasing NREM sleep. In the acute phase, this is followed by restorative sleep compensation, meaning increased REM and overall sleep time. However, if stress stimuli become repetitive/chronic, such as in the case of COVID-19, sleep restoration will be reduced, and hyperactivity of the HPA axis will result in fragmented sleep (Zhuo et al. 2020). Over time, this will excite the sympathetic catecholamine system, raising cortisol levels and resulting in persistent insomnia and other sleep disturbances in this group (Levenson et al. 2015).

There is a clear overlap between sleep problems and psychological distress. Other mental health problems reported during the pandemic, such as anxiety and depression, are associated with sleep disorders, and the two behave bidirectionally (Kim et al. 2021). Psychological distress can lead to insufficient sleep, and insufficient sleep can further increase levels of distress. Both pose risk factors for suicidal ideation, attempt, and completion, and suicidal ideation has been directly linked to the severity of insomnia (Killgore et al. 2020). Systems that aim to reduce psychological distress among HCWs may help reduce sleep loss, while screening for insomnia may also reduce the risk of suicidal ideation and other psychological disorders (Fig. 5.2).

5.5 Impact of Sleep Dysfunction

Sleep problems are common, highly treatable conditions that often go untreated despite their health and safety effects (Altevogt and Colten 2006). In studying HCWs, a preexisting trend of increased sleeplessness makes it difficult to quantify the impact of COVID-19 directly. However, it is accepted that a lack of quality sleep at this time has been harmful at all levels, including personal, socioeconomic, and organizational.

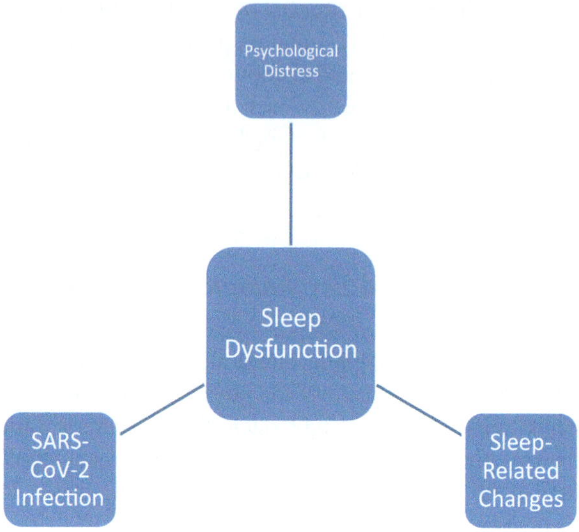

Fig. 5.2 Mechanisms involved in the development of sleep dysfunction in HCWs during the COVID-19 pandemic

5.5.1 The Personal Impact of Sleeplessness

The personal impact of sleep dysfunction among HCWs includes reduced alertness, altered cognitive and psychomotor performance, emotional dysregulation, increased fatigue, and a lower quality of life (Bollu and Kaur 2019). It is also associated with increased morbidity and mortality, adversely affecting most body systems (AlDabal and BaHammam 2011), and increasing the risk of cardiovascular disease, hypertension, T2DM, asthma, and thyroid disorders. Individuals who sleep less than 6 h per day are thought to have a 13% higher mortality risk than those sleeping 7–9 h every night (Cappuccio et al. 2010).

5.5.2 Socioeconomic Impact of Sleeplessness

While most HCWs only experience a mild-to-moderate disruption in sleep, this can have a knock-on effect at a social level. Sleep deprivation can cause the overall health of the frontline workforce to suffer, forcing a deterioration of population health, and, in turn, an increase in healthcare spending for organizations and individuals alike. There is a link between sleep problems and financial suffering in the general population (Partinen et al. 2021), and this may apply to HCWs too. Before the outbreak, it is thought that poor sleep costs the UK economy as much as £50bn, and 200,000 working days, per year (Hafner et al. 2017). It is without a doubt that in the face of COVID-19, this figure will have increased substantially.

5.5.3 *Organizational Impact of Sleeplessness*

In an organizational context, sleeplessness can lead to decreased productivity and increased absenteeism. Personal effects of insufficient sleep, such as reduced attention and altered cognitive processing, can preclude medical error (Medic et al. 2017), and screening positive for a sleep disorder has been associated with a raised incidence of adverse safety outcomes, both in the workplace and out (Weaver et al. 2018). This makes sleeplessness a likely risk factor for occupational and patient safety and suggests that it can affect the well-being of both the workforce and the public. Finally, COVID-19 is fuelling a staffing crisis; it has left a self-reported 50% of all UK HCWs in a deteriorating mental state, with up to 21% stating they are likely to quit because of the pandemic. In England alone, this could result in the loss of approximately 300,000 healthcare workers (Patel and Thomas 2021). This highlights the need for institutions to take the matter seriously and act to preserve and protect their workforce (Fig. 5.3).

5.6 Management

Preserving sleep quality and quantity in HCWs can support psychological well-being, improve work performance, and uphold patient care. Unfortunately, maintaining a healthy level of sleep in this population group, especially during a pandemic, poses a huge logistical challenge. Nevertheless, prevention and intervention measures should be prioritized, and considerations should be made at both an individual and institutional level.

5.6.1 *Preventing the Development of Sleep Problems*

To prevent the onset of sleep dysfunction in the first place, HCWs can be grouped according to the relative risk of developing sleep problems, considering factors at each ecological level (World Health Organization 2022). Screening needs to be

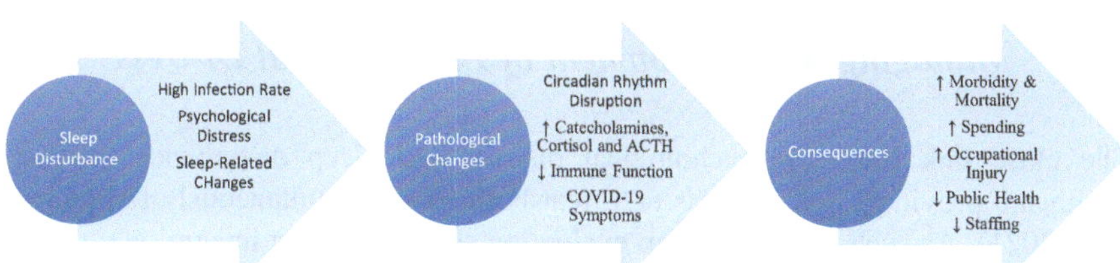

Fig. 5.3 Short- and long-term implications of sleep dysfunction in HCWs during the COVID-19 pandemic

implemented systematically, using a top-down approach, and the application of a validated screening tool may help in achieving this. Appropriate tiered and tailored measures can then be taken to try and minimize risk.

5.6.2 Managing the Development of Sleep Problems

Although some sleep conditions are more difficult to treat, most can be routinely managed. Early recognition is key and any subsequent intervention strategies should be personalized, timely, and evidence-based to prevent the onset of chronic or recalcitrant sleep dysfunction in HCWs. Practical and easy-to-implement strategies identified during the COVID-19 pandemic include:

1. Personal—Expressing stress and concerns to family, practicing sleep hygiene, effective sleep schedules, regular daytime exercise, mindfulness, chronotherapy, relaxation techniques, and yoga. Following a stepwise approach, pharmacological interventions can also be trialed (Altena et al. 2020).
2. Societal—Mitigating the risk of COVID-19 transmission, reducing stigma, and discouraging negative media coverage relating to HCWs. Public authorities may also wish to support health professionals in delivering sleep-related help, encourage employers to address sleep issues, and adjust public policy to support flexible and healthy shift working (Geoffroy et al. 2020).
3. Institutional—Administrative support, workplace napping, safe workplace conditions, removal of clinical care barriers, and provision of PPE to reduce fear of COVID-19. According to the World Health Organisation, a reduction of workplace distress is also necessary and includes decreasing work hours, increasing workforce numbers, and training managers and workers alike in mental health literacy and awareness (World Health Organization 2022). Easy access to mental health services is equally important; counseling, CBT for insomnia, and mindfulness must be considered for HCWs and can be delivered in-person, via a dedicated hotline or smartphone application (Geoffroy et al. 2020). Above all, employers need to understand the importance of sleep, the effect of sleeplessness on service delivery, and their role in preserving it (Blake et al. 2020; Redeker et al. 2019).

5.6.3 Managing the Development of Psychological Distress

The close link between psychological distress and sleep disturbance suggests healthcare providers could assess and treat both issues simultaneously (Alimoradi et al. 2021). As such, a formal sleep evaluation should form an integral part of any mental health assessment, and vice versa.

5.6.4 Sleep in a Post-COVID World

Real-world change is needed urgently to protect HCWs, both in future crises but also in their present recovery from the effects of the pandemic. Sleep dysfunction in HCWs must be acknowledged as a public health crisis and sleep quality in this group must be prioritized and addressed on a global scale.

References

Al Maqbali M, Al Sinani M, Al-Lenjawi B (2021) Prevalence of stress, depression, anxiety and sleep disturbance among nurses during the COVID-19 pandemic: a systematic review and meta-analysis. J Psychosom Res 141:110343

AlDabal L, BaHammam AS (2011) Metabolic, endocrine, and immune consequences of sleep deprivation. Open Respir Med J 5:31

Alimoradi Z, Broström A, Tsang HW, Griffiths MD, Haghayegh S, Ohayon MM, Lin CY, Pakpour AH (2021) Sleep problems during COVID-19 pandemic and its' association to psychological distress: a systematic review and meta-analysis. EClinicalMedicine 36:100916

Altena E, Baglioni C, Espie CA, Ellis J, Gavriloff D, Holzinger B, Schlarb A, Frase L, Jernelöv S, Riemann D (2020) Dealing with sleep problems during home confinement due to the COVID-19 outbreak: practical recommendations from a task force of the European CBT-I Academy. J Sleep Res 29(4):e13052

Altevogt BM, Colten HR (eds) (2006) Sleep disorders and sleep deprivation: an unmet public health problem. National Academies Press, Washington, DC

Bangasser DA, Valentino RJ (2014) Sex differences in stress-related psychiatric disorders: neurobiological perspectives. Front Neuroendocrinol 35(3):303–319

Blake H, Bermingham F, Johnson G, Tabner A (2020) Mitigating the psychological impact of COVID-19 on healthcare workers: a digital learning package. Int J Environ Res Public Health 17(9):2997

Bollu PC, Kaur H (2019) Sleep medicine: insomnia and sleep. Mo Med 116(1):68

Booker LA, Magee M, Rajaratnam SM, Sletten TL, Howard ME (2018) Individual vulnerability to insomnia, excessive sleepiness and shift work disorder amongst healthcare shift workers. A systematic review. Sleep Med Rev 41:220–233

Borrmann H, McKeating JA, Zhuang X (2021) The circadian clock and viral infections. J Biol Rhythm 36(1):9–22

Brum MCB, Dantas Filho FF, Schnorr CC, Bertoletti OA, Bottega GB, da Costa Rodrigues T (2020) Night shift work, short sleep and obesity. Diabetol Metab Syndr 12(1):1–9

Busch IM, Moretti F, Mazzi M, Wu AW, Rimondini M (2021) What we have learned from two decades of epidemics and pandemics: a systematic review and meta-analysis of the psychological burden of frontline healthcare workers. Psychother Psychosom 90(3):178–190

Cappuccio FP, D'Elia L, Strazzullo P, Miller MA (2010) Sleep duration and all-cause mortality: a systematic review and meta-analysis of prospective studies. Sleep 33(5):585–592

Centers for Disease Control and Prevention (2015) The social-ecological model: a framework for prevention. Centers for Disease Control and Prevention

Chirico F, Ferrari G, Nucera G, Szarpak L, Crescenzo P, Ilesanmi O (2021) Prevalence of anxiety, depression, burnout syndrome, and mental health disorders among healthcare workers during the COVID-19 pandemic: a rapid umbrella review of systematic reviews. J Health Soc Sci 6(2): 209–220

Chiu HY, Chang LY, Hsieh YJ, Tsai PS (2016) A meta-analysis of diagnostic accuracy of three screening tools for insomnia. J Psychosom Res 87:85–92

Crook H, Raza S, Nowell J, Young M, Edison P (2021) Long covid—mechanisms, risk factors, and management. BMJ 374:n1648

de Pablo GS, Vaquerizo-Serrano J, Catalan A, Arango C, Moreno C, Ferre F, Shin JI, Sullivan S, Brondino N, Solmi M, Fusar-Poli P (2020) Impact of coronavirus syndromes on physical and mental health of health care workers: systematic review and meta-analysis. J Affect Disord 275: 48–57

Dosil M, Ozamiz-Etxebarria N, Redondo I, Picaza M, Jaureguizar J (2020) Psychological symptoms in health professionals in Spain after the first wave of the COVID-19 pandemic. Front Psychol 11:606121

Dragioti E, Tsartsalis D, Mentis M, Mantzoukas S, Gouva M (2022) Impact of the COVID-19 pandemic on the mental health of hospital staff: an umbrella review of 44 meta-analyses. Int J Nurs Stud 131:104272

Dutta A, Sharma A, Torres-Castro R, Pachori H, Mishra S (2021) Mental health outcomes among health-care workers dealing with COVID-19/severe acute respiratory syndrome coronavirus 2 pandemic: a systematic review and meta-analysis. Indian J Psychiatry 63(4):335

Ertel KA, Berkman LF, Buxton OM (2011) Socioeconomic status, occupational characteristics, and sleep duration in African/Caribbean immigrants and US White health care workers. Sleep 34(4): 509–518

Fabbri M, Beracci A, Martoni M, Meneo D, Tonetti L, Natale V (2021) Measuring subjective sleep quality: a review. Int J Environ Res Public Health 18(3):1082

Gaber TAK, Ashish A, Unsworth A (2021) Persistent post-covid symptoms in healthcare workers. Occup Med (Lond) 71(3):144–146

Geoffroy PA, Le Goanvic V, Sabbagh O, Richoux C, Weinstein A, Dufayet G, Lejoyeux M (2020) Psychological support system for hospital workers during the Covid-19 outbreak: rapid design and implementation of the Covid-Psy hotline. Front Psychiatry 11:511

Gupta R, Pandi-Perumal SR (2021) SARS-CoV-2 infection: paving way for sleep disorders in long term! Sleep Vigilance 5(1):1–2

Hafner M, Stepanek M, Taylor J, Troxel WM, Van Stolk C (2017) Why sleep matters—the economic costs of insufficient sleep: a cross-country comparative analysis. Rand Health Q 6(4):11

Hassan EM, Mahmoud HN (2021) Impact of multiple waves of COVID-19 on healthcare networks in the United States. PLoS One 16(3):e0247463

Hu N, Deng H, Yang H, Wang C, Cui Y, Chen J, Wang Y, He S, Chai J, Liu F, Zhang P (2022) The pooled prevalence of the mental problems of Chinese medical staff during the COVID-19 outbreak: a meta-analysis. J Affect Disord 303:323–330

Jahrami H, BaHammam AS, Bragazzi NL, Saif Z, Faris M, Vitiello MV (2021) Sleep problems during the COVID-19 pandemic by population: a systematic review and meta-analysis. J Clin Sleep Med 17(2):299–313

Jahrami HA, Alhaj OA, Humood AM, Alenezi AF, Fekih-Romdhane F, AlRasheed MM, Saif ZQ, Bragazzi NL, Pandi-Perumal SR, BaHammam AS, Vitiello MV (2022) Sleep disturbances during the COVID-19 pandemic: a systematic review, meta-analysis, and meta-regression. Sleep Med Rev 62:101591

Katsarou M-S, Iasonidou E, Osarogue A, Kalafatis E, Stefanatou M, Pappa S, Gatzonis S, Verentzioti A, Gounopoulos P, Demponeras C, Konstantinidou E, Drakoulis N, Asimakos A, Antonoglou A, Mavronasou A, Spetsioti S, Kotanidou A, Katsaounou P (2022) The Greek Collaborative Long COVID Study: non-hospitalized and hospitalized patients share similar symptom patterns. J Pers Med 12:987

Killgore WD, Cloonan SA, Taylor EC, Fernandez F, Grandner MA, Dailey NS (2020) Suicidal ideation during the COVID-19 pandemic: the role of insomnia. Psychiatry Res 290:113134

Kim H, Hegde S, LaFiura C, Raghavan M, Luong E, Cheng S, Rebholz CM, Seidelmann SB (2021) COVID-19 illness in relation to sleep and burnout. BMJ Nutr Prev Health 4(1):132

LeBlanc M, Beaulieu-Bonneau S, Mérette C, Savard J, Ivers H, Morin CM (2007) Psychological and health-related quality of life factors associated with insomnia in a population-based sample. J Psychosom Res 63(2):157–166

Levenson JC, Kay DB, Buysse DJ (2015) The pathophysiology of insomnia. Chest 147(4): 1179–1192

Mahmud S, Hossain S, Muyeed A, Islam MM, Mohsin M (2021) The global prevalence of depression, anxiety, stress, and, insomnia and its changes among health professionals during COVID-19 pandemic: a rapid systematic review and meta-analysis. Heliyon 7(7):e07393

Marvaldi M, Mallet J, Dubertret C, Moro MR, Guessoum SB (2021) Anxiety, depression, trauma-related, and sleep disorders among healthcare workers during the COVID-19 pandemic: a systematic review and meta-analysis. Neurosci Biobehav Rev 126:252–264

Medic G, Wille M, Hemels ME (2017) Short- and long-term health consequences of sleep disruption. Nat Sci Sleep 9:151

Moll SE (2014) The web of silence: a qualitative case study of early intervention and support for healthcare workers with mental ill-health. BMC Public Health 14(1):1–13

Nguyen LH, Drew DA, Graham MS, Joshi AD, Guo CG, Ma W, Mehta RS, Warner ET, Sikavi DR, Lo CH, Kwon S (2020) Risk of COVID-19 among front-line health-care workers and the general community: a prospective cohort study. Lancet Public Health 5(9):e475–e483

Papoutsi E, Giannakoulis VG, Ntella V, Pappa S, Katsaounou P (2020) Global burden of COVID-19 pandemic on healthcare workers. ERJ Open Res 6(2):00195-2020

Pappa S, Ntella V, Giannakas T, Giannakoulis VG, Papoutsi E, Katsaounou P (2020) Prevalence of depression, anxiety, and insomnia among healthcare workers during the COVID-19 pandemic: a systematic review and meta-analysis. Brain Behav Immun 88:901–907

Pappa S, Giannakoulis VG, Papoutsi E, Katsaounou P (2021a) Author reply–Letter to the editor "The challenges of quantifying the psychological burden of COVID-19 on healthcare workers". Brain Behav Immun 92:209

Pappa S, Barnett J, Berges I, Sakkas N (2021b) Tired, worried and burned out, but still resilient: a cross-sectional study of mental health workers in the UK during the COVID-19 pandemic. Int J Environ Res Public Health 18(9):4457

Pappa S, Athanasiou N, Sakkas N, Patrinos S, Sakka E, Barmparessou Z, Tsikrika S, Adraktas A, Pataka A, Migdalis I, Gida S (2021c) From recession to depression? Prevalence and correlates of depression, anxiety, traumatic stress and burnout in healthcare workers during the COVID-19 pandemic in Greece: a multi-center, cross-sectional study. Int J Environ Res Public Health 18(5):2390

Pappa S, Sakkas N, Sakka E (2022a) A year in review: sleep dysfunction and psychological distress in healthcare workers during the COVID-19 pandemic. Sleep Med 91:237–245

Pappa S, Barmparessou Z, Athanasiou N, Sakka E, Eleftheriou K, Patrinos S, Sakkas N, Pappas A, Kalomenidis I, Katsaounou P (2022b) Depression, insomnia and post-traumatic stress disorder in COVID-19 survivors: role of gender and impact on quality of life. J Pers Med 12(3):486

Pappa S, Chen J, Barnett J, Chang A, Dong RK, Xu W, Yin A, Chen BZ, Delios AY, Chen RZ, Miller S, Wan X, Zhang SX (2022c) A systematic review and meta-analysis of the mental health symptoms during the Covid-19 pandemic in Southeast Asia. Psychiatry Clin Neurosci 76(2): 41–50

Partinen M, Bjorvatn B, Holzinger B, Chung F, Penzel T, Espie CA, Morin CM, ICOSS-Collaboration Group (2021) Sleep and circadian problems during the coronavirus disease 2019 (COVID-19) pandemic: the International COVID-19 Sleep Study (ICOSS). J Sleep Res 30(1):e13206

Pataka A, Kotoulas S, Sakka E, Katsaounou P, Pappa S (2021) Sleep dysfunction in COVID-19 patients: prevalence, risk factors, mechanisms, and management. J Pers Med 11(11):1203

Patel P, Thomas C (2021) Recover, reward, renew: a post-pandemic plan for the healthcare workforce. IPPR

Power N, Perreault M, Ferrari M, Boudreau P, Boivin DB (2022) Sleep of healthcare workers during the COVID-19 pandemic and the role of atypical work schedules: a scoping review. J Biol Rhythm 37(4):358–384
Qiu D, Yu Y, Li RQ, Li YL, Xiao SY (2020) Prevalence of sleep disturbances in Chinese healthcare professionals: a systematic review and meta-analysis. Sleep Med 67:258–266
Que J, Shi LE, Deng J, Liu J, Zhang L, Wu S, Gong Y, Huang W, Yuan K, Yan W, Sun Y (2020) Psychological impact of the COVID-19 pandemic on healthcare workers: a cross-sectional study in China. Gen Psychiatr 33(3):e100259
Redeker NS, Caruso CC, Hashmi SD, Mullington JM, Grandner M, Morgenthaler TI (2019) Workplace interventions to promote sleep health and an alert, healthy workforce. J Clin Sleep Med 15(4):649–657
Reicher S, Drury J (2021) Pandemic fatigue? How adherence to covid-19 regulations has been misrepresented and why it matters. BMJ 372:n137
Roth T (2007) Insomnia: definition, prevalence, etiology, and consequences. J Clin Sleep Med 3(5 suppl):S7–S10
Sahebi A, Abdi K, Moayedi S, Torres M, Golitaleb M (2021) The prevalence of insomnia among health care workers amid the COVID-19 pandemic: an umbrella review of meta-analyses. J Psychosom Res 149:110597
Salehinejad MA, Azarkolah A, Ghanavati E, Nitsche MA (2022) Circadian disturbances, sleep difficulties and the COVID-19 pandemic. Sleep Med 91:246–252
Serrano-Ripoll MJ, Zamanillo-Campos R, Castro A, Fiol-de Roque MA, Ricci-Cabello I (2021) Insomnia and sleep quality in healthcare workers fighting against COVID-19: a systematic review of the literature and meta-analysis. Actas Espanolas de Psiquiatria 49(4):155–179
Shah AS, Wood R, Gribben C, Caldwell D, Bishop J, Weir A, Kennedy S, Reid M, Smith-Palmer A, Goldberg D, McMenamin J (2020) Risk of hospital admission with coronavirus disease 2019 in healthcare workers and their households: nationwide linkage cohort study. BMJ 371:m3582
Shi L, Lu ZA, Que JY, Huang XL, Liu L, Ran MS, Gong YM, Yuan K, Yan W, Sun YK, Shi J (2020) Prevalence of and risk factors associated with mental health symptoms among the general population in China during the coronavirus disease 2019 pandemic. JAMA Netw Open 3(7):e2014053
Taylor S, Landry CA, Rachor GS, Paluszek MM, Asmundson GJ (2020) Fear and avoidance of healthcare workers: an important, under-recognized form of stigmatization during the COVID-19 pandemic. J Anxiety Disord 75:102289
Weaver MD, Vetter C, Rajaratnam SM, O'Brien CS, Qadri S, Benca RM, Rogers AE, Leary EB, Walsh JK, Czeisler CA, Barger LK (2018) Sleep disorders, depression and anxiety are associated with adverse safety outcomes in healthcare workers: a prospective cohort study. J Sleep Res 27(6):e12722
World Health Organization (2022) WHO guidelines on mental health at work. World Health Organization, Geneva
Xiao H, Zhang Y, Kong D, Li S, Yang N (2020) The effects of social support on sleep quality of medical staff treating patients with coronavirus disease 2019 (COVID-19) in January and February 2020 in China. Med Sci Monit 26:e923549–e923541
Zhang H, Li W, Li H, Zhang C, Luo J, Zhu Y, Wu H, Zhu Z, Xu Y, Wang J, Li C (2021) Prevalence and dynamic features of psychological issues among Chinese healthcare workers during the COVID-19 pandemic: a systematic review and cumulative meta-analysis. Gen Psychiatr 34(3): e100344
Zhou Y, Wang W, Sun Y, Qian W, Liu Z, Wang R, Qi L, Yang J, Song X, Zhou X, Zeng L (2020) The prevalence and risk factors of psychological disturbances of frontline medical staff in China under the COVID-19 epidemic: workload should be concerned. J Affect Disord 277:510–514
Zhuo K, Gao C, Wang X, Zhang C, Wang Z (2020) Stress and sleep: a survey based on wearable sleep trackers among medical and nursing staff in Wuhan during the COVID-19 pandemic. Gen Psychiatr 33(3):e100260

Chapter 6
Sleep Disturbances Among Patients with COVID-19 Infection

Suhaila Ghuloum, Sulaiman Alshakhs, and Hassen Al-Amin

Abstract The COVID-19 pandemic had global implications for people's physical and mental health. Sleep is one of the determinants of health and well-being and sleep disturbance and mental illness are closely correlated. The COVID-19 infection, as well as the restrictions associated with the control of its spread, have resulted in a higher prevalence of sleep disturbances and mental health conditions, specifically anxiety and depression. Some evidence links the severity of the infection with the onset and severity of sleep disturbance. Preventive measures, early recognition, and management contribute to a better prognosis in those infected with the virus. This chapter explores the impact of the pandemic on sleep patterns, quality, and duration. We analyze the literature evidence for a correlation between infection and sleep disturbance and the theories behind that. We allude briefly to other sleep disorders that will be addressed in more depth in other chapters. This chapter ends with a description of key management considerations, pharmacological and non-pharmacological.

Keywords Sleep · COVID-19 · Insomnia · Sleep quality · Somnolence

6.1 Introduction

Sleep plays an essential role in promoting health, regeneration, and immunomodulation. Normal sleep strengthens the synaptic connections in neurons, maintaining brain functions like memory consolidation and learning (Tononi and Cirelli 2014). Sleep duration, timing, and quality are essential in these roles (Czeisler 2015). Sleep disturbances, on the other hand, are both a precipitating factor for mental illness and a symptom. It is associated with anxiety, depression, and dementia

S. Ghuloum (✉) · S. Alshakhs
Mental Health Services, Hamad Medical Corporation, Doha, Qatar
e-mail: sghuloum@hamad.qa

H. Al-Amin
Department of Psychiatry, Weill Cornell Medicine, Ar-Rayyan, Qatar

A. S. BaHammam et al. (eds.), *COVID-19 and Sleep: A Global Outlook*, Progress in Sleep Research, https://doi.org/10.1007/978-981-99-0240-8_6

and is a risk factor for suicide (Porras-Segovia et al. 2019). This becomes more significant during any pandemic, where the prevalence of sleep disorders is increased, to reduce the risk of suicide (Sher 2020). Therefore, early recognition and treatment of sleep problems among those with mental illness and the general population are essential to prevent psychological and physical morbidities during pandemics. Beyond mental health consequences, sleep disturbances are also a risk factor for several medical conditions, including cardiovascular disease, obesity, and overall increased mortality (Czeisler 2015; Sofi et al. 2014). Good quality of sleep promotes a better immune system and can subsequently promote better recovery from infection (Mahmoudi et al. 2021).

6.2 Background

The COVID-19 pandemic may be the most challenging health crisis of this century. Its global health and economic impacts affected all aspects of life, including people's physical and mental health. While necessary measures were implemented to control the spread of the virus, the restrictions imposed on daily life caused other social, health, and financial consequences beyond the risk of infection. It is therefore not surprising to see a rise in mental health disorders worldwide. The increased prevalence of depression, anxiety, and psychological distress have been well-reported in the literature since the pandemic's start (Xiong et al. 2020). Sleep, an essential component of maintaining physical and mental well-being, was particularly affected. The sleep–wake cycle is regulated by a circadian rhythm influenced by daylight exposure and daily activities and routines. These include regular waking times, work, leisure activities, exercise, etc. (Morin et al. 2020). The imposed lockdowns around the world confined most people to home, disrupting their daily activities and often leaving them with little if any exposure to sunlight thus resulting in increased rates of insomnia and other sleep disorders; disruptions to both sleep pattern and quality were reported (Morin et al. 2020; Cellini et al. 2020). While this is true for many people, patients infected with COVID-19 were particularly susceptible.

The prevalence of sleep problems during the pandemic increased steadily. A study conducted in China during the early days of the pandemic reported that 18.2% of people had decreased sleep quality (Huang and Zhao 2020). A larger systematic review that spanned a wider data range estimated the prevalence of sleep problems at 35.7% (Jahrami et al. 2021). This systematic review and meta-analysis reported that most studies exploring sleep problems were conducted on the general population, followed by healthcare practitioners, and lastly, on patients infected with the virus. Yet, those infected were the most affected, with a 74.8% prevalence of sleep problems. Older age and male gender had a significantly higher prevalence. This is strikingly higher than their reported 40% prevalence among the general public and healthcare practitioners (Jahrami et al. 2021). A later systematic review on the long-term sleep problems after COVID-19 infection reported considerable variation in the literature from no effect of COVID-19 on sleep 5 months after recovery to 85% of patients having sleep disturbances within a similar time frame.

The impact of the severity of the acute viral infection on sleep disturbances also varied; some studies reported sleep to be independent of the severity of the infection, while others showed a higher prevalence of disturbed sleep in the more acutely ill. Whatever the extent of the effect, it seems to improve with time after recovery (Schou et al. 2021). Some factors contributing to poor sleep in this population include hospitalization duration, low absolute lymphocyte count, increased neutrophil to lymphocyte ratio, and a history of mood disorder. Immunological and other factors are associated with long-haulers who continue to have clinical sequelae after passing through the index infection; many of those individuals also report sleep complaints (Bhat and Chokroverty 2022).

Several theories attempted to explain the relationship between disturbed sleep and the severity of COVID-19 infection. The virus attacks the respiratory system to cause cough and shortness of breath, which are associated with sleep disruption (Ferrando et al. 2016). This was more pronounced before the introduction of vaccines when severe respiratory symptoms were more common among those infected. The other physiological effects of the virus, such as fever and pain, are also linked to sleep problems. Medications used to treat infected patients may also cause impaired sleep (Tsang et al. 2004). In the following sections, we will discuss the biopsychosocial and medical aspects related to reported sleep problems (insomnia, quality, duration, and others) related to the COVID-19 pandemic. We will also briefly cover the available treatment options.

6.3 Insomnias

Insomnia, "a difficulty in initiating, maintaining, and consolidating sleep..." is the most common of sleep disorders (de Sousa Martins e et al. 2020). Insomnia is associated with hypertension, diabetes, and cardiovascular disease (Sofi et al. 2014; Fernandez-Mendoza and Vgontzas 2013). It is also a common symptom of depression and anxiety disorders, both occurring at a higher rate during the pandemic. Published literature from previous viral outbreaks such as swine flu, severe acute respiratory syndrome coronavirus (SARS), and Middle East respiratory syndrome coronavirus (MERS) provided evidence of insomnia developing because of isolation and quarantine (Hossain et al. 2020; Sanghera et al. 2020). Among the factors contributing to the increased prevalence of insomnia during the pandemic are the uncertainties of the news of the spread of the virus and updates on social media, the lockdown periods, and also the people's concerns for their health and of their loved ones (Pappa et al. 2020; Voitsidis et al. 2020; Kokou-Kpolou et al. 2020). Specific population groups involved as front-liners in managing the pandemic, for instance, healthcare professionals, working longer hours and more overnight shifts, were more susceptible to insomnia (Cénat et al. 2021).

Patients with the infection who were quarantined or hospitalized, and those exposed to the virus and isolated, were particularly at risk of difficulty falling asleep (early insomnia) or waking early in the morning (late insomnia). Earlier in the pandemic, data from China showed a significantly higher prevalence of insomnia among medically isolated patients than in others. A systematic review of 31 studies, with a total sample size of 5153 patients with COVID-19, showed an overall prevalence of insomnia of 34%; there was no significant gender difference (Deng et al. 2021). However, others reported more severe sleep problems in women (31%) than in men (16%) (Wichniak et al. 2021). The International COVID-19 Sleep Study (ICOSS) is a collaborative study among 14 countries across four continents that explored the pandemic's impact on sleep. They concluded that being infected with the virus was significantly associated with poor sleep quality, early insomnia, late insomnia, increased use of hypnotics, daytime sedation, and fatigue. This was reported in all the participating countries, where more than 20% of participants reported poorer sleep quality. They assume the sleepiness caused by the infection may be biological (Partinen et al. 2021).

Impaired consciousness, ranging from somnolence to delirium and coma, has been reported in 15% of hospitalized COVID-19 patients. Mechanisms include direct infection-related causes and general factors associated with increasing the risk of delirium in hospitalized patients, such as prolonged hospitalization, pain, and metabolic abnormalities. Sleep deprivation is both a risk factor for delirium and a consequence. Several studies explored the use of Melatonin and Melatonin receptor agonists and their association with reducing ICU admissions, reducing the risk of delirium, and improving sleep quality in COVID-19 patients (Zambrelli et al. 2020).

6.4 Sleep Duration

Sleep has a bidirectional association with infectious processes. Poor sleep impairs the body's normal immune response to infective agents thus increasing the risk of getting infected and experiencing a more severe illness (de Sousa Martins e et al. 2020). On the other hand, the infection and restrictions associated with the pandemic have physical and mental health consequences that affect sleep. In a study that examined observational and genetic data, short sleep duration was associated with acceleration of immune senescence and obesity, both associated with the susceptibility to and severity of COVID-19 infection (Liu et al. 2022). Another study in China found that the risk of severe infection was six to eight times higher in those who had decreased sleep in the week before their infection (Huang et al. 2020). Conversely, patients with an acute COVID-19 infection had an increased risk of sleep disturbance. Among those, the severity of sleep quality was associated with a prolonged hospital stay and greater immune disturbances as measured by laboratory markers (Zhang et al. 2021).

Systematic reviews on children and adolescents during the COVID-19 pandemic reported increased early insomnia and also longer sleep duration that was correlated with increased levels of anxiety and depression (Chawla et al. 2021; Richter et al. 2022). Similarly, in adults, the restrictions imposed by the COVID-19 pandemic showed a similar pattern of increased sleep duration that was also attributed to the consequent mismatch between the biological internal rhythm and change in daily activities (Blume et al. 2020).

6.5 Sleep Quality

Beyond insomnia, patients with COVID-19 are also likely to have poor sleep quality. An observational study in Wuhan, China, found an association with time to seek medical help. Patients who sought medical care more than 3 days post the onset of symptoms reported a subjectively worse sleep quality and early insomnia (Sun et al. 2021). Sleep problems, broadly categorized, were also associated with older age, having a comorbid chronic illness, and subjective perception of the severity of the infection (Jiang et al. 2021). Zhang et al. published one of the few papers focusing on the effect of sleep quality in hospitalized patients with COVID-19 on their immune function recovery and prognosis. Two weeks of poor sleep quality during hospitalization was associated with an increased need for ICU hospitalization, slower recovery from lymphopenia, reflective of immune function status, and a longer duration of hospitalization by an average of 8 days (Zhang et al. 2020a). Researchers in Turkey surveyed 189 patients hospitalized with COVID-19 to study the relationship between sleep quality and mental health status. They found that patients with poor sleep quality required longer periods of hospitalization, had more clinical deterioration, and had a higher rate of depression than those with good sleep quality; the rate of anxiety was not affected between the two groups (Akıncı and Melek 2021). The isolation imposed on infected patients, stress, psychological distress, and the severity of dyspnea contribute to poor sleep quality in this population (Gungor et al. 2021). Stress disrupts sleep quality; it activates the hypothalamic-pituitary-adrenal (HPA) axis, increases cortisol release, and subsequently suppresses the body's immune response. Poor sleep quality could increase the risk of upper respiratory tract infections, while good sleep quality enhances the immune system (Zhang et al. 2020a).

Patients infected with COVID-19 complained of sleep disturbances regardless of the presence of infective symptoms, need for hospitalization, and intensive care unit (ICU) admission. Mahmoudi et al. explored the association between mental health, sleep problems, Post Traumatic Stress Disorder (PTSD), and health-related quality of life (HRQoL) in COVID-19 patients. They concluded that infected patients with sleep problems and PTSD had worse HRQoL (Mahmoudi et al. 2021). Compared to those with good sleep, patients with COVID-19 admitted to ICU with poorer sleep quality had worse outcomes in terms of a longer hospital stay, lower HRQoL, and lower recovery (Zhang et al. 2020a). A significantly longer period of hospitalization among those with poorer sleep quality compared to those with better sleep was also reported by a study in Turkey; those patients were also found to have a significantly higher incidence of depression (Ak and Bas 2021).

6.6 Other Sleep Disorders in COVID-19

6.6.1 Restless Leg Syndrome (RLS)

The association between RLS and COVID-19 infections has not been much reported beyond a few retrospective studies and case reports (Tony et al. 2020; Goldstein et al. 2021). However, the higher prevalence of stress, anxiety, and depression, and the medication used to treat them, are all contributing factors to either causing RLS or aggravating the condition in those with preexisting RLS. The discomfort associated with RLS can be relieved with movement. The confinement related to hospitalization and restricted movement is expected to worsen the symptoms (Bhat and Chokroverty 2022).

6.6.2 Obstructive Sleep Apnea (OSA)

Obstructive sleep apnea causes intermittent partial or total airway obstruction, resulting in difficulty breathing and sleep arousal. The disorder is associated with other medical comorbidities, including hypertension, obesity, depression, asthma, and others. Many of these were also identified as medical conditions associated with higher vulnerability and worse outcomes in patients infected with COVID-19 (Miller and Cappuccio 2021). Sleep deprivation, interruption, and intermittent hypoxia in OSA result in an inflammatory response that ultimately weakens the body's immune response to the viral infection. Treatment with Continuous Positive Airway Pressure (CPAP) improves the well-being of patients with OSA. However, CPAP is considered a high-risk aerosol-generating procedure and can place those in the vicinity at increased risk of viral infection. The reader is advised to refer to chap. 14 of this book for more detailed information on OSA and COVID-19.

6.6.3 Somnolence

Somnolence, or excessive daytime sleepiness, is the inability to remain awake during wakefulness, resulting in sleep. In comparing hospitalized and non-hospitalized patients with COVID-19 among a UK biobank cohort, the hospitalized group had more insomnia, excessive daytime sleepiness, and daytime napping. The authors report that daytime sleepiness was often associated with higher COVID-19 hospitalization, and sometimes daytime napping was associated with susceptibility to the virus and a higher risk of hospitalization, daytime napping being a sign as opposed to a cause of severe infection. They make the argument that metabolic pathways may be the link between somnolence and COVID-19 infections. Somnolence is significantly associated with KSR 2 gene, which is also linked to obesity and insulin resistance; both conditions are reported to contribute to the severity of COVID-19 (Liu et al. 2022). OSA is the most common cause of somnolence; the association of OSA with COVID-19 is briefly described above and will be addressed in more depth in another chapter of this book.

6.7 Narcolepsy

Narcolepsy is a sleep disorder characterized by increased daytime sleepiness, cataplexy, hallucinations, and sleep paralysis. The condition has been reported in association with previous pandemics. Increased incidence was reported in people infected with Influenza-A H1N1 in 2009 (Sarkanen et al. 2018). Current evidence supports Narcolepsy as a product of genetic and triggering environmental factors affecting Orexin neurons in the lateral hypothalamus. The immune response to certain infective pathogens has been suggested to detriment the function of Orexin neurons thus contributing to the development of Narcolepsy (Schirinzi et al. 2021; Bassetti et al. 2019). Earlier, during the COVID-19 pandemic, a paper was published suggesting the possibility of COVID-19 triggering Narcolepsy (Schirinzi et al. 2021). Since then, there has not been more reported incidence of Narcolepsy caused by this virus, except for isolated case reports (Deshpande et al. 2022). It is worthwhile investigating a possible link between Narcolepsy and COVID-19 in wider scale research projects.

6.8 Management Considerations

While sleep-related issues are widely reported among those acutely infected with COVID-19, clinical trials investigating improving sleep for those patients are scarce. Therefore, physicians managing sleep-related issues using specific patient-tailored approaches should follow clinical judgment per the most updated guidelines. Screening for sleep patterns, duration, quality, and insomnia is important in assessing patients with infectious diseases. It would allow targeted interventions that enhance the recovery process. Special consideration should be given to scrupulously investigating the underlying sleep complaints associated with COVID-19 in a patient-specific manner. The management of restless legs syndrome developing due to COVID-19 infection would not be identical to managing more commonly reported sleep consequences of COVID-19, such as insomnia (Tony et al. 2020; Mohiuddin et al. 2021).

Pharmacological options to manage sleep disturbances among COVID-19 patients are limited compared to the general population, especially given the increased risk of respiratory depression in symptomatic patients, which limits the use of several anxiolytics and sleep-aid medications for risks of respiratory depression. The natural hormone Melatonin, which helps regulate the sleep–wake cycle, is not new when managing respiratory infections and illnesses. Melatonin has been used as an adjunct medication in the Ebola outbreak and was shown to alleviate radiation-induced lung injury in mice models (Wu et al. 2019; Reiter et al. 2020). One clinical trial found that Melatonin improved sleep quality and oxygen saturation among those hospitalized for COVID-19 when given in addition to the standard treatment as opposed to standard therapy alone (Mousavi et al. 2022). It has also been recommended for use to prevent and treat sleep disturbances and delirium in ICU patients with COVID-19 (Zhang et al. 2020b). This is due to its good safety profile in therapeutic doses and based on previous research that associated melatonin with improved sleep and shorter duration of ICU stay as well as lower prevalence of delirium among patients in ICU (Zhang et al. 2019).

Non-pharmacological approaches to managing sleep disturbances in patients with COVID-19 have also been used. First developed by Edmund Jacobson in 1934, progressive muscle relaxation (PMR) attempts to help stressed people relieve distress by teaching them how to reduce muscle tension (Conrad and Roth 2007). It has since been reintroduced in more abbreviated ways and has shown usefulness in relieving anxiety, somatic pain, and disturbed sleep (Conrad and Roth 2007; Seyedi Chegeni et al. 2018). In a randomized controlled trial on COVID-19 patients admitted to the hospital, daily PMR sessions for five consecutive days were reported to improve the quality of sleep and anxiety symptoms (Liu et al. 2020). Another study demonstrated that PMR among hospitalized COVID-19 patients decreased sleep, anxiety, and mood disturbances (Xiao et al. 2020). One advantage of this technique is that it can be delivered online; this is particularly advantageous to patients with COVID-19 who are often required to self-isolate. Another advantage is that non-pharmacological methods help patients avoid medication side effects, especially among vulnerable populations with multiple comorbidities and increased sensitivity to medication interactions. However, a possible challenge towards adapting PMR is its need for patients to be able and willing to actively engage for 15–60 min (depending on the protocol used) and for multiple sessions; this might not be feasible for patients in severe symptomatic distress or altered cognition as a result of their primary COVID-19 illness.

References

Ak T, Bas HM (2021) Relationship between sleep quality and the psychological status of patients hospitalised with COVID-19. Sleep Med 80:167–170

Akıncı T, Melek Başar H (2021) Relationship between sleep quality and the psychological status of patients hospitalised with COVID-19. Sleep Med 80(January):167–170. https://doi.org/10.1016/j.sleep.2021.01.034

Bassetti CLA, Adamantidis A, Burdakov D et al (2019) Narcolepsy—clinical spectrum, aetiopathophysiology, diagnosis and treatment. Nat Rev Neurol 15(9):519–539. https://doi.org/10.1038/s41582-019-0226-9

Bhat S, Chokroverty S (2022) Sleep disorders and COVID-19. Sleep Med 91:253–261. https://doi.org/10.1016/j.sleep.2021.07.021

Blume C, Schmidt MH, Cajochen C (2020) Effects of the COVID-19 lockdown on human sleep and rest-activity rhythms. Curr Biol 30(14):R795–R797. https://doi.org/10.1016/j.cub.2020.06.021

Cellini N, Canale N, Mioni G, Costa S (2020) Changes in sleep pattern, sense of time and digital media use during COVID-19 lockdown in Italy. J Sleep Res 29(4):1–5. https://doi.org/10.1111/jsr.13074

Cénat JM, Blais-Rochette C, Kokou-Kpolou CK et al (2021) Prevalence of symptoms of depression, anxiety, insomnia, posttraumatic stress disorder, and psychological distress among populations affected by the COVID-19 pandemic: a systematic review and meta-analysis. Psychiatry Res 295:113599. https://doi.org/10.1016/J.PSYCHRES.2020.113599

Chawla N, Tom A, Sen MS, Sagar R (2021) Psychological impact of COVID-19 on children and adolescents: a systematic review. Indian J Psychol Med 43(4):294–299. https://doi.org/10.1177/02537176211021789

Conrad A, Roth WT (2007) Muscle relaxation therapy for anxiety disorders: it works but how? J Anxiety Disord 21(3):243–264. https://doi.org/10.1016/j.janxdis.2006.08.001

Czeisler CA (2015) Duration, timing and quality of sleep are each vital for health, performance and safety. Sleep Health 1(1):5–8. https://doi.org/10.1016/j.sleh.2014.12.008

de Sousa Martins e Silva E, Ono BHVS, Souza JC (2020) Sleep and immunity in times of COVID-19. Rev Assoc Med Bras 66(Suppl 2):143–147. https://doi.org/10.1590/1806-9282.66.S2.143

Deng J, Zhou F, Hou W et al (2021) The prevalence of depression, anxiety, and sleep disturbances in COVID-19 patients: a meta-analysis. Ann N Y Acad Sci 1486(1):90–111. https://doi.org/10.1111/NYAS.14506

Deshpande P, Thomas M, Tran L, Sampath R, Clark DR (2022) A rare case of narcolepsy following SARS-CoV-2 infection. ATS J:A4589–A4589. https://doi.org/10.1164/ajrccm-conference.2022.205.1_meetingabstracts.a4589

Fernandez-Mendoza J, Vgontzas AN (2013) Insomnia and its impact on physical and mental health. Curr Psychiatry Rep 15(12):418. https://doi.org/10.1007/s11920-013-0418-8

Ferrando M, Bagnasco D, Roustan V, Canonica GW, Braido F, Baiardini I (2016) Sleep complaints and sleep breathing disorders in upper and lower obstructive lung diseases. J Thorac Dis 8(8):E716–E725. https://doi.org/10.21037/jtd.2016.07.82

Goldstein CA, Rizvydeen M, Conroy DA et al (2021) The prevalence and impact of pre-existing sleep disorder diagnoses and objective sleep parameters in patients hospitalized for COVID-19. J Clin Sleep Med 17(5):1039–1050. https://doi.org/10.5664/JCSM.9132/SUPPL_FILE/JCSM.9132.DS001.PDF

Gungor S, Tosun B, Unal N, Dusak I (2021) Evaluation of dyspnea severity and sleep quality in patients with novel coronavirus. Int J Clin Pract 75(10):1–10. https://doi.org/10.1111/ijcp.14631

Hossain MM, Sultana A, Purohit N (2020) Mental health outcomes of quarantine and isolation for infection prevention: a systematic umbrella review of the global evidence. Epidemiol Health 42:1–11. https://doi.org/10.4178/epih.e2020038

Huang Y, Zhao N (2020) Generalized anxiety disorder, depressive symptoms and sleep quality during COVID-19 outbreak in China: a web-based cross-sectional survey. Psychiatry Res 288 (January):2–3

Huang B, Niu Y, Zhao W, Bao P, Li D (2020) Reduced sleep in the week prior to diagnosis of COVID-19 is associated with the severity of COVID-19. Nat Sci Sleep 12:999–1007. https://doi.org/10.2147/NSS.S263488

Jahrami H, BaHammam AS, Bragazzi NL, Saif Z, Faris M, Vitiello MV (2021) Sleep problems during the COVID-19 pandemic by population: a systematic review and meta-analysis. J Clin Sleep Med 17(2):299–313. https://doi.org/10.5664/JCSM.8930

Jiang ZY, Zhu PP, Wang LY et al (2021) Psychological distress and sleep quality of COVID-19 patients in Wuhan, a lockdown city as the epicenter of COVID-19. J Psychiatr Res 136:595–602. https://doi.org/10.1016/j.jpsychires.2020.10.034

Kokou-Kpolou CK, Megalakaki O, Laimou D, Kousouri M (2020) Insomnia during COVID-19 pandemic and lockdown: prevalence, severity, and associated risk factors in French population. Psychiatry Res 290:113128. https://doi.org/10.1016/j.psychres.2020.113128

Liu K, Chen Y, Wu D, Lin R, Wang Z, Pan L (2020) Effects of progressive muscle relaxation on anxiety and sleep quality in patients with COVID-19. Complement Ther Clin Pract 39:101132. https://doi.org/10.1016/j.ctcp.2020.101132

Liu Z, Luo Y, Su Y et al (2022) Associations of sleep and circadian phenotypes with COVID-19 susceptibility and hospitalization: an observational cohort study based on the UK Biobank and a two-sample Mendelian randomization study. Sleep 45(6):1–10. https://doi.org/10.1093/sleep/zsac003

Mahmoudi H, Saffari M, Movahedi M et al (2021) A mediating role for mental health in associations between COVID-19-related self-stigma, PTSD, quality of life, and insomnia among patients recovered from COVID-19. Brain Behav 11(5):1–8. https://doi.org/10.1002/brb3.2138

Miller MA, Cappuccio FP (2021) A systematic review of COVID-19 and obstructive sleep apnoea. Sleep Med Rev 55:101382. https://doi.org/10.1016/J.SMRV.2020.101382

Mohiuddin O, Khan AA, Shah SMI et al (2021) New-onset restless leg syndrome in a covid-19 patient: a case report with literature review. Pan Afr Med J 38:318. https://doi.org/10.11604/pamj.2021.38.318.28836

Morin CM, Carrier J, Bastien C, Godbout R (2020) Sleep and circadian rhythm in response to the COVID-19 pandemic. Can J Public Health 111(5):654–657. https://doi.org/10.17269/s41997-020-00382-7

Mousavi SA, Heydari K, Mehravaran H et al (2022) Melatonin effects on sleep quality and outcomes of COVID-19 patients: an open-label, randomized, controlled trial. J Med Virol 94(1):263–271. https://doi.org/10.1002/jmv.27312

Pappa S, Ntella V, Giannakas T, Giannakoulis VG, Papoutsi E, Katsaounou P (2020) Prevalence of depression, anxiety, and insomnia among healthcare workers during the COVID-19 pandemic: a systematic review and meta-analysis. Brain Behav Immun 88:901. https://doi.org/10.1016/J.BBI.2020.05.026

Partinen M, Holzinger B, Morin CM et al (2021) Sleep and daytime problems during the COVID-19 pandemic and effects of coronavirus infection, confinement and financial suffering: a multinational survey using a harmonised questionnaire. BMJ Open 11(12):e050672. https://doi.org/10.1136/BMJOPEN-2021-050672

Porras-Segovia A, Pérez-Rodrígeuz MM, López-Esteban P et al (2019) Contribution of sleep deprivation to suicidal behaviour: a systematic review, vol 44. Elsevier Ltd. https://doi.org/10.1016/j.smrv.2018.12.005

Reiter RJ, Ma Q, Sharma R (2020) Treatment of ebola and other infectious diseases: melatonin “goes viral”. Melatonin Res 3(1):43–57. https://doi.org/10.32794/mr11250047

Richter SA, Ferraz-Rodrigues C, Schilling LB, Camargo NF, Nunes ML (2022) Effects of the COVID-19 pandemic on sleep quality in children and adolescents: a systematic review and meta-analysis. J Sleep Res. https://doi.org/10.1111/jsr.13720

Sanghera J, Pattani N, Hashmi Y et al (2020) The impact of SARS-CoV-2 on the mental health of healthcare workers in a hospital setting—a systematic review. J Occup Health 62(1):1–16. https://doi.org/10.1002/1348-9585.12175

Sarkanen TO, Alakuijala APE, Dauvilliers YA, Partinen MM (2018) Incidence of narcolepsy after H1N1 influenza and vaccinations: systematic review and meta-analysis. Sleep Med Rev 38: 177–186. https://doi.org/10.1016/j.smrv.2017.06.006

Schirinzi T, Landi D, Liguori C (2021) COVID-19: dealing with a potential risk factor for chronic neurological disorders. J Neurol 268(4):1171–1178. https://doi.org/10.1007/s00415-020-10131-y

Schou TM, Joca S, Wegener G, Bay-Richter C (2021) Psychiatric and neuropsychiatric sequelae of COVID-19 – a systematic review. Brain Behav Immun 97(March):328–348. https://doi.org/10.1016/j.bbi.2021.07.018

Seyedi Chegeni P, Gholami M, Azargoon A, Hossein Pour AH, Birjandi M, Norollahi H (2018) The effect of progressive muscle relaxation on the management of fatigue and quality of sleep in patients with chronic obstructive pulmonary disease: a randomized controlled clinical trial. Complement Ther Clin Pract 31:64–70. https://doi.org/10.1016/j.ctcp.2018.01.010

Sher L (2020) COVID-19, anxiety, sleep disturbances and suicide. Sleep Med 70(January):124. https://doi.org/10.1016/j.sleep.2020.04.019

Sofi F, Cesari F, Casini A, Macchi C, Abbate R, Gensini GF (2014) Insomnia and risk of cardiovascular disease: a meta-analysis. Eur J Prev Cardiol 21(1):57–64. https://doi.org/10.1177/2047487312460020

Sun L, Yi B, Pan X et al (2021) PTSD symptoms and sleep quality of COVID-19 patients during hospitalization: an observational study from two centers. Nat Sci Sleep 13(April):1519–1531. https://doi.org/10.2147/NSS.S317618

Tononi G, Cirelli C (2014) Sleep and the price of plasticity: from synaptic and cellular homeostasis to memory consolidation and integration. Neuron 81(1):12–34. https://doi.org/10.1016/j.neuron.2013.12.025

Tony AA, Tony EA, Ali SB, Ezzeldin AM, Mahmoud AA (2020) COVID-19-associated sleep disorders: a case report. Neurobiol Sleep Circadian Rhythm 9:100057. https://doi.org/10.1016/J.NBSCR.2020.100057

Tsang HWH, Scudds RJ, Chan EYL (2004) Psychosocial impact of SARS. Emerg Infect Dis 10(7): 1326–1327

Voitsidis P, Gliatas I, Bairachtari V et al (2020) Insomnia during the COVID-19 pandemic in a Greek population. Psychiatry Res 289(January):113076. https://doi.org/10.1016/j.psychres.2020.113076

Wichniak A, Kania A, Siemiński M, Cubała WJ (2021) Melatonin as a potential adjuvant treatment for COVID-19 beyond sleep disorders. Int J Mol Sci 22(16):8623. https://doi.org/10.3390/IJMS22168623

Wu X, Ji H, Wang Y et al (2019) Melatonin alleviates radiation-induced lung injury via regulation of miR-30e/NLRP3 axis. Oxidative Med Cell Longev 2019:4087298. https://doi.org/10.1155/2019/4087298

Xiao CX, Lin YJ, Lin RQ, Liu AN, Zhong GQ, Lan CF (2020) Effects of progressive muscle relaxation training on negative emotions and sleep quality in COVID-19 patients: a clinical observational study. Medicine (Baltimore) 99(47):e23185. https://doi.org/10.1097/MD.0000000000023185

Xiong J, Lipsitz O, Nasri F, Lui LMW, Gill H, Phan L (2020) Impact of COVID-19 pandemic on mental health in the general population: a systematic review. J Affect Disord 277 (January):55–64

Zambrelli E, Canevini M, Gambini O, D'Agostino A (2020) Delirium and sleep disturbances in COVID-19: a possible role for melatonin in hospitalized patients? Sleep Med 70:111. https://doi.org/10.1016/j.sleep.2020.04.006

Zhang Q, Gao F, Zhang S, Sun W, Li Z (2019) Prophylactic use of exogenous melatonin and melatonin receptor agonists to improve sleep and delirium in the intensive care units: a systematic review and meta-analysis of randomized controlled trials. Sleep Breath 23(4): 1059–1070. https://doi.org/10.1007/s11325-019-01831-5

Zhang J, Xu D, Xie B, Zhang Y, Huang H, Liu H (2020a) Poor-sleep is associated with slow recovery from lymphopenia and an increased need for ICU care in hospitalized patients with COVID-19: a retrospective cohort study. Brain Behav Immun 88:50–58

Zhang R, Wang X, Ni L et al (2020b) COVID-19: melatonin as a potential adjuvant treatment. Life Sci 250(March):117583. https://doi.org/10.1016/j.lfs.2020.117583

Zhang L, Li T, Chen L et al (2021) Association of sleep quality before and after sars-cov-2 infection with clinical outcomes in hospitalized patients with covid-19 in China. EXCLI J 20:894–906. https://doi.org/10.17179/excli2021-3451

Chapter 7
Sleep Patterns and Sleep Disturbances During the Lockdown Periods

Federico Salfi and Michele Ferrara

Abstract Since December 2019, the spread of acute respiratory syndrome (COVID-19) due to the SARS-CoV-2 virus has posed unprecedented challenges to the international healthcare systems. The uncontrollable increase in contagions, hospitalizations, and deaths forced governments around the globe to implement severe restriction policies, known as *lockdowns*. During the lockdown periods, people were forced to stay at home, face-to-face social interactions were forbidden, and most school and work activities were mandatorily closed. This stressful scenario dramatically affected the daily routine of the world population, impacting the sleep health and habits of hundreds of millions of people.

In this chapter, we provided a general overview of the multifaceted changes in human sleep due to the lockdown periods. We described the negative consequences on sleep quality and insomnia symptoms among the general population, outlining possible demographic and psychological risk factors. We discussed the pervasive changes in sleep duration and sleep schedules, the global reduction in *social jet lag*, and how all these factors interacted with the different circadian typologies. Finally, we discussed the consequences of working and schooling adjustments during the self-confinement period, as well as the repercussions of reduced daylight exposure and the concomitant increase in screen time worldwide. Besides the obvious negative impact on sleep health, the lockdown periods gave birth to the broadest natural experiment in human history, representing a unique opportunity to deepen our knowledge about the role of the societal and environmental contexts on sleep behavior.

Keywords COVID-19 lockdown · Sleep health · Social jet lag · Chronotype · Zeitgebers

---

F. Salfi · M. Ferrara (✉)
Department of Biotechnological and Applied Clinical Sciences, University of L'Aquila, L'Aquila, Italy
e-mail: Federico.salfi@graduate.univaq.it; Michele.ferrara@univaq.it

A. S. BaHammam et al. (eds.), *COVID-19 and Sleep: A Global Outlook*, Progress in Sleep Research, https://doi.org/10.1007/978-981-99-0240-8_7

7.1 Lockdown and Sleep: The Perfect Storm

The SARS-CoV-2 virus started to spread in China in the last months of 2019. This virus led to an acute respiratory syndrome associated with a potentially life-threatening pneumonia disease (COVID-19). In the subsequent months, the contagions increased worldwide, giving rise to a global pandemic (Platto et al. 2021). Governments around the world promptly reacted to this situation by applying unprecedented containment measures to counteract the increasing contagion, hospitalization, and death rates. The set of extreme restraining measures became known as *lockdown*, consisting of home confinement, social distancing, and the mandatory closure of schools and most work activities. The pervasive impact of these measures on the lifestyle of the general population was associated with generalized alterations in main *zeitgebers*. The nationwide lockdown periods led to reduced sunlight exposure, social interactions, and physical activity (Altena et al. 2020). Meanwhile, the daily use of smartphones and computers dramatically increased to compensate for the limited face-to-face interactions, spend longer free time, and relieve boredom (Trott et al. 2022). In this scenario, millions of workers and students began to work from home (Eurofound 2020; Brynjolfsson et al. 2020), while fears of contagion, economic and employment uncertainties, and worries about the well-being of loved ones burdened everyday life, leading to increased distress and anxiety levels (Rajkumar 2020). It did not take long to realize that all the abovementioned factors would have represented the perfect storm to impact the sleep health/habits of the world population (O'regan et al. 2021).

7.2 Sleep Disturbances During the Lockdown

Since the early months of the pandemic, a considerable effort by the international scientific community has been conducted to understand the consequences of the lockdown periods on human sleep. However, the restraining measures led to the closure of research laboratories worldwide, and the only way to collect data during social distancing consisted of online surveys. Despite the intrinsic limitations of self-reported questionnaires (selection and response biases), the massive literature developed during the first pandemic phase provided a crucial contribution to the advancement of knowledge about the lockdown consequences.

Italy was the first Western country to deal with the pandemic and to apply a total lockdown, and the first evidence of the effect of the self-isolation period on sleep was addressed by an Italian study (Cellini et al. 2020). Through an online survey, the authors evaluated the sleep quality of 1310 young Italian citizens, retrospectively comparing the second week of lockdown with the previous month. The investigation showed decreased sleep quality, with an increase in poor sleep rates from 40.5% to 52.4% based on the Pittsburgh Sleep Quality Index (PSQI) (Buysse et al. 1989). In the subsequent months, the detrimental effect of the self-isolation period was systematically confirmed in Italy (Gualano et al. 2020; Casagrande et al. 2020). One of the most extensive studies (Salfi et al. 2021c) on 13,989 participants reported alarming rates of poor sleepers (61.1%) and people with at least subthreshold insomnia (52.8%), based on PSQI and Insomnia Severity Index (ISI) (Bastien et al. 2001), respectively. As the virus spread worldwide, international scientific interest increased in studying the repercussion of stay-at-home orders on sleep. Blume and colleagues (2020) showed reduced sleep quality in three European countries (Austria, Germany, and Switzerland). Increased sleep disturbances were reported in Greece (Voitsidis et al. 2020), France (Kokou-Kpolou et al. 2020), Spain (Dal Santo et al. 2021), Belgium (Cellini et al. 2021), India (Gupta et al. 2020), China (Li et al. 2020; Lin et al. 2021), Brazil (Taporoski et al. 2022), the US (Mandelkorn et al. 2021), Argentina (Valiensi et al. 2022), as well as by multinational studies (Mandelkorn et al. 2021; Yuksel et al. 2021).

In response to this emergent literature, some studies summarized the results coming from all over the globe. The first systematic review and meta-analysis on this topic (Jahrami et al. 2021) was carried out on 44 papers involving a total of 54,231 participants from 13 countries. This study estimated a pooled prevalence rate of sleep problems among all populations of 35.7%. A more recent systematic review and meta-analysis (Jahrami et al. 2022), covering the first 6 months of the pandemic, analyzed 250 papers comprising 493,475 participants from 49 countries. This study revealed a global prevalence of sleep disturbances of 40.5%, showing that sleep disturbances were higher during lockdown (42.5%) compared to no lockdown periods (38.0%). Finally, to provide a global overview of insomnia symptoms, a systematic review, and individual participant data meta-analysis were specifically performed on research involving the ISI (AlRasheed et al. 2022). This study analyzed 48 studies from 25 countries comprising 133,006 respondents, showing that the pooled estimate of insomnia symptoms (subthreshold and clinically significant) was 52.6%. Specifically, 16.7% of the population suffered from clinically significant insomnia, of which 13.8% suffered from moderate insomnia, and 2.5% suffered from severe insomnia.

The same meta-analyses tried to clarify potential risk factors for the development of sleep disturbances. One of these studies focused on possible differences between genders, reporting similar prevalence rates of sleep problems during the lockdown periods (Alimoradi et al. 2022). Notably, this evidence was supported by other meta-analyses that confirmed the absence of a moderating role of sex in the global prevalence rates of sleep issues (Alimoradi et al. 2021; Jahrami et al. 2021, 2022) and insomnia symptoms (AlRasheed et al. 2022). These findings are inconsistent with the pre-pandemic literature as sleep problems such as insomnia were typically more common among women (Zhang and Wing 2006). However, the closure of the gender gap could be specifically ascribable to the confinement situation, as shown by a longitudinal study on 2701 Italian adults highlighting a specific gender-related time course of sleep disturbances during prolonged self-isolation (Salfi et al. 2020). Comparing the third and the seventh weeks of lockdown, men reported decreased sleep quality and exacerbated insomnia symptoms, while women relieved insomnia severity. In this scenario, the higher prevalence of moderate/severe insomnia conditions among females was no longer present with the extension of the confinement period.

Meta-analytic studies also confuted a possible role of age in predicting sleep problems during the first pandemic phase (Alimoradi et al. 2021, 2022; AlRasheed et al. 2022; Jahrami et al. 2022). Again, this finding was unexpected as aging is typically associated with increased sleep disturbances (Li et al. 2018; Patel et al. 2018). The absence of moderation effects of age could be driven by a specific vulnerability to sleep problems of the youngest people (Amicucci et al. 2021) and the student population under confinement (Jahrami et al. 2022). During the first contagion wave, healthcare professionals were the frontline workers in dealing with the emergency. The increased stressful workload, accompanied by higher contagion risk and irregular work schedules, could have led them to experience acute sleep disturbances. Some meta-analyses confirmed this idea, showing that sleep problems were most prevalent in healthcare workers (Alimoradi et al. 2021; Cénat et al. 2021; Jahrami et al. 2022). However, the literature in the field is heterogenous, and some reports failed to highlight statistically significant differences (AlRasheed et al. 2022), showing only numerically higher rates of sleep disturbances in healthcare professionals compared to the general population (Jahrami et al. 2021; Alimoradi et al. 2022).

The raised sleep issues under confinement were complemented by higher levels of mental health problems and decreased psychological well-being (Rajkumar 2020; Vindegaard and Benros 2020). Considering the bidirectionality between sleep and mental health (Alvaro et al. 2013), psychological distress could have played a crucial role in fostering sleep disturbances during the first contagion wave of COVID-19. A systematic review and meta-analysis of 177 papers comprising 345,270 participants from 39 countries addressed this topic, evaluating the relationship between sleep problems and psychological distress during the first months of the pandemic (Alimoradi et al. 2021). The study revealed a moderate positive association between sleep disturbances and the severity of depressive and anxiety symptoms, emphasizing the need for effective programs treating mental health to improve pandemic-related sleep problems and *vice versa*.

Overall, the literature supported the evidence of a general increase in sleep disturbances in the worldwide population due to the exceptional measures applied to contrast the virus spread. In this scenario, the concomitant psychological distress could have significantly contributed to the development and exacerbation of sleep problems.

7.3 The Sleep Duration Paradox

Although the stressful situation of self-confinement led to increased sleep problems among the general population, several studies reported a clear dissociation between trends of quality and quantity of sleep during the lockdowns. Before the COVID-19 outbreak, modern societies were already dealing with a sleep-loss epidemic (CDC

2015). The social and technological revolutions in most industrialized nations were associated with a decline in sleep duration across the last decades (Bixler 2009). In this view, the loosening of social obligations during the lockdown seemed to paradoxically unlock more time for sleep (Kantermann 2020). One of the first studies to address this issue was by Leone and colleagues (2020), who analyzed data from 1021 respondents that completed questionnaires before and during the home confinement period. This study showed longer sleep duration during lockdown weekdays, with only 37.3% of participants not reaching the recommended 7 h of sleep (Hirshkowitz et al. 2015) during the quarantine compared to 60.2% during the pre-pandemic period. Consistently, a larger multinational investigation (Global Chrono Corona Survey) on 7517 respondents from 40 countries described longer sleep duration on workdays by 26 min than before COVID-19-mandated social restrictions (Korman et al. 2020). This phenomenon was systematically reported by several studies on adult populations (Blume et al. 2020; Wright et al. 2020; Cellini et al. 2020), and similar results were reported by research focused on younger people. A study involving 17,000 school-aged children and adolescents in the UK during the first national lockdown showed that participants reported longer sleep duration than in 2019, with a maximal improvement in younger secondary school students (+45 min) (Illingworth et al. 2022). These results were objectively confirmed by research that analyzed large amounts of smartphone users' data. Robbins and coworkers (2021) examined 2.9 million nights of sleep recordings from different continents, suggesting the lockdown periods were associated with a 20-min sleep extension worldwide compared with the previous year.

Further confirmations of the lockdown-induced sleep extension came from longitudinal studies carried out during different pandemic phases. Salfi and colleagues (2022a) longitudinally surveyed 1062 Italian people from the first lockdown to 2 years later, providing a compelling countercheck of the *sleep duration paradox*. The authors showed that, as the COVID-19 mitigation strategies were gradually loosened, sleep disturbances improved, while sleep duration decreased (−20 min). Reduced sleep times after the lockdowns were confirmed by other studies using wearable sleep/activity trackers (Massar et al. 2021) and by investigations that analyzed large datasets from thousands of smartphone users (Yuan et al. 2022).

In sum, this literature supported the idea that lockdowns provided people worldwide increased opportunities to sleep, suggesting the existence of a societal sleep deficit during pre-pandemic times. However, it was a transitory phenomenon that tended to disappear with the gradual resolution of the emergency.

7.4 Changes in Sleep Schedule and Social Jet lag

One of the most striking consequences of the lockdown periods was a pervasive shift in sleep schedules. The international literature consistently found delayed sleep timing, perhaps because of the relaxed social time pressure under self-confinement. An Italian survey of approximately 14,000 respondents showed that 59% of the sample reported delayed bedtimes and 63% delayed wake-up times during lockdown (Salfi et al. 2021c). The same studies that demonstrated the sleep extension effect also shed light on when people slept during the lockdown. In a sample of young workers, Cellini and colleagues (2020) demonstrated later bedtimes (+41 min) and even later wake-up times (+73 min). Similar results were obtained by investigations on a large sample of children and adolescents (Illingworth et al. 2022), as well as among university student populations (Wright et al. 2020; Marelli et al. 2021). Later sleep timing than in 2019 was confirmed by a global analysis of almost 65,000 users of the "*Sleep As Android*" smartphone application, with the extent of changes strictly linked with the progression of the emergency in each country (Yuan et al. 2022).

During home confinement, people almost completely limited their daily social activity, and millions of workers and schoolers began to work from home with more flexibility in working hours. In this view, the lockdown represented a unique opportunity to reduce the discrepancy between the social/working clocks and the endogenous day and night rhythms (*social jet lag*) (Wittmann et al. 2006). Several studies demonstrated that the changes in sleep schedule were significantly different between weekdays and weekends, and the most evident variations were found on weekdays (Korman et al. 2020; Leone et al. 2020; Blume et al. 2020; Wright et al. 2020). A large international study collecting data from 40 countries showed delayed midsleep time on workdays and free days by 50 and 22 min, respectively. These outcomes were objectively confirmed by analyzing the nocturnal sleep patterns of ~113,000 consumer sleep tracker users from 20 countries (Ong et al. 2021). This study found later midsleep times, particularly on weekdays, whose extent was greater with increasing stringency of confinement measures. Based on this evidence, the current literature is consistent in indicating that the lockdown periods led to better regularity of sleep timing worldwide, alleviating the social jet lag phenomenon (Korman et al. 2020; Leone et al. 2020; Blume et al. 2020; Wright et al. 2020).

The confirmation of the lockdown-related delay of sleep times and reduced social jet lag also came from studies evaluating people longitudinally after lifting the stay-at-home orders. Salfi and colleagues (2021b) evaluated the variations of sleep schedules in a large Italian sample (2013 adults), showing that the participants went to sleep 36 min earlier and woke up 56 min earlier during a subsequent period of lighter restraining measures compared to the first lockdown. Moreover, a study using wearable sleep trackers showed an immediate increase in social jet lag as a consequence of the cessation of the isolation measures (Massar et al. 2021).

In conclusion, studies across multiple societies and different population groups showed a substantial shift to later sleep schedules during the lockdown, and the effect was more prominent on weekdays. This effect temporarily allowed people to comply with their circadian rhythm, leading to a transitory reduction of social jet lag.

7.5 The Chronotype Matter

The lifted societal demands under home confinement led to a worldwide reduction in social jet lag (Korman et al. 2020; Leone et al. 2020; Blume et al. 2020; Wright et al. 2020). This phenomenon is intrinsically linked with the circadian typology concept, with evening-types experiencing the most pronounced misalignment between social and biological clocks in modern society (Roenneberg et al. 2019). In this view, the lockdown periods could have unevenly affected the sleep patterns of different chronotypes.

The first to address this topic was a large Italian study that surveyed 13,989 adults (Salfi et al. 2021c). The sample was divided into morning-type (MT, 23.3%), intermediate-type (IT, 65.6%), and evening-type people (ET, 11.1%) according to the Morningness-Eveningness questionnaire-reduced version (MEQr) criteria (Adan and Almirall 1991). The study found a higher prevalence of delayed bedtime (75.3%) and wake-up time (77.1%) in ET compared with MT people (44.3% and 49.8%, respectively). In stark contrast with pre-pandemic literature (Adan et al. 2012), ET slept longer than MT. However, a higher rate of ET respondents reported a negative impact of the confinement measures on sleep than the MT group (69.0% vs. 50.4%), and eveningness was associated with lower sleep quality and more severe insomnia symptoms. The vulnerability of late chronotypes during the lockdown periods was subsequently confirmed by a multinational investigation of 19,267 adults from 15 countries (Merikanto et al. 2022). The authors showed specific delayed sleep timing and increased sleep duration in the ET group under confinement compared with a retrospective pre-lockdown assessment. This effect led to the disappearance of the well-known differences in sleep duration between chronotypes. On the other hand, the same study highlighted that sleep problems (e.g., insomnia symptoms, nightmares, and daytime sleepiness) primarily increased among evening-types. A specific benefit on sleep duration of *night owls* was confirmed by another investigation on 610 US adults (Bottary et al. 2022) that found a stronger lockdown-related decrease in social sleep restriction (the difference between weekend and weekday sleep duration) among the ET population. Remarkably, the only investigation providing an objective sleep evaluation reported similar results (Pépin et al. 2021). Using data from 599 adults collected by a commercial EEG wearable headband, the authors compared confinement and the pre-lockdown period, demonstrating a larger shift to later sleep timing among the ET people and no difference in sleep duration. Furthermore, eveningness preference was associated with a higher increase in REM sleep as a consequence of the longer sleep duration. Remarkably, looking at subsequent pandemic phases, a longitudinal study on 1062 Italian adults showed that the time course of sleep duration changes differed between chronotypes (Salfi et al. 2022a). The authors showed that the gradual loosening of restraining measures led ET people to sleep less and less than morning-types across the subsequent 2 years from lockdown.

Overall, the literature described distinct changes in sleep patterns according to circadian typology. The lockdown period allowed ET people to sleep longer and more aligned with their endogenous clock (decreased social jet lag), suggesting a preexisting sleep deficit in this population and greater susceptibility to pandemic-related societal changes. Meanwhile, these benefits were over-weighted by a concomitant worsening of sleep disturbances. On the other hand, the MT population was characterized by a more stable sleep schedule, with morningness suggested as a protective factor against the development of sleep disturbances during home confinement.

7.6 The Impact of Working/Schooling Adjustments

The societal changes imposed by the COVID-19 outbreak radically disrupted the labor market worldwide. Millions of people suspended their work, while most of the general population began working from home for the first time (Brynjolfsson et al. 2020; Eurofound 2020). Home working removed the need to commute to the workplace and could be characterized by higher flexibility of working schedules. This situation drastically affected the daily routine as well as the sleep rhythms of workers, giving rise to an unprecedented natural experiment to understand how sleep changes when work hours and work environments change.

In one study, Leone and colleagues analyzed a subgroup of their sample that reported working from home, compared with people continuing to reach the workplace during the stay-at-home orders (Leone et al. 2020). The authors showed a specific benefit of working from home, as this condition was associated with longer sleep time, lower social jet lag, and delayed sleep timing compared to the pre-pandemic assessment. Moreover, the study found an increased prevalence of remote workers reaching the minimum recommended 7 h of sleep on weekdays during the lockdown (from 39% to 65%). In contrast, this prevalence was stable among those who continued to work outside (from 27% to 32%). A subsequent investigation provided similar results by surveying a large Italian sample (Salfi et al. 2021c). The authors demonstrated better sleep quality, lower insomnia severity, longer sleep duration (+27 min), and later bedtime and wake-up time among 3536 adults working from home compared with 1675 respondents who reached the workplace. Raman and Coogan (2021) confirmed these findings in a group of 797 Irish adults, showing that remote working during restrictions led to a more delayed midsleep time, longer workday sleep duration, and a more marked reduction of social jet lag than "essential" workers who continued to attend their workplace. However, the benefits of remote working may not have involved all circadian typologies in the same way, considering that the effect of working schedules on sleep quality and duration seems to depend on individual chronotypes (Juda et al. 2013; Vetter et al. 2015). In line with this idea, an investigation on 675 office workers and 265 home workers demonstrated that the effects of the working adjustments due to the COVID-19 outbreak were not generalizable to the whole population, highlighting better sleep quality, reduced insomnia symptoms, and longer sleep duration specifically among the *night owls* (Salfi et al. 2022b).

Another main consequence of the confinement measures was a pervasive upheaval in the school community. In-presence lessons were suspended, and remote learning became the norm for millions of students. Although homeschooling continued to be characterized by fixed schedules, removing morning commutes could have facilitated young people to follow their endogenous circadian rhythm, typically oriented to the eveningness (Roenneberg et al. 2019), with possible implications for their sleep health. This idea finds its roots in a large body of evidence demonstrating the beneficial effects of later school start time on the younger population's sleep (e.g., Alfonsi et al. 2020).

A Brazilian longitudinal study addressed this topic, comparing sleep measures collected during the lockdown with measures collected 1 year before among a sample of 259 high school adolescents (Santos and Louzada 2022). The authors found that students during remote classes slept later, spent more time in bed (+152 min), and reported lower daily sleepiness compared to the pre-pandemic assessment. Moreover, the nap habits decreased during the lockdown as a consequence of increased sleep duration at nighttime. Consistently, another longitudinal study on 94 high school students showed substantially delayed sleep schedules during the pandemic. Moreover, participants who slept less than 7.4 h before the COVID-19 outbreak reported longer sleep duration (+30 min) and improved sleep quality during school closures (Genta et al. 2021). Notably, similar findings were obtained by a study that evaluated sleep patterns using wrist actigraphy and sleep diaries, confirming later bedtime and wake-up time, longer sleep duration (+22 min), and lower sleepiness during school days from home (Stone et al. 2021).

In conclusion, working and schooling adjustments due to the COVID-19 outbreak compromised the role of one of the crucial *social zeitgebers*, leading to a general delay in sleep schedules. However, this scenario allowed both workers and students to sleep longer and better, suggesting a detrimental effect of working/schooling obligations in pre-pandemic times, particularly among the so-called *night owls*.

7.7 The Role of Daylight Exposure

The sun's daily cycle substantially affects the circadian clock, sleep, and alertness. Ambient light is the most important *zeitgeber* and plays a crucial role in human physiology, entraining the circadian cycle to local time (Blume et al. 2019). Moreover, environmental light is intimately involved in the daily regulation of melatonin, a key sleep-promoting pineal gland hormone (Brown 1994). High-intensity daylight exposure is beneficial for sleep, and some studies showed that low light levels could impact sleep quality (Boubekri et al. 2014; Figueiro et al. 2017), lead to shorter sleep duration (Boubekri et al. 2014), cause longer sleep-onset latency (Figueiro et al. 2017), and interfere with sleep architecture (Wams et al. 2017).

During lockdown periods, the freedom of movement was substantially reduced as people were allowed to leave their homes for limited purposes, e.g., shopping for necessities and reaching the workplace. Consequently, one of the main implications of the stay-at-home orders was a large-scale reduction of daylight exposure. This situation could have interfered with sleep health and the circadian clock, and international agencies immediately recommended ensuring adequate daylight exposure during self-isolation (Morin et al. 2020; Altena et al. 2020). However, under strict confinement, many citizens could have hardly followed the advice, especially those living in homes with limited outside areas and small windows. Despite the significance of this topic, few studies were performed in this field. Blume and colleagues (2020) addressed this issue by surveying 435 adults and retrospectively comparing the lockdown scenario with a pre-confinement period. Their study showed decreased sleep quality under quarantine, but higher levels of daylight exposure buffered the sleep quality impairment and were associated with longer sleep time. An online survey performed during New Zealand's lockdown on 723 adults confirmed these results, reporting that people spent 1 h less per day under the open sky. The reduced daylight exposure predicted worsened sleep quality compared with a pre-pandemic assessment (Gibson et al. 2022). Similar findings were obtained by the Global Chrono Corona Survey (Korman et al. 2022). The authors demonstrated a 1-h median decrease in outdoor light exposure among 7517 respondents during social restrictions. This variation was significantly associated with reduced sleep quality than the pre-lockdown period.

Besides sunlight, artificial light also plays a role in regulating sleep rhythms, with positive effects of high levels of office lights in the morning (Figueiro et al. 2017). However, indoor home light levels are generally lower than those for commercial office or school spaces, and some evidence suggested that staying in a bright room indoors may counteract sleep disturbances and sleep-related impairments under confinement (Figueiro et al. 2021).

In sum, the literature consistently demonstrated that reduced daylight exposure due to home confinement was an important contributory factor in explaining the raised sleep disturbances worldwide. Spending more time outdoors, when possible, represented an effective strategy to mitigate the detrimental effects of home confinement.

7.8 The Screen Time Effect

The imposition of social distancing measures and the limitations of outdoor activities during the lockdowns had an inevitable consequence around the world: the massive use of digital devices. Self-confined people substantially increased the time spent on social networks and video calling to compensate for the limited face-to-face interactions. The world population has begun to spend more and more time facing television or using the Internet to occupy their growing free time and fight boredom. Furthermore, hundreds of millions of people began to work or attend school from home, leading to an unprecedented daily use of computers and tablets. All these factors led to a worldwide increase in screen time. A systematic review and meta-analysis of 89 studies (Trott et al. 2022) tried to summarize this phenomenon, indicating that the total screen time and leisure screen time (non-work/non-academic) of the adult population increased by 1 h/day and 0.7 h/day during the pandemic, respectively.

Notwithstanding that the use of computers, smartphones, tablets, and televisions may have helped to deal with the stressful confinement situation, the pre-pandemic literature consistently described a negative impact of evening screen exposure on sleep health. Several studies showed alerting effects of lights emitted by modern electronic devices by dampening melatonin release (Cajochen et al. 2011; Green et al. 2017).

Salfi and coworkers (2021a) were the first to address a possible impact of the increased evening screen exposure during the lockdown on sleep. The authors longitudinally collected information on 2123 participants during the 3rd and the 7th week of home confinement, showing that the variations of screen habits before bedtime played a critical role in the time course of sleep problems under self-isolation. The study demonstrated that increased screen time led to lower sleep quality, more severe insomnia symptoms, shorter sleep duration, prolonged sleep-onset latency, and delayed bedtime and get-up time. On the other hand, people who decreased their exposure to electronic devices reported the opposite pattern of outcomes. In line with these results, a recent systematic review of 18 articles confirmed the detrimental role of screen exposure on sleep duration and quality during the COVID-19 pandemic (Drumheller and Fan 2022). Notably, the pandemic-related upsurge in screen time was reported in all age groups (Trott et al. 2022), and some studies confirmed the association between the overuse of electronic devices and sleep problems also in the younger population (Bruni et al. 2021).

Another way that evening exposure to electronic devices could have impaired sleep is through the arousing and exciting effects of some screen-mediated contents (Higuchi et al. 2005). During the lockdown periods, the media have played an essential role in advising people about adequate prevention behaviors. However, constant overexposure to ever-changing news and the overload of information consisting of potentially traumatic content and dramatic news may have contributed to deteriorating sleep by fueling anxiety and distress (Gao et al. 2020). In this view, an online survey of 1005 adults representative of the French population found that overexposure to media content about COVID-19 emergency was linked with aggravated and severe sleep problems with daytime impairment and/or sleeping pill use during the self-isolation situation (Léger et al. 2020).

Overall, the current literature supported the assumption that the large-scale increased screen times during the self-isolation period negatively impacted the general population's sleep health, also affecting their sleep schedule. The light *per se* and the COVID-19-related contents could have interfered with sleep patterns, simultaneously intervening in physiological and cognitive mechanisms.

References

Adan A, Almirall H (1991) Horne & Östberg morningness-eveningness questionnaire: a reduced scale. Pers Individ Dif 12:241–253. https://doi.org/10.1016/0191-8869(91)90110-W

Adan A, Archer SN, Hidalgo MP et al (2012) Circadian typology: a comprehensive review. Chronobiol Int 29:1153–1175. https://doi.org/10.3109/07420528.2012.719971

Alfonsi V, Scarpelli S, D'Atri A et al (2020) Later school start time: the impact of sleep on academic performance and health in the adolescent population. Int J Environ Res Public Health 17:2574. https://doi.org/10.3390/ijerph17072574

Alimoradi Z, Broström A, Tsang HWH et al (2021) Sleep problems during COVID-19 pandemic and its' association to psychological distress: a systematic review and meta-analysis. EClinicalMedicine 36:100916. https://doi.org/10.1016/j.eclinm.2021.100916

Alimoradi Z, Gozal D, Tsang HWH et al (2022) Gender-specific estimates of sleep problems during the COVID-19 pandemic: systematic review and meta-analysis. J Sleep Res 31:1–16. https://doi.org/10.1111/jsr.13432

AlRasheed MM, Fekih-Romdhane F, Jahrami H et al (2022) The prevalence and severity of insomnia symptoms during COVID-19: a global systematic review and individual participant data meta-analysis. Sleep Med 100:7–23. https://doi.org/10.1016/j.sleep.2022.06.020

Altena E, Baglioni C, Espie CA et al (2020) Dealing with sleep problems during home confinement due to the COVID-19 outbreak: practical recommendations from a task force of the European CBT-I Academy. J Sleep Res 29:e13052. https://doi.org/10.1111/jsr.13052

Alvaro PK, Roberts RM, Harris JK (2013) A systematic review assessing bidirectionality between sleep disturbances, anxiety, and depression. Sleep 36:1059–1068. https://doi.org/10.5665/sleep.2810

Amicucci G, Salfi F, D'atri A et al (2021) The differential impact of COVID-19 lockdown on sleep quality, insomnia, depression, stress, and anxiety among late adolescents and elderly in Italy. Brain Sci 11:1136. https://doi.org/10.3390/brainsci11101336

Bastien CH, Vallières A, Morin CM (2001) Validation of the Insomnia Severity Index as an outcome measure for insomnia research. Sleep Med 2:297–307. https://doi.org/10.1016/S1389-9457(00)00065-4

Bixler E (2009) Sleep and society: an epidemiological perspective. Sleep Med 10:S3–S6. https://doi.org/10.1016/j.sleep.2009.07.005

Blume C, Garbazza C, Spitschan M (2019) Effects of light on human circadian rhythms, sleep and mood. Somnologie 23:147–156. https://doi.org/10.1007/S11818-019-00215-X

Blume C, Schmidt MH, Cajochen C (2020) Effects of the COVID-19 lockdown on human sleep and rest-activity rhythms. Curr Biol 30:R795. https://doi.org/10.1016/j.cub.2020.06.021

Bottary R, Fields EC, Kensinger EA, Cunningham TJ (2022) Age and chronotype influenced sleep timing changes during the first wave of the COVID-19 pandemic. J Sleep Res 31:e13495. https://doi.org/10.1111/JSR.13495

Boubekri M, Cheung IN, Reid KJ et al (2014) Impact of windows and daylight exposure on overall health and sleep quality of office workers: a case-control pilot study. J Clin Sleep Med 10:603–611. https://doi.org/10.5664/JCSM.3780

Brown GM (1994) Light, melatonin and the sleep-wake cycle. J Psychiatry Neurosci 19:345

Bruni O, Giallonardo M, Sacco R et al (2021) The impact of lockdown on sleep patterns of children and adolescents with ADHD. J Clin Sleep Med 17:1759–1765. https://doi.org/10.5664/jcsm.9296

Brynjolfsson E, Horton JJ, Ozimek A, et al (2020) COVID-19 and remote work: an early look at US data. https://doi.org/10.3386/W27344

Buysse DJ, Reynolds CF, Monk TH et al (1989) The Pittsburgh sleep quality index: a new instrument for psychiatric practice and research. Psychiatry Res 28:193–213. https://doi.org/10.1016/0165-1781(89)90047-4

Cajochen C, Frey S, Anders D et al (2011) Evening exposure to a light-emitting diodes (LED)-backlit computer screen affects circadian physiology and cognitive performance. J Appl Physiol 110:1432–1438. https://doi.org/10.1152/japplphysiol.00165.2011

Casagrande M, Favieri F, Tambelli R, Forte G (2020) The enemy who sealed the world: effects quarantine due to the COVID-19 on sleep quality, anxiety, and psychological distress in the Italian population. Sleep Med 75:12–20. https://doi.org/10.1016/j.sleep.2020.05.011

CDC Center for Disease Control and Prevention (2015) Insufficient sleep is a public health problem. https://www.cdc.gov/sleep/index.html. Accessed 30 Sept 2022

Cellini N, Canale N, Mioni G, Costa S (2020) Changes in sleep pattern, sense of time and digital media use during COVID-19 lockdown in Italy. J Sleep Res 29:e13074. https://doi.org/10.1111/JSR.13074

Cellini N, Conte F, de Rosa O et al (2021) Changes in sleep timing and subjective sleep quality during the COVID-19 lockdown in Italy and Belgium: age, gender and working status as modulating factors. Sleep Med 77:112–119. https://doi.org/10.1016/j.sleep.2020.11.027

Cénat JM, Blais-Rochette C, Kokou-Kpolou CK et al (2021) Prevalence of symptoms of depression, anxiety, insomnia, posttraumatic stress disorder, and psychological distress among populations affected by the COVID-19 pandemic: a systematic review and meta-analysis. Psychiatry Res 295:113599. https://doi.org/10.1016/j.psychres.2020.113599

Dal Santo F, González-Blanco L, Rodríguez-Revuelta J et al (2021) Early impact of the COVID-19 outbreak on sleep in a large Spanish sample. Behav Sleep Med 20:100–115. https://doi.org/10.1080/15402002.2021.1890597

Drumheller K, Fan C-W (2022) Unprecedented times and uncertain connections: a systematic review examining sleep problems and screentime during the COVID-19 pandemic. Sleep Epidemiol 2:100029. https://doi.org/10.1016/j.sleepe.2022.100029

Eurofound (2020) Living, working and COVID-19, COVID-19 series. Publications Office of the European Union, Luxembourg. https://www.eurofound.europa.eu/publications/report/2020/living-working-and-covid-19. Accessed 30 Sept 2022

Figueiro MG, Steverson B, Heerwagen J et al (2017) The impact of daytime light exposures on sleep and mood in office workers. Sleep Health 3:204–215. https://doi.org/10.1016/j.sleh.2017.03.005

Figueiro MG, Jarboe C, Sahin L (2021) The sleep maths: a strong correlation between more daytime light and better night-time sleep. Light Res Technol 53:423–435. https://doi.org/10.1177/14771535211005832

Gao J, Zheng P, Jia Y et al (2020) Mental health problems and social media exposure during COVID-19 outbreak. PLoS One 15:e0231924. https://doi.org/10.1371/journal.pone.0231924

Genta FD, Neto GBR, Sunfeld JPV et al (2021) COVID-19 pandemic impact on sleep habits, chronotype, and health-related quality of life among high school students: a longitudinal study. J Clin Sleep Med 17:1371–1377. https://doi.org/10.5664/JCSM.9196

Gibson R, Shetty H, Carter M, Münch M (2022) Sleeping in a bubble: factors affecting sleep during New Zealand's COVID-19 lockdown. Sleep Adv 3:zpac017. https://doi.org/10.1093/sleepadvances/zpac017

Green A, Cohen-Zion M, Haim A, Dagan Y (2017) Evening light exposure to computer screens disrupts human sleep, biological rhythms, and attention abilities. Chronobiol Int 34:855–865. https://doi.org/10.1080/07420528.2017.1324878

Gualano MR, Lo Moro G, Voglino G et al (2020) Effects of Covid-19 lockdown on mental health and sleep disturbances in Italy. Int J Environ Res Public Health 17:4779. https://doi.org/10.3390/ijerph17134779

Gupta R, Grover S, Basu A et al (2020) Changes in sleep pattern and sleep quality during COVID-19 lockdown. Indian J Psychiatry 62:370–378. https://doi.org/10.4103/psychiatry.IndianJPsychiatry_523_20

Higuchi S, Motohashi Y, Liu Y, Maeda A (2005) Effects of playing a computer game using a bright display on presleep physiological variables, sleep latency, slow wave sleep and REM sleep. J Sleep Res 14:267–273. https://doi.org/10.1111/J.1365-2869.2005.00463.X

Hirshkowitz M, Whiton K, Albert SM et al (2015) National Sleep Foundation's updated sleep duration recommendations: final report. Sleep Health 1:233–243. https://doi.org/10.1016/j.sleh.2015.10.004

Illingworth G, Mansfield KL, Espie CA et al (2022) Sleep in the time of COVID-19: findings from 17000 school-aged children and adolescents in the UK during the first national lockdown. Sleep Adv 3:zpab021. https://doi.org/10.1093/sleepadvances/zpab021

Jahrami H, BaHammam AS, Bragazzi NL et al (2021) Sleep problems during the COVID-19 pandemic by population: a systematic review and meta-analysis. J Clin Sleep Med 17:299–313. https://doi.org/10.5664/jcsm.8930

Jahrami H, Alhaj OA, Humood AM et al (2022) Sleep disturbances during the COVID-19 pandemic: a systematic review, meta-analysis, and meta-regression. Sleep Med Rev 62: 101591. https://doi.org/10.1016/j.smrv.2022.101591

Juda M, Vetter C, Roenneberg T (2013) Chronotype modulates sleep duration, sleep quality, and social jet lag in shift-workers. J Biol Rhythm 28:141–151. https://doi.org/10.1177/0748730412475042

Kantermann T (2020) Behavior: how a global social lockdown unlocks time for sleep. Curr Biol 30: R822–R823. https://doi.org/10.1016/j.cub.2020.06.037

Kokou-Kpolou CK, Megalakaki O, Laimou D, Kousouri M (2020) Insomnia during COVID-19 pandemic and lockdown: prevalence, severity, and associated risk factors in French population. Psychiatry Res 290:113128. https://doi.org/10.1016/j.psychres.2020.113128

Korman M, Tkachev V, Reis C et al (2020) COVID-19-mandated social restrictions unveil the impact of social time pressure on sleep and body clock. Sci Rep 10:1–10. https://doi.org/10.1038/s41598-020-79299-7

Korman M, Tkachev V, Reis C et al (2022) Outdoor daylight exposure and longer sleep promote wellbeing under COVID-19 mandated restrictions. J Sleep Res 31:e13471. https://doi.org/10.1111/JSR.13471

Léger D, Beck F, Fressard L et al (2020) Poor sleep associated with overuse of media during the COVID-19 lockdown. Sleep 43:zsaa125. https://doi.org/10.1093/sleep/zsaa125

Leone MJ, Sigman M, Golombek DA (2020) Effects of lockdown on human sleep and chronotype during the COVID-19 pandemic. Curr Biol 30:R930–R931. https://doi.org/10.1016/j.cub.2020.07.015

Li J, Vitiello MV, Gooneratne NS (2018) Sleep in normal aging. Sleep Med Clin 13:1. https://doi.org/10.1016/j.jsmc.2017.09.001

Li Y, Qin Q, Sun Q et al (2020) Insomnia and psychological reactions during the COVID-19 outbreak in China. J Clin Sleep Med 16:1417–1418. https://doi.org/10.5664/JCSM.8524

Lin LY, Wang J, Ou-Yang XY et al (2021) The immediate impact of the 2019 novel coronavirus (COVID-19) outbreak on subjective sleep status. Sleep Med 77:348–354. https://doi.org/10.1016/j.sleep.2020.05.018

Mandelkorn U, Genzer S, Choshen-Hillel S et al (2021) Escalation of sleep disturbances amid the COVID-19 pandemic: a cross-sectional international study. J Clin Sleep Med 17:45–53. https://doi.org/10.5664/jcsm.8800

Marelli S, Castelnuovo A, Somma A et al (2021) Impact of COVID-19 lockdown on sleep quality in university students and administration staff. J Neurol 268:8–15. https://doi.org/10.1007/s00415-020-10056-6

Massar SAA, Ng ASC, Soon CS et al (2021) Reopening after lockdown: the influence of working-from-home and digital device use on sleep, physical activity, and wellbeing following COVID-19 lockdown and reopening. Sleep 45:zsab250. https://doi.org/10.1093/sleep/zsab250

Merikanto I, Kortesoja L, Benedict C et al (2022) Evening-types show highest increase of sleep and mental health problems during the COVID-19 pandemic—multinational study on 19 267 adults. Sleep 45:zsab216. https://doi.org/10.1093/sleep/zsab216

Morin CM, Carrier J, Bastien C, Godbout R (2020) Sleep and circadian rhythm in response to the COVID-19 pandemic. Can J Public Health 111:654–657. https://doi.org/10.17269/s41997-020-00382-7

O'regan D, Jackson ML, Young AH, Rosenzweig I (2021) Understanding the Impact of the COVID-19 pandemic, lockdowns and social isolation on sleep quality. Nat Sci Sleep 13: 2053–2064. https://doi.org/10.2147/nss.s266240

Ong JL, Lau TY, Karsikas M et al (2021) A longitudinal analysis of COVID-19 lockdown stringency on sleep and resting heart rate measures across 20 countries. Sci Rep 11:14413. https://doi.org/10.1038/s41598-021-93924-z

Patel D, Steinberg J, Patel P (2018) Insomnia in the elderly: a review. J Clin Sleep Med 14:1017–1024. https://doi.org/10.5664/jcsm.7172

Pépin JL, Bailly S, Mordret E et al (2021) Greatest changes in objective sleep architecture during COVID-19 lockdown in night owls with increased REM sleep. Sleep 44:zsab075. https://doi.org/10.1093/sleep/zsab075

Platto S, Wang Y, Zhou J, Carafoli E (2021) History of the COVID-19 pandemic: origin, explosion, worldwide spreading. Biochem Biophys Res Commun 538:14–23. https://doi.org/10.1016/j.bbrc.2020.10.087

Rajkumar RP (2020) COVID-19 and mental health: a review of the existing literature. Asian J Psychiatr 52:102066. https://doi.org/10.1016/j.ajp.2020.102066

Raman S, Coogan AN (2021) Effects of societal-level COVID-19 mitigation measures on the timing and quality of sleep in Ireland. Sleep Med. https://doi.org/10.1016/j.sleep.2021.02.024

Robbins R, Affouf M, Weaver MD et al (2021) Estimated sleep duration before and during the COVID-19 pandemic in major metropolitan areas on different continents: observational study of smartphone app data. J Med Internet Res 23:e20546. https://doi.org/10.2196/20546

Roenneberg T, Pilz LK, Zerbini G, Winnebeck EC (2019) Chronotype and social jetlag: a (self-) critical review. Biology (Basel) 8:1–19. https://doi.org/10.3390/biology8030054

Salfi F, Lauriola M, Amicucci G et al (2020) Gender-related time course of sleep disturbances and psychological symptoms during the COVID-19 lockdown: a longitudinal study on the Italian population. Neurobiol Stress 13:100259. https://doi.org/10.1016/j.ynstr.2020.100259

Salfi F, Amicucci G, Corigliano D et al (2021a) Changes of evening exposure to electronic devices during the COVID-19 lockdown affect the time course of sleep disturbances. Sleep 44:zsab080. https://doi.org/10.1093/sleep/zsab080

Salfi F, D'Atri A, Tempesta D, Ferrara M (2021b) Sleeping under the waves: a longitudinal study across the contagion peaks of the COVID-19 pandemic in Italy. J Sleep Res 30:e13313. https://doi.org/10.1111/jsr.13313

Salfi F, Lauriola M, D'Atri A et al (2021c) Demographic, psychological, chronobiological, and work-related predictors of sleep disturbances during the COVID-19 lockdown in Italy. Sci Rep 11:11416. https://doi.org/10.1038/s41598-021-90993-y

Salfi F, Amicucci G, Corigliano D et al (2022a) Two years after lockdown: longitudinal trajectories of sleep disturbances and mental health over the COVID-19 pandemic and the effects of age, gender, and chronotype. J Sleep Res e13767. https://doi.org/10.1111/jsr.13767

Salfi F, D'Atri A, Amicucci G et al (2022b) The fall of vulnerability to sleep disturbances in evening chronotypes when working remotely and its implications on depression. Sci Rep 12:12249. https://doi.org/10.1038/s41598-022-16256-6

Santos JS, Louzada FM (2022) Changes in adolescents' sleep during COVID-19 outbreak reveal the inadequacy of early morning school schedules. Sleep Sci 15:74–79. https://doi.org/10.5935/1984-0063.20200127

Stone JE, Phillips AJK, Chachos E et al (2021) In-person vs home schooling during the COVID-19 pandemic: differences in sleep, circadian timing, and mood in early adolescence. J Pineal Res 71:e12757. https://doi.org/10.1111/jpi.12757

Taporoski TP, Beijamini F, Gómez LM et al (2022) Subjective sleep quality before and during the COVID-19 pandemic in a Brazilian rural population. Sleep Health 8:167–174. https://doi.org/10.1016/j.sleh.2021.11.007

Trott M, Driscoll R, Irlado E, Pardhan S (2022) Changes and correlates of screen time in adults and children during the COVID-19 pandemic: a systematic review and meta-analysis. EClinicalMedicine 48:101452. https://doi.org/10.1016/j.eclinm.2022.101452

Valiensi SM, Folgueira AL, Garay A (2022) Early impact on sleep and mental health during the mandatory social isolation of COVID-19 outbreak: an observational cross-sectional study carried out in Argentina. Sleep Sci 15:41–48. https://doi.org/10.5935/1984-0063.20200121

Vetter C, Fischer D, Matera JL, Roenneberg T (2015) Aligning work and circadian time in shift workers improves sleep and reduces circadian disruption. Curr Biol 25:907–911. https://doi.org/10.1016/j.cub.2015.01.064

Vindegaard N, Benros ME (2020) COVID-19 pandemic and mental health consequences: systematic review of the current evidence. Brain Behav Immun 89:531–542. https://doi.org/10.1016/j.bbi.2020.05.048

Voitsidis P, Gliatas I, Bairachtari V et al (2020) Insomnia during the COVID-19 pandemic in a Greek population. Psychiatry Res 289:113076. https://doi.org/10.1016/j.psychres.2020.113076

Wams EJ, Woelders T, Marring I et al (2017) Linking light exposure and subsequent sleep: a field polysomnography study in humans. Sleep 40:zsx165. https://doi.org/10.1093/sleep/zsx165

Wittmann M, Dinich J, Merrow M, Roenneberg T (2006) Social jetlag: misalignment of biological and social time. Chronobiol Int 23:497–509. https://doi.org/10.1080/07420520500545979

Wright KP, Linton SK, Withrow D et al (2020) Sleep in university students prior to and during COVID-19 Stay-at-Home orders. Curr Biol 30:R797–R798. https://doi.org/10.1016/j.cub.2020.06.022

Yuan RK, Zitting KM, Maskati L, Huang J (2022) Increased sleep duration and delayed sleep timing during the COVID-19 pandemic. Sci Rep 12:10937. https://doi.org/10.1038/s41598-022-14782-x

Yuksel D, McKee GB, Perrin PB et al (2021) Sleeping when the world locks down: correlates of sleep health during the COVID-19 pandemic across 59 countries. Sleep Health 7:134–142. https://doi.org/10.1016/j.sleh.2020.12.008

Zhang B, Wing Y-K (2006) Sex differences in insomnia: a meta-analysis. Sleep 29:85–93. https://doi.org/10.1093/sleep/29.1.85

Chapter 8
Impact of COVID-19 Mobility Restrictions on Sleep and Well-Being

Stijn A. A. Massar, Ju Lynn Ong, and Michael W. L. Chee

Abstract The COVID-19 pandemic disrupted lives around the world. Mobility restrictions imposed to stem the spread of the disease severely disrupted the daily routines of billions of people. We review how mobility restrictions have affected sleep, well-being, and performance among general populations around the world. Overall, clear shifts toward later sleep timing, and longer sleep duration have been observed in response to mobility restrictions. At the same time sleep quality deteriorated. Within these broad shifts in sleep characteristics, there was heterogeneity in response among individuals. Elderly people, generally experienced small changes in sleep during movement restrictions while school-aged children and adolescents showed the largest change. Individuals who have an evening chronotype showed larger sleep changes than morning-type persons. Evening chronotypes also showed the most increase in insomnia and depression symptoms. Finally, we discuss how hybrid work, blended learning, and routine-setting to reinforce social time cues could be beneficial adjustments to improve sleep in the post-pandemic world.

Keywords Mobility restrictions · Sleep timing · Sleep duration · Covid-19 · Sleep quality · Chronotype

8.1 Introduction

The COVID-19 pandemic resulted in global disruption with largely negative effects. However, measures taken to reduce the spread of infection provided an unprecedented opportunity for many to work from home, spend time with family, and have more time for sleep. COVID-19 mobility restrictions curtailed or altogether stopped commuting, as well as in-person shopping and socializing, disrupting daily schedules linked to sleep timing, duration, and regularity. Alongside, the multitude of

---

S. A. A. Massar · J. L. Ong · M. W. L. Chee (✉)
Sleep and Cognition Laboratory, Centre for Sleep and Cognition, Yong Loo Lin School of Medicine, National University of Singapore, Singapore, Singapore
e-mail: Michael.chee@nus.edu.sg

A. S. BaHammam et al. (eds.), *COVID-19 and Sleep: A Global Outlook*, Progress in Sleep Research, https://doi.org/10.1007/978-981-99-0240-8_8

uncertainties wrought by the pandemic, together with isolation associated with hospitalization, quarantines, and loss of opportunities for in-person work and social interactions affected sleep quality. In this chapter, we review how the pandemic and its subsequent unwinding influenced sleep timing, duration, and quality of persons in different countries. We reviewed how persons of different ages were affected. We also comment on how those with preexisting insomnia and evening chronotype were impacted. We close by suggesting community-wide actions to improve sleep in the aftermath of the pandemic.

8.2 Mobility Restrictions Resulted in Later and Longer Sleep in Adults

Commuting consumes significant time each workday. Pre-pandemic, an average of 20 min was spent commuting in the United States, Finland, Spain, and Sweden. In countries where urban population density is higher, like in Japan, China, and Korea, it might take 40–60 min to get to work (OECD Social Policy Division 2016). With the mobility restrictions resulting in the closures of public transportation, shops, workplaces, and schools, time was saved with the implementation of work-from-home procedures. Globally, data from both surveys and objective tracking found that although bedtimes were delayed, wake times were delayed further, leading to an overall increase in sleep duration. This arrangement also allowed individuals—particularly evening types, to sleep closer to their preferred time. Against these benefits, weaker social time cues, less morning light, together with increased evening light exposure might have contributed to poorer sleep quality and later bedtime (Leone et al. 2020). Reversion of sleep schedules to pre-pandemic ones occurred swiftly with the lifting of movement restrictions and return to the office, highlighting the strength of social influences on sleep patterns (Massar et al. 2022).

8.2.1 Evidence from Surveys

Sleep researchers were quick to launch large-scale online surveys soon after national lockdowns. Cellini et al. (2020) conducted a survey in the second week of lockdown (Mar 17–23, 2020) on 501 workers and 809 university students in Italy, comparing lockdown with pre-lockdown sleep. They observed that while bedtimes were delayed by ~41 min in both workers and students, wake times were delayed further, by 1 h and 13 min in workers, and ~45 min in students. This resulted in overall longer time spent in bed (~26 min in workers, ~5 min in students).

In Austria, Germany, and Switzerland, Blume et al. (2020) administered sleep questionnaires between Mar 23 and Apr 26 and found that median social jet lag (difference in the mid-sleep time on free and workdays) reduced by 13 min while social sleep restriction (difference in sleep duration on free and workdays) decreased by 25 min compared to recollections of sleep patterns pre-lockdown. This was indicative of a reduced mismatch between external (social) and internal (biological) sleep–wake timing.

By the end of March, well over 100 countries had either instituted full or partial lockdowns. Self-reported later but longer sleep was also observed around the world, e.g., in the US (Gao and Scullin 2020), Argentina (Leone et al. 2020), India (Sinha et al. 2020), Scotland (Janssen et al. 2020), Greece, Switzerland, Austria, Germany, France, and Brazil (Trakada et al. 2020), as well as regions in the Middle East and North Africa (Cheikh Ismail et al. 2021). In a survey of 14,000 individuals across 11 countries, a multinational research team found that the most significant changes to sleep patterns occurred in the first 14 days of lockdown, stabilizing at 1 month and 2 months, as individuals adapted to a new routine (AMHSI Research Team et al. 2020).

8.2.2 Evidence from Consumer Sleep Technology

Using consumer sleep technology (CST) observations of pre-pandemic sleep patterns provided objective evidence for how sleep shifted during the lockdown and following the lifting of restrictions. In China, where the outbreak started, crowdsourced data from the SleepAsAndroid app of 563 users were analyzed from Jan 2011 to Mar 2020 (Lee et al. 2020). Although sleep timings were manually entered into the app in the majority, logging was temporally more proximate to sleep than in survey-based reports. Sleep patterns in China changed drastically from Jan 23, 2020, when strict quarantine laws were implemented. Weekday sleep duration increased by an average of 20 min while bedtimes were delayed by 30 min. In the US, a similar digital diary-based Sleep Cycle app showed significant increases in estimated sleep duration at the onset of the pandemic in major cities of London, Los Angeles, New York City, Seoul, and Stockholm (Robbins et al. 2021).

Consumer sleep trackers enable objective recording of sleep over months or longer. Our work assessed ~113,000 Oura users across 20 countries from Jan to Jul 2020 and found that stricter lockdown measures (defined by the Oxford Government Response Tracker (Hale et al. 2021)) resulted in delayed mid-sleep times (+0.09–+0.58 h) and reduced mid-sleep variability (−0.12 to −0.26 h) compared with an equivalent period in 2019 (Ong et al. 2021a). Seasonality effects also appeared to be modulated during the lockdown, with an apparent advance in the onset of summertime in the Northern hemisphere and an apparent prolongation of summertime in the South. In Singapore, where seasonal variation is minimal, we showed similarly graded responses to lockdown severity utilizing Fitbit data from a separate study. Comparing equivalent 3-week periods before the pandemic, with increased restrictions, and during the lockdown, we observed greater delays in bedtimes and waketimes, increased sleep duration, and decreased social jet lag and social sleep restriction, particularly on weekdays (Ong et al. 2021b). We further found that upon lifting of restrictions, sleep was affected by post-lockdown work arrangements, with those who returned to the office reverting to earlier and shorter sleep (Massar et al. 2022).

In the US, data from 163,524 active Fitbit users across six major US cities similarly showed that compared to 2019 patterns, sleep duration in 2020 was increased in most age groups. The sleep phase was delayed, and bedtime variability was reduced owing to decreased weekday-weekend differences (Rezaei and Grandner 2021).

Using data from 599 individuals in France who wore the EEG-based Dreem headband before and during the lockdown, Pepin et al. found that although participants slept slightly more during the lockdown (~8 min), they had a less deep sleep, more light sleep, and longer REM sleep (<5 min) (Pepin et al. 2021). In a follow-up study ~7 months later during a period of partial lockdown where movement within 10 km was allowed, no significant changes to sleep schedules were observed in comparison to pre-lockdown. Interestingly, there was a return to 2019 sleep patterns occurring within 2 weeks of socioeconomic activity resumption (Pepin et al. 2022).

8.3 Mobility Restrictions Were Often Associated with Poorer Quality Sleep

Subjective sleep quality is another dimension through which sleep is assessed. It could be affected by uncertainty about the pandemic's trajectory and worries about one's job and contraction of COVID-19 by oneself or loved ones. The increased engagement with social media and news feeds close to bedtime could have further delayed bedtime, giving rise to longer sleep onset latency and worsening sleep quality. Additional factors negatively impacting sleep quality include a reduction in morning light exposure, an increase in evening light exposure, and a reduction in opportunities for physical activity (Wang and Boros 2019; Figueiro et al. 2017).

In Italy, a significant increase in the PSQI score was observed during the lockdown, with no difference between students and workers (Cellini et al. 2020). The proportion of poor sleepers with PSQI >5 increased from 40.5% to 52.4%. It was proposed that the subjective elongation of time from boredom or the feeling of being "stuck" contributed to decreased sleep quality in working adults. In Austria, Germany, and Switzerland, overall sleep quality decreased, but only by 0.25 points on a scale of 0–25 (Blume et al. 2020). The drop could have been attenuated by the reduced social jet lag during this period. Poor sleep quality and increased sleep disturbances were also reported elsewhere, e.g., in China (Huang and Zhao 2020), the Middle East and North Africa regions (Cheikh Ismail et al. 2021; Salehinejad et al. 2020), and in large multi-country surveys (Jahrami et al. 2021; Jahrami et al. 2022; Mandelkorn et al. 2021), with trends in 2021 being worse than in 2020 (Jahrami et al. 2022). However, a few reports also indicated no change (Leone et al. 2020; AlRasheed et al. 2021) or even trends toward improvement (Gao and Scullin 2020), indicating heterogeneity of changes to sleep quality.

8.4 Pandemic Effects on Sleep in Older Adults

During the pandemic, elderly persons were flagged as being at higher risk of severe symptoms, leading a majority to adhere to strict isolation that reduced social interaction, including those with family members (Wu 2020). Elderly persons were also less likely to adopt the digital tools that allowed vital social interactions despite physical isolation (Song et al. 2021). Unsurprisingly many elderly reported increased social isolation and loneliness, and a deterioration in the quality-of-life (Sayin Kasar and Karaman 2021).

These adversities notwithstanding, older adults seemed to show less change in sleep compared to younger adults or children/adolescents. A study by Rezaei and Grandner (2021) examined the sleep records of 163,524 Fitbit users in the US. While younger adults (18–29-year-old) showed increased sleep duration (7.7–11.1 min) and delayed bedtime (23.7–24 min), the elderly (65+) had much more modest changes (sleep duration: −1.1 to 0.8 min; bedtime delay: 8.4–11.1 min). Similarly, Amicucci et al. (2021) found that the elderly (65–75-year-old) reported mostly unchanged sleep timing (56.1% unchanged bedtime, 49.4% unchanged wake time), while most younger participants (18–20-year-old) reported delayed bed (74.6%) and wake times (74.2%). Sleep quality ratings were poor for both groups (elderly: 63.7% poor; young: 60% poor), but insomnia ratings were lower for older (60.5% reporting no insomnia symptoms) compared to younger participants (47.6% having no symptoms). Further, the elderly reported lower depression and less stress than younger participants. It therefore seems that despite their susceptibility to adverse health outcomes, elderly persons may have experienced less impact on their sleep routines. It was possible because of greater flexibility in time use, to begin with, for example, not having to deal with work or childcare issues (Rezaei and Grandner 2021). Elderly persons may also have had more life experience to develop effective coping mechanisms to adapt to the pandemic (Amicucci et al. 2021).

8.5 Pandemic Effects on the Sleep of Children and Adolescents

Another vulnerable age group was school-aged children and adolescents (Meherali et al. 2021). With school closures and a shift to home-based learning this younger age group experienced a major shift in their routines. Physical activity and social interactions were drastically restricted due to pandemic safety measures (Bates et al. 2020; Rossi et al. 2021; Stavridou et al. 2020). Moreover, studies have reported a large increase in time spent using electronic devices in this age group (e.g., computers, tablets, smartphones) (Drouin et al. 2020; Dutta et al. 2020). Early in the pandemic, it was recognized that these factors could strongly affect children and adolescents' sleep during restrictions (Becker and Gregory 2020).

Studies examining children and adolescent sleep have reported a sizable shift toward later and longer sleep during restrictions. A survey study among Italian school children observed a delay in bedtime by 1 h and 18 min and a delayed wake time by 1 h and 50 min, resulting in 32 min increase in TIB on average (Cellini et al. 2021). Similarly, studies in adolescents found delayed bedtimes (by 15–36 min) and wake times (by ~90 min), leading to an increase in sleep opportunity of 1 h or more (Albrecht et al. 2022; Saxvig et al. 2022). These shifts were accompanied by a reduction in social jet lag from 2 h 37 min to 1 h 53 min in 2020 (Saxvig et al. 2022) which indicates better alignment with preferred sleep–wake timing on free days. Studies examining a wide age range of children and adolescents demonstrated that the shifts in sleep schedule were smaller in younger children (preschool age), and were most pronounced in adolescents (Bruni et al. 2022; Illingworth et al. 2022; Lim et al. 2021).

Reports of sleep quality changes during restrictions have been mixed. Some studies observed a slight improvement in self or parent-scored sleep quality (Albrecht et al. 2022; Lim et al. 2021), while others reported increased difficulties falling and staying asleep, increased anxiety at bedtime, increased sleep onset latency, and awakenings (Becker and Gregory 2020; Bruni et al. 2022). The changes in sleep quality were also associated with quality-of-life scores, happiness levels, social relationships, and loneliness. Less happy respondents, reported problems getting along with household members and were lonely and more likely to report worse sleep quality (Albrecht et al. 2022; Illingworth et al. 2022).

Adolescents, unlike college students, are still very much constrained by school start times, which can begin as early as 7.30 am in Singapore. During the lockdown period, remote/home-based learning arrangements allowed students to wake much later as there was no longer a need to commute and get ready for school, and, consequently, sleep duration increased, particularly in older adolescents/high school students with later chronotypes. As with working adults, the reopening of schools also saw the prevalence of insufficient sleep start to increase, with a nearly fourfold increase for ninth-grade students in China (Lian et al. 2021).

In the US, a cross-sectional, nationwide sample of 5245 adolescents showed a close link between instructional mode and sleep patterns, with the earliest bed/wake times and shortest sleep duration observed for nights preceding in-person instruction, followed by online instruction, and then no instruction (Meltzer et al. 2021). The authors also found that hybrid (online/in-person) instruction resulted in increased night-to-night variability that could affect mood and behavior in the long term.

Despite encouraging extensions to sleep duration, lockdowns were also generally associated with increased depression, anxiety, loneliness, and poor sleep quality (Naff et al. 2022). This was particularly true for the most vulnerable groups: persons of lower socioeconomic status, those with preexisting disabilities and mental health issues, and communities of color. Poor outcomes were compounded by reduced access to school-based counselors during school closures. Remote learning arrangements also significantly affected learning outcomes. In the Netherlands, despite favorable factors, having a relatively short lockdown, equitable funding and widespread broadband access, learning losses equivalent to one-fifth of a school year were observed (Engzell et al. 2021). Persons from less-educated homes lost up to 60% more. Coupled with the stress and isolation of online learning, and lack of student engagement, there was a profound sense of learning loss—not just academically but also socially and psychologically.

8.6 Sleep-Related Characteristics

Another relevant dimension to assess inter-individual differences is how prior individual sleep-related characteristics influence an individual's response to pandemic restrictions. In the next sections, we outline two such characteristics: preexisting insomnia and chronotype.

8.6.1 Preexisting Insomnia

The prevalence of insomnia symptoms increased during the pandemic (for review see Mahmud et al. 2021; Cheshmehzangi et al. 2022). This could be related to anxiety and stress brought about by the sudden challenges of pandemic life (e.g., home confinement, fear of contagion, economic hardship), and could have been exacerbated by the loss of daily routines and reduced exposure to environmental cues for circadian entrainment. Given these, we might expect patients with preexisting insomnia to be especially affected by mobility restrictions. Instead, findings have been mixed. In a survey across 67 countries, Meaklim et al. (2021) reported that, out of 804 participants with preexisting insomnia, 41% reported no change in symptom severity during the lockdown. 42.5% reported deterioration of insomnia symptoms, while 16.5% reported improvement. Those with improved symptoms had lower stress and anxiety scores. Similarly, Kocevska et al. (2020) found that in participants without prior insomnia, there was an increase in symptoms during the pandemic, while those with prior insomnia improved. In both studies, improvements in symptoms were related to features of better mental well-being (i.e., lower anxiety and stress, lower negative affect). Further studies are needed to delineate the factors that could have led to increased vulnerability versus improvement (e.g., sociodemographic, economic, and/or cultural/geographic factors) (Bhat and Chokroverty 2022).

8.6.2 Chronotype

While mobility restrictions, by their very nature, restricted freedom, they also provided some flexibility for time allocation. As many workplaces adapted to remote work arrangements, office hours became less strictly stipulated. Moreover, not having to commute freed up substantial time. Globally, sleep patterns shifted to later timings and longer duration, during restrictions, and people relied less on alarm clocks for waking up (Korman et al. 2020), allowing many to schedule their sleep and wake closer to their biological circadian preferences. Several studies found that individuals with an evening chronotype showed the largest shifts in sleep timing and duration (Pepin et al. 2021; Marelli et al. 2021; Oved et al. 2021), and the greatest reduction in social sleep restriction (measured as the extension of sleep duration on weekends versus weekdays) (Bottary et al. 2022). Further support for this idea comes from evidence that the association between chronotype on the one hand and sleep quality and well-being on the other is moderated by work-from-home status (Salfi et al. 2022). Pre-pandemic, persons with an evening chronotype generally showed worse ratings of sleep quality and depression. However, this association was less strong in individuals who worked from home during the pandemic, as compared to those who continued to (or returned to) work in-person (Salfi et al. 2022).

In theory, the increased opportunity to obtain sufficient sleep and/or sleep according to one's circadian preference under mobility restrictions would benefit health and well-being. However, empirical findings suggest the opposite. A survey among university students in Bangladesh found that even though evening chronotypes showed the largest delay in sleep timing, this shift was associated with poorer mood and well-being (Hasan et al. 2022). Likewise, a large-scale study across 15 countries showed that evening chronotypes showed the greatest increase in sleep problems, insomnia symptoms, and nightmares, along with poorer mental health (anxiety, depression, PTSD symptoms) during the pandemic (Merikanto et al. 2022). In our work, we found that, for most people, actual sleep timing was later than their self-reported preferred sleep timing, both during the lockdown and after reopening (Massar et al. 2022). It is possible that the reduced exposure to natural and social time cues suppressed circadian entrainment and caused misalignment between internal and external night-day rhythms (Kahawage et al. 2022a, b). Alternatively, negative outcomes of pandemic restrictions in evening chronotypes could arise from socioeconomic differences. Merikanto et al. (2022) found that more women were evening types than morning types, and evening types were younger, had lower education, and had a higher percentage of financial hardship. All these factors were markers for mental health problems during the pandemic.

8.7 Lessons for Post-pandemic Sleep Health

8.7.1 Hybrid Work

Improvements in video conferencing and existing long-distance collaborative work have made hybrid working possible for some years now, but the pandemic obligated its widespread adoption. As the foregoing material has illustrated, hybrid work has meant time savings for some and greater flexibility in time allocation allowing for sleep at more preferred times (Salfi et al. 2022). Hybrid work is fundamentally asynchronous in that not all members of a work team need to be contactable or engaged at any given time. This can allow workers to schedule activities more according to their optimal timings with potential benefits for deeper focus and working with less interruption (Volk et al. 2017). On the other hand, loss of regularity of schedules can have collateral effects on sleep, and erosion of boundaries between work and non-work whereby employees are worn down by implied obligations to answer calls, respond to emails or complete tasks outside traditional work hours and across the entire week. One study reported that employees who experience more blurring of work-life boundaries show higher emotional exhaustion and can less effectively maintain healthy lifestyles (e.g., by disrupting sufficient opportunities for relaxation and sleep) (Pluut and Wonders 2020). This was in turn associated with reduced work performance. Whether remote work can be effectively sustained further depends on organizational support and management practices (e.g., effective communication and coordination within virtual teams) (Vyas 2022; Shockley et al. 2021; Breideband et al. 2022; Kniffin et al. 2021), individual qualities (e.g., self-discipline/conscientiousness versus procrastination) (Wang et al. 2021; Rocchi and Bernacchio 2022), as well as on situational factors (e.g., family structure, care duties, housing situation) (Vicari et al. 2022; Costa et al. 2022).

To what extent remote work will continue to be adopted beyond the pandemic remains to be seen. A large-scale Gallup poll conducted in the US in September 2021 showed that 45% of full-time employees worked from home either all (25%) or part of the time (20%) (Saad and Wigert 2022). While these numbers have come down slightly, there continues to be vigorous debate about the extent to which hybrid work will remain with some business leaders decrying it while others embrace it (Savage 2021; Wingard 2022). At the same time, there is a growing call for continued remote or hybrid work arrangements among employees in many sectors. As work arrangements in some countries start to return to a pre-pandemic status quo, policymakers in other countries have instated legislation to protect and regulate the need for remote work beyond the pandemic (Senatori and Spinelli 2021; Baazil and Cras 2022). The eventual success of hybrid work will require policymakers, businesses, and employees to rethink how we can work more productively while staying well and sleeping better (Lin et al. 2021).

8.7.2 School Arrangements

While prolonged home-based/asynchronous learning has disadvantages because of attendant social isolation and caregiving issues, creating an environment where students can have an opportunity to better align their biological and social clocks could benefit academic outcomes and well-being (van der Vinne et al. 2015). One way this could be achieved is by delaying school start times. Proponents of later school start times have long advocated this for its effects on sleep health, cognition, and well-being (Minges and Redeker 2016; Winnebeck et al. 2020; Lo et al. 2018) and these lockdowns have provided a front seat into how sleep health can improve in this group if students are allowed to sleep in more. In addition, hybrid or blended learning methods that incorporate both online and face-to-face instruction can improve student success and satisfaction (Means et al. 2013) while preserving a sense of community (Rovai and Jordan 2004). While fully online learning is convenient and advantageous for motivated/advanced students to learn at a quicker pace, the lack of interaction between peers and between instructor and student can cause those less motivated to disengage and perform poorly (Artino Jr. and Stephens 2009). The most successful forms of blended learning also typically involve strong institutional support and alignment with faculty and student goals in the design of infrastructure and course planning (Moskal et al. 2013), but much of this has been catalyzed by the pandemic. This can be further improved now that there are opportunities to reconnect in the classroom for synchronous discussion and hands-on collaborative work. Still, others can leverage upon institutions that have successfully transformed their educational programs and provide training and toolkits for others to adopt (Moskal et al. 2013). These flexible learning environments will also conceivably help students prepare for hybrid models of work when they leave the school system.

Motivated students who exercise good time management can benefit significantly from later start times and online learning at their own pace or in virtual groups mimicking what they are likely to encounter later in life when hybrid work becomes more common. Like with other instances where social disparities are widened by crises, those in less conducive environments or those with poorer self-regulatory ability, risk falling by the wayside without traditional in-person delivery of education, and measures to enable these persons to adapt to future modes of education and work are needed.

8.7.3 Setting Time Cues

A particular challenge for maintaining healthy sleep rhythms is when accustomed social time cues are removed. During the pandemic, having to get up at a relatively constant time to engage in prework preparation and then the commute disappeared together with work-related mealtimes, returning home, and socializing with friends and/or family. Failure to replace such zeitgebers will result in irregular rest-activity-work and feeding rhythms that could be deleterious to sleep, health, and mental well-being. Such practices could include sleep/waking at the same time every day, eating meals at regular times, setting aside specific times daily for social activities, e.g., calls with a friend, ensuring natural light exposure (preferably in the morning), and avoiding evening light exposure (e.g., through bright indoor lighting or electronic screens) (Murray et al. 2021). Difficulty in retaining a scheduled routine not only affects circadian regularity, but it might also compound it by inducing "bedtime procrastination" (Oliveira et al. 2022). Rebuilding regular routines, on the other hand, could help to restore a sense of self-control and predictability which can benefit mental health (Hou et al. 2020).

8.8 Conclusion

The pandemic provided an unprecedented opportunity to study how alterations to social routines can affect sleep behavior. Mobility restrictions are now minimal in most countries, and life is slowly returning to normal. The key question is whether we fully reinstate pre-pandemic routines. Work-from-home and home-based learning arrangements that eliminate commute time allowed for longer sleep that was better aligned with an individual's chronotype. This had benefits on well-being. Flexible or staggered arrangements could boost sleep in the evening types while reducing overcrowding on public transportation and roads. In tandem, employers should recognize that flexible work hours do not equate to unlimited work hours. Delayed school start times particularly in older adolescents can result in longer sleep. These structural measures that provide environments conducive to sleep are more effective for shifting population behavior compared to efforts primarily directed at instilling personal agency (Chaput 2019).

References

Albrecht JN, Werner H, Rieger N et al (2022) Association between homeschooling and adolescent sleep duration and health during COVID-19 pandemic high school closures. JAMA Netw Open 5(1):e2142100

AlRasheed MM, Alkadir AM, Bin Shuqiran KI, Al-Aqeel S, Jahrami HA, BaHammam AS (2021) The impact of quarantine on sleep quality and psychological distress during the COVID-19 pandemic. Nat Sci Sleep 13:1037–1048

AMHSI Research Team, Milken Research Team, Roitblat Y et al (2020) Stay-at-home circumstances do not produce sleep disorders: an international survey during the COVID-19 pandemic. J Psychosom Res 139:110282

Amicucci G, Salfi F, D'Atri A, Viselli L, Ferrara M (2021) The differential impact of COVID-19 lockdown on sleep quality, insomnia, depression, stress, and anxiety among late adolescents and elderly in Italy. Brain Sci 11(10):1336

Artino AR Jr, Stephens JM (2009) Academic motivation and self-regulation: a comparative analysis of undergraduate and graduate students learning online. Internet High Educ 12(3–4):146–151

Baazil D, Cras PF (2022) Dutch house approves to make work from home a legal right. Bloomberg

Bates LC, Zieff G, Stanford K et al (2020) COVID-19 impact on behaviors across the 24-hour day in children and adolescents: physical activity, sedentary behavior, and sleep. Children (Basel) 7(9):138

Becker SP, Gregory AM (2020) Editorial perspective: Perils and promise for child and adolescent sleep and associated psychopathology during the COVID-19 pandemic. J Child Psychol Psychiatry 61(7):757–759

Bhat S, Chokroverty S (2022) Sleep disorders and COVID-19. Sleep Med 91:253–261

Blume C, Schmidt MH, Cajochen C (2020) Effects of the COVID-19 lockdown on human sleep and rest-activity rhythms. Curr Biol 30(14):R795–R7R7

Bottary R, Fields EC, Kensinger EA, Cunningham TJ (2022) Age and chronotype influenced sleep timing changes during the first wave of the COVID-19 pandemic. J Sleep Res 31(2):e13495

Breideband T, Sukumar P, Mark G, Caruso M, D'Mello S, Striegel AD (2022) Home-life and work rhythm diversity in distributed teamwork: a study with information workers during the COVID-19 pandemic. Proc ACM Hum Comput Interact 6:CSCW1

Bruni O, Malorgio E, Doria M et al (2022) Changes in sleep patterns and disturbances in children and adolescents in Italy during the Covid-19 outbreak. Sleep Med 91:166–174

Cellini N, Canale N, Mioni G, Costa S (2020) Changes in sleep pattern, sense of time and digital media use during COVID-19 lockdown in Italy. J Sleep Res 29(4):e13074

Cellini N, Di Giorgio E, Mioni G, Di Riso D (2021) Sleep and psychological difficulties in Italian school-age children during COVID-19 lockdown. J Pediatr Psychol 46(2):153–167

Chaput JP (2019) The integration of pediatric sleep health into public health in Canada. Sleep Med 56:4–8

Cheikh Ismail L, Osaili TM, Mohamad MN et al (2021) Assessment of eating habits and lifestyle during the coronavirus 2019 pandemic in the Middle East and North Africa region: a cross-sectional study. Br J Nutr 126(5):757–766

Cheshmehzangi A, Chen H, Su Z, Zou T, Xiang YT, Dawodu A (2022) How does the COVID-19 fuel insomnia? Brain Behav Immun Health 21:100426

Costa C, Teodoro M, Mento C et al (2022) Work performance, mood and sleep alterations in home office workers during the COVID-19 pandemic. Int J Environ Res Public Health 19(4):1990

Drouin M, McDaniel BT, Pater J, Toscos T (2020) How parents and their children used social media and technology at the beginning of the COVID-19 pandemic and associations with anxiety. Cyberpsychol Behav Soc Netw 23(11):727–736

Dutta K, Mukherjee R, Sen D, Sahu S (2020) Effect of COVID-19 lockdown on sleep behavior and screen exposure time: an observational study among Indian school children. Biol Rhythm Res 53(4):628–639

Engzell P, Frey A, Verhagen MD (2021) Learning loss due to school closures during the COVID-19 pandemic. Proc Natl Acad Sci U S A 118(17):e2022376118

Figueiro MG, Steverson B, Heerwagen J et al (2017) The impact of daytime light exposures on sleep and mood in office workers. Sleep Health 3(3):204–215

Gao C, Scullin MK (2020) Sleep health early in the coronavirus disease 2019 (COVID-19) outbreak in the United States: integrating longitudinal, cross-sectional, and retrospective recall data. Sleep Med 73:1–10

Hale T, Angrist N, Goldszmidt R et al (2021) A global panel database of pandemic policies (Oxford COVID-19 Government Response Tracker). Nat Hum Behav 5(4):529–538

Hasan MM, Jankowski KS, Khan MHA (2022) Morningness-eveningness preference and shift in chronotype during COVID-19 as predictors of mood and well-being in university students. Pers Individ Dif 191:111581

Hou WK, Lai FT, Ben-Ezra M, Goodwin R (2020) Regularizing daily routines for mental health during and after the COVID-19 pandemic. J Glob Health 10(2):020315

Huang Y, Zhao N (2020) Generalized anxiety disorder, depressive symptoms and sleep quality during COVID-19 outbreak in China: a web-based cross-sectional survey. Psychiatry Res 288: 112954

Illingworth G, Mansfield KL, Espie CA, Fazel M, Waite F (2022) Sleep in the time of COVID-19: findings from 17000 school-aged children and adolescents in the UK during the first national lockdown. Sleep Adv 3(1):zpab021

Jahrami H, BaHammam AS, Bragazzi NL, Saif Z, Faris M, Vitiello MV (2021) Sleep problems during the COVID-19 pandemic by population: a systematic review and meta-analysis. J Clin Sleep Med 17(2):299–313

Jahrami HA, Alhaj OA, Humood AM et al (2022) Sleep disturbances during the COVID-19 pandemic: a systematic review, meta-analysis, and meta-regression. Sleep Med Rev 62:101591

Janssen X, Fleming L, Kirk A et al (2020) Changes in physical activity, sitting and sleep across the COVID-19 national lockdown period in Scotland. Int J Environ Res Public Health 17(24):9362

Kahawage P, Bullock B, Meyer D et al (2022a) Social rhythm disruption is associated with greater depressive symptoms in people with mood disorders: findings from a multinational online survey during COVID-19. Can J Psychiatr 2022:7067437221097905

Kahawage P, Crowe M, Gottlieb J et al (2022b) Adrift in time: the subjective experience of circadian challenge during COVID-19 amongst people with mood disorders. Chronobiol Int 39(1):57–67

Kniffin KM, Narayanan J, Anseel F et al (2021) COVID-19 and the workplace: implications, issues, and insights for future research and action. Am Psychol 76(1):63–77

Kocevska D, Blanken TF, Van Someren EJW, Rosler L (2020) Sleep quality during the COVID-19 pandemic: not one size fits all. Sleep Med 76:86–88

Korman M, Tkachev V, Reis C et al (2020) COVID-19-mandated social restrictions unveil the impact of social time pressure on sleep and body clock. Sci Rep 10(1):22225

Lee PH, Marek J, Nalevka P (2020) Crowdsourced smartphone data reveal altered sleep/wake pattern in quarantined Chinese during the COVID-19 outbreak. Chronobiol Int 37(8): 1181–1190

Leone MJ, Sigman M, Golombek DA (2020) Effects of lockdown on human sleep and chronotype during the COVID-19 pandemic. Curr Biol 30(16):R930–R9R1

Lian Q, Zuo X, Zhong X et al (2021) The effect of COVID-19 school closures on adolescent sleep duration: an uncontrolled before-after study. BMC Public Health 21(1):1528

Lim MTC, Ramamurthy MB, Aishworiya R et al (2021) School closure during the coronavirus disease 2019 (COVID-19) pandemic - impact on children's sleep. Sleep Med 78:108–114

Lin Y, Islan W, Sayal L, Vadaketh S, Simms A, da Silva M (2021) Making hybrid work human: 2021 global survey report: economist impact

Lo JC, Lee SM, Lee XK et al (2018) Sustained benefits of delaying school start time on adolescent sleep and well-being. Sleep 41(6):zsy052

Mahmud S, Hossain S, Muyeed A, Islam MM, Mohsin M (2021) The global prevalence of depression, anxiety, stress, and, insomnia and its changes among health professionals during COVID-19 pandemic: a rapid systematic review and meta-analysis. Heliyon 7(7):e07393

Mandelkorn U, Genzer S, Choshen-Hillel S et al (2021) Escalation of sleep disturbances amid the COVID-19 pandemic: a cross-sectional international study. J Clin Sleep Med 17(1):45–53

Marelli S, Castelnuovo A, Somma A et al (2021) Impact of COVID-19 lockdown on sleep quality in university students and administration staff. J Neurol 268(1):8–15

Massar SAA, Ng ASC, Soon CS et al (2022) Reopening after lockdown: the influence of working-from-home and digital device use on sleep, physical activity, and wellbeing following COVID-19 lockdown and reopening. Sleep 45(1):zsab250

Meaklim H, Junge MF, Varma P, Finck WA, Jackson ML (2021) Pre-existing and post-pandemic insomnia symptoms are associated with high levels of stress, anxiety, and depression globally during the COVID-19 pandemic. J Clin Sleep Med 17(10):2085–2097

Means B, Toyama Y, Murphy R, Baki M (2013) The effectiveness of online and blended learning: a meta-analysis of the empirical literature. Teach Coll Rec 115(3)

Meherali S, Punjani N, Louie-Poon S et al (2021) Mental health of children and adolescents amidst COVID-19 and past pandemics: a rapid systematic review. Int J Environ Res Public Health 18(7):3432

Meltzer LJ, Saletin JM, Honaker SM et al (2021) COVID-19 instructional approaches (in-person, online, hybrid), school start times, and sleep in over 5,000 U.S. adolescents. Sleep 44(12)

Merikanto I, Kortesoja L, Benedict C et al (2022) Evening-types show highest increase of sleep and mental health problems during the COVID-19 pandemic-multinational study on 19 267 adults. Sleep 45(2):zsab216

Minges KE, Redeker NS (2016) Delayed school start times and adolescent sleep: a systematic review of the experimental evidence. Sleep Med Rev 28:86–95

Moskal P, Dziuban C, Hartman J (2013) Blended learning: a dangerous idea? Internet High Educ 18:15–23

Murray G, Gottlieb J, Swartz HA (2021) Maintaining daily routines to stabilize mood: theory, data, and potential intervention for circadian consequences of COVID-19. Can J Psychiatr 66(1):9–13

Naff D, Williams S, Furman-Darby J, Yeung M (2022) The mental health impacts of COVID-19 on PK–12 students: a systematic review of emerging literature. AERA Open. https://doi.org/10.1177/23328584221084722

OECD Social Policy Division (2016) Time spent travelling to and from work

Oliveira A, Pereira B, Rosario P, Magalhaes P (2022) Bedtime procrastination during the second COVID-19 lockdown in Portugal. Sleep Sci 15(1):105–111

Ong JL, Lau T, Karsikas M, Kinnunen H, Chee MWL (2021a) A longitudinal analysis of COVID-19 lockdown stringency on sleep and resting heart rate measures across 20 countries. Sci Rep 11(1):14413

Ong JL, Lau T, Massar SAA et al (2021b) COVID-19-related mobility reduction: heterogenous effects on sleep and physical activity rhythms. Sleep 44(2):zsaa179

Oved S, Mofaz M, Lan A et al (2021) Differential effects of COVID-19 lockdowns on well-being: interaction between age, gender and chronotype. J R Soc Interface 18(179):20210078

Pepin JL, Bailly S, Mordret E et al (2021) Greatest changes in objective sleep architecture during COVID-19 lockdown in night owls with increased REM sleep. Sleep 44(9):zsab075

Pepin JL, Bailly S, Mignot E et al (2022) Digital markers of sleep architecture to characterize the impact of different lockdown regimens on sleep health during the COVID-19 pandemic. Sleep 45(6):zsac074

Pluut H, Wonders J (2020) Not able to lead a healthy life when you need it the most: dual role of lifestyle behaviors in the association of blurred work-life boundaries with well-being. Front Psychol 11:607294

Rezaei N, Grandner MA (2021) Changes in sleep duration, timing, and variability during the COVID-19 pandemic: large-scale Fitbit data from 6 major US cities. Sleep Health 7(3):303–313

Robbins R, Affouf M, Weaver MD et al (2021) Estimated sleep duration before and during the COVID-19 pandemic in major metropolitan areas on different continents: observational study of smartphone app data. J Med Internet Res 23(2):e20546

Rocchi M, Bernacchio C (2022) The virtues of COVID-19 pandemic: how working from home can make us the best (or the worst) version of ourselves. Bus Soc Rev. https://doi.org/10.1111/basr.12282

Rossi L, Behme N, Breuer C (2021) Physical activity of children and adolescents during the COVID-19 pandemic-a scoping review. Int J Environ Res Public Health 18(21):11440

Rovai AP, Jordan HM (2004) Blended learning and sense of community: a comparative analysis with traditional and fully online graduate courses. Int Rev Res Open Distance Learn 5(2):1–13

Saad L, Wigert B (2022) Remote work persisting and trending permanent. Gallup News

Salehinejad MA, Majidinezhad M, Ghanavati E et al (2020) Negative impact of COVID-19 pandemic on sleep quantitative parameters, quality, and circadian alignment: implications for health and psychological well-being. EXCLI J 19:1297–1308

Salfi F, D'Atri A, Amicucci G et al (2022) Demographic, psychological, chronobiological, and work-related predictors of sleep disturbances during the COVID-19 lockdown in Italy. Sci Rep 11:11416

Savage M (2021) What bosses really think about remote work. BBC Worklife

Saxvig IW, Pallesen S, Sivertsen B, Hysing M, Evanger LN, Bjorvatn B (2022) Sleep during COVID-19-related school lockdown, a longitudinal study among high school students. J Sleep Res 31(2):e13499

Sayin Kasar K, Karaman E (2021) Life in lockdown: social isolation, loneliness and quality of life in the elderly during the COVID-19 pandemic: a scoping review. Geriatr Nurs 42(5):1222–1229

Senatori I, Spinelli C (2021) (Re-)regulating remote work in the post-pandemic scenario: lessons from the Italian experience. Italian Labour Law e-Journal 14(1):209–260

Shockley KM, Allen TD, Dodd H, Waiwood AM (2021) Remote worker communication during COVID-19: the role of quantity, quality, and supervisor expectation-setting. J Appl Psychol 106(10):1466–1482

Sinha M, Pande B, Sinha R (2020) Impact of COVID-19 lockdown on sleep-wake schedule and associated lifestyle related behavior: a national survey. J Public Health Res 9:1826

Song Y, Qian C, Pickard S (2021) Age-related digital divide during the COVID-19 pandemic in China. Int J Environ Res Public Health 18(21):11285

Stavridou A, Stergiopoulou AA, Panagouli E et al (2020) Psychosocial consequences of COVID-19 in children, adolescents and young adults: a systematic review. Psychiatry Clin Neurosci 74(11): 615–616

Trakada A, Nikolaidis PT, Andrade MDS et al (2020) Sleep during "lockdown" in the COVID-19 pandemic. Int J Environ Res Public Health 17(23):9094

van der Vinne V, Zerbini G, Siersema A et al (2015) Timing of examinations affects school performance differently in early and late chronotypes. J Biol Rhythm 30(1):53–60

Vicari B, Zoch G, Bächmann A-C (2022) Childcare, work or worries? What explains the decline in parents' well-being at the beginning of the COVID-19 pandemic in Germany? J Fam Res 34(1): 310–332

Volk S, Pearsall MJ, Christian MS, Becker WJ (2017) Chronotype diversity in teams: toward a theory of team energetic asynchrony. Acad Manag Rev 42(4):683–702

Vyas L (2022) "New normal" at work in a post-COVID world: work–life balance and labor markets. Polic Soc 41(1):155–167

Wang F, Boros S (2019) The effect of physical activity on sleep quality: a systematic review. Eur J Phys 23(1):11–18

Wang B, Liu Y, Qian J, Parker SK (2021) Achieving effective remote working during the COVID-19 pandemic: a work design perspective. Appl Psychol 70(1):16–59

Wingard J (2022) Elon Musk's big bet against remote work: will Tesla win? Forbescom

Winnebeck EC, Vuori-Brodowski MT, Biller AM et al (2020) Later school start times in a flexible system improve teenage sleep. Sleep 43(6):zsz307

Wu B (2020) Social isolation and loneliness among older adults in the context of COVID-19: a global challenge. Glob Health Res Policy 5:27

Chapter 9
Sleep Disturbances and COVID-19 Vaccines

Daniela Santoro Rosa, Silvia Beatriz Boscardin, Gabriel Natan Pires, Monica Levy Andersen, and Sergio Tufik

Abstract Sleep and the immune system have a bidirectional relationship, as sleep can strengthen the immune system to achieve balanced and effective immune function, while immune system activation may affect sleep. One of the most relevant topics in this interplay is the effects of sleep disturbances on the immune response to vaccination. The effect of sleep disturbances (including sleep deprivation, sleep restriction, insomnia, and shift work) on the vaccine response has been tested for several diseases, including flu, hepatitis A/B, and meningitis, and in most cases, it leads to a reduced post-vaccination antibody titer. Given these results and considering the COVID-19 pandemic, it has been wondered if sleep disturbances could reduce antibody titers or otherwise reduce the vaccinal response against SARS-CoV-2. The results available so far demonstrate that obstructive sleep apnea among older adults does not lead to reduced antibody titers after vaccination against COVID-19. Regarding circadian rhythmicity, mixed findings were observed when considering the time of the day of COVID-19 vaccination. No studies have been performed regarding sleep deprivation, sleep restriction, or insomnia. In summary, substantial evidence demonstrates that sleep plays an important role in immune system homeostasis and its effects on infection control and vaccine effectiveness. Although the theoretical background suggested sleep disturbances might reduce immune response after vaccination against COVID-19, the current results do not allow concluding it. The effects of sleep on the outcome of COVID-19 and SARS-CoV2 immune responses are still to be better explored, especially concerning

D. S. Rosa
Departamento de Microbiologia, Imunologia e Parasitologia, Universidade Federal de São Paulo (UNIFESP), São Paulo, Brasil

S. B. Boscardin
Departamento de Parasitologia, Universidade de São Paulo, São Paulo, Brasil

G. N. Pires (✉) · M. L. Andersen · S. Tufik
Departamento de Psicobiologia, Universidade Federal de São Paulo (UNIFESP), São Paulo, Brasil
e-mail: Gabriel.Pires@sono.org.br

A. S. BaHammam et al. (eds.), *COVID-19 and Sleep: A Global Outlook*, Progress in Sleep Research, https://doi.org/10.1007/978-981-99-0240-8_9

conditions that are more classically related to reduced vaccination response (including sleep deprivation and insomnia).

Keywords Sleep · COVID-19 · SARS-CoV-2 · Vaccines · Antibodies

9.1 Sleep and the Immune System

Sleep science has evolved dramatically in the past few decades, demonstrating the profound importance of sleep for nearly every system in the body. In addition, immune system homeostasis is critical for overall health. Sleep and the immune system have a bidirectional relationship (Besedovsky et al. 2019; Irwin 2019), as sleep can strengthen the immune system to achieve balanced and effective immune function, while immune system activation may affect sleep. The mechanisms by which sleep and its disturbances influence the immune response are not yet fully understood but are the subject of continuous investigation.

The hypothalamic-pituitary-adrenal (HPA) axis through the production of glucocorticoids and the central nervous system (CNS) through the production of catecholamines are the main effector systems that link sleep and immunity. However, other neuroendocrine signals, such as melatonin, prolactin, and growth hormone (GH), also regulate the immune response (Dantzer 2018). Physiological concentrations of glucocorticoids are essential for immune homeostasis, but, in excess, they can be potent immunosuppressors. Cortisol (or corticosterone in rodents), produced by the adrenal glands, is the main glucocorticoid involved in the regulation of the immune response. While, under physiological conditions, cortisol inhibits the transcription of pro-inflammatory genes (Petta et al. 2016), its excess inhibits leukocyte proliferation and function, inducing apoptosis (Quatrini and Ugolini 2021).

In humans, sleep disturbances cause persistent activation of the HPA axis, which, in turn, can cause immune cells to become resistant to glucocorticoids. For example, in individuals with insomnia, it is possible to notice an increase in the activation of the HPA axis and, consequently, an increase in inflammatory biomarkers (Floam et al. 2015; Irwin and Piber 2018). According to this notion, persistent sleep disturbance leads to sustained activation of the inflammatory response, which can be damaging to the host (Irwin 2019). In fact, in a meta-analysis of ~50,000 adults, sleep disturbances were associated with higher levels of C-reactive protein (CRP) and IL-6 (Irwin et al. 2016). Inflammatory markers such as IL-6, TNF-α, and C-reactive protein (CRP) appear as potential biomarkers that associate sleep duration and mortality (Hall et al. 2015).

As stated above, sleep affects a wide spectrum of immunological parameters, and, thus, it is not surprising that lack of sleep has been associated with alterations in the innate and adaptive responses, influencing susceptibility to pathogens, the course of an infection, as well as the immune response to vaccination (Besedovsky et al. 2019; Garbarino et al. 2021). In addition to medical concerns, factors primarily associated with the modern 24/7 society contribute to the current phenomenon of chronic sleep deprivation (Zomers et al. 2017).

Studies in animal models have shown that sleep manipulation influences the course of an infection. By prolonging the duration of sleep in *Drosophila* through a genetic approach, it was possible to observe greater resistance to bacterial infection and survival (Kuo and Williams 2014). In animal models of sepsis, sleep disruption caused by fragmentation of sleep, intermittent hypoxia, or rapid eye movement (REM) sleep deprivation led to high mortality after septic challenge. Similarly, mice subjected to sleep deprivation and subsequent infection with a malaria parasite had higher mortality. In the case of *Plasmodium* infection, the mechanisms involved include exacerbated production of glucocorticoids, inhibition of follicular helper T (Tfh) cell differentiation, and consequently reduced production of specific antibodies. The combination of sleep deprivation and infection resulted in an additive effect on glucocorticoids synthesis, and chemical inhibition of this exacerbated glucocorticoid synthesis reduced mortality, and restored immune effector functions, suggesting a role of HPA axis hyperactivation in impairing the host immune response under sleep deprivation (Fernandes et al. 2020). It has already been described that prolonged sleep deprivation can lead to the translocation of commensal bacteria from the intestine (e.g., *P. aeruginosa*, *K. pneumoniae*, and *S. aureus*) to other tissues thus causing generalized infections (Krueger and Opp 2016).

In humans, many studies evaluated sleep time and susceptibility to infection and convincingly showed an association between the two variables (Besedovsky et al. 2019; Bryant et al. 2004). A study using a rhinovirus infection model indicated that individuals who slept ≤6 h per night (short sleepers) were at higher risk of developing clinical symptoms of upper respiratory infection (Prather et al. 2015). Short sleepers (<5 h) were more likely to report colds and infections, including influenza, pneumonia, and ear infections (Prather and Leung 2016). In adolescents, reduced sleep duration was associated with a greater number of disease-related events, such as gastroenteritis, cold, and flu (Orzech et al. 2014). There is epidemiological evidence to suggest effects on morbidity and mortality in individuals who are sleep deprived. Participants with less than 7 h of sleep were 2.94 times more likely to develop a cold when infected with rhinovirus than those with 8 h or more of sleep. Furthermore, a shorter duration of sleep in the weeks preceding exposure to a rhinovirus was associated with lower resistance to the disease (Cohen et al. 2009). Similarly, patients with sleep disturbances exhibited a 1.2-fold higher risk of herpes zoster infection than did the comparison cohort (Chung et al. 2016).

A cohort of nearly 57,000 women, who reported sleeping less than 6 h a night, had a significantly higher risk of developing pneumonia compared to those who slept 8 h a night (Patel et al. 2012). From an evolutionary point of view, increased sleep time in mammals has a protective effect, as it is associated with a greater number of immune cells with consequent infection control (Preston et al. 2009).

9.2 Sleep and Vaccination

Vaccination is the most effective and long-lasting strategy to prevent diseases at all stages of life. The goal of a vaccine is to induce a specific immune response after the administration of an attenuated, inactivated pathogen or its subunits (peptide, protein, DNA, and mRNA). Vaccines induce pathogen-specific humoral and cellular immune responses by mimicking infection without causing pathology. In almost all vaccines, neutralizing antibodies are the best immunological correlates of protection (Pollard and Bijker 2021). Figure 9.1 depicts the main vaccine mechanisms of action.

In contrast to the abundance of data on the impact of natural and experimentally induced sleep disturbances on immunological parameters, few studies have investigated its direct effect on the immune response to vaccination (Besedovsky et al. 2019). In this context, the first study in humans examined the influence of sleep restriction (4 h per night) on influenza vaccination. Analysis of the humoral immune response showed a more than 50% reduction in specific antibody titers in the sleep-restricted group compared to subjects with regular sleep (Spiegel et al. 2002). Another study involving adults with and without insomnia observed that insomniacs developed significantly lower levels of antibodies against influenza 4 weeks after vaccination, demonstrating the impact of chronic sleep deprivation on the development of the specific humoral immune response (Taylor et al. 2017). Other studies pointed out that a single night without sleep after vaccination against hepatitis A (HAV), hepatitis B (HBV), and H1N1 (swine flu) significantly reduced the specific

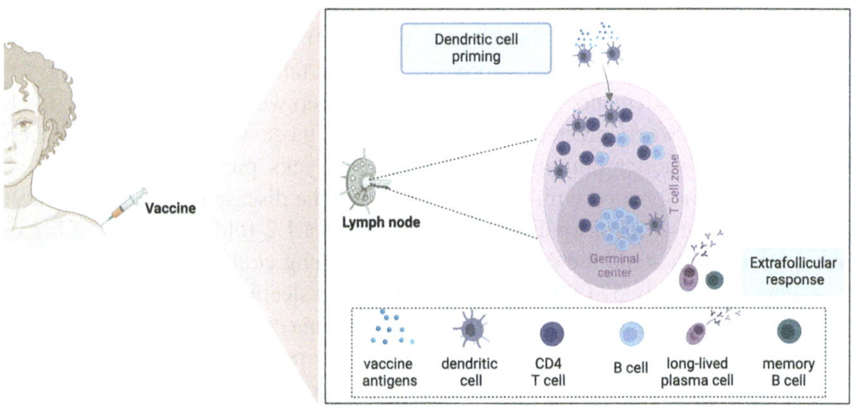

Fig. 9.1 Vaccine mechanism of action. After vaccine administration into the muscle, the vaccine antigen is taken up by dendritic cells, which become activated. Then, dendritic cells migrate to draining lymph nodes and present the antigen for T cells. The interaction between $CD4^+$ T and B cells occurs (help effect), leading to the maturation of the antibody response (increase antibody affinity and induce different antibody isotypes). Over the next weeks, plasma cells secrete vaccine-specific antibodies that can be detected in serum. Long-lived plasma cells and memory B cells are generated

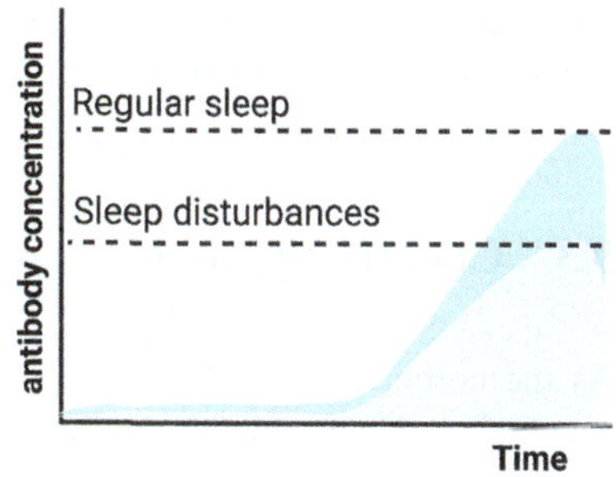

Fig. 9.2 Vaccine-induced antibody titers in individuals with regular sleep compared to individuals with sleep disturbances

antibody response to these vaccines (Lange et al. 2003, 2011; Benedict et al. 2011). In addition, sleep has been considered to influence the number of HAV-specific IFNγ-producing $CD4^+$ T cells (Lange et al. 2011). Another study using the hepatitis B vaccine reported that shorter sleep was associated with a lower antibody response and predicted a decrease in clinical protection against hepatitis B 6 months after vaccination (Prather et al. 2012). It seems that the pro-inflammatory environment induced during sleep acts as an adjuvant that facilitates the transfer of antigenic information from antigen-presenting cells (e.g., dendritic cells) to antigen-specific T cells. More recently, short sleep two nights before influenza vaccination is associated with a lower humoral response (Prather et al. 2021). Using night shift workers as a natural model of sleep restriction, our group also observed a reduction in the humoral immune response after immunization against meningococcus C. Furthermore, the night workers presented a decrease in the duration of the N3 stage and REM sleep, increased inflammatory mediators (TNF-α and IL-6), reduced $CD4^+$ T lymphocytes and plasmacytoid dendritic cells, reduced prolactin levels, and increased regulatory T cells and IL-10 levels (Ruiz et al. 2020). Altogether, these results suggest that sleep provides essential support for the development of a protective vaccine immune response.

In addition to sleep, circadian rhythms regulate different aspects of the immune response. The central circadian clock is located in the suprachiasmatic nucleus (SCN) of the hypothalamus and synchronizes the synthesis and secretion of catecholamines, which leads to the control of immune cell trafficking. In turn, the immune system feeds back through cytokine and chemokine production (Imeri and Opp 2009). Several immune cells and molecules (e.g., cytokines and inflammatory markers) exhibit circadian rhythmicity (levels fluctuate cyclically over 24 h) (for a detailed review, see Haspel et al. 2020). For this reason, it has been suggested that morning vaccination induces a higher antibody response than afternoon vaccination (Phillips et al. 2008; Long et al. 2016). On the other hand, some studies did not find a clear association between these parameters (Kurupati et al. 2017; Anna et al. 2022).

Sleep disturbances, such as sleep restriction, and sleep fragmentation, among others, seem to reduce the vaccine-specific humoral response (Fig. 9.2). Together, these findings link sleep duration with vaccine-induced effector immune response and suggest that sleeping the nights before and after immunization is essential to ensure vaccine effectiveness.

9.3 COVID-19 Pandemic and Sleep

As the coronavirus disease 2019 (COVID-19) pandemic continues, its causative agent, the severe acute respiratory syndrome coronavirus 2 (SARS-CoV-2), has presented numerous and continuing challenges to the scientific community. Despite the new therapies and vaccines that have been and are being evaluated, the contagion returns wave after wave. The pandemic altered the daily life of many people, causing stress, anxiety, and grief, but also affecting sleep patterns and quality. It is estimated that around one in ten people has reported sleep problems during the COVID-19 pandemic (Jahrami et al. 2021).

A meta-analysis indicated a prevalence of 31% of sleep problems among healthcare workers, 18% among the general population, and 57% among COVID-19 patients (Alimoradi et al. 2021). For this reason, portmanteaus, such as Coronasomnia or COVID-somnia, have been used to encompass the myriad of sleep problems. For instance, it has been raised as main conditions: insomnia, nightmares, changes in the sleep–wake cycle (later bedtimes, increased time in bed, the emergence of napping), complaints in sleep patterns, and excessive daytime sleepiness due to reduced nighttime sleep (Gupta and Pandi-Perumal 2020; Bhat and Chokroverty 2022). Furthermore, sleep problems were more common among healthcare workers working on the front line due to increased workload, duration of shifts, and changes in shifts (Bhat and Chokroverty 2022; Lai et al. 2020; Ishikura et al. 2021).

SARS-CoV2 infection can produce systemic inflammation with massive infiltration of pro-inflammatory monocytes and neutrophils (Ramos-Casals et al. 2021) and the release of inflammatory mediators (e.g., TNF-α, IL-6, and CRP) that are known to affect sleep, and vice versa (Irwin 2019; Ibarra-Coronado et al. 2015). As stated above, sleep disturbances also induce an inflammatory environment. Hence, it is plausible to suggest a mechanistic link between sleep disorders and COVID-19 infection. So far, sleep disorders, poor sleep quality, and sleep deprivation have been associated with negative COVID-19 outcomes (Li et al. 2021; Mello et al. 2020). For example, obstructive sleep apnea (OSA) is a risk factor for more severe outcomes from SARS-CoV-2 infections, such as admission to the ICU, mechanical ventilation, and death (Tufik et al. 2020; McSharry et al. 2020; Strausz et al. 2021). These associations may be more relevant among older adults (Pires et al. 2021a), who are at increased risk of severe COVID-19 and present a high prevalence of OSA (Tufik et al. 2010; Pires et al. 2021b). In addition to inflammation, COVID-19 and OSA share common risk factors, such as cardiovascular disease, obesity, diabetes, hypertension, and male sex (Salles and Mascarenhas Barbosa 2020). In individuals with untreated OSA, the risk of developing COVID-19 was more than double (Najafi et al. 2021). Another study found that among patients with COVID-19, OSA was

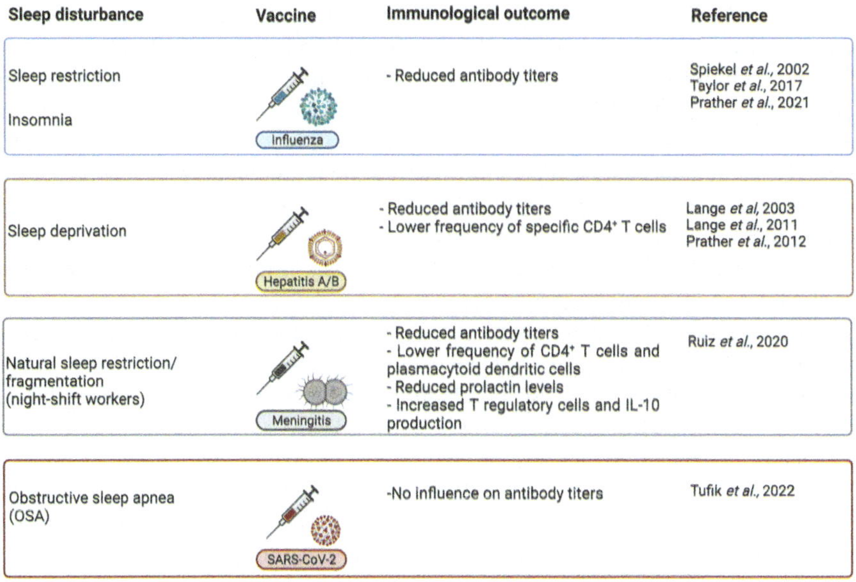

Fig. 9.3 Influence of different sleep disturbances on vaccine-specific immune responses

associated with an increased risk of hospitalization and approximately double the risk of developing respiratory failure (Maas et al. 2021). There is an increased mortality rate in patients with confirmed cases of COVID-19 and OSA (11.7% in participants with OSA vs. 6.9% in non-OSA) (Cade et al. 2020).

Since good sleep builds and improves the immune system, both quantitatively and qualitatively, this could be extrapolated to COVID-19 vaccinations. Protection induced by currently available vaccines against the SARS-CoV-2 virus is based primarily on the production of neutralizing antibodies (Jeyanathan et al. 2020). As stated above, sleep disturbances can affect immune responses after vaccination. Therefore, sleep may function as a natural immune adjuvant for COVID-19 vaccination (Benedict and Cedernaes 2021). For instance, until now, only one study has evaluated the influence of OSA on the antibody response after COVID-19 vaccination. In this work, anti-SARS-CoV-2 antibodies were analyzed among older adults who underwent full-night type-I polysomnography (PSG) and received the CoronaVac (Sinovac Biotech) or ChAdOx1 nCoV-19 (AstraZeneca/Fiocruz) COVID-19 vaccines. The results demonstrated that the increase in OSA severity did not lead to a decrease in anti-SARS-CoV-2 IgG levels, and, therefore, the efficiency of COVID-19 vaccines is not reduced from mild to severe OSA (Tufik et al. 2022). As observed in COVID-19/OSA, other sleep disturbances have also been associated with reduced immune responses to different vaccines (Fig. 9.3).

Regarding circadian rhythmicity, mixed findings were observed when considering the time of the day of COVID-19 vaccination (Zhang et al. 2021; Barnoud et al. 2021; Matryba et al. 2022; Wang et al. 2022). One study did not find significant differences in antibody titers between morning and afternoon vaccination after two doses of the mRNA-based BNT162b2 vaccine (Matryba et al. 2022). On the other hand, others observed significant differences after the administration of an inactivated vaccine (BBIBP-CorV, Sinopharm) (Zhang et al. 2021), mRNA- or adenovirus-based vaccines (Wang et al. 2022) during morning or afternoon. Interestingly, participants who received the inactivated vaccine in the morning presented a higher antibody response, a twofold higher level of neutralizing antibodies, and a higher frequency of antibody-secreting cells, as well as higher frequencies of monocytes, dendritic cells, memory B cells, and Tfh cells (Zhang et al. 2021; Barnoud et al. 2021). Another study carried out with healthcare workers who received either the mRNA-based BNT162b2 or the adenoviral ChAdOx1 nCoV-19 (AstraZeneca) showed that participants vaccinated late in the afternoon presented higher anti-Spike antibody levels (Wang et al. 2022). These contradictory results may reflect differences in vaccine type (inactivated whole virus, mRNA, adenovirus), antigen in the vaccine (T-dependent or independent), number of days after vaccination, number of doses administered before sample collection, time of sample collection, age, health status, sleep patterns, and shiftwork of participants (Anna et al. 2022). Additional studies will be necessary to better understand the relationship between the time of vaccine administration and the induction of an immune response.

In summary, substantial evidence demonstrates that sleep plays an important role in immune system homeostasis and its effects on infection control and vaccine effectiveness. The effects of sleep on the outcome of COVID-19 and SARS-CoV2 immune responses are still poorly explored.

Acknowledgments GNP is a shareholder at SleepUp© and founder of P&P Metanálises. The other authors have no conflicts of interest to disclose. This work was supported by grants from the Associação Fundo de Incentivo à Pesquisa (AFIP), the Coordenação de Aperfeiçoamento de Pessoal de Nível Superior (CAPES), and the Conselho Nacional de Desenvolvimento Científico e Tecnológico (CNPq). MLA and DRS are CNPq fellowship recipients.

References

Alimoradi Z, Brostrom A, Tsang HWH, Griffiths MD, Haghayegh S, Ohayon MM et al (2021) Sleep problems during COVID-19 pandemic and its' association to psychological distress: a systematic review and meta-analysis. EClinicalMedicine 36:100916

Anna C, Whittaker S, Drayson M (2022) Time of day of vaccination does not relate to antibody response to thymus-independent vaccinations. Vaccine 11:100178

Barnoud C, Wang C, Scheiermann C (2021) Timing vaccination against SARS-CoV-2. Cell Res 31(11):1146–1147

Benedict C, Cedernaes J (2021) Could a good night's sleep improve COVID-19 vaccine efficacy? Lancet Respir Med 9(5):447–448

Benedict C, Hallschmid M, Lassen A, Mahnke C, Schultes B, Schioth HB et al (2011) Acute sleep deprivation reduces energy expenditure in healthy men. Am J Clin Nutr 93(6):1229–1236

Besedovsky L, Lange T, Haack M (2019) The sleep-immune crosstalk in health and disease. Physiol Rev 99(3):1325–1380

Bhat S, Chokroverty S (2022) Sleep disorders and COVID-19. Sleep Med 91:253–261

Bryant PA, Trinder J, Curtis N (2004) Sick and tired: does sleep have a vital role in the immune system? Nat Rev Immunol 4(6):457–467

Cade BE, Dashti HS, Hassan SM, Redline S, Karlson EW (2020) Sleep apnea and COVID-19 mortality and hospitalization. Am J Respir Crit Care Med 202(10):1462–1464

Chung WS, Lin HH, Cheng NC (2016) The incidence and risk of Herpes Zoster in patients with sleep disorders: a population-based cohort study. Medicine (Baltimore) 95(11):e2195

Cohen S, Doyle WJ, Alper CM, Janicki-Deverts D, Turner RB (2009) Sleep habits and susceptibility to the common cold. Arch Intern Med 169(1):62–67

Dantzer R (2018) Neuroimmune interactions: from the brain to the immune system and vice versa. Physiol Rev 98(1):477–504

Fernandes ER, Barbosa ML, Amaral MP, de Souza Apostolico J, Sulczewski FB, Tufik S et al (2020) Sleep disturbance during infection compromises Tfh differentiation and impacts host immunity. iScience 23(10):101599

Floam S, Simpson N, Nemeth E, Scott-Sutherland J, Gautam S, Haack M (2015) Sleep characteristics as predictor variables of stress systems markers in insomnia disorder. J Sleep Res 24(3): 296–304

Garbarino S, Lanteri P, Bragazzi NL, Magnavita N, Scoditti E (2021) Role of sleep deprivation in immune-related disease risk and outcomes. Commun Biol 4(1):1304

Gupta R, Pandi-Perumal SR (2020) COVID-somnia: how the pandemic affects sleep/wake regulation and how to deal with it? Sleep Vigil 4(2):51–53

Hall MH, Smagula SF, Boudreau RM, Ayonayon HN, Goldman SE, Harris TB et al (2015) Association between sleep duration and mortality is mediated by markers of inflammation and health in older adults: the Health, Aging and Body Composition Study. Sleep 38(2):189–195

Haspel JA, Anafi R, Brown MK, Cermakian N, Depner C, Desplats P et al (2020) Perfect timing: circadian rhythms, sleep, and immunity - an NIH workshop summary. JCI Insight 5(1):e131487

Ibarra-Coronado EG, Pantaleon-Martinez AM, Velazquez-Moctezuma J, Prospero-Garcia O, Mendez-Diaz M, Perez-Tapia M et al (2015) The bidirectional relationship between sleep and immunity against infections. J Immunol Res 2015:678164

Imeri L, Opp MR (2009) How (and why) the immune system makes us sleep. Nat Rev Neurosci 10(3):199–210

Irwin MR (2019) Sleep and inflammation: partners in sickness and in health. Nat Rev Immunol 19(11):702–715

Irwin MR, Piber D (2018) Insomnia and inflammation: a two hit model of depression risk and prevention. World Psychiatry 17(3):359–361

Irwin MR, Olmstead R, Carroll JE (2016) Sleep disturbance, sleep duration, and inflammation: a systematic review and meta-analysis of cohort studies and experimental sleep deprivation. Biol Psychiatry 80(1):40–52

Ishikura IA, Rosa DS, Hachul H, Pires GN, Tufik S, Andersen ML (2021) Sleep deficit in COVID-19 health-care workers may increase the infection risk. Scand J Public Health 49(7):697–699

Jahrami H, BaHammam AS, Bragazzi NL, Saif Z, Faris M, Vitiello MV (2021) Sleep problems during the COVID-19 pandemic by population: a systematic review and meta-analysis. J Clin Sleep Med 17(2):299–313

Jeyanathan M, Afkhami S, Smaill F, Miller MS, Lichty BD, Xing Z (2020) Immunological considerations for COVID-19 vaccine strategies. Nat Rev Immunol 20(10):615–632

Krueger JM, Opp MR (2016) Sleep and microbes. Int Rev Neurobiol 131:207–225

Kuo TH, Williams JA (2014) Increased sleep promotes survival during a bacterial infection in Drosophila. Sleep 37(6):1077–86, 86A–86D

Kurupati RK, Kossenkoff A, Kannan S, Haut LH, Doyle S, Yin X et al (2017) The effect of timing of influenza vaccination and sample collection on antibody titers and responses in the aged. Vaccine 35(30):3700–3708

Lai J, Ma S, Wang Y, Cai Z, Hu J, Wei N et al (2020) Factors associated with mental health outcomes among health care workers exposed to coronavirus disease 2019. JAMA Netw Open 3(3):e203976

Lange T, Perras B, Fehm HL, Born J (2003) Sleep enhances the human antibody response to hepatitis A vaccination. Psychosom Med 65(5):831–835
Lange T, Dimitrov S, Bollinger T, Diekelmann S, Born J (2011) Sleep after vaccination boosts immunological memory. J Immunol 187(1):283–290
Li P, Zheng X, Ulsa MC, Yang HW, Scheer F, Rutter MK et al (2021) Poor sleep behavior burden and risk of COVID-19 mortality and hospitalization. Sleep 44(8):zsab138
Long JE, Drayson MT, Taylor AE, Toellner KM, Lord JM, Phillips AC (2016) Morning vaccination enhances antibody response over afternoon vaccination: a cluster-randomised trial. Vaccine 34(24):2679–2685
Maas MB, Kim M, Malkani RG, Abbott SM, Zee PC (2021) Obstructive sleep apnea and risk of COVID-19 infection, hospitalization and respiratory failure. Sleep Breath 25(2):1155–1157
Matryba P, Gawalski K, Ciesielska I, Horvath A, Bartoszewicz Z, Sienko J et al (2022) The influence of time of day of vaccination with BNT162b2 on the adverse drug reactions and efficacy of humoral response against SARS-CoV-2 in an observational study of young adults. Vaccines (Basel) 10(3):443
McSharry D, Lam MT, Malhotra A (2020) OSA as a probable risk factor for severe COVID-19. J Clin Sleep Med 16(9):1649
Mello MT, Silva A, Guerreiro RC, da Silva FR, Esteves AM, Poyares D et al (2020) Sleep and COVID-19: considerations about immunity, pathophysiology, and treatment. Sleep Sci 13(3): 199–209
Najafi A, Sadeghniiat-Haghighi K, Akbarpour S, Samadi S, Rahimi B, Alemohammad ZB (2021) The effect of apnea management on novel coronavirus infection: a study on patients with obstructive sleep apnea. Sleep Health 7(1):14–18
Orzech KM, Acebo C, Seifer R, Barker D, Carskadon MA (2014) Sleep patterns are associated with common illness in adolescents. J Sleep Res 23(2):133–142
Patel SR, Malhotra A, Gao X, Hu FB, Neuman MI, Fawzi WW (2012) A prospective study of sleep duration and pneumonia risk in women. Sleep 35(1):97–101
Petta I, Dejager L, Ballegeer M, Lievens S, Tavernier J, De Bosscher K et al (2016) The interactome of the glucocorticoid receptor and its influence on the actions of glucocorticoids in combatting inflammatory and infectious diseases. Microbiol Mol Biol Rev 80(2):495–522
Phillips AC, Gallagher S, Carroll D, Drayson M (2008) Preliminary evidence that morning vaccination is associated with an enhanced antibody response in men. Psychophysiology 45(4):663–666
Pires GN, Ishikura IA, Xavier SD, Petrella C, Piovezan RD, Andersen ML et al (2021a) Sleep in older adults and its possible relations with COVID-19. Front Aging Neurosci 13:647875
Pires GN, Ishikura IA, Xavier SD, Petrella C, Piovezan RD, Xerfan EMS et al (2021b) Sleep in older adults and its possible relations with COVID-19. Front Aging Neurosci 13:647875
Pollard AJ, Bijker EM (2021) A guide to vaccinology: from basic principles to new developments. Nat Rev Immunol 21(2):83–100
Prather AA, Leung CW (2016) Association of insufficient sleep with respiratory infection among adults in the United States. JAMA Intern Med 176(6):850–852
Prather AA, Hall M, Fury JM, Ross DC, Muldoon MF, Cohen S et al (2012) Sleep and antibody response to hepatitis B vaccination. Sleep 35(8):1063–1069
Prather AA, Janicki-Deverts D, Hall MH, Cohen S (2015) Behaviorally assessed sleep and susceptibility to the common cold. Sleep 38(9):1353–1359
Prather AA, Pressman SD, Miller GE, Cohen S (2021) Temporal links between self-reported sleep and antibody responses to the influenza vaccine. Int J Behav Med 28(1):151–158
Preston BT, Capellini I, McNamara P, Barton RA, Nunn CL (2009) Parasite resistance and the adaptive significance of sleep. BMC Evol Biol 9:7
Quatrini L, Ugolini S (2021) New insights into the cell- and tissue-specificity of glucocorticoid actions. Cell Mol Immunol 18(2):269–278
Ramos-Casals M, Brito-Zeron P, Mariette X (2021) Systemic and organ-specific immune-related manifestations of COVID-19. Nat Rev Rheumatol 17(6):315–332

Ruiz FS, Rosa DS, Zimberg IZ, Dos Santos Quaresma MV, Nunes JO, Apostolico JS et al (2020) Night shift work and immune response to the meningococcal conjugate vaccine in healthy workers: a proof of concept study. Sleep Med 75:263–275

Salles C, Mascarenhas Barbosa H (2020) COVID-19 and obstructive sleep apnea. J Clin Sleep Med 16(9):1647

Spiegel K, Sheridan JF, Van Cauter E (2002) Effect of sleep deprivation on response to immunization. JAMA 288(12):1471–1472

Strausz S, Kiiskinen T, Broberg M, Ruotsalainen S, Koskela J, Bachour A et al (2021) Sleep apnoea is a risk factor for severe COVID-19. BMJ Open Respir Res 8(1):e000845

Taylor DJ, Kelly K, Kohut ML, Song KS (2017) Is insomnia a risk factor for decreased influenza vaccine response? Behav Sleep Med 15(4):270–287

Tufik S, Santos-Silva R, Taddei JA, Bittencourt LR (2010) Obstructive sleep apnea syndrome in the Sao Paulo Epidemiologic Sleep Study. Sleep Med 11(5):441–446

Tufik S, Gozal D, Ishikura IA, Pires GN, Andersen ML (2020) Does obstructive sleep apnea lead to increased risk of COVID-19 infection and severity? J Clin Sleep Med 16(8):1425–1426

Tufik SAM, Rosa DS, Tufik SB, Pires GN (2022) Effects of obstructive sleep apnea on SARS-CoV-2 antibody response after vaccination against COVID-19 in older adults. Nat Sci Sleep 14:1203–1211

Wang W, Balfe P, Eyre DW, Lumley SF, O'Donnell D, Warren F et al (2022) Time of day of vaccination affects SARS-CoV-2 antibody responses in an observational study of health care workers. J Biol Rhythm 37(1):124–129

Zhang H, Liu Y, Liu D, Zeng Q, Li L, Zhou Q et al (2021) Time of day influences immune response to an inactivated vaccine against SARS-CoV-2. Cell Res 31(11):1215–1217

Zomers ML, Hulsegge G, van Oostrom SH, Proper KI, Verschuren WMM, Picavet HSJ (2017) Characterizing adult sleep behavior over 20 years-the population-based doetinchem cohort study. Sleep 40(7):zsx085

Chapter 10
Structural Inequity and Racial/Ethnic Disparities in Sleep Health During the COVID-19 Pandemic

Dana M. Alhasan, Symielle A. Gaston, and Chandra L. Jackson

Abstract The onset of the coronavirus disease of 2019 (COVID-19) pandemic and resulting global lockdowns, hospitalizations, and increased unemployment rates have disproportionately affected minoritized racial/ethnic groups, which likely exacerbated existing racial/ethnic disparities in sleep health around the world. In this chapter, we summarize research findings that describe various dimensions of sleep health by race/ethnicity, globally, before and during the ongoing pandemic. Understanding that structural inequities impact sleep-health disparities, we also describe the existing literature on racial/ethnic disparities in sleep during the COVID-19 pandemic from a socioecological perspective at the individual, household, community, and societal levels. We also identify salient research gaps, such as limited objectively measured sleep data, a lack of intersectionality being investigated, and few public health interventions employed. Considering these literature gaps and limitations, we provide future research directions, promote the need to employ an intersectional perspective, as well as, a health equity or structural racism lens, and discuss the importance of multilevel, multi-pronged interventions that can improve sleep health across socially disadvantaged groups.

Keywords Sleep · COVID-19 · Racial groups · Health equity · Racism

D. M. Alhasan · S. A. Gaston
Epidemiology Branch, National Institute of Environmental Health Sciences, Department of Health and Human Services, National Institutes of Health, Research Triangle Park, NC, USA

C. L. Jackson (✉)
Epidemiology Branch, National Institute of Environmental Health Sciences, Department of Health and Human Services, National Institutes of Health, Research Triangle Park, NC, USA

Intramural Program, National Institute on Minority Health and Health Disparities, Department of Health and Human Services, National Institutes of Health, Bethesda, MD, USA
e-mail: Chandra.Jackson@nih.gov

A. S. BaHammam et al. (eds.), *COVID-19 and Sleep: A Global Outlook*, Progress in Sleep Research, https://doi.org/10.1007/978-981-99-0240-8_10

10.1 Overview of Racial/Ethnic Disparities in Sleep Health Before the COVID-19 Pandemic

A health disparity is considered "a health difference that adversely affects defined disadvantaged populations, based on one or more health outcomes" (Duran and Perez-Stable 2019). Health outcomes are typically considered preventable and socially unjust because differential exposure to environmental and social problems can lead to observable biological differences (Braveman 2014). As designated by the National Institutes of Health, groups disproportionately affected by health disparities in the United States (US) include (but are not limited to): Blacks/African Americans; Hispanics/Latinos; Native Americans/Alaskan Natives; Asian Americans; Native Hawaiians and other Pacific Islanders; socioeconomically disadvantaged populations; underserved rural populations; and minoritized sexual or gender groups (Duran and Perez-Stable 2019). Health outcomes referenced in the definition of health disparities include higher disease incidence or prevalence as well as the earlier onset and/or more aggressive progression; greater global burden of disease, such as disability-adjusted life years as measured by population health metrics; premature or excessive mortality from specific health conditions; poorer health behaviors and clinical outcomes that have been related to the aforementioned information; and worse outcomes on even self-reported measures that have been validated and reflect daily functioning or condition-specific symptoms (Duran and Perez-Stable 2019). To achieve the goal of eliminating health disparities, health equity must be achieved, which is considered "the assurance of the conditions needed for optimal health among all people regardless of social group membership (Jones 2000)." This goal would require: "(1) valuing all individuals and populations equally—that is, there are no invisible, undervalued, or disposable people; (2) recognizing and rectifying historical injustices; and (3) providing resources according to need—not equally, but according to need (Jones 2000)."

It is important to note that race is a social construct that assigns categories to individuals who appear phenotypically (e.g., skin color; hair texture; facial features) similar while ethnicity is based on shared culture or customs often common to a particular geographic area (Flanagin et al. 2021). Hence, race/ethnicity is not a marker that denotes genetic differences across populations, but rather a marker of relative social disadvantage or advantage resulting from historical and contemporary forms of structural racism (Bailey et al. 2017). Related to relative disadvantage and advantage, studies of sleep disparities by race/ethnicity often reflect deleterious manifestations of disadvantage or marginalization among minoritized racial groups (Johnson et al. 2019; Jackson et al. 2015).

A review summarizing the published scientific literature on racial/ethnic disparities in sleep health prior to the COVID-19 pandemic indicates that—in the US—minoritized racial/ethnic groups compared to White Americans are generally less likely to get the recommended duration of sleep (Johnson et al. 2019; Liu et al. 2016; McElfish et al. 2021). Lower sleep efficiency also appears more prevalent among minoritized racial/ethnic groups (Williams et al. 2015). Furthermore, minoritized groups generally spend less time in slow wave sleep, which is considered the most physiologically restorative stage (Mezick et al. 2008). Studies have also found a higher likelihood of circadian misalignment, more excessive daytime sleepiness, as well as more variability in sleep timing among minoritized racial/ethnic groups (Patel et al. 2015). Minoritized racial/ethnic groups are also generally less likely to report sleep complaints although objectively measured sleep data commonly indicates worse sleep compared to their White counterparts (Jackson et al. 2018, 2020; Grandner et al. 2010). Furthermore, Black or African Americans, Hispanics/Latinos, and Asians (who appear particularly susceptible) have an especially high prevalence of obstructive sleep apnea although the condition is largely underdiagnosed, untreated, and more severe among these groups compared to their White counterparts (Chen et al. 2015). Although the findings are mixed and methodological limitations can influence data, White adults appear generally more likely to have insomnia (Jean-Louis et al. 2007).

10.1.1 Documented Early Life Sleep Disparities Before the COVID-19 Pandemic

Before the COVID-19 pandemic, it has been shown that racial/ethnic disparities in sleep emerge in infancy (Yu et al. 2021) and persist over the life course (Johnson et al. 2019; Taveras et al. 2010). For example, Black and Hispanic/Latino infants compared to White infants have been shown to exceed a threefold and 2.5-fold higher odds of not getting the recommended amount of at least 12 h of sleep, respectively (Taveras et al. 2010). Nightly and 24-h sleep duration that was confirmed to increase between 1 and 6 months of age increased less among non-Hispanic (NH) Black and Hispanic/Latino compared to NH-White infants in a study of 306 Asian, NH-Black, Hispanic/Latino, and NH-White infants in the USs (Yu et al. 2021). Differences were largely explained by markers of socioeconomic status (i.e., annual household income, and maternal educational attainment). A systematic review of US literature investigating the relationships between race/ethnicity and various sleep variables (i.e., bedtime, bedtime consistency, use of a bedtime routine, nocturnal sleep duration and quality, sleep efficiency) among preschool children identified nine studies (Smith et al. 2019). The authors found that White children in the study tended to go to bed earlier, have more regular bedtime consistency, and were more likely to use a bedtime routine than children of other races. The authors also found that White children compared to children of color had longer nocturnal sleep duration and spent more time in Stage 1 sleep (i.e., non-rapid eye movement sleep). Sleep efficiency was lower among minoritized compared to White children. Moreover, compared to White children who were 5 years old, their Black counterparts took longer naps and were more likely to take naps. Some studies related to total sleep over 24 h found similar sleep duration for White and Black children but higher sleep duration for Hispanic/Latino children and children of "other" races while other studies reported that White children—daily—tended to sleep longer than Black children.

A literature review examined the results of 23 studies investigating racial/ethnic sleep disparities among US school-aged children and adolescents (aged 6–19 years) before the COVID-19 pandemic (Guglielmo et al. 2018). Each eligible study found disparities for at least one sleep characteristic, including sleep duration, wake time, bedtime/sleep onset, efficiency, daytime sleepiness, sleep/wake problems, quality, fragmentation, and night-to-night variability. Most of these studies used self-reported sleep data ($n = 17$) while some used objective measures ($n = 4$) (i.e., polysomnography and wrist actigraphy), and both subjective and objective measures were used in two studies. Most studies investigated sleep duration and generally found that White children slept longer than Black and Hispanic/Latino children. Although data on Asian children were limited, the review found that Asian children slept less than White children but longer than Black and Hispanic/Latino children. Similarly, studies with measured wake and bed times also documented racial/ethnic disparities in that minoritized groups had later bedtimes than White children. Conversely, the included studies that measured sleep/wake problems reported more frequent sleep/wake problems among White adolescents than their minoritized counterparts. When results were stratified by children (aged 6–13 years) and adolescents (aged 14–19 years), racial/ethnic disparities were also found. The literature review found that sleep tended to worsen as children's age increased. For example, the percentage of children's sleep insufficiency increased by 15% across racial/ethnic groups, which underscores the importance of intervening early in life. Because several of the studies are older (i.e., nine published before 2010), sleep disparities may have worsened among current children and adolescents over time due to, for instance, new technologies (e.g., handheld electronic devices), and ongoing public health crises (e.g., COVID-19 pandemic, climate change, state-sanctioned war) (Guglielmo et al. 2018).

10.2 Racial/Ethnic Disparities in Sleep During the COVID-19 Pandemic

Adults and children around the world have had to adapt to changes in their typical daily lives during the coronavirus disease of 2019 (COVID-19) pandemic lockdowns, which started to varying extents in almost all countries across the world in March 2020 (Onyeaka et al. 2021). There were documented changes in where individuals lived and how people worked as well as engaged in recreational activities, which could all affect sleep health (Violant-Holz et al. 2020; Stockwell et al. 2021). There is evidence—even from early in the pandemic—that the health and social consequences of the virus such as joblessness, hospitalizations, and an increased death rate have disproportionately burdened disadvantaged communities (Luck et al. 2022). A recent meta-analysis of 250 studies from 49 countries that did not assess racial/ethnic disparities reported higher levels of sleep disturbances during versus before the pandemic, disproportionately impacting those infected with COVID-19, older adults, children, and healthcare workers (Jahrami et al. 2022). It is plausible that minoritized racial/ethnic groups would be more socially vulnerable to COVID-19 due to historical and contemporary forms of, for instance, racial residential and labor market segregation, which have resulted in minoritized groups being more likely to be essential workers and in positions that were low-wage and without worker protections (e.g., paid sick leave) (Krieger 2020; Williams et al. 2022; Jackson and Johnson 2020). Minoritized groups are also more likely to use public transportation and live in substandard multifamily units with poor ventilation, exacerbating existing health conditions including poor sleep and consequently existing inequities (Abraham et al. 2021). The COVID-19 pandemic's negative impact across communities globally reveals health, social, and economic inequities disproportionately burdening minoritized racial/ethnic groups. The observed racial/ethnic inequities are reinforced by structural racism, a system of policies, norms or customs, and practices that advantage one group over a different group (Bailey et al. 2017).

The objective of this chapter is to summarize—from a socioecological perspective—the existing literature on racial/ethnic disparities in sleep health during the COVID-19 pandemic. We summarize findings from studies published by May 2022, and generally among adult populations. Data that were not published in the English language as well as results focused on special populations (e.g., pregnant women) were not included. In this review of the literature and consistent with the terminology used in the United States, the terms "White" and "non-Hispanic White"; "African American," "Black," "non-Hispanic Black," and "Black/African American"; and "Hispanic/Latino(a)" and "Latino(a)(x)" will, hereafter, be used interchangeably.

Studies investigating racial/ethnic disparities in sleep during the COVID-19 pandemic were primarily conducted in the United States (followed by the United Kingdom and Canada) and surveyed mainly adult participants. Studies were largely cross-sectional, relied on self-reported sleep measures, and had limited generalizability due to, for instance, nonrandom sampling and online survey modalities within target populations, which may not capture a representative sample including socially under-resourced individuals (Cobb et al. 2022; Yip et al. 2021; Simonelli et al. 2021; Zhang et al. 2022a). Nonetheless, studies overall suggested there was persistent and exacerbated racial/ethnic sleep-health disparities during the period of the pandemic studied (Cheng et al. 2022; Troxel et al. 2022; John-Henderson et al. 2021; Bann et al. 2021). The literature we identified among US populations assessed either sleep health within understudied populations or racial/ethnic disparities across multiple sleep dimensions, including sleep duration, sleep quality, sleep disturbances, and sleep disorders such as insomnia along with its severity. Among a sample of White ($n = 148$) and Black ($n = 48$) adults enrolled in the longitudinal Sleep to Prevent Evolving Affecting Disorders trial who previously received insomnia treatment (2016–2017) in Michigan, data suggested that racial/ethnic disparities in insomnia grew disproportionately among Black compared to White adults during the pandemic (Cheng et al. 2022). Conversely, a descriptive study of White and Hispanic/Latino men and women ($n = 2040$) suggested Hispanic/Latino race/ethnicity as a protective factor against the progression of worse sleep quality over the pandemic (French et al. 2022). Similarly, another study found that associations between more COVID-19 experiences and insomnia severity were stronger among White adults compared to People of Color (Dzierzewski et al. 2022).

Longitudinal descriptive studies among minoritized populations offered additional insight into the impacts of the COVID-19 pandemic on sleep health over time among understudied populations in the United States. Using data from surveys administered among American Indian/Alaskan Native (AI/AN) adolescents in California between March 2019 and 2020 as well as May to September 2021, changes in self-reported sleep included delays in bed- and wake times, increases in sleep duration, and increases in sleep disturbance (Troxel et al. 2022). Across four waves of data collection from August to November of 2020 among Blackfeet AI/AN adults in Montana, sleep-health scores assessed through the sleep health scale known as RU-SATED decreased from August to October followed by improvement in sleep health from October to November (John-Henderson et al. 2021). While sleep duration appeared to increase, detrimental impacts on sleep quality may persist.

Cross-sectional studies consistently found that minoritized groups compared to NH-White groups encountered greater COVID-19-related psychosocial stress, which may contribute to poor sleep, although these studies were inconsistent regarding sleep outcomes. For example, using population-based data from the Pew Research Center, a study found no varying associations between COVID-19 stressors and sleep troubles by race/ethnicity, although the study reported Black and Latino adults had higher levels of COVID-19 threat than White adults (Cobb et al. 2022). The COVID-19 Southern Cities Study consisting of a population of 1688 male and female residents of four major cities in the southern United States also reported no differences in self-reported trouble sleeping May to June 2020 among minoritized groups compared to NH-White adults (Goldmann et al. 2021). Similarly, a large cross-sectional study of US adults ($n = 4542$; 93% White and 7% Black) conducted from April to May 2020 found no differences in self-reported changes in sleep duration during the pandemic among Black compared with NH-White adults. However, the investigators did find disproportionate changes to employment and household income among Black compared with NH-White adults (Sparks et al. 2022). Comparably, a more diverse cross-sectional study among 473 US women who self-identified as AI/AN, Asian/Pacific Islander, Black/African American, Hispanic/Latina, Multiracial, NH-White, or some other race/ethnicity also reported that select stressors (e.g., food insecurity, financial hardship) were more common among minoritized compared to NH-White women (Stockman et al. 2021). Further, there were racial/ethnic differences in sleep. Compared to NH-White women, Asian/Pacific Islander and AI/AN women had higher odds, Hispanic/Latina women had similar odds, and Black/African American women had lower odds of self-reported sleep loss due to worrying about COVID-19 between May and June 2020 (Stockman et al. 2021). Racial/ethnic differences in sleep quality, as well as sleep duration, were also found among 547 Asian, Black/African American, Latinx, and White young adults who completed the Pittsburgh Sleep Quality Index (PSQI) during April 2020. Specifically, Black/African American young adults had the shortest sleep duration and poorer sleep quality compared to Asian young adults (Yip et al. 2021). Structural inequities, such as being employed as an essential worker or COVID-19 victimization distress, were also considered as potential explanations for the observed disparities in this study (Yip et al. 2021).

Similarly, changes in sleep duration during COVID-19 (compared to pre-COVID) were found to be larger among ethnic minorities compared to White adults across four age cohorts in the UK (Bann et al. 2021). In a separate longitudinal study of participants aged 16 years and older in the UK, Black and Asian participants were also found to have higher odds of sleep loss compared to White counterparts; however, the association became inversed after adjustment for sociodemographic and mental health characteristics (Falkingham et al. 2022).

There was also one notable cross-sectional descriptive study of sleep among a visible minoritized population in Israel (Ghanamah and Eghbaria-Ghanamah 2021). Among 382 visible minoritized children (i.e., Arab Palestinian) aged 5–11 years in Israel, parents of approximately 40% of the children reported new sleep problems during the COVID-19 pandemic as of December 2020. While the authors suggested potential cultural differences to compare between Israeli Jewish and Palestinian Arab children in future research (Ghanamah and Eghbaria-Ghanamah 2021), structural inequity is also likely to play a role given over 60 laws have shown to negatively impact Arab Palestinians (Adalah: The Legal Center for Arab Minority Rights in Israel 2019).

10.3 Sleep Disparities During the COVID-19 Pandemic from a Socioecological Perspective

10.3.1 Individual Level

Multiple social categories (e.g., race, ethnicity, gender, sexual orientation, socioeconomic status, immigration status, disability) intersect at the individual level from structural systems of oppression (e.g., racism, sexism, heterosexism) impacting health (Bowleg 2012). Therefore, other social determinants of health along with race/ethnicity should be considered to adequately address COVID-19-related sleep-health disparities. Consistent with literature regarding intersectionality, multiple groups experiencing marginalization including minoritized racial/ethnic groups, sexual orientation groups, and women may act as modifiers of associations between COVID-19-related stressors and sleep health (Bann et al. 2021; Falkingham et al. 2022; King and Devonish 2021; Osiogo et al. 2021; Pedrozo-Pupo et al. 2022). Age has also often acted as a modifier and/or risk factor for subjectively poorer sleep health during the COVID-19 pandemic in multiple studies (Simonelli et al. 2021;

Bann et al. 2021; King and Devonish 2021; Osiogo et al. 2021). For example, associations between economic hardship related to COVID-19 and worsening sleep quality is worse among women compared to men and among younger compared to older adults (Almeida et al. 2021; Wielgoszewska et al. 2022).

10.3.1.1 Sex/Gender and Sexual Orientation

Most studies asking about gender use self-reported data and sex choice options (i.e., female or male) where generally it is unclear how participants perceive this question (i.e., biological sex or the social construct of gender). Therefore, we combine sex and gender when summarizing these results when applicable. Although some US studies were racially/ethnically diverse, few considered racial/ethnic differences while investigating associations either with or by sex/gender or by sexual orientation. Nonetheless, most studies suggested greater severity in poor subjective sleep quality and more insomnia symptoms among women compared with men as well as sex/gender as a modifier between COVID-19-related stressors and sleep outcomes (French et al. 2022; Dzierzewski et al. 2022; Fuller-Rowell et al. 2021; Knell et al. 2020; Bigalke et al. 2020). This is in line with data elucidating the burden of poor sleep among minoritized populations and populations with intersectional identities. A study that comprised an understudied population of Latinx transgender women and men found sleep problems to be highly prevalent (67%) and associated with COVID-19-related stressors (e.g., lost job, verbal arguments with a partner) (MacCarthy et al. 2020). In another study of racially/ethnically diverse sexually minoritized men, reports of poorer sleep quality during COVID-19 were also prevalent (43%) (Millar et al. 2021). Further, a study of Latino sexually minoritized men similarly reported a high prevalence of sleep problems among the sample, although sleep problems did not vary by immigration/nativity status (Harkness et al. 2021). It will be important nonetheless to continue investigations among intersectional groups of multiple marginalized identities and assess racial/ethnic differences.

Studies outside the United States also examined sex/gender and sexual orientation as modifiers of sleep during the COVID-19 pandemic. While these studies included racially/ethnically diverse participants, intersectionality was not assessed (i.e., differences between multiple intersecting social categories) (Simonelli et al. 2021; Bann et al. 2021; Falkingham et al. 2022; King and Devonish 2021; Osiogo et al. 2021; Pedrozo-Pupo et al. 2022; Menouni et al. 2022; Ding et al. 2022; Cellini et al. 2021; Kolakowsky-Hayner et al. 2021; Taporoski et al. 2022; Terán-Pérez et al. 2021; Tsukamoto et al. 2021; Monterrosa-Castro and Monterrosa-Blanco 2021). For example, a nationally representative study in the UK examining changes in sleep problems by gender during COVID-19 found that women reported more sleep loss than men (Ding et al. 2022). Studies in other countries (e.g., Japan, Mexico, Brazil) similarly reported worse sleep (e.g., insomnia symptoms, poor sleep quality) concerning COVID-19-related stressors among women compared to men (Cellini et al. 2021; Kolakowsky-Hayner et al. 2021; Taporoski et al. 2022; Terán-Pérez et al.

2021; Tsukamoto et al. 2021), although these studies were not nationally representative. A Swiss study also found that sexually minoritized men compared to heterosexual men demonstrated lower sleep quality (Marmet et al. 2021a). These studies importantly investigated and demonstrated sex/gender and sexual orientation differences in the impacts of COVID-19 on sleep health across the world, and more information about differential impacts can be gained by additionally considering potential racial/ethnic differences.

10.3.1.2 Socioeconomic Status (SES)

While almost all studies included descriptions of the SES of participants, racial/ethnic differences in relationships between SES and sleep during the COVID-19 pandemic were understudied. Further, SES measurement varied by study limiting comparability: educational attainment, income, and privileged status or class, to name a few examples. While there were mixed results, studies generally suggested poorer subjective sleep health among lower compared with higher SES individuals (Simonelli et al. 2021; Bann et al. 2021; Pedrozo-Pupo et al. 2022; Menouni et al. 2022; Agberotimi et al. 2020; Marmet et al. 2021b; Palla et al. 2021). For example, lower SES was associated with lower sleep quality among Swiss men (mean age = 29 years), yet insomnia prevalence was reported as higher among privileged compared with underprivileged men and women in Pakistan (Marmet et al. 2021b; Palla et al. 2021). In a Nigerian sample, the prevalence of insomnia did not vary by SES; however, clinical insomnia symptoms were higher among participants with "standard" compared with "above" and "below" standard incomes, suggesting a possible U-shaped relationship (Agberotimi et al. 2020). Future research to elucidate associations across various populations of the world while also considering race/ethnicity is important given that lower SES is often disproportionately more prevalent among minoritized adults (Johnson et al. 2019). Longitudinal studies to capture the SES impact of COVID-19 are also warranted to decipher the temporary versus long-term impacts of the COVID-19 pandemic on SES among minoritized populations.

10.3.1.3 Health Conditions

Given that certain health conditions (e.g., obesity, type II diabetes, and hypertension) and comorbidities increase the risk for COVID-19, a wealth of studies examined the impact of sleep during the COVID-19 pandemic among those with health conditions and comorbidities. Yet, few of these studies addressed racial/ethnic disparities despite the wide documentation of comorbidities among minoritized racial/ethnic groups. Health conditions that increase COVID-19 are also disproportionately higher among minoritized racial/ethnic groups because of differences in environments and structural racism. More so, intersecting identities are also disproportionately impacted by health conditions further exacerbating health disparities. For example, Black women have the highest prevalence of obesity in the United States compared to other women and men of racial/ethnic groups (Hales et al. 2020).

Despite the lack of examination of racial/ethnic differences, cross-sectional studies reported more frequent insomnia symptoms among those with COVID-19, obesity, diabetes, epilepsy, and other diseases. For instance, 25% of people with epilepsy reported increases in seizures, and 63% of them attributed the increase primarily to disrupted sleep (Casassa et al. 2021). Using a racially diverse sample made up of White (52.6%), Black/African American (26.5%), "other" (11.8%), Asian (3.3%), and Native American (0.4%), as well as Latinx ethnicity (12.9%), another study in the United States found that workers with preexisting behavioral disorders compared to nonworkers reported altered sleep habits (Cook et al. 2022). Moreover, a racially/ethnically diverse sample among participants with obesity (54.5% of the sample were NH-White, 21.5% NH-Black, 19.8% Hispanic/Latino, and 4.2% "other") reported that substance versus non-substance-users had more difficulty falling asleep (OR = 1.64 [1.14–2.34]) (Almandoz et al. 2021).

Most studies outside of the United States reported sleep prevalence among a sample of COVID-19 patients. For example, three studies conducted in China found that: (1) Chinese patients with COVID-19 reported high symptoms of insomnia (Gu et al. 2021); (2) COVID-19 patients with moderate (9.7) or severe (11.0) compared to mild (8.0) symptoms scored higher on PSQI (p-value = 0.088, 0.017, respectively) (Jiang et al. 2021); and (3) female compared to male COVID-19 patients were more likely to have insomnia (Hu et al. 2020). Studies in other countries also reported a high prevalence of insomnia among COVID-19 patients, such as in Saudi Arabia (Alodhayani et al. 2021), India (Chakrabarti 2021), Bangladesh (Nabi et al. 2022), and Colombia (Pedrozo-Pupo et al. 2022). Likewise, studies also reported high sleep disturbances among COVID-19 patients, such as in Pakistan (Imran et al. 2022), Egypt (Abdelghani et al. 2022), and Italy (Aragona et al. 2022). Among hemodialysis patients in Morocco, 49% reported difficulty falling asleep during the COVID-19 pandemic (Alafifi et al. 2022).

10.3.2 Household Structure and Family

Studies have also supported the importance of considering and including measurement of family structure (e.g., single-parent household), domestic conflict/violence, work-family conflict, household demands, and other social environmental characteristics when considering the detrimental impact of COVID-19-related stressors (e.g., economic instability and hardship, social isolation) (Almeida et al. 2021; Yuksel et al. 2021; Antino et al. 2022; Wang et al. 2022a; Galindo-Aldana et al. 2022).

Generally, studies reported poor sleep health among children and parents/guardians. Using data from 9438 parental reports pre-COVID from a national survey and data from 2365 parental reports during COVID, a study found that children had more sleep problems during COVID compared to children's sleep problems pre-COVID in Australia (Olive et al. 2021). Likewise, parents also had significantly poorer sleep quality during COVID compared to parents pre-COVID. While the study accounted for Aboriginal and Torres Strait Islander Peoples's Status, the authors did not explicitly study differences by racial identity (Olive et al. 2021). Another study in Chile also identified decreases in sleep quality among both parents and children (Olhaberry et al. 2022).

Studies also highlighted poor sleep health among various household structures (e.g., single-parent households). For example, among 122 parents/guardians of Asian, Black/African American, White, and "other" race/ethnicity US children aged 6–14 years, children in families highly impacted by COVID-19 had lower parent-reported sleep duration compared to those in families who experienced medium impacts (Sprague et al. 2022). Further, Black children and children from single-parent households were more likely to belong to the high-impact group, suggesting the potential for racial/ethnic disparities (Sprague et al. 2022). Among 210 American Indians, a longitudinal study found that childhood trauma predicted greater declines in sleep quality associated with the onset of the pandemic partly due to psychological stress related to COVID-19 (John-Henderson 2020). Another longitudinal study among 124 parent/guardian-children dyad living in the southeastern part of the United States (primarily European descent) found that deprivation (defined as experiences of neglect that reflect a child's lack of fundamental physical and emotional care from a caregiver) predicted increased sleep problems among boys during the pandemic (Zhang et al. 2022b). Further, while a quasi-experimental study among 281 families living in historically disinvested neighborhoods found no differences in parent sleep or child sleep after school closure as compared to before school closure, racial/ethnic differences were reported among children and parents (Ursache et al. 2022). On average, Black vs Latinx parents reported more sleep for children, and Spanish-speaking vs English-speaking parents reported lower sleep problems for themselves but higher mental health problems for children (Ursache et al. 2022).

Family-related stressors may be associated with poorer subjective sleep quality among older-aged family members as well. For example, a study of college students in China reported that students in single-parent households or who experienced domestic violence had poorer subjective sleep quality compared to students who experienced neither (Wang et al. 2022a). Regarding resilience factors, a study in Europe found family social support linked to better sleep quality among adults aged 18–30 years (Zsido et al. 2022).

10.3.2.1 Caregivers

Studies also assessed sleep prevalence during the COVID-19 pandemic among caregivers and generally reported poor sleep health among family members, yet few considered racial/ethnic sleep disparities. For example, among a sample of 3509 living in Pennsylvania, family caregivers compared to non-caregivers reported higher sleep disturbances as well as higher levels of anxiety, depression, fatigue and lower social participation, lower financial well-being and increased food insecurity, and increased financial worries (Beach et al. 2021). Although there were no statistically significant different associations between NH-Black, NH-"Other", and Hispanic/Latino compared with NH-White in sleep disturbances, the sample was predominantly NH-White (~90%) (Beach et al. 2021). Similarly, another study among 835 participants found that higher caregiver burden was associated with higher sleep-related impairment and sleep disturbance scores, yet Black and Asian compared to White caregivers had lower mean sleep-related impairment scores (Greaney et al. 2022). Finally, another study among 300 caregivers, of whom 58% were Hispanic/Latino, found that caregiver financial strain was associated with inconsistent sleep routines, poorer mental health as well as changes in children's problematic and prosocial behaviors (Reich et al. 2021). Another interesting finding to note is that the majority of their sample included mothers (89%), where women are commonly in caregiver roles (Reich et al. 2021).

Studies outside the United States also found higher poor sleep prevalence among caregivers and other family members during the pandemic. For example, a longitudinal study in Italy found that caregivers compared to non-caregivers had a higher frequency of stress 1 year after the pandemic, and females compared to male caregivers had poorer sleep quality (Busse et al. 2022). Further, a study in India found high levels of insomnia among parents of children with pediatric kidney disease (87.2%) (Sharma et al. 2021). Finally, a study in Belgium found that COVID-19-specific stressors and parental psychological experiences were associated with higher levels of sleep problems, anxiety, and depression among parents of children with a chronic disease (Wauters et al. 2022). The study in Belgium also reported that parents of Arabic-speaking, Asian, Black, and mixed ethnicity compared to White reported more sleep problems; however, the sample of the minoritized racial/ethnic groups was small where White parents made up 96% of the sample (Wauters et al. 2022). Nonetheless, these findings may indicate the disproportionate impact of COVID-19 stressors on minoritized racial/ethnic groups (Wauters et al. 2022). While few studies explicitly studied racial/ethnic sleep disparities, overall studies reported higher caregiver burden was associated with higher sleep-related impairment and more sleep disturbances.

10.3.3 Community

Studies conducted worldwide have demonstrated that community disruptions, such as social isolation, loneliness, and changes in social environments resultant from imposed shutdowns impact multiple dimensions of subjective sleep disturbances (i.e., sleep efficiency, sleep duration, sleep problems, sleep quality, insomnia symptoms) with very few suggesting no associations with sleep duration and sleep problems, including a study of objectively measured sleep in Japan (Simonelli et al. 2021; López-Moreno et al. 2020; Sepúlveda-Loyola et al. 2020; Talbot et al. 2021; Viola and Nunes 2022; Wang et al. 2020; O'Sullivan et al. 2021; Liao et al.

2022; de Almondes et al. 2022; Muhammad et al. 2021; Lewis 2020; Haydon and Salvatore 2021; Grey et al. 2022; Ataka et al. 2022). These studies have been mostly cross-sectional and among adult populations but also conducted across continents of Asia, Europe, and North America. Social support and capital were also frequently investigated worldwide and suggested to act as protective factors/buffers as well as modifiers against associations between COVID-19-related stressors with sleep health (e.g., quality) (Talbot et al. 2021; Wright et al. 2021; Grey et al. 2020; Xiao et al. 2020a, b). Lower support was also associated with subjective sleep problems (Grey et al. 2020; Lee and Waters 2021; Sahin et al. 2022). Despite these important results, studies while inclusive of diverse populations, did not assess potential racial/ ethnic differences, offering an area for further research. Similarly, it will be important to investigate longitudinal associations as a review suggested that sleep disturbances related to social isolation may persist beyond the COVID-19 pandemic (O'Regan et al. 2021).

10.3.4 *Society*

Resources, policies, and other upstream determinants as a result of structural racism and other systems of oppression may contribute to racial/ethnic sleep disparities. Availability and access to resources (e.g., personal protective equipment and healthcare facilities) were suggested to explain differences in insomnia symptoms across Latin American countries, for example (Zhang et al. 2022a). These are crucial to study as they may inform upstream interventions aimed at addressing root causes to help achieve health equity.

10.3.4.1 Racism and COVID-19-Related Discrimination

We consider discrimination a societal problem even though most studies assessed discrimination at the interpersonal level. Studies of racial/ethnic discrimination and sleep during the COVID-19 pandemic occurred in Canada and the United States, but only US studies considered potential variations in relationships by race/ethnicity. A Canadian study did not find associations between racial/ethnic discrimination and decreased sleep duration during the first wave of the COVID-19 pandemic among a population demographically representative of the Canadian population (Currie and Higa 2021). However, racial/ethnic discrimination was associated with worsened sleep problems during the pandemic among US Black men and women in rural Georgia as well as US Asian adults (Lee and Waters 2021; Adesogan et al. 2021). Racial/ethnic discrimination was also associated with higher odds of poor sleep quality only among Black and East Asian adults in a sample of White, Black, Hispanic/Latino, East Asian, South Asian, and Southeast Asian adults in the United States (Shi et al. 2022).

Further, in the United States, Anti-Asian, anti-Black/African American, and anti-Latinx sentiments were either consistent or exacerbated during the COVID-19 pandemic. Specifically in the United States, anti-Asian racism and hate crimes surged where approximately one-half of hate incidents included anti-China or anti-immigrant rhetoric (Yellow Horse et al. 2021). Everyday forms of COVID-19-related discrimination were more strongly associated with more self-reported sleep troubles among Asian and Black adults versus other groups while major forms of COVID-19-related discrimination were more strongly associated with more self-reported sleep troubles among White compared with Black adults (Le et al. 2022). Additionally, a study of COVID-related stigma in China, which reported individuals with more discrimination experiences and higher compared with lower perceived stigma, reported poorer sleep quality but did not consider ethnic background as a potential modifier (Fu et al. 2020). Racism and discrimination in the social environment appeared to contribute to compromised sleep quality and non-recommended sleep duration in multiple studies, suggesting structural racism and related factors either pre- or during the pandemic, in part, may explain racial/ethnic disparities in sleep health (Cheng et al. 2022; Currie and Higa 2021; Yip et al. 2022). Both COVID-19-related discrimination and higher vigilance have also been found to be higher among Black, Hispanic/Latino, and Asian compared to White adults in the COVID-19 Southern Cities Study in the United States, and there were differences in additive associations (Le et al. 2022).

10.3.4.2 Labor and Employment

While employment is similar to SES and usually measured at the individual level, we consider it a societal problem given its ties to structural inequity, especially during global lockdowns (e.g., impact on essential workers who tend to be minoritized). As previously mentioned, economic instability and financial hardships related to COVID-19 were prevalent, associated with sleep health, and often found to disproportionately affect minoritized racial/ethnic groups (Cheng et al. 2022; Goldmann et al. 2021; Stockman et al. 2021; Lima et al. 2021). Across literature to date including reviews of the literature, longitudinal, cross-sectional, and descriptive studies, economic instability was consistently associated with poorer sleep health and possible sleep disorders across geographic areas (i.e., Brazil, US, and United Kingdom) (Cobb et al. 2022; Simonelli et al. 2021; Almeida et al. 2021; Millar et al. 2021; Marmet et al. 2021b; Yuksel et al. 2021; Antino et al. 2022; Wright et al. 2021; Lee et al. 2021; Samuel et al. 2022). Although, two studies suggested there was no association between changes in employment and sleep among US and UK populations (Wielgoszewska et al. 2022; Baird et al. 2022). Two reviews summarized evidence about indicators of economic instability (e.g., unemployment, financial stress, food insecurity) during times of crises, including the COVID-19 pandemic with some discussion of the relationship to disparities (Simonelli et al. 2021; Lee et al. 2021). According to the reviews, articles published by early 2020 supported that job insecurity, fear of unemployment, food insecurity, and financial stress related to public health crises and COVID-19, specifically, were associated with subjectively and objectively measured sleep health including lower sleep efficiency, poor sleep quality, and insomnia symptoms (Simonelli et al. 2021; Lee et al. 2021). Further, economic instability may mediate racial/ethnic disparities in sleep duration because of structural racism (Simonelli et al. 2021). However, most studies did not include investigations of economic instability and sleep health within the context of racial/ethnic disparities. Nonetheless, one longitudinal study among a predominantly Black/African American sample of older adults who reported financial strain also reported that changes in employment status were associated with increases in poor sleep quality from 2018 to March–May 2020 (Baird et al. 2022). A US study restricted to 466 Black fathers found that working in high-risk areas compared to low-risk areas had more sleep disturbances but not sleep quality (Cooper et al. 2021). Since economic impacts are disparate and are consistently associated with poor sleep health, future research among diverse populations is warranted to elucidate economic hardships related to COVID-19 in association with racial/ethnic disparities in sleep health.

Healthcare Workers

The majority of articles reported sleep prevalence among healthcare workers compared to other types of workers. More frequent insomnia symptoms and sleep disturbances (as well as poor mental health outcomes) were reported among various healthcare providers: ICU nurses, reproductive healthcare workers, and outpatient providers. Few of these articles were stratified by race/ethnicity; for a further examination of sleep disturbances among healthcare workers during the COVID-19 pandemic, please refer Chap. 5 (Pappa, Sofia). Essential workers, who experience disparate exposures, are more likely to be minoritized. Therefore, future research needs to assess differences by race/ethnicity. Racism and other social factors are likely major drivers of existing racial disparities. Future research should be conducted at all levels to inform interventions.

Among 14,600 healthcare workers in the United States using the HERO registry made up of NH-White, NH-Black, Asian, and "other" races, 41% reported trouble sleeping (Forrest et al. 2021). Few studies investigated differences by race/ethnicity. Yet, these few studies documented racial/ethnic differences. For example, one study found that Hispanic/Latino ethnicity had increased odds of insomnia compared to NH-White (Stewart et al. 2021).

Studies outside of the United States also did not consider race/ethnicity. For example, we identified two cross-sectional studies conducted in Turkey that found high rates of poor sleep health among healthcare workers (Yilmaz et al. 2021; Şayık et al. 2022). Among 600 healthcare workers, insomnia was higher for those who worked in clinics compared to intensive care units, women compared to men, and nurses compared to physicians (Yilmaz et al. 2021). Another study in Turkey found that emergency department personnel helped cope with stress via adequate sleep (Demir and Ataman 2021). Two studies conducted in Italy also reported a high prevalence of sleep problems (Magnavita et al. 2021) and insomnia (Italia et al. 2021). Specifically, among hospital workers, women compared to men had higher odds of insomnia (OR = 2.20 [1.48–3.28]) in adjusted models (Italia et al. 2021). Furthermore, psycho-oncologists with compared with those without peritraumatic distress (e.g., anxiety, depression, cognitive change, loss of social functioning in the previous week) in Italy were found to use sleep remedies (OR = 3.79 [1.41–10.21]) (Costantini et al. 2022). A high prevalence of insomnia was also reported among healthcare workers in Jilin, China (Dong et al. 2021), Taiwan (Lu et al. 2021) as well as sleep disturbances among frontline nurses in Oman (Al Maqbali 2021), healthcare workers in Iran (Amra et al. 2021), and doctors in India (Chaudhuri et al. 2022).

10.3.4.3 Neighborhood Built Environment

Despite the vast literature demonstrating the influence of the neighborhood environment on health, sleep health, and sleep-health disparities, few studies examined the neighborhood environments' influence on sleep health during the COVID-19 pandemic. Among these few studies, they varied in which neighborhood features were assessed and rarely examined differences by race/ethnicity. We identified three studies conducted in the United States (Testa et al. 2022; Rassu et al. 2021; Finucane et al. 2022). One study found that residents living in more compared to less dangerous neighborhoods had poorer sleep quality (Testa et al. 2022). Residents living in neighborhoods that became more dangerous after the pandemic compared to residents living in neighborhoods that are the same danger level had poorer sleep quality (relative risk ratio = 2.81 [1.31–6.04]). Black, Hispanic/Latino, and "other" race/ethnicity residents living in neighborhoods that became more dangerous after the pandemic compared to White residents living in neighborhoods that are the same danger level also had poorer sleep quality; although, the measure of association was only significant among the Hispanic/Latino group (Testa et al. 2022). In contrast, another study among participants with chronic low back pain did not find any significant differences in reported sleep quality between neighborhood deprivation tertiles (measured via the Area Deprivation Index) since the pandemic (Rassu et al. 2021). While this study did not report findings by race, their sample was made up of 54.5% White or Caucasian, 36.4% Black or African American, 3% "other" race, 2% Asian, and 2% more than one race (Rassu et al. 2021). The third study among residents in predominantly Black, low-income neighborhoods in Pennsylvania found that business/school closures were associated with poorer sleep quality among residents living in areas with lower neighborhood walkability but not among residents living in higher neighborhood walkability (Finucane et al. 2022). Neighborhood walkability as a significant modifier between closure experiences and sleep quality indicates the value of physical resources available to meet the demands of stress during closures. A study in China found that the prevalence of insomnia was higher in urban (29.5%) compared to rural (25.35%) residents (Liu et al. 2021). Given the more significant impact on urban residents' work and businesses resulting from precautionary strategies (e.g., social distancing, lockdown), this may explain the higher prevalence of insomnia observed. The documented rural global disparities also indicate the need to understand the depth of racial/ethnic sleep disparities. Health outcomes, including sleep health, follow place-based patterns and depend on the neighborhood in which one lives, works, and plays (Williams and Cooper 2020).

10.3.4.4 Education Access and Quality

Education access was disrupted by the COVID-19 pandemic where most governments worldwide closed schools at almost all levels (United Nations Educational, Scientific and Cultural Organization n.d.). Both changes to education access/modalities (e.g., in-person, online) concerning sleep as well as sleep health among students across ages (e.g., secondary school, college) were widely studied across the world. Yet, there were few investigations of teachers and racial/ethnic differences (Fuller-Rowell et al. 2021; Ursache et al. 2022; Benham 2021; Charles et al. 2021; Gruber et al. 2020; Hendriksen et al. 2021; Liu et al. 2022; Meltzer et al. 2021; Souto-Manning and Melvin 2022; Zhang et al. 2021; Chandler et al. 2021; Sarsak 2022; Fila-Witecka et al. 2021; Niba Rawlings et al. 2022; Wang et al. 2022b; Widyastari et al. 2022; García and Garvey 2022; Caldwell et al. 2021; Staller et al. 2022; Wieman et al. 2022; Prabhat et al. 2022). Overall, studies assessed subjective sleep and suggested higher sleep disturbances, increases in insomnia symptoms, inconsistent associations with both excessive daytime sleepiness and sleep quality (although most suggested a decrease in sleep quality), worse sleep efficiency, and poorer sleep hygiene but longer or no changes in sleep duration during the pandemic among students (Ursache et al. 2022; Benham 2021; Charles et al. 2021; Gruber et al. 2020; Hendriksen et al. 2021; Liu et al. 2022; Meltzer et al. 2021; Souto-Manning and Melvin 2022; Zhang et al. 2021; Chandler et al. 2021; Sarsak 2022; Fila-Witecka et al. 2021; Niba Rawlings et al. 2022; Wang et al. 2022b; Widyastari et al. 2022; García and Garvey 2022; Caldwell et al. 2021; Staller et al. 2022; Wieman et al. 2022; Prabhat et al. 2022; Kurz et al. 2022). A sample in the UK consisting of a racially/ethnically diverse population reported that almost one-half of the 107 adolescent participants had worsened sleep quality (Marques and Braidwood 2021).

Racial/ethnic differences were assessed in a few US studies. One study compared a sample of college students after shutdowns were imposed to another sample 6 months later and found no differences in sleep problems. However, White students reported more sleep problems compared to Black/African American students (Charles et al. 2021). Among 263 students (mostly aged 18–25 years) at a southeastern college, sleep problems increased more among White compared to Black students (Fuller-Rowell et al. 2021). Another study among early childhood education teachers in the New York City area found sleep disturbances were higher among Latinx compared to multiracial teachers (Souto-Manning and Melvin 2022). Evidence suggests that the sleep quality of both students and teachers may have been negatively impacted by the COVID-19 pandemic. Given the racial/ethnic differences in sleep health observed in US studies, it will be important to extend the vast literature on student and teacher populations during COVID-19 to assess possible racial/ethnic disparities, even if retrospective.

10.4 Gaps in the Literature on Sleep Disparities During the COVID-19 Pandemic

Few studies were longitudinal, assessed objective sleep, investigated intersectionality (i.e., belonging to multiple groups experiencing marginalization), and employed an intervention. Further, most studies that considered racial/ethnic disparities were conducted among the United States, the United Kingdom, and Canadian populations with the majority of studies in the United States. Since most studies were conducted during the early phases of the COVID-19 pandemic (e.g., March–May 2020), there is limited data on exacerbated racial/ethnic sleep disparities throughout the pandemic.

10.5 Future Research Directions

There is a need for future research to incorporate lessons learned and close gaps in knowledge in the field. Few studies examined racial/ethnic differences despite the known adverse impacts of structural racism on health, which were exacerbated during the COVID-19 pandemic. For example, despite the known heterogeneity among Asian Americans, there remains limited disaggregated data among this group. Yet anti-Asian racism surged during the COVID-19 pandemic, which demonstrates the urgency to adequately study minoritized racial/ethnic groups. Future researchers can employ a Syndemic framework to account for compounding and exacerbating adverse effects from interactions of health conditions, including poor sleep health, with inequality caused by structural racism (e.g., poverty, stigmatization, stress) among Asian Americans (Saw et al. 2022). By understanding that race/ethnicity is a social construct and a proxy measure for racism, studies can appropriately study racial/ethnic differences through stratification as opposed to adjustment where White is the "referent" population. Incorporating descriptive measures to assess the burden of both the exposure and outcome can provide salient information regarding health disparities and thus can better inform resource allocation (Ward et al. 2019). Similarly, other appropriate public health measures aimed at assessing public health impact, such as the population-attributable risk, should be included (Cox and Li 2012). There is also a need for longitudinal studies that use an exposomic, life course approach given that sleep disparities emerge in infancy. Furthermore, future research should employ an intersectional lens as well as a structural racism framework. A previous study conducted by chapter authors demonstrated how sleep disparities may, at least in part, be explained by the neighborhood environment across multiple identities where age, sex/gender, and race/ethnicity impact each other to explain sleep disparities (Alhasan et al. 2020). Studying multiple intersecting identities, and employing an intersectional perspective in general, can disentangle the complex ways identities experience sleep disparities and thus help us move towards achieving health equity. In addition to intersectionality and structural racism as well as other systems of oppression, it is increasingly evident for future research to also consider a global perspective. The COVID-19 pandemic is ongoing and transmission rates can only be reduced in countries if they are reduced globally. There is a need for global cooperation to end vaccine apartheid (Bajaj et al. 2022).

Regardless of the adequate effort to document racial/ethnic sleep disparities during COVID-19, there is a wealth of evidence demonstrating the detrimental impact of structural racism on health and thus future directions must also include conducting interventions aimed at addressing root causes. Interventions to improve population health and address health disparities should center efforts around assuring the material and social conditions (especially those known to modulate sleep and circadian rhythms) for optimal health among all individuals regardless of a social identity group. Interventions can also modify adverse physical and social exposures of the neighborhood environment to improve sleep health across marginalized groups. For example, because environmental toxicants from swine industrial livestock have been shown to impact sleep (MacNell et al. 2021), interventions in the physical environment that could improve sleep include air pollution mitigation strategies. Furthermore, the implementation of the World Health Organization guidelines has the potential to prevent deaths directly related to air particulate matter ($PM_{2.5}$) (WHO 2021).

10.6 Conclusion

Prior to the pandemic, minoritized racial/ethnic groups were disproportionately burdened by insufficient sleep, poor sleep quality, and sleep disorders including sleep apnea (Johnson et al. 2019). Social and environmental factors (e.g., residential segregation, labor discrimination, household crowding, social disorder, air pollution) likely contributing to these stark sleep disparities have further exacerbated sleep disparities during the ongoing pandemic (Jackson and Johnson 2020). Health conditions, worker strain, and employment burden have further exacerbated sleep throughout the pandemic. While a potentially traumatizing public health crisis like the COVID-19 pandemic and associated global lockdown impacted sleep universally, studies included assessed sleep health during the beginning of the pandemic (March–May 2020) where it is possible that novel stressors may have not further exacerbated considerable disparities in sleep or severe inequities, and implications for sleep may not have been fully realized. Privileged, affluent groups may have temporarily experienced poor sleep health during this time frame, which may dissipate as conditions for these groups return to "normal". However, poorer sleep health among marginalized groups is likely to remain without interventions to address the social disadvantage. Despite limited data assessing racial/ethnic disparities in sleep during the ongoing pandemic, there remains an urgent need to intervene and improve health. Specifically, upstream interventions addressing the root causes of structural racism may help close the gap of racial/ethnic inequities. To address racial/ethnic sleep disparities, research can apply a health equity lens by, for instance, promoting universal healthcare available to all citizens, and governments can implement structural interventions aimed at achieving and establishing equity, such as livable wages (Jackson and Johnson 2020). In turn, addressing sleep disparities will subsequently improve and promote overall health.

Conflicts of Interest

None declared.
Sources of Funding This work was funded by the Intramural Program at the NIH, National Institute of Environmental Health Sciences (Z1AES103325-01).

References

Abdelghani M, Alsadik M, Abdelmoaty A, Atwa S, Said A, Hassan M (2022) Sleep disturbances following recovery from COVID-19: a comparative cross-sectional study in Egypt. East Mediterr Health J 28(1):14–22

Abraham P, Williams E, Bishay AE, Farah I, Tamayo-Murillo D, Newton IG (2021) The roots of structural racism in the United States and their manifestations during the COVID-19 pandemic. Acad Radiol 28(7):893–902

Adesogan O, Lavner JA, Carter SE, Beach SRH (2021) COVID-19 stress and the health of Black Americans in the rural south. Clin Psychol Sci. https://doi.org/10.1177/2167702621104937

Agberotimi SF, Akinsola OS, Oguntayo R, Olaseni AO (2020) Interactions between socioeconomic status and mental health outcomes in the Nigerian context amid COVID-19 pandemic: a comparative study. Front Psychol 11:559819

Al Maqbali M (2021) Sleep disturbance among frontline nurses during the COVID-19 pandemic. Sleep Biol Rhythms 19(4):467–473

Alafifi R, Maazouzi C, Chaanoun K, Elkhayat SS, Mtioui N, Zamd M et al (2022) POS-922 psychological impact of Covid 19 in hemodialysis patients. Kidney Int Rep 7(2):S402

Alhasan DM, Gaston SA, Jackson WB II, Williams PC, Kawachi I, Jackson CL (2020) Neighborhood social cohesion and sleep health by age, sex/gender, and race/ethnicity in the United States. Int J Environ Res Public Health 17(24):9475

Almandoz JP, Xie L, Schellinger JN, Mathew MS, Bismar N, Ofori A et al (2021) Substance use, mental health and weight-related behaviours during the COVID-19 pandemic in people with obesity. Clin Obes 11(2):e12440

Almeida WDS, Szwarcwald CL, Malta DC, Barros MBA, Souza Júnior PRB, Azevedo LO et al (2021) Changes in Brazilians' socioeconomic and health conditions during the COVID-19 pandemic. Rev Bras Epidemiol 23:e200105

Alodhayani AA, Almutairi KM, Alshobaili FA, Alotaibi AF, Alkhaldi G, Vinluan JM et al (2021) Predictors of mental health status among quarantined COVID-19 patients in Saudi Arabia. Healthcare (Basel) 9(10):1271

Amra B, Salmasi M, Soltaninejad F, Sami R, Nickpour M, Mansourian M et al (2021) Healthcare workers' sleep and mood disturbances during COVID-19 outbreak in an Iranian referral center. Sleep Breath 25(4):2197–2204

Antino M, Ruiz-Zorrilla P, Sanz-Vergel AI, Leon-Perez JM, Rodriguez-Munoz A (2022) The role of job insecurity and work-family conflict on mental health evolution during COVID-19 lockdown. Eur J Work Organ Psychol. https://doi.org/10.1080/1359432X.2022.2049251

Aragona M, Tumiati MC, Ferrari F, Viale S, Nicolella G, Barbato A et al (2022) Psychopathological effects of the Coronavirus (Sars-CoV-2) imposed lockdown on vulnerable patients in treatment in a mental health outpatient department for migrants and individuals in poor socioeconomic conditions. Int J Soc Psychiatry 68(1):203–209

Ataka T, Kimura N, Eguchi A, Matsubara E (2022) Changes in objectively measured lifestyle factors during the COVID-19 pandemic in community-dwelling older adults. BMC Geriatr 22(1):326

Bailey ZD, Krieger N, Agénor M, Graves J, Linos N, Bassett MT (2017) Structural racism and health inequities in the USA: evidence and interventions. Lancet 389(10077):1453–1463

Baird MD, Dubowitz T, Cantor J, Troxel WM (2022) Examining the impact of employment status on sleep quality during the COVID-19 pandemic in two low-income neighborhoods in Pittsburgh, PA. Sleep 45(3):zsab303

Bajaj SS, Maki L, Cody Stanford F (2022) Vaccine apartheid: global cooperation and equity. Lancet 399(10334):1452–1453

Bann D, Villadsen A, Maddock J, Hughes A, Ploubidis GB, Silverwood R et al (2021) Changes in the behavioural determinants of health during the COVID-19 pandemic: gender, socioeconomic and ethnic inequalities in five British cohort studies. J Epidemiol Community Health 75(12): 1136–1142

Beach SR, Schulz R, Donovan H, Rosland AM (2021) Family caregiving during the COVID-19 pandemic. Gerontologist 61(5):650–660

Benham G (2021) Stress and sleep in college students prior to and during the COVID-19 pandemic. Stress Health 37(3):504–515

Bigalke JA, Greenlund IM, Carter JR (2020) Sex differences in self-report anxiety and sleep quality during COVID-19 stay-at-home orders. Biol Sex Differ 11(1):56

Bowleg L (2012) The problem with the phrase women and minorities: intersectionality-an important theoretical framework for public health. Am J Public Health 102(7):1267–1273

Braveman P (2014) What are health disparities and health equity? We need to be clear. Public Health Rep 129(S2):5–8

Busse C, Barnini T, Zucca M, Rainero I, Mozzetta S, Zangrossi A et al (2022) Depression, anxiety and sleep alterations in caregivers of persons with dementia after 1-year of COVID-19 pandemic. Front Psych 13:826371

Caldwell HAT, Hancock Friesen CL, Kirk SFL (2021) The effects of the COVID-19 pandemic on health and well-being of children and youth in Nova Scotia: youth and parent perspectives. Front Pediatr 9:725439

Casassa C, Moss R, Goldenholz DM (2021) Epilepsy during the COVID-19 pandemic lockdown: a US population survey. Epileptic Disord 23(2):257–267

Cellini N, Conte F, De Rosa O, Giganti F, Malloggi S, Reyt M et al (2021) Changes in sleep timing and subjective sleep quality during the COVID-19 lockdown in Italy and Belgium: age, gender and working status as modulating factors. Sleep Med 77:112–119

Chakrabarti S (2021) Mental health in hospitalised COVID 19 patients in quarantine during second wave in a South Indian private teaching hospital. J Multidiscip Healthc 14:2777–2789

Chandler AJ, Arent MA, Cintineo HP, Torres-McGehee TM, Winkelmann ZK, Arent SM (2021) The impacts of COVID-19 on collegiate student-athlete training, health, and well-being. Transl J Am Coll Sport Med 6(4):11

Charles NE, Strong SJ, Burns LC, Bullerjahn MR, Serafine KM (2021) Increased mood disorder symptoms, perceived stress, and alcohol use among college students during the COVID-19 pandemic. Psychiatry Res 296:113706

Chaudhuri A, Sood R, Paul S, Saha E (2022) A study of perceived stress level and sleep quality among physicians in a hospital of West Bengal, during the second wave of the present coronavirus disease-19 pandemic. Nat J Physiol Pharm Pharmacol 12(1):21–25

Chen X, Wang R, Zee P, Lutsey PL, Javaheri S, Alcantara C et al (2015) Racial/ethnic differences in sleep disturbances: the multi-ethnic study of atherosclerosis (MESA). Sleep 38(6):877–888

Cheng P, Casement MD, Cuellar R, Johnson DA, Kalmbach D, Cuamatzi Castelan A et al (2022) Sleepless in COVID-19: racial disparities during the pandemic as a consequence of structural inequity. Sleep 45(1):zsab242

Cobb RJ, Sheehan CM, Nguyen AW, Johnson D (2022) COVID-19 hardships and self-reported sleep quality among American adults in March and April 2020: results from a nationally representative panel study. Sleep Health 8(3):288–293

Cook JA, Swarbrick M, Boss KA, Steigman PJ, Nemec P, Jonikas JA et al (2022) The importance of employment to workers with preexisting behavioral health disorders during the COVID-19 pandemic. Psychiatr Rehabil J 45(1):11–17

Cooper SM, Thomas A, Bamishigbin O (2021) Black American fathers employed in higher-risk contexts for contracting COVID-19: implications for individual wellbeing and work-family spillover. Am J Mens Health 15(2):15579883211005617

Costantini A, Mazzotti E, Serpentini S, Piattelli A, Scarponi D, De Benedetta G et al (2022) COVID-19 pandemic distress among a sample of Italian psycho-oncologists: risk of isolation and loneliness. Tumori 108(1):77–85

Cox C, Li X (2012) Model-based estimation of the attributable risk: a loglinear approach. Comput Stat Data Anal 56(12):4180–4189

Currie CL, Higa EK (2021) The impact of racial and non-racial discrimination on health behavior change among visible minority adults during the COVID-19 pandemic. J Racial Ethn Health Disparities:1–9

de Almondes KM, Castro EAS, Paiva T (2022) Sleep habits, quality of life and psychosocial aspects in the older age: before and during COVID-19. Front Neurosci 16:694894

Demir M, Ataman A (2021) Emergency medicine physicians' approaches to coping with stress in COVID-19 pandemic. Eurasian J Emerg Med 20(4):241–254

Ding X, Brazel DM, Mills MC (2022) Gender differences in sleep disruption during COVID-19: cross-sectional analyses from two UK nationally representative surveys. BMJ Open 12(4): e055792

Discriminatory Laws in Israel. Adalah: The Legal Center for Arab Minority Rights in Israel (2019). https://www.adalah.org/en/law/index

Dong HS, Gao JJ, Dong YX, Han CX, Sun L (2021) Prevalence of insomnia and anxiety among healthcare workers during the COVID-19 pandemic in Jilin Province. Braz J Med Biol Res 54(9):e10602

Duran DG, Perez-Stable EJ (2019) Novel approaches to advance minority health and health disparities research. Am J Public Health 109(S1):S8–S10

Dzierzewski JM, Dautovich ND, Ravyts SG, Perez E, Soto P, Donovan EK (2022) Insomnia symptoms during the COVID-19 pandemic: an examination of biopsychosocial moderators. Sleep Med 91:175–178

Falkingham JC, Evandrou M, Qin M, Vlachantoni A (2022) Prospective longitudinal study of 'Sleepless in Lockdown': unpacking differences in sleep loss during the coronavirus pandemic in the UK. BMJ Open 12(1):e053094

Fila-Witecka K, Senczyszyn A, Kołodziejczyk A, Ciułkowicz M, Maciaszek J, Misiak B et al (2021) Lifestyle changes among Polish university students during the COVID-19 pandemic. Int J Environ Res Public Health 18(18):9571

Finucane ML, Beckman R, Ghosh-Dastidar M, Dubowitz T, Collins RL, Troxel W (2022) Do social isolation and neighborhood walkability influence relationships between COVID-19 experiences and wellbeing in predominantly Black urban areas? Landsc Urban Plan 217:104264

Flanagin A, Frey T, Christiansen SL (2021) Updated guidance on the reporting of race and ethnicity in medical and science journals. JAMA 326(7):621–627

Forrest CB, Xu H, Thomas LE, Webb LE, Cohen LW, Carey TS et al (2021) Impact of the early phase of the COVID-19 pandemic on US healthcare workers: results from the HERO Registry. J Gen Intern Med 36(5):1319–1326

French MT, Mortensen K, Timming AR (2022) Changes in self-reported health, alcohol consumption, and sleep quality during the COVID-19 pandemic in the United States. Appl Econ Lett 29(3):219–225

Fu W, Wang C, Zou L, Guo Y, Lu Z, Yan S et al (2020) Psychological health, sleep quality, and coping styles to stress facing the COVID-19 in Wuhan, China. Transl Psychiatry 10(1):225

Fuller-Rowell TE, Nichols OI, Doan SN, Adler-Baeder F, El-Sheikh M (2021) Changes in depressive symptoms, physical symptoms, and sleep-wake problems from before to during the COVID-19 pandemic among emerging adults: inequalities by gender, socioeconomic position, and race. Emerg Adulthood 9(5):492–505

Galindo-Aldana GM, Padilla-López LA, Torres-González C, García-León IA, Padilla-Bautista JA, Alvarez-Núñez DN (2022) Effects of socio-familial behavior on sleep quality predictive risk factors in individuals under social isolation. Int J Environ Res Public Health 19(6):3702

García IJ, Garvey AM (2022) Economic and social interactions in business students during COVID-19 confinement: relationship with sleep disturbance. Behav Sci (Basel) 12(4):100

Ghanamah R, Eghbaria-Ghanamah H (2021) Impact of COVID-19 pandemic on behavioral and emotional aspects and daily routines of Arab Israeli children. Int J Environ Res Public Health 18(6)

Goldmann E, Hagen D, Khoury EE, Owens M, Misra S, Thrul J (2021) An examination of racial and ethnic disparities in mental health during the Covid-19 pandemic in the U.S. South. J Affect Disord 295:471–478

Grandner MA, Patel NP, Gehrman PR, Xie D, Sha D, Weaver T, Gooneratne N (2010) Who gets the best sleep? Ethnic and socioeconomic factors related to sleep complaints. Sleep Med 11(5): 470–478. https://doi.org/10.1016/j.sleep.2009.10.006. PMCID: PMC2861987

Greaney ML, Kunicki ZJ, Drohan MM, Nash CC, Cohen SA (2022) Sleep quality among informal caregivers during the COVID-19 pandemic: a cross-sectional study. Gerontol Geriatr Med 8: 23337214211057387

Grey I, Arora T, Thomas J, Saneh A, Tohme P, Abi-Habib R (2020) The role of perceived social support on depression and sleep during the COVID-19 pandemic. Psychiatry Res 293:113452

Grey I, Arora T, Sanah A (2022) Generalized anxiety mediates the relationship between loneliness and sleep quality amongst young adults during the COVID-19 pandemic. Psychol Rep 2022: 332941221079723

Gruber R, Saha S, Somerville G, Boursier J, Wise MS (2020) The impact of COVID-19 related school shutdown on sleep in adolescents: a natural experiment. Sleep Med 76:33–35

Gu Y, Zhu Y, Xu F, Xi J, Xu G (2021) Factors associated with mental health outcomes among patients with COVID-19 treated in the Fangcang Shelter Hospital in China. Asia Pac Psychiatry 13(2):e12443

Guglielmo D, Gazmararian JA, Chung J, Rogers AE, Hale L (2018) Racial/ethnic sleep disparities in US school-aged children and adolescents: a review of the literature. Sleep Health 4(1):68–80

Hales CM, Carroll MD, Fryar CD, Ogden CL (2020) Prevalence of obesity and severe obesity among adults: United States, 2017–2018. NCHS Data Brief, no 360. National Center for Health Statistics, Hyattsville, MD

Harkness A, Weinstein ER, Mayo D, Rodriguez-Diaz C, Safren SA (2021) Latinx sexual minority men's behavioral, psychosocial, and medical experiences during COVID-19: differences across immigration statuses. Ann LGBTQ Public Popul Health 2(2):104–115

Haydon KC, Salvatore JE (2021) A prospective study of mental health, well-being, and substance use during the initial COVID-19 pandemic surge. Clin Psychol Sci 10(1):58–73

Hendriksen PA, Garssen J, Bijlsma EY, Engels F, Bruce G, Verster JC (2021) COVID-19 lockdown-related changes in mood, health and academic functioning. Eur J Investig Health Psychol Educ 11(4):1440–1461

Hu Y, Chen Y, Zheng Y, You C, Tan J, Hu L et al (2020) Factors related to mental health of inpatients with COVID-19 in Wuhan, China. Brain Behav Immun 89:587–593

Imran N, Aamer I, Afzal H, Hashmi A, Shabbir B, Asif A (2022) Psychiatric impact on COVID-19 patients isolated in a tertiary care hospital in Pakistan. EMHJ 28(1):5–13

Italia S, Costa C, Briguglio G, Mento C, Muscatello MRA, Alibrandi A et al (2021) Quality of life, insomnia and coping strategies during COVID-19 pandemic in hospital workers. A cross-sectional study. Int J Environ Res Public Health 18(23):12466

Jackson CL, Johnson DA (2020) Sleep disparities in the era of the COVID-19 pandemic highlight the urgent need to address social determinants of health like the virus of racism. J Clin Sleep Med 16(8):1401–1402

Jackson CL, Redline S, Emmons KM (2015) Sleep as a potential fundamental contributor to disparities in cardiovascular health. Annu Rev Public Health 36:417–440

Jackson CL, Patel SR, Jackson WB II, Lutsey PL, Redline S (2018) Agreement between self-reported and objectively measured sleep duration among white, black, Hispanic, and Chinese adults in the United States: multi-ethnic study of atherosclerosis. Sleep 41(6):zsy057

Jackson CL, Ward JB, Johnson DA, Sims M, Wilson J, Redline S (2020) Concordance between self-reported and actigraphy-assessed sleep duration among African-American adults: findings from the Jackson Heart Sleep Study. Sleep 43(3):zsz246

Jahrami HA, Alhaj OA, Humood AM et al (2022) Sleep disturbances during the COVID-19 pandemic: a systematic review, meta-analysis, and meta-regression. Sleep Med Rev 62:101591

Jean-Louis G, Magai C, Consedine NS, Pierre-Louis J, Zizi F, Casimir GJ et al (2007) Insomnia symptoms and repressive coping in a sample of older Black and White women. BMC Womens Health 7:1

Jiang Z, Zhu P, Wang L, Hu Y, Pang M, Ma S et al (2021) Psychological distress and sleep quality of COVID-19 patients in Wuhan, a lockdown city as the epicenter of COVID-19. J Psychiatr Res 136:595–602

John-Henderson NA (2020) Childhood trauma as a predictor of changes in sleep quality in American Indian adults during the COVID-19 pandemic. Sleep Health 6(6):718–722

John-Henderson NA, Oosterhoff B, Hall B, Johnson L, Lafromboise ME, Malatare M et al (2021) Covid-19 and changes in sleep health in the Blackfeet Community. Sleep Med 85:87–93

Johnson DA, Jackson CL, Williams NJ, Alcántara C (2019) Are sleep patterns influenced by race/ethnicity - a marker of relative advantage or disadvantage? Evidence to date. Nat Sci Sleep 11:79–95

Jones CP (2000) Levels of racism: a theoretic framework and a gardener's tale. Am J Public Health 90(8):1212–1215

King J, Devonish D (2021) Mental health and COVID-19 perceptions in a predominant black population in the Eastern Caribbean: an exploratory study of residents of Barbados. J Mental Health Training Educ Pract. https://doi.org/10.1108/JMHTEP-07-2021-0085

Knell G, Robertson MC, Dooley EE, Burford K, Mendez KS (2020) Health behavior changes during COVID-19 pandemic and subsequent "stay-at-home" orders. Int J Environ Res Public Health 17(17):6268

Kolakowsky-Hayner SA, Goldin Y, Kingsley K, Alzueta E, Arango-Lasprilla JC, Perrin PB et al (2021) Psychosocial impacts of the COVID-19 quarantine: a study of gender differences in 59 countries. Medicina (Kaunas) 57(8):789

Krieger N (2020) ENOUGH: COVID-19, structural racism, police brutality, plutocracy, climate change- and time for health justice, democratic governance, and an equitable, sustainable future. Am J Public Health 110(11):1620–1623

Kurz D, Braig S, Genuneit J, Rothenbacher D (2022) Lifestyle changes, mental health, and health-related quality of life in children aged 6–7 years before and during the COVID-19 pandemic in South Germany. Child Adolesc Psychiatry Mental Health 16(1):20

Le PD, Misra S, Hagen D, Wang SM, Li T, Brenneke SG et al (2022) Coronavirus disease (COVID-19) related discrimination and mental health in five U.S. southern cities. Stigma Health. https://doi.org/10.1037/sah0000351

Lee S, Waters SF (2021) Asians and Asian Americans' experiences of racial discrimination during the COVID-19 pandemic: impacts on health outcomes and the buffering role of social support. Stigma Health 6(1):70–78

Lee S, Deason K, Rancourt D, Gray HL (2021) Disentangling the relationship between food insecurity and poor sleep health. Ecol Food Nutr 60(5):580–595

Lewis K (2020) COVID-19: preliminary data on the impact of social distancing on loneliness and mental health. J Psychiatr Pract 26(5):400–404

Liao G, Huang B, Lee PM, Zhao S, Chan CK, Tai LB et al (2022) Differences in sleep patterns and mental health problems during different periods of COVID-19 outbreak among community-dwelling older men in Hong Kong. Int J Public Health 67:1604363

Lima MG, Barros MBA, Szwarcwald CL, Malta DC, Romero DE, Werneck AO et al (2021) Association of social and economic conditions with the incidence of sleep disorders during the COVID-19 pandemic. Cad Saude Publica 37(3):e00218320

Liu Y, Wheaton AG, Chapman DP, Cunningham TJ, Lu H, Croft JB (2016) Prevalence of healthy sleep duration among adults — United States, 2014. MMWR 65(6):134–165

Liu L, Xue P, Li SX, Zhang J, Zhou J, Zhang W (2021) Urban-rural disparities in mental health problems related to COVID-19 in China. Gen Hosp Psychiatry 69:119–120

Liu Y, Frazier PA, Porta CM, Lust K (2022) Mental health of US undergraduate and graduate students before and during the COVID-19 pandemic: differences across sociodemographic groups. Psychiatry Res 309:114428

López-Moreno M, López MTI, Miguel M, Garcés-Rimón M (2020) Physical and psychological effects related to food habits and lifestyle changes derived from Covid-19 home confinement in the Spanish population. Nutrients 12(11):3445

Lu MY, Ahorsu DK, Kukreti S, Strong C, Lin YH, Kuo YJ et al (2021) The prevalence of post-traumatic stress disorder symptoms, sleep problems, and psychological distress among COVID-19 frontline healthcare workers in Taiwan. Front Psychiatry 12:705657

Luck AN, Preston SH, Elo IT, Stokes AC (2022) The unequal burden of the Covid-19 pandemic: capturing racial/ethnic disparities in US cause-specific mortality. SSM Popul Health 17:101012

MacCarthy S, Izenberg M, Barreras JL, Brooks RA, Gonzalez A, Linnemayr S (2020) Rapid mixed-methods assessment of COVID-19 impact on Latinx sexual minority men and Latinx transgender women. PLoS One 15(12):e0244421

MacNell NS, Jackson CL, Heaney CD (2021) Relation of repeated exposures to air emissions from swine industrial livestock operations to sleep duration and awakenings in nearby residential communities. Sleep Health 7(5):528–534

Magnavita N, Soave PM, Antonelli M (2021) A one-year prospective study of work-related mental health in the intensivists of a COVID-19 hub hospital. Int J Environ Res Public Health 18(18): 9888

Marmet S, Wicki M, Gmel G, Gachoud C, Bertholet N, Studer J (2021a) Sexual minority orientation is associated with greater psychological impact due to the COVID-19 crisis-evidence from a longitudinal cohort study of young Swiss men. Front Public Health 9:692884

Marmet S, Wicki M, Gmel G, Gachoud C, Daeppen JB, Bertholet N et al (2021b) The psychological impact of the COVID-19 crisis is higher among young Swiss men with a lower socioeconomic status: evidence from a cohort study. PLoS One 16(7):e0255050

Marques SS, Braidwood R (2021) Impact of the coronavirus lockdown on older adolescents engaged in a school-based stress management program: changes in mental health, sleep, social support, and routines. Child Sch 43(4):198–208

McElfish PA, Narcisse MR, Selig JP, Felix HC, Scott AJ, Long CR (2021) Effects of race and poverty on sleep duration: analysis of patterns in the 2014 native Hawaiian and Pacific Islander National Health Interview Survey and General National Health Interview Survey Data. J Racial Ethn Health Disparities 8(4):837–843

Meltzer LJ, Saletin JM, Honaker SM, Owens JA, Seixas A, Wahlstrom KL et al (2021) COVID-19 instructional approaches (in-person, online, hybrid), school start times, and sleep in over 5,000 U.S. adolescents. Sleep 44(12)

Menouni A, Berni I, Abchouch T, Khouchoua M, Filali-Zegzouti Y, Janssen P et al (2022) Mental health among the Moroccan population during SARS-CoV-2 outbreak: MAROCOVID study. J Affect Disord 308:343–352

Mezick EJ, Matthews KA, Hall M, Strollo PJ Jr, Buysse DJ, Kamarck TW et al (2008) Influence of race and socioeconomic status on sleep: Pittsburgh SleepSCORE project. Psychosom Med 70(4):410–416

Millar BM, Adebayo T, Dellucci TV, Behar E, Starks TJ (2021) Keeps me awake at night: the potential of the COVID-19 pandemic to affect sleep quality among sexual minority men in the U.S.A. Psychol Sex Orientat Gend Divers 8(2):213–219

Monterrosa-Castro Á, Monterrosa-Blanco A (2021) Prevalence of sleep disorders in menopausal women during the COVID-19 pandemic in Colombia. Revista Colombiana de Obstetricia y Ginecologia 72(2):365–373

Muhammad DG, Abubakar IA, Gbonjubola YT, Ahmad HY, Sadiq AAA, Abdullahi S (2021) Impact of Covid-19 on lifestyle related behaviors of Nigerians: a cross sectional survey. Acta Medica Martiniana 21(3):118–128

Nabi SG, Rashid MU, Sagar SK, Ghosh P, Shahin M, Afroz F et al (2022) Psychological impact of COVID-19 pandemic: a cross-sectional study of hospitalized COVID-19 patients in an urban setting, Bangladesh. Heliyon 8(3):e09110

Niba Rawlings N, Akwah EA, Musisi J, Awanchiri K, Babirye R, Emalieu D et al (2022) Perceived risks of COVID-19, attitudes towards preventive guidelines and impact of the lockdown on students in Uganda: a cross-sectional study. PLoS One 17(4):e0266249

O'Regan D, Jackson ML, Young AH, Rosenzweig I (2021) Understanding the impact of the COVID-19 pandemic, lockdowns and social isolation on sleep quality. Nat Sci Sleep 13: 2053–2064

O'Sullivan R, Burns A, Leavey G, Leroi I, Burholt V, Lubben J et al (2021) Impact of the covid-19 pandemic on loneliness and social isolation: a multi-country study. Int J Environ Res Public Health 18(19):9982

Olhaberry M, Sieverson C, Franco P, Romero M, Tagle T, Iribarren D et al (2022) The impact of COVID-19 on experiences of pregnancy and/or early parenting in Chile. Infant Ment Health J 43(1):8–23

Olive LS, Sciberras E, Berkowitz TS, Hoare E, Telford RM, O'Neil A et al (2021) Child and parent physical activity, sleep, and screen time during COVID-19 and associations with mental health: implications for future psycho-cardiological disease? Front Psych 12:774858

Onyeaka H, Anumudu CK, Al-Sharify ZT, Egele-Godswill E, Mbaegbu P (2021) COVID-19 pandemic: a review of the global lockdown and its far-reaching effects. Sci Prog 104(2): 368504211019854

Osiogo F, Shalaby R, Adegboyega S, Hrabok M, Gusnowski A, Vuong W et al (2021) COVID-19 pandemic: demographic and clinical correlates of disturbed sleep among 6,041 Canadians. Int J Psychiatry Clin Pract 25(2):164–171

Palla AH, Amin F, Durrani N, Ur-Rehman N, Ikram R (2021) Covid-19 pandemic-prevalence of depression, anxiety and insomnia during Covid-19 lockdown and its associated factors in Karachi, Pakistan. Pak J Pharm Sci 34(4 Suppl):1597–1605

Patel SR, Sotres-Alvarez D, Castaneda SF, Dudley KA, Gallo LC, Hernandez R et al (2015) Social and health correlates of sleep duration in a US Hispanic population: results from the Hispanic community health study/study of Latinos. Sleep 38(10):1515–1522

Pedrozo-Pupo JC, Caballero-Domínguez CC, Campo-Arias A (2022) Prevalence and variables associated with insomnia among COVID-19 survivors in Colombia. Acta Biomed 93(1): e2022019

Prabhat A, Kumar A, Bhardwaj SK (2022) Prolonged lockdown due to COVID-19 alters sleep-wake timings and negatively affects self-esteem, personality, depression and anxiety in college-going Indian students. Sleep Vigil 6(1):199–210

Rassu FS, McFadden M, Aaron RV, Wegener ST, Ephraim PL, Lane E et al (2021) The relationship between neighborhood deprivation and perceived changes for pain-related experiences among

US patients with chronic low back pain during the COVID-19 pandemic. Pain Med 22(11): 2550–2565
Reich SM, Dahlin M, Tulagan N, Kerlow M, Cabrera N, Piroutek MJ et al (2021) Caregivers' experiences during the COVID-19 pandemic and their children's behavior. J Fam Issues. https://doi.org/10.1177/0192513X21105551
Sahin F, Karadag F, Kucukkarapinar M (2022) A cross-sectional study investigating mental health among Turkish citizens during the COVID-19 pandemic: the importance of perceived social support. Psychiatry Behav Sci 12(1):14–25
Samuel LJ, Dwivedi P, Hladek M, Cudjoe TKM, Drazich BF, Li Q et al (2022) The effect of COVID-19 pandemic-related financial challenges on mental health and well-being among U.S. older adults. J Am Geriatr Soc 70(6):1629–1641
Sarsak HI (2022) Psychological impact of the COVID-19 pandemic on medical and rehabilitation sciences university students in Saudi Arabia. Work 71(3):473–480
Saw A, Yi SS, Ethoan LN, Tsoh JY, Yellow Horse AJ, Kwon SC et al (2022) Improving Asian American health during the Syndemic of COVID-19 and racism. EClinicalMedicine 45:101313
Şayık D, Açıkgöz A, Kaya S (2022) Anxiety and sleep quality levels of emergency medical personnel and factors affecting them during the coronavirus disease-2019 pandemic. J Turk Sleep Med 9:27–36
Sepúlveda-Loyola W, Rodríguez-Sánchez I, Pérez-Rodríguez P, Ganz F, Torralba R, Oliveira DV et al (2020) Impact of social isolation due to COVID-19 on health in older people: mental and physical effects and recommendations. J Nutr Health Aging 24(9):938–947
Sharma R, Kumar K, Pilania R, Dawman L, Kaur N, Sharma R et al (2021) Alarming rates of psychological problems among caregivers of pediatric kidney patients admitted during the COVID-19 pandemic lockdown. Ind Psychiatry J 30(1):84–89
Shi L, Zhang D, Martin E, Chen Z, Li H, Han X et al (2022) Racial discrimination, mental health and behavioral health during the COVID-19 pandemic: a national survey in the United States. J Gen Intern Med 37(10):2496–2504
Simonelli G, Petit D, Delage JP, Michaud X, Lavoie MD, Morin CM et al (2021) Sleep in times of crises: a scoping review in the early days of the COVID-19 crisis. Sleep Med Rev 60:101545
Smith JP, Hardy ST, Hale LE, Gazmararian JA (2019) Racial disparities and sleep among preschool aged children: a systematic review. Sleep Health 5(1):49–57
Souto-Manning M, Melvin SA (2022) Early childhood teachers of color in New York City: heightened stress, lower quality of life, declining health, and compromised sleep amidst COVID-19. Early Child Res Q 60:34–48
Sparks JR, Kebbe M, Flanagan EW, Beyl RA, Altazan AD, Yang S et al (2022) Impact of COVID-19 stay-at-home orders on health behaviors and anxiety in Black and White Americans. J Racial Ethn Health Disparities 9(5):1932–1936
Sprague NL, Rundle AG, Ekenga CC (2022) The COVID-19 pandemic as a threat multiplier for childhood health disparities: evidence from St. Louis, MO. J Urban Health 99(2):208–217
Staller N, Kalbacher L, Randler C (2022) Impact of pandemic lockdown on learning behaviour and sleep quality in German students: results of an online survey before and during the pandemic. Somnologie 26(2):98–105
Stewart N, Koza A, Dhaon S, Shoushtari C, Martinez M, Arora V (2021) Sleep loss in healthcare workers during the COVID-19 pandemic. Sleep 44(722)
Stockman JK, Wood BA, Anderson KM (2021) Racial and ethnic differences in COVID-19 outcomes, stressors, fear, and prevention behaviors among US women: Web-based cross-sectional study. J Med Internet Res 23(7):e26296
Stockwell S, Trott M, Tully M, Shin J, Barnett Y, Butler L et al (2021) Changes in physical activity and sedentary behaviours from before to during the COVID-19 pandemic lockdown: a systematic review. BMJ Open Sport Exerc Med 7(1):e000960
Talbot J, Charron V, Konkle AT (2021) Feeling the void: lack of support for isolation and sleep difficulties in pregnant women during the COVID-19 pandemic revealed by Twitter data analysis. Int J Environ Res Public Health 18(2):393

Taporoski TP, Beijamini F, Gómez LM, Ruiz FS, Ahmed SS, von Schantz M et al (2022) Subjective sleep quality before and during the COVID-19 pandemic in a Brazilian rural population. Sleep Health 8(2):167–174

Taveras EM, Gillman MW, Kleinman K, Rich-Edwards JW, Rifas-Shiman SL (2010) Racial/ethnic differences in early-life risk factors for childhood obesity. Pediatrics 125(4):686–695

Terán-Pérez G, Portillo-Vásquez A, Arana-Lechuga Y, Sánchez-Escandón O, Mercadillo-Caballero R, González-Robles RO et al (2021) Sleep and mental health disturbances due to social isolation during the COVID-19 pandemic in Mexico. Int J Environ Res Public Health 18(6):2804

Testa A, Fahmy C, Hill TD (2022) Perceptions of neighborhood dangerousness and changes in sleep quality during the COVID-19 pandemic: assessing the mediating role of changes in health behaviors. Prev Med 156:106991

Troxel WM, Palimaru AI, Klein DJ, Dong L, Dickerson DL, Brown RA et al (2022) Changes in sleep-wake patterns and disturbances before and during COVID-19 in urban American Indian/Alaska native adolescents. Behav Sleep Med 20(3):343–356

Tsukamoto R, Kataoka Y, Mino K, Ishibashi N, Shibata M, Matsuo H et al (2021) Gender differences in anxiety among COVID-19 inpatients under isolation: a questionnaire survey during the first and second waves of the COVID-19 pandemic in Japan. Front Public Health 9:708965

United Nations Educational, Scientific and Cultural Organization. Education: from school closure to recovery. https://www.unesco.org/en/covid-19/education-response

Ursache A, Barajas-Gonzalez RG, Adhikari S, Kamboukos D, Brotman LM, Dawson-McClure S (2022) A quasi-experimental study of parent and child well-being in families of color in the context of COVID-19 related school closure. SSM Popul Health 17:101053

Viola TW, Nunes ML (2022) Social and environmental effects of the COVID-19 pandemic on children. J Pediatr 98(Suppl 1):S4–S12

Violant-Holz V, Gallego-Jimenez MG, Gonzalez-Gonzalez CS, Munoz-Violant S, Rodriguez MJ, Sansano-Nadal O et al (2020) Psychological health and physical activity levels during the COVID-19 pandemic: a systematic review. Int J Environ Res Public Health 17(24):9419

Wang S, Zhang Y, Ding W, Meng Y, Hu H, Liu Z et al (2020) Psychological distress and sleep problems when people are under interpersonal isolation during an epidemic: a nationwide multicenter cross-sectional study. Eur Psychiatry 63(1):e77

Wang R, He L, Xue B, Wang X, Ye S (2022a) Sleep quality of college students during COVID-19 outbreak in China: a cross-sectional study. Altern Ther Health Med 28(3):58–64

Wang D, Zhao J, Zhai S, Huang S, Yang Z, Pan Y et al (2022b) Longitudinal trajectories of insomnia symptoms among college students during the COVID-19 lockdown in China. J Psychosom Res 157:110795

Ward JB, Gartner DR, Keyes KM, Fliss MD, McClure ES, Robinson WR (2019) How do we assess a racial disparity in health? Distribution, interaction, and interpretation in epidemiological studies. Ann Epidemiol 29:1–7

Wauters A, Vervoort T, Dhondt K, Soenens B, Vansteenkiste M, Morbee S et al (2022) Mental health outcomes among parents of children with a chronic disease during the COVID-19 pandemic: the role of parental burn-out. J Pediatr Psychol 47(4):420–431

WHO (2021) What are the WHO Air quality guidelines? Improving health by reducing air pollution. https://www.who.int/news-room/feature-stories/detail/what-are-the-who-air-quality-guidelines

Widyastari DA, Kesaro S, Rasri N, Saonuam P, Katewongsa P (2022) Learning methods during school closure and its correlation with anxiety and health behavior of Thai students. Front Pediatr 10:815148

Wielgoszewska B, Maddock J, Green MJ, Di Gessa G, Parsons S, Griffith GJ et al (2022) The UK Coronavirus Job Retention Scheme and diet, physical activity, and sleep during the COVID-19 pandemic: evidence from eight longitudinal population surveys. BMC Med 20(1):147

Wieman ST, Fields JS, Arditte Hall KA, MacDonald HZ, Liverant GI (2022) Effects of the COVID-19 pandemic on anhedonia, reward exposure and responsiveness, and sleep in college students. J Am Coll Health:1–5
Williams DR, Cooper LA (2020) COVID-19 and health equity-a new kind of "herd immunity". JAMA 323(24):2478–2480
Williams NJ, Grandner MA, Snipes A, Rogers A, Williams O, Airhihenbuwa C et al (2015) Racial/ethnic disparities in sleep health and health care: importance of the sociocultural context. Sleep Health 1(1):28–35
Williams PC, Binet A, Alhasan DM, Riley NM, Jackson CL (2022) Urban planning for health equity must employ an intersectionality framework. J Am Plan Assoc. https://doi.org/10.1080/01944363.2022.2079550
Wright L, Steptoe A, Fancourt D (2021) Are adversities and worries during the COVID-19 pandemic related to sleep quality? Longitudinal analyses of 46,000 UK adults. PLoS One 16(3):e0248919
Xiao H, Zhang Y, Kong D, Li S, Yang N (2020a) Social capital and sleep quality in individuals who self-isolated for 14 days during the coronavirus disease 2019 (COVID-19) outbreak in January 2020 in China. Med Sci Monit 26:e923921
Xiao H, Zhang Y, Kong D, Li S, Yang N (2020b) The effects of social support on sleep quality of medical staff treating patients with coronavirus disease 2019 (COVID-19) in January and February 2020 in China. Med Sci Monit 26:e923549
Yellow Horse AJ, Jeung R, Lim R, Tang B, Im M, Higashiyama L, et al (2021) Stop AAPI Hate National Report: 3/19/20–6/30/21
Yilmaz Y, Erdogan A, Bahadir E (2021) Fear, anxiety, burnout, and insomnia levels of healthcare workers during COVID-19 pandemic in Turkey. Psychiatr Danub 33:S350–S356
Yip T, Feng Y, Fowle J, Fisher CB (2021) Sleep disparities during the COVID-19 pandemic: an investigation of AIAN, Asian, Black, Latinx, and White young adults. Sleep Health 7(4): 459–467
Yip T, Chung K, Chae DH (2022) Vicarious racism, ethnic/racial identity, and sleep among Asian Americans. Cultur Divers Ethnic Minor Psychol. https://doi.org/10.1037/cdp0000534
Yu X, Quante M, Rueschman M, Ash T, Kaplan ER, Guo N et al (2021) Emergence of racial/ethnic and socioeconomic differences in objectively measured sleep-wake patterns in early infancy: results of the Rise & SHINE study. Sleep 44(3):zsaa193
Yuksel D, McKee GB, Perrin PB, Alzueta E, Caffarra S, Ramos-Usuga D et al (2021) Sleeping when the world locks down: correlates of sleep health during the COVID-19 pandemic across 59 countries. Sleep Health 7(2):134–142
Zhang Y, Wang D, Zhao J, Chen XY, Chen H, Ma Z et al (2021) Insomnia and other sleep-related problems during the remission period of the COVID-19 pandemic: a large-scale survey among college students in China. Psychiatry Res 304:114153
Zhang SX, Batra K, Xu W, Liu T, Dong RK, Yin A et al (2022a) Mental disorder symptoms during the COVID-19 pandemic in Latin America - a systematic review and meta-analysis. Epidemiol Psychiatr Sci 31:e23
Zhang L, Cui Z, Sasser J, Carvalho C, Oshri A (2022b) Family stress during the pandemic worsens the effect of adverse parenting on adolescent sleep quality. Child Abuse Negl 123:105390
Zsido AN, Arato N, Inhof O, Matuz-Budai T, Stecina DT, Labadi B (2022) Psychological well-being, risk factors, and coping strategies with social isolation and new challenges in times of adversity caused by the COVID-19 pandemic. Acta Psychol 225:103538

Chapter 11
The Effect of Cognitive-Behavioral Therapy and Coping Strategies on Sleep Disturbances During the COVID-19 Pandemic

Horacio Balam Álvarez-García, Isaías Vicente Lugo-González, and Ulises Jiménez Correa

Abstract The COVID-19 pandemic brought about several health problems, due to the disease itself, but also due to the indirect effects of the health measures adopted to reduce the risk of contagion or to the management of infected people. Sleep was affected worldwide, with alterations reported both in health personnel and in the general population that had to abide by the confinement measures. Insomnia emerged as the main sleep disorder, which had a direct impact on the execution of daily activities, reduced productivity, decreased quality of life, and increased the risk of developing cardiovascular and respiratory problems and mental disorders (anxiety and depression). In this context, several research groups have implemented specific cognitive-behavioral interventions for insomnia to prevent its chronicity. Within this chapter, the general panorama of sleep disturbances, psychiatric comorbidities, and stress levels that affected the population is addressed, followed by a conceptualization of the model of coping with stress, the repercussions on health, and thus understanding the role of coping strategies to manage stress. Finally, the cognitive-behavioral interventions used during confinement to improve sleep indicators in the population are mentioned.

H. B. Álvarez-García (✉)
Sleep Disorders Clinic, Research Division, School of Medicine, National Autonomous University of Mexico, UNAM, Ciudad de México, Mexico

I. V. Lugo-González
Interdisciplinary Research Unit in Health Sciences and Education (IRUHSE), Faculty of Higher Studies Iztacala, National Autonomous University of Mexico, UNAM, Ciudad de México, Mexico

U. J. Correa
Sleep Disorders Clinic, Research Division, School of Medicine, National Autonomous University of Mexico, UNAM, Ciudad de México, Mexico

Postgraduate Program in Behavioral Neuroscience, Psychology Faculty, National Autonomous University of Mexico, UNAM, Ciudad de México, Mexico

A. S. BaHammam et al. (eds.), *COVID-19 and Sleep: A Global Outlook*, Progress in Sleep Research, https://doi.org/10.1007/978-981-99-0240-8_11

Keywords Sleep disorders · Cognitive-behavioral therapy · Coping · Sleep quality · Insomnia

11.1 Introduction

Sleep is a very important physiological process for maintaining physical and mental health and quality of life (Silva et al. 2020). The sleep–wake cycle was affected at the beginning of the COVID-19 pandemic in March 2020 when quarantine was established as a measure to reduce the risk of contagion. The population had to abide by the measures of physical confinement and social distancing proposed by the World Health Organization to reduce the number of infections (Onyeaka et al. 2021). However, confinement limited exposure to sunlight in the morning and increased exposure to artificial light at night. This explains the increase in sleep latency and the phase shift of the sleep cycle concerning the light-dark cycle, leading to the alteration of homeostatic and circadian factors (Borbély 1982).

At the same time, confinement and social distancing caused changes in work hours, eating, and sleeping habits (Di Renzo et al. 2020). Online work and school voided the need to get up early to get to school or work on time, and the quarantine caused a decrease in the time spent on physical exercise and loss of exposure to environmental indicators of time, known as zeitgebers (e.g., the noise of the first daytime activities). In addition, by decreasing physical interaction with other people, the risk of developing symptoms of mental illness such as anxiety and depression increased (Gloster et al. 2020). It has been suggested that living in strict quarantine, excessive exposure to information related to the pandemic, job loss, and drastic lifestyle changes contributed to emotional stress in entire populations around the world.

11.2 Epidemiological Data on Sleep and Mental Health During the COVID-19 Pandemic

The first studies clearly showed the effect of the pandemic on characteristics of mental health such as sleep and mood. The COVID-19 pandemic caused a worsening of sleep quality associated with a negative mood (Targa et al. 2021). Lin et al. (2021) studied the early impact of the pandemic on sleep-disordered and psychological symptoms in healthcare workers and adults in the general population, finding prevalences of 20%, 15.8%, 18.5%, and 24.5% for insomnia, acute stress, anxiety, and depression, respectively.

Zhou et al. (2020) found a prevalence of insomnia symptoms of 23.2% in a sample of adolescents and young adults; mainly insomnia for the female sex, having anxiety or depression. They found however that family support was a protective factor against insomnia.

Li et al. (2020) found that during the pandemic, the prevalence of insomnia increased from 26.2% to 36.7%, anxiety from 16.1% to 26.5%, and depression from 22.7% to 31.2%. Insomnia symptoms worsened by 12.5%, even though time in bed (485.5 $\pm$ 76.2 vs. 531.5 $\pm$ 94.2 min) and total sleep time (432.8 $\pm$ 65.6 vs. 466.9 $\pm$ 96.6) also increased (Li et al. 2020). In addition, the time for going to bed and getting up was delayed. In Italy, 57% of respondents reported complaints of poor sleep quality, anxiety, and stress, mainly among females and the youngest people (Casagrande et al. 2020).

Another study reported prevalences of anxiety of 35.1%, symptoms of depression of 20.1%, and poor quality sleep of 18.2%, mainly in young people, hospital staff, and those who consumed more information about the pandemic (Huang and Zhao 2020).

In France, Peretti-Wate, et al. (2020) applied a survey on coronavirus and confinement; finding an increase in depression symptoms from 7% to 8.8% and sleep disorders from 49.7% to 68.7%; primarily in the 18–25 age group.

In another investigation, Deng et al. (2021) implemented the COVID Collateral impacts (COCOS) study and evaluated symptoms of depression, generalized anxiety, and sleep disorders. They found a prevalence of 24.7%, 23.2%, and 42.2%, respectively; of which 17.4% reported moderate or severe insomnia (Deng et al. 2021). In a systematic review and meta-analysis, Zhang et al. (2020) studied the prevalence of depression, anxiety, and symptoms of sleep disorders; 31 studies were identified with a sample of 5153 participants; identifying a combined prevalence of 45%, 47%, and 34% for depression, anxiety, and insomnia or poor sleep quality, respectively. They found no significant differences among genders in these prevalences.

11.3 Circadian Rhythms Disorders During the COVID-19 Pandemic

Since the beginning of the COVID-19 pandemic, the possibility has been raised that the combination of exposure to artificial lighting late into the night along with the loss of the need to get up early to go to work or school could lead to Delayed Sleep Phase Syndrome. As an initial effort, the International COVID-19 Sleep Study (ICOSS) was devised, to explore direct and indirect effects of the COVID-19 pandemic on various aspects of sleep and circadian rhythms in adults (Partinen et al. 2021). In this sense, in different reports, it was confirmed that the change in work schedules derived from confinement led to an increase in the incidences of delayed sleep phase syndrome. Epstein et al. (2022) reported two cases that after receiving sleep restriction therapy and modification of bedtime and wake-up times had an improvement in the quality and total time of sleep.

It has been reported that sleep onset, offset, and mid-time of the sleep phase were delayed on working days but not on free days; sleep duration was increased on work days and social jet lag was reduced during the lockdown (Tahara et al. 2021). Participants also slept longer and later during lockdown weekdays and exhibited lower levels of social Jet Lag (Leone et al. 2020). Although this would seem to be an overall improvement in sleep conditions, chronotype was also delayed under lockdown (Leone et al. 2020).

In an online study, Bottary et al. (2022) found data suggesting delayed sleep phase syndrome, reduced social jet lag, and social sleep restriction. They also found a pronounced shift towards later sleep combined with an increase in sleep duration; these changes were more pronounced in adolescents than in young adults and seemed to occur mostly during weekdays. Teens also reported an improvement in daytime sleepiness and subjective sleep quality, while young adults reported an increase in sleep difficulties associated with sleep onset, nocturnal and early morning awakenings, and nightmares.

11.4 Parasomnias

In an epidemiological study, it was found that adolescents and young adults presented alterations in their sleep patterns in a peculiar way. Young adults presented difficulty falling asleep and had increased nightmares and frequent awakenings during the night (Ramos Socarras et al. 2021; Drager et al. 2022). In addition, it has been reported that the increase in the frequency of vivid dreams during the pandemic (Fränkl et al. 2021) was significantly associated with female sex, nightmares, awakenings, symptoms of REM Sleep Behavior Disorder, and repetitive annoying thoughts.

Compared to the first wave of COVID-19, there was a decrease in recall ability for dreams, nightmares, lucid dreams, and emotional stress during the second wave of the pandemic; showing that the strongest impact of the pandemic was during the first wave. This is very important because the negative emotional tone of nightmares has a strong relationship with emotional processing problems as well as symptoms of mental illness such as anxiety and depression and post-traumatic stress disorder (Scarpelli et al. 2021).

11.5 Insomnia, Anxiety, and Depression in Health Staff

Since the spread of the SARS-CoV-2 virus became a public health problem, protective measures have impacted the physical and mental health of the entire population (Umakanthan et al. 2020). However, health staff in COVID-19 areas had to comply with greater protection measures to reduce the risk of contagion, which increased their stress levels. In addition, there was an increase in the number of hours of the workday, causing more accelerated physical and mental exhaustion

(Walton et al. 2020). This directly impacted the sleep–wake cycle, causing symptoms of sleep disorders and anxiety (Xiao et al. 2020).

In this context, it was reported that worldwide the health personnel who treated infected patients in the first months of the pandemic presented anxiety (23.2%) and depression (22.8%); while insomnia had a prevalence of 38.9% (Pappa et al. 2020). In contrast, among the general population, depression was the most prevalent disorder with 33.7%, followed by anxiety at 31.9% and stress symptoms at 29.6% (Salari et al. 2020). Sleep disturbances occurred in 34% (Deng et al. 2021), of which insomnia had the highest incidence at 36.7% (Morin et al. 2021).

11.6 Long COVID-19 and Sleep

Concerned to REM sleep, Heidbreder et al. (2021) found REM sleep without muscle atony in 36% of a sample of patients who were studied with video polysomnography 60 days after being diagnosed with COVID-19. These studies confirm alterations related to the onset and continuity of sleep, as well as REM sleep parasomnias.

11.7 Sleep-Disordered Breathing

It has recently been observed that patients with higher body mass index or diabetes have a higher risk of developing COVID-19, and the probability of being hospitalized in intensive care units is doubled (Chung et al. 2021). In addition, there have been reports of patients with transient snoring beginning with Coronavirus 19 infection (COVID-19 Transient Snoring, COTS); in these cases, the snoring was not associated with sleep apnea or other previously diagnosed sleep disorders. The cause and pathophysiological course of COTS are unknown, although it is important to note that with sleep hygiene, sleep position therapy, and limiting caffeine consumption, a significant decrease was achieved in a couple of months (Riad et al. 2021).

11.8 Movement Disorders Related to Sleep

Confinement increased the symptoms of bruxism and temporomandibular disorder, negatively affecting sleep quality; these symptoms had a significant correlation with symptoms of anxiety, depression, and stress (Peixoto et al. 2021). In addition, it has been reported that one of the neurological sequelae of COVID-19 could be symptoms of restless legs syndrome (Mohiuddin et al. 2021). For example, Nakamura et al. (2021) in a case report describe a 72-year-old male patient with a variant of restless legs syndrome, which began after COVID-19 infection; in the anal region without causal evidence identified anatomically, structurally, or functionally.

11.9 Insomnia

Complications in convalescence from COVID-19 have been reported to include interstitial lung disease, cardiac pathology, fatigue, neuro-psychiatric symptoms (such as depression, anxiety, post-traumatic stress disorder, and cognitive deficits), and insomnia (Altena et al. 2020; Efstathiou et al. 2022). Compared to matched controls, patients with COVID-19 have increased sleep disturbances and decreased sleep quality during and after infection. Patients with long COVID-19 were characterized by increased awake time and decreased total sleep and deep sleep. These findings show that COVID-19 may present with long-lasting symptoms such as autonomic and neurologic disturbances such as COVID-19 brain fog (Mekhael et al. 2022).

With a sample of 478 patients, after 4 months of hospitalization 51% declare at least one symptom that did not exist before COVID-19 infection: fatigue in 31%, cognitive symptoms in 21%, new onset dyspnea in 16%; and insomnia in 53.6% (Morin et al. 2021). Insomnia was more frequent among non-intubated than intubated patients (57.6% vs. 44% respectively; Morin et al. 2021).

Studying a sample of 251 COVID-19 survivors, 1 month after discharge from the hospital, 41% of the patients presented insomnia; insomnia prevalence decreased to 25% after 3 months post-discharge evaluation (Bourmistrova et al. 2022).

In another study in a sample of 402 patients who were assessed 1 month after hospital treatment, 40% reported insomnia, which decreased significantly 3 months later (Mazza et al. 2020).

In a study that aimed to assess the short-term consequences of SARS-CoV-2 infection-related pneumonia, 4 months after hospital discharge 52%of patients had fatigue in 52%, 36% had effort dyspnea, 10% had anorexia, 14% had dysgeusia or anosmia, 21% had anxiety, and 31% had insomnia (Boari et al. 2021).

Considering the above, treatment should be implemented as soon as possible to avoid complications of COVID-19; such as insomnia, mental health symptoms, and even misuse or abuse of pharmacological treatments for insomnia.

11.10 Stress: General Considerations

Psychological stress can be understood as a series of emotional, cognitive, behavioral, and physiological responses derived from the interaction between an individual and their environment. In particular, individual evaluates their resources concerning the situations they are experiencing, and if the situation exceeds the individual's resources to cope, they will have difficulty facing it (Lazarus and Folkman 1984).

This initial evaluation process is aimed at discriminating between situations that are irrelevant, benign/positive, or that imply some threat or challenge. In this sense, only situations that imply a threat or challenge for the individual (loss or damage, threat, and challenge) would be considered stressful (Cao et al. 2018). A stressful situation reflects all those environmental, social, and internal stimuli that challenge the adaptive capacities and resources of an individual (Monroe and Slavich 2016). Under this concept, many of the situations generated by the COVID-19 pandemic are stressful (Burtscher et al. 2020).

Caballo and Anguiano (2002) mention that there are three types of events that can function as stressors:

- Major changes that affect a large number of people: natural disasters, wars, or pandemics are generally considered stressful due to the negative effects that they produce and the limited control that can be had over them, regardless of their duration.
- Major changes that affect a single person or a small number of people: natural disasters, wars, or epidemics can also be included here because their effects can occur in a smaller number of people. However, the death of relatives, the diagnosis of an illness, the threat to life, a divorce, or loss of employment are more clearly conceptualized in this category.
- The setbacks of daily life: This category includes all the situations we face due to the way we live and where we live and work, for example, the constant demands of a chronic illness, family problems, excessive responsibilities in the home or at work, and feeling lonely, among others.

Based on the previous descriptions, it is clear that the setbacks of daily life can be less impactful in negative terms than a natural disaster, experiencing a pandemic, or experiencing the death of loved ones (Reynoso and Avila 2014). However, given the extended duration and the frequency and intensity with which they occur (chronic and acute demands), they can lead to greater difficulties in adapting to living conditions and managing emotions and can have repercussions on health (Ben-Zur 2019; Biggs et al. 2017).

11.11 Stress and Repercussions on Health

Since the beginning of research on psychological stress, it has been linked to various health problems such as infectious diseases, cancer, and cardiovascular diseases (Lazarus 2000). For example, living with high levels of stress is a risk factor for alterations in blood pressure, heart rate, insulin resistance, tissue damage, and visceral adiposity (Brady and Matthews 2006; Bruce et al. 2009, 2015; Mooy et al. 2000), as well as hyperglycemia in those living with diabetes (Surwit et al. 1992; Wales 1995).

In addition, over the last two decades studies have continued to examine the relationship and effects of stress on autonomic processes (Kim et al. 2018; Lucini et al. 2007; Uusitalo et al. 2011) inflammation, and immunity (Duric et al. 2016; Graham et al. 2006; Salmoirago-Blotcher et al. 2019; Segerstrom and Miller 2004), cardiovascular, metabolic, cancer, pain-related and respiratory diseases

(Christiansen et al. 2021; Grinde 2017; Johansen et al. 2017; Oren and Martinez 2020; Schnall et al. 2017), and psychological and psychiatric disorders (Maydych 2019; Zhaoyang et al. 2020). These studies have focused both on the general population and on populations with specific occupations in the context of work (Ganster and Rosen 2013; West et al. 2016).

Another problem associated with stress has been insomnia (Basta et al. 2007); this includes insomnia from stress related to the COVID-19 pandemic. For example, it was reported that in different countries in North America, Europe, and Asia, insomnia problems increased due to COVID-19 confinement and stress (Gualano et al. 2020; Marelli et al. 2021; Stanton et al. 2020).

Although there are no figures in Latin America on the prevalence of sleep disorders derived from the COVID-19 pandemic, there have been clear increases in psychological problems including anxiety, depression, stress, post-traumatic stress symptoms (rumination, avoidance, and physiological hyperarousal), consumption of psychoactive substances and disruption of daily routines, which have been associated with the development of sleep disorders (Caqueo-Urízar et al. 2020; Cortés-Álvarez et al. 2020; González et al. 2020; Orellana and Orellana 2020; Escobar-Córdoba et al. 2021).

Some of the explanations that link stress with the development of various diseases are the result of the activation of the inflammatory-immune system response and adaptation processes of the Central Nervous System (Graham et al. 2006). However, in addition to these mechanisms, the execution of unhealthy behaviors (tobacco, alcohol or drug use, sleep and eating disorders, and sedentary lifestyle) and ineffective behaviors to manage the stressful demands likely also have a role (O'Connor et al. 2009; Orpana et al. 2007).

11.12 Stress Coping Strategies

Upon coming into contact with a stressful situation, individuals are forced to deal with it through various cognitive and behavioral resources, known as coping strategies. In other words, coping strategies are the behavioral repertoire that mediates the relationship between the perception of stress and the process of adaptation (Rodríguez et al. 2016).

Lazarus and Folkman (1984) proposed two general forms of coping:

- Emotion-oriented coping strategies: Cognitive and behavioral resources that are implemented to reduce the degree of emotional distress, such as avoidance, minimization, distraction, and selective attention, among others.
- Coping strategies aimed at the problem: Cognitive and behavioral resources that are implemented to change the nature of the problem or its effects, they include the definition of the problem, seeking a solution for it, and considering different options based on the costs and benefits.

When talking about stress, it is therefore always necessary to talk about the coping process. When facing a situation, the individual's primary determines whether the situation involves loss/damage/threat/challenge and if so a stress response will be favored. Subsequently, the individual will carry out a secondary evaluation in which they define what can be done in the situation and what the results would be, ending with the implementation of some form of coping. This process will be regulated by the forecast, ambiguity, and controllability of the situation, as well as its possible consequences (Rodríguez et al. 2016).

11.13 Psychological Treatment of Sleep Disorders

Within the context of the pandemic, the lifestyle change was reflected in a delay in the sleep phase, increased periods of wakefulness after midnight, and decreased total sleep time (Morin et al. 2020). According to the expected consensus in the face of the health emergency, psychological interventions became more relevant, with behavioral recommendations being a viable treatment option to reduce the adverse effects of confinement (Crew et al. 2020; De Almondes et al. 2021). Techniques derived from Cognitive-Behavioral Therapy for Insomnia (CBT-I) began to be applied as preventive or therapeutic measures (Altena et al. 2020) and to avoid the side effects of pharmacological treatment.

It is important to mention that CBT-I is an empirically validated treatment with high efficacy, and it is considered the first-line treatment of choice (Haynes et al. 2018). This treatment lasts between six and eight sessions, administered weekly, in individual or group format, in a health services office with a clinical psychologist, online, or even using cell phone apps (Trauer et al. 2015; Álvarez-García et al. 2020). Usually, the CBT-I protocol consists of sleep restriction, stimulus control therapy, relaxation techniques, sleep hygiene, and cognitive restructuring (Dopheide 2020).

11.14 Sleep Disorders and Coping Strategies During the Pandemic

The first interventions were implemented to increase nighttime sleep; this is why progressive muscle relaxation was outlined as an ideal procedure to increase sleep time. Through guided instruction, the participants tensed and relaxed muscle groups for periods of 10–15 s; the training was given in an intensive format of five weekly sessions, obtaining an increase in the quality of sleep of the participants (Liu et al. 2020).

However, later studies showed that the application of multicomponent treatments presented better results since they increased sleep quality and total sleep time and decreased sleep onset latency and insomnia symptoms. An especially important aspect of these interventions is that they were administered in group formats or through teleconsultation, which reduced the risk of contagion for both the health personnel and the patients (Becker 2022).

Likewise, it was observed that people who presented coping strategies aimed at the problem have better predictors of mental health (Budimir et al. 2021). This is highly relevant since health workers are the most vulnerable population. Therefore, education in this coping style makes it possible to reduce the psychological consequences derived from work overload and the high risk of contagion and death (Liu et al. 2021). Finally, it is reported that the strategies used mostly were acceptance, active coping, positive reframing, use of emotional support, planning, and use of instrumental support (Jewell et al. 2021).

11.15 Discussion

The COVID-19 pandemic has significantly affected sleep, causing high prevalences of mental morbidity and symptoms of sleep disorders such as parasomnias, circadian rhythm disorders, sleep-disordered breathing, sleep-related movement disorders, and, most frequently, insomnia. These issues are due to multiple environmental factors, including the loss of time indicators, psychosocial factors such as loss of interaction in school, sports, or work environments, inadequate sleep hygiene, and especially, cardio-respiratory and neurological (central and peripheral) consequences.

As discussed in this chapter, one of the main treatment options for sleep disorders (mainly insomnia) is CBT. This treatment has been shown to have a positive effect on mental health and coping with stress (Friedrich et al. 2018), and improvements in sleep quality led to improvements in mental health (Scott et al. 2021). Therefore, it is very important to implement population strategies that aim to improve sleep quality in conjunction with mental health. It is also of great importance to identify vulnerable groups and design CBT-I protocols that include training in coping skills.

11.16 Research Agenda

There are sleep disorders that include negative emotional content, such as REM sleep parasomnias (whose prevalence increased during the pandemic). It is therefore necessary to investigate the possible beneficial effects of coping skills training on the symptoms of nightmares, sleep paralysis, and REM sleep behavior disorder. Its hypothesis deserves future research.

It would also be necessary to identify and resolve the main difficulties in implementing coping skills training and CBT.

11.17 Conclusion

It is very important to identify the specific needs of each patient in terms of sleep disorders and coping styles to personalize this area of CBT (beyond insomnia), improving the patient's prognosis, avoiding pharmacological treatment (and its side effects) to the extent possible, and reducing the risk of developing symptoms of mental illness.

References

Altena E, Baglioni C, Espie CA, Ellis J, Gavriloff D, Holzinger B, Schlarb A, Frase L, Jernelöv S, Riemann D (2020) Dealing with sleep problems during home confinement due to the COVID-19 outbreak: practical recommendations from a task force of the European CBT-I Academy. J Sleep Res 29(4):e13052

Álvarez-García HB, Jiménez-Correa U, de Almondes KM (2020) Effectiveness of a brief behavioral intervention for insomnia (BBII) during the COVID-19 pandemic: Mexican case report. Sleep Sci (Sao Paulo, Brazil) 13(3):210–213

Basta M, Chrousos GP, Vela-Bueno A, Vgontzas AN (2007) Chronic insomnia and stress system. Sleep Med Clin 2(2):279–291

Becker PM (2022) Overview of sleep management during COVID-19. Sleep Med 91:211–218

Ben-Zur H (2019) Transactional model of stress and coping. In: Zeigler-Hill V, Shackelford TK (eds) Encyclopedia of personality and individual differences. Springer, Switzerland

Biggs A, Brough P, Drummond S (2017) Lazarus and Folkman's psychological stress and coping theory. In: Cooper C, Campbell J (eds) The handbook of stress and health: a guide to research and practice. Wiley Blackwell

Boari G, Bonetti S, Braglia-Orlandini F, Chiarini G, Faustini C, Bianco G, Santagiuliana M, Guarinoni V, Saottini M, Viola S, Ferrari-Toninelli G, Pasini G, Bonzi B, Desenzani P, Tusi C, Malerba P, Zanotti E, Turini D, Rizzoni D (2021) Short-term consequences of SARS-CoV-2-related pneumonia: a follow up study. High Blood Pressure Cardiovasc Prev 28(4): 373–381

Borbély AA (1982) A two process model of sleep regulation. Hum Neurobiol 1(3):195–204

Bottary R, Fields EC, Kensinger EA, Cunningham TJ (2022) Age and chronotype influenced sleep timing changes during the first wave of the COVID-19 pandemic. J Sleep Res 31(2):e13495

Bourmistrova NW, Solomon T, Braude P, Strawbridge R, Carter B (2022) Long-term effects of COVID-19 on mental health: a systematic review. J Affect Disord 299:118–125

Brady SS, Matthews KA (2006) Chronic stress influences ambulatory blood pressure in adolescents. Ann Behav Med 31(1):80–88

Bruce MA, Beech BM, Sims M, Brown TN, Wyatt SB, Taylor HA, Williams DR, Crook E (2009) Social environmental stressors, psychological factors, and kidney disease. J Investig Med 57(4): 583–589

Bruce MA, Griffith DM, Thorpe RJ Jr (2015) Stress and the kidney. Adv Chronic Kidney Dis 22(1): 46–53

Budimir S, Probst T, Pieh C (2021) Coping strategies and mental health during COVID-19 lockdown. J Ment Health 30(2):156–163

Burtscher J, Burtscher M, Millet GP (2020) (Indoor) isolation, stress, and physical inactivity: vicious circles accelerated by COVID-19? Scand J Med Sci Sports 30(8):1544–1545

Caballo VE, Anguiano SA (2002) Estrés, estilos de personalidad y contratiempos cotidianos. In: Reynoso-Erazo L, Seligson I (eds) Psicología y Salud. Facultad de Psicología U.N.A.M. CONACyT, México

Cao X, Masood A, Luqman A, Ali A (2018) Excessive use of mobile social networking sites and poor academic performance: antecedents and consequences from stressor-strain-outcome perspective. Comput Hum Behav 85:163–174

Caqueo-Urízar A, Urzúa A, Aragón-Caqueo D, Charles CH, El-Khatib Z, Otu A, Yaya S (2020) Mental health and the COVID-19 pandemic in Chile. Psychol Trauma Theory Res Pract Policy 12(5):521–523

Casagrande M, Favieri F, Tambelli R, Forte G (2020) The enemy who sealed the world: effects quarantine due to the COVID-19 on sleep quality, anxiety, and psychological distress in the Italian population. Sleep Med 75:12–20

Christiansen J, Lund R, Qualter P, Andersen CM, Pedersen SS, Lasgaard M (2021) Loneliness, social isolation, and chronic disease outcomes. Ann Behav Med 55(3):203–215

Chung F, Waseem R, Pham C, Penzel T, Han F, Bjorvatn B, Morin CM, Holzinger B, Espie CA, Benedict C, Cedernaes J, Saaresranta T, Wing YK, Nadorff MR, Dauvilliers Y, De Gennaro L, Plazzi G, Merikanto I, Matsui K, Leger D et al (2021) The association between high risk of sleep apnea, comorbidities, and risk of COVID-19: a population-based international harmonized study. Sleep Breath 25(2):849–860

Cortés-Álvarez NY, Piñeiro-Lamas R, Vuelvas-Olmos CR (2020) Psychological effects and associated factors of COVID-19 in a Mexican sample. Disaster Med Public Health Prep 14(3): 413–424

Crew EC, Baron KG, Grandner MA, Ievers-Landis CE, McCrae CS, Nadorff MR, Nowakowski S, Ochsner Margolies S, Hansen K (2020) The Society of Behavioral Sleep Medicine (SBSM) COVID-19 Task Force: objectives and summary recommendations for managing sleep during a pandemic. Behav Sleep Med 18(4):570–572

De Almondes KM, Marín Agudelo HA, Jiménez-Correa U (2021) Impact of sleep deprivation on emotional regulation and the immune system of healthcare workers as a risk factor for COVID 19: practical recommendations from a task force of the Latin American Association of Sleep Psychology. Front Psychol 12:564227

Deng J, Zhou F, Hou W, Silver Z, Wong CY, Chang O, Huang E, Zuo QK (2021) The prevalence of depression, anxiety, and sleep disturbances in COVID-19 patients: a meta-analysis. Ann N Y Acad Sci 1486(1):90–111

Di Renzo L, Gualtieri P, Pivari F, Soldati L, Attinà A, Cinelli G, Leggeri C, Caparello G, Barrea L, Scerbo F, Esposito E, De Lorenzo A (2020) Eating habits and lifestyle changes during COVID-19 lockdown: an Italian survey. J Transl Med 18(1):229

Dopheide JA (2020) Insomnia overview: epidemiology, pathophysiology, diagnosis and monitoring, and nonpharmacologic therapy. Am J Manag Care 26(4 Suppl):S76–S84

Drager LF, Pachito DV, Moreno C, Tavares AR, Conway SG, Assis M, Sguillar DA, Moreira GA, Bacelar A, Genta PR (2022) Insomnia episodes, new-onset pharmacological treatments, and other sleep disturbances during the COVID-19 pandemic: a nationwide cross-sectional study in Brazilian health care professionals. J Clin Sleep Med 18(2):373–382

Duric V, Clayton S, Leong ML, Yuan LL (2016) Comorbidity factors and brain mechanisms linking chronic stress and systemic illness. Neural Plast 2016

Efstathiou V, Stefanou MI, Demetriou M, Siafakas N, Makris M, Tsivgoulis G, Zoumpourlis V, Kympouropoulos SP, Tsoporis JN, Spandidos DA, Smyrnis N, Rizos E (2022) Long COVID and neuropsychiatric manifestations (Review). Exp Ther Med 23(5):363

Epstein LJ, Cai A, Klerman EB, Czeisler CA (2022) Resolving delayed sleep-wake phase disorder with a pandemic: two case reports. J Clin Sleep Med 18(1):315–318

Escobar-Córdoba F, Ramírez-Ortiz J, Fontecha-Hernández J (2021) Effects of social isolation on sleep during the COVID-19 pandemic. Sleep Sci (Sao Paulo, Brazil) 14(Spec 1):86–93

Fränkl E, Scarpelli S, Nadorff MR, Bjorvatn B, Bolstad CJ, Chan NY, Chung F, Dauvilliers Y, Espie CA, Inoue Y, Leger D, Macêdo T, Matsui K, Merikanto I, Morin CM, Mota-Rolim S, Partinen M, Penzel T, Plazzi G, Sieminski M et al (2021) How our dreams changed during the COVID-19 pandemic: effects and correlates of dream recall frequency - a multinational study on 19,355 adults. Nat Sci Sleep 13:1573–1591

Friedrich A, Claßen M, Schlarb AA (2018) Sleep better, feel better? Effects of a CBT-I and HT-I sleep training on mental health, quality of life and stress coping in university students: a randomized pilot controlled trial. BMC Psychiatry 18(1):268

Ganster DC, Rosen CC (2013) Work stress and employee health: a multidisciplinary review. J Manag 39:1085–1122

Gloster AT, Lamnisos D, Lubenko J, Presti G, Squatrito V, Constantinou M, Nicolaou C, Papacostas S, Aydın G, Chong YY, Chien WT, Cheng HY, Ruiz FJ, Garcia-Martin MB, Obando-Posada DP, Segura-Vargas MA, Vasiliou VS, McHugh L, Höfer S, Baban A, Dias Neto D, Nunes da Silva A, Monestès JL, Alvarez-Galvez J, Paez-Blarrina M, Montesinos F, Valdivia-Salas S, Ori D, Kleszcz B, Lappalainen R, Ivanović I, Gosar D, Dionne F, Merwin RM, Kassianos AP, Karekla M (2020) Impact of COVID-19 pandemic on mental health: an international study. PLoS One 15:e0244809

González RLP, Martínez ARJ, Hernández GMA, De la Roca-Chiapas JM (2020) Psychological distress and signs of post-traumatic stress in response to the COVID-19 health emergency in a Mexican sample. Psychol Res Behav Manag 13:589–597

Graham JE, Christian LM, Kiecolt-Glaser JK (2006) Stress, age, and immune function: toward a lifespan approach. J Behav Med 29(4):389–400

Grinde B (2017) Stress and chronic fatigue syndrome. In: Cooper C, Campbell J (eds) The handbook of stress and health: a guide to research and practice. Wiley Blackwell

Gualano MR, Lo Moro G, Voglino G, Bert F, Siliquini R (2020) Effects of Covid-19 lockdown on mental health and sleep disturbances in Italy. Int J Environ Res Public Health 17(13):4779

Haynes J, Talbert M, Fox S, Close E (2018) Cognitive behavioral therapy in the treatment of insomnia. South Med J 111(2):75–80

Heidbreder A, Sonnweber T, Stefani A, Ibrahim A, Cesari M, Bergmann M, Brandauer E, Tancevski I, Löffler-Ragg J, Högl B (2021) Video-polysomnographic findings after acute COVID-19: REM sleep without atonia as sign of CNS pathology? Sleep Med 80:92–95

Huang Y, Zhao N (2020) Generalized anxiety disorder, depressive symptoms and sleep quality during COVID-19 outbreak in China: a web-based cross-sectional survey. Psychiatry Res 288: 112954

Jewell C, Vandivort C, Patterson B, Schnapp BH (2021) Coping strategies utilized by emergency department providers during the COVID-19 pandemic. WMJ 120(4):262–267

Johansen C, Katajavaara I, Lim B, Bidstrup P, Dalton S (2017) Stress and cancer. In: Cooper C, Campbell J (eds) The handbook of stress and health: a guide to research and practice. Wiley Blackwell

Kim HG, Cheon EJ, Bai DS, Lee YH, Koo BH (2018) Stress and heart rate variability: a meta-analysis and review of the literature. Psychiatry Investig 15(3):235–245

Lazarus RS (2000) Estrés y emoción: Manejo e implicaciones en nuestra salud. Descleé de Brower

Lazarus RS, Folkman S (1984) Stress, appraisal and coping. Springer Publishing Company

Leone MJ, Sigman M, Golombek DA (2020) Effects of lockdown on human sleep and chronotype during the COVID-19 pandemic. Curr Biol 30(16):R930–R931

Li Y, Qin Q, Sun Q, Sanford LD, Vgontzas AN, Tang X (2020) Insomnia and psychological reactions during the COVID-19 outbreak in China. J Clin Sleep Med 16(8):1417–1418

Lin LY, Wang J, Ou-Yang XY, Miao Q, Chen R, Liang FX, Zhang YP, Tang Q, Wang T (2021) The immediate impact of the 2019 novel coronavirus (COVID-19) outbreak on subjective sleep status. Sleep Med 77:348–354

Liu K, Chen Y, Wu D, Lin R, Wang Z, Pan L (2020) Effects of progressive muscle relaxation on anxiety and sleep quality in patients with COVID-19. Complement Ther Clin Pract 39:101132

Liu Z, Liu R, Zhang Y, Zhang R, Liang L, Wang Y, Wei Y, Zhu R, Wang F (2021) Association between perceived stress and depression among medical students during the outbreak of COVID-19: the mediating role of insomnia. J Affect Disord 292:89–94

Lucini D, Riva S, Pizzinelli P, Pagani M (2007) Stress management at the worksite: reversal of symptoms profile and cardiovascular dysregulation. Hypertension 49(2):291–297

Marelli S, Castelnuovo A, Somma A, Castronovo V, Mombelli S, Bottoni D, Leitner C, Fossati A, Ferini-Strambi L (2021) Impact of COVID-19 lockdown on sleep quality in university students and administration staff. J Neurol 268(1):8–15

Maydych V (2019) The interplay between stress, inflammation, and emotional attention: relevance for depression. Front Neurosci 13:384

Mazza MG, De Lorenzo R, Conte C, Poletti S, Vai B, Bollettini I, Melloni E, Furlan R, Ciceri F, Rovere-Querini P, COVID-19 BioB Outpatient Clinic Study group, Benedetti F (2020) Anxiety and depression in COVID-19 survivors: role of inflammatory and clinical predictors. Brain Behav Immun 89:594–600

Mekhael M, Lim CH, El Hajjar AH, Noujaim C, Pottle C, Makan N, Dagher L, Zhang Y, Chouman N, Li DL, Ayoub T, Marrouche N (2022) Studying the effect of long COVID-19 infection on sleep quality using wearable health devices: observational study. J Med Internet Res 24(7):e38000

Mohiuddin O, Khan AA, Shah S, Malick M, Memon SF, Jalees S, Yasmin F (2021) New-onset restless leg syndrome in a COVID-19 patient: a case report with literature review. Pan Afr Med J 38:318

Monroe S, Slavich G (2016) Psychological stressors: overview. In: Fink G (ed) Stress: concepts, cognition, emotion, and behavior. Elsevier

Mooy JM, de Vries H, Grootenhuis PA, Bouter LM, Heine RJ (2000) Major stressful life events in relation to prevalence of undetected type 2 diabetes: the Hoorn Study. Diabetes Care 23(2): 197–201

Morin CM, Carrier J, Bastien C, Godbout R, Canadian Sleep and Circadian Network (2020) Sleep and circadian rhythm in response to the COVID-19 pandemic. Can J Public Health 111(5): 654–657

Morin CM, Bjorvatn B, Chung F, Holzinger B, Partinen M, Penzel T, Ivers H, Wing YK, Chan NY, Merikanto I, Mota-Rolim S, Macêdo T, De Gennaro L, Léger D, Dauvilliers Y, Plazzi G, Nadorff MR, Bolstad CJ, Sieminski M, Benedict C et al (2021) Insomnia, anxiety, and depression during the COVID-19 pandemic: an international collaborative study. Sleep Med 87:38–45

Nakamura I, Itoi T, Inoue T (2021) Case report of restless anal syndrome as restless legs syndrome variant after COVID-19. BMC Infect Dis 21:993

O'Connor DB, Conner M, Jones F, McMillan B, Ferguson E (2009) Exploring the benefits of conscientiousness: an investigation of the role of daily stressors and health behaviors. Ann Behav Med 37(2):184–196

Onyeaka H, Anumudu CK, Al-Sharify ZT, Egele-Godswill E, Mbaegbu P (2021) COVID-19 pandemic: a review of the global lockdown and its far-reaching effects. Sci Prog 104(2): 368504211019854

Orellana CI, Orellana LM (2020) Predictores de síntomas emocionales durante la cuarentena domiciliar por pandemia de COVID-19 en El Salvador. Actualidades en Psicología 34:103–120

Oren E, Martinez FD (2020) Stress and asthma: physiological manifestations and clinical implications. Ann Allergy 125:372–373

Orpana HM, Lemyre L, Kelly S (2007) Do stressors explain the association between income and declines in self-rated health? A longitudinal analysis of the National Population Health Survey. Int J Behav Med 14(1):40–47

Pappa S, Ntella V, Giannakas T, Giannakoulis VG, Papoutsi E, Katsaounou P (2020) Prevalence of depression, anxiety, and insomnia among healthcare workers during the COVID-19 pandemic: a systematic review and meta-analysis. Brain Behav Immun 88:901–907

Partinen M, Bjorvatn B, Holzinger B, Chung F, Penzel T, Espie CA, Morin CM, ICOSS-Collaboration Group (2021) Sleep and circadian problems during the coronavirus disease 2019 (COVID-19) pandemic: the International COVID-19 Sleep Study (ICOSS). J Sleep Res 30(1):e13206

Peixoto KO, Resende C, Almeida EO, Almeida-Leite CM, Conti P, Barbosa G, Barbosa JS (2021) Association of sleep quality and psychological aspects with reports of bruxism and TMD in Brazilian dentists during the COVID-19 pandemic. J Appl Oral Sci 29:e20201089

Peretti-Watel P, Alleaume C, Léger D, Beck F, Verger P, COCONEL Group (2020) Anxiety, depression and sleep problems: a second wave of COVID-19. Gen Psychiatry 33(5):e100299

Ramos Socarras L, Potvin J, Forest G (2021) COVID-19 and sleep patterns in adolescents and young adults. Sleep Med 83:26–33

Reynoso L, Avila MR (2014) Estrés y enfermedad crónica. In: Reynoso L, Becerra AL (eds) Medicina Conductual: teoría y práctica. Ed. UNAM-Qartuppi, México

Riad A, Kassem I, Badrah M, Klugar M (2021) COVID-19 transient snoring (CVTS): clinical and laboratory description. J Med Virol 93(4):1890–1892

Rodríguez A, Zúñiga P, Contreras K, Gana S, Fortunato J (2016) Revisión Teórica de las Estrategias de Afrontamiento (Coping) según el Modelo Transaccional de Estrés y Afrontamiento de Lazarus & Folkman en Pacientes Quirúrgicos Bariátricos, Lumbares y Plásticos. Revista El Dolor 66:14–23

Salari N, Hosseinian-Far A, Jalali R, Vaisi-Raygani A, Rasoulpoor S, Mohammadi M, Rasoulpoor S, Khaledi-Paveh B (2020) Prevalence of stress, anxiety, depression among the general population during the COVID-19 pandemic: a systematic review and meta-analysis. Glob Health 16(1):57

Salmoirago-Blotcher E, Hovey KM, Andrews CA, Allison M, Brunner RL, Denburg NL, Eaton C, Garcia L, Sealy-Jefferson SM, Zaslavsky O, Kang J, López L, Post SG, Tindle H, Wassertheil-Smoller S (2019) Psychological traits, heart rate variability, and risk of coronary heart disease in healthy aging women-the women's health initiative. Psychosom Med 81(3):256–264

Scarpelli S, Alfonsi V, Gorgoni M, Musetti A, Filosa M, Quattropani MC, Lenzo V, Vegni E, Borghi L, Margherita G, Freda MF, Saita E, Cattivelli R, Castelnuovo G, Manari T, Plazzi G, De Gennaro L, Franceschini C (2021) Dreams and nightmares during the first and second wave of the COVID-19 infection: a longitudinal study. Brain Sci 11(11):1375

Schnall PL, Dobson M, Landsbergis P (2017) Work, stress, and cardiovascular disease. In: Cooper C, Campbell J (eds) The handbook of stress and health: a guide to research and practice. Wiley Blackwell

Scott AJ, Webb TL, Martyn-St James M, Rowse G, Weich S (2021) Improving sleep quality leads to better mental health: a meta-analysis of randomised controlled trials. Sleep Med Rev 60: 101556

Segerstrom SC, Miller GE (2004) Psychological stress and the human immune system: a meta-analytic study of 30 years of inquiry. Psychol Bull 130(4):601–630

Silva E, Ono B, Souza JC (2020) Sleep and immunity in times of COVID-19. Rev Assoc Med Bras (1992) 66(Suppl 2):143–147

Stanton R, To QG, Khalesi S, Williams SL, Alley SJ, Thwaite TL, Fenning AS, Vandelanotte C (2020) Depression, anxiety and stress during COVID-19: associations with changes in physical activity, sleep, tobacco and alcohol use in Australian adults. Int J Environ Res Public Health 17(11):4065

Surwit RS, Schneider MS, Feinglos MN (1992) Stress and diabetes mellitus. Diabet Care 15(10): 1413–1422

Tahara Y, Shinto T, Inoue K, Roshanmehr F, Ito A, Michie M, Shibata S (2021) Changes in sleep phase and body weight of mobile health App users during COVID-19 mild lockdown in Japan. Int J Obes 45(10):2277–2280

Targa A, Benítez ID, Moncusí-Moix A, Arguimbau M, de Batlle J, Dalmases M, Barbé F (2021) Decrease in sleep quality during COVID-19 outbreak. Sleep Breath 25(2):1055–1061

Trauer JM, Qian MY, Doyle JS, Rajaratnam SM, Cunnington D (2015) Cognitive behavioral therapy for chronic insomnia: a systematic review and meta-analysis. Ann Intern Med 163(3): 191–204

Umakanthan S, Sahu P, Ranade AV, Bukelo MM, Rao JS, Abrahao-Machado LF, Dahal S, Kumar H, Kv D (2020) Origin, transmission, diagnosis and management of coronavirus disease 2019 (COVID-19). Postgrad Med J 96(1142):753–758

Uusitalo A, Mets T, Martinmäki K, Mauno S, Kinnunen U, Rusko H (2011) Heart rate variability related to effort at work. Appl Ergon 42(6):830–838

Wales JK (1995) Does psychological stress cause diabetes? Diabet Med 12(2):109–112

Walton M, Murray E, Christian MD (2020) Mental health care for medical staff and affiliated healthcare workers during the COVID-19 pandemic. Eur Heart J Acute Cardiovasc Care 9(3): 241–247

West CP, Dyrbye LN, Erwin PJ, Shanafelt TD (2016) Interventions to prevent and reduce physician burnout: a systematic review and meta-analysis. Lancet (London, England) 388(10057): 2272–2281

Xiao H, Zhang Y, Kong D, Li S, Yang N (2020) The effects of social support on sleep quality of medical staff treating patients with coronavirus disease 2019 (COVID-19) in January and February 2020 in China. Med Sci Monit 26:e923549

Zhang C, Yang L, Liu S, Ma S, Wang Y, Cai Z, Du H, Li R, Kang L, Su M, Zhang J, Liu Z, Zhang B (2020) Survey of insomnia and related social psychological factors among medical staff involved in the 2019 novel coronavirus disease outbreak. Front Psychiatry 11:306

Zhaoyang R, Scott SB, Smyth JM, Kang JE, Sliwinski MJ (2020) Emotional responses to stressors in everyday life predict long-term trajectories of depressive symptoms. Ann Behav Med 54(6): 402–412

Zhou SJ, Wang LL, Yang R, Yang XJ, Zhang LG, Guo ZC, Chen JC, Wang JQ, Chen JX (2020) Sleep problems among Chinese adolescents and young adults during the coronavirus-2019 pandemic. Sleep Med 74:39–47

Chapter 12
Sleep and Circadian Rhythm in Post-COVID-19 Patients

Reaz Mahmud

Abstract The current COVID-19 pandemic caused enormous suffering to people in the disease state and caused significant mortality; the morbidity of people after COVID-19 is also alarming. It is named post-COVID conditions by WHO and CDC USA. Its estimated prevalence ranges from 10% to 50%, and circadian rhythm sleep disturbance is one of the essential symptoms of the post-COVID state. The circadian sleep disturbance may appear isolated or comorbid, associated with other symptoms like chronic fatigue syndrome, depression, post-traumatic stress disorder, and anxiety. The relationship of the circadian rhythm disturbance with the comorbid condition may be bidirectional. The circadian rhythm modulates gene expression, body metabolism, immune status, and endocrine function of the body, as well as a person's behavior. So, the consequences of sleep rhythm disturbance may be enormous.

The pathophysiology of sleep-rhythm disturbance in the post-COVID state is mainly unknown and is still under research. Persistence of the organism or prolonged and dysregulated inflammation also explains sleep, circadian rhythm, and other neuropsychiatric disorders in the post-COVID period. The countermeasures against this disorder are still under research. Sleep extension, appropriate bedtime recommendations, maintaining sleep hygiene and reciprocity between physical activity and the circadian system, time-of-day-restricted feeding (TRF), chronobiology, and drugs, like melatonin, may be an approach to treat such conditions. The concept of chronology is the newer one; it identifies the body's circadian rhythm of various biologic responses and modifies the treatment to gain optimum benefit.

Keywords Post-COVID-19 condition · Sleep · Circadian rhythm · Sleep hygiene · Chronobiology

---

R. Mahmud (✉)
Department of Neurology, Dhaka Medical College, Dhaka, Bangladesh

A. S. BaHammam et al. (eds.), *COVID-19 and Sleep: A Global Outlook*, Progress in Sleep Research, https://doi.org/10.1007/978-981-99-0240-8_12

12.1 Introduction

The COVID-19 infection, and countermeasures containing the disease, change the lifestyle of most people in the world. Compulsory home confinement changes the daily rhythm and behavior of a significant number of the world population, influencing people's circadian rhythm. Due to the lockdown, more people became flexible about bedtimes and arousal time. It may sometime improve or deteriorate social jet lag. The pandemic-induced negative mental well-being further exaggerates the situation. During the lockdown, there was a shift to a later bedtime, reduced night-sleep time, and increased daytime sleeping (Gupta et al. 2020) among the general population. The infection with COVID-19 further escalates the sleep problem. A study found that 57% (vs. 18% in the general population) of COVID-19-infected patients complained of sleep problems (Alimoradi et al. 2021). Among COVID-19-infected patients, young adolescent people, and healthcare workers, the prevalence is high (Jahrami et al. 2022). Many patients remain unwell even after the apparent cure of COVID-19. Various nomenclature has been used to define conditions, like long COVID, long haul COVID, post-COVID syndrome, and post-acute COVID-19. Sleep problem was also reported among the wide range of symptoms, like fatigue, cough, and dyspnea (Mahmud et al. 2021). Even 6 months after the COVID-19 infection, the patients suffered from fatigue, anxiety, and sleep problems (Huang et al. 2021). It may be a part of chronic fatigue syndrome, psychiatric disorders (depression, anxiety, adjustment disorder), or an isolated one. Three forms of sleep problems, insomnia, hypersomnia, and sleep cycle or circadian rhythm disorders, are commonly found in different works of literature. Approximately one-third of the patients had increased sleep, and one-third reported decreased sleep quality (Tsang et al. 2021). Whether sleep or circadian rhythm disorders are the risk factors for the post-COVID-19 condition or socioeconomic consequences, personal stress, or the neurologic effect of COVID-19 cause the sleep or circadian rhythm in the post-COVID state is a matter of debate. Some literature emphasizes the bidirectional relationship between sleep, circadian rhythms disorders, inflammation, and immunity (Giri et al. 2021).

12.2 Circadian Rhythm Sleep–Wake Disorders

Circadian Rhythm Sleep–Wake Disorders are the groups of sleep disorders resulting from alteration, disruption of the core clock, or misaligning endogenous clock with the body's sleep–wake schedule. There are two categories of circadian Rhythm Sleep–Wake Disorders; transient disorders (caused by extrinsic factors, like shift work and jet lag) and chronic disorders (caused by intrinsic factors). The subclassification of chronic disorders includes (1) Advanced sleep–wake phase disorder (extremely early bird) (2) Delayed sleep–wake phase disorder (extreme night owl), (3) Non-24-h sleep–wake rhythm disorder (drifting circadian rhythm), and Irregular sleep–wake rhythm disorder (no rhythm) (American Academy of Sleep Medicine 2014).

Among the sleep–wake rhythm disorders, the most common is delayed sleep–wake phase disorder. Its prevalence varies according to age; in adolescents and young adults, 3.3% and 4.6%, and in adults, 0.2–1.7% (Lee 2020). In the COVID-19 pandemic era, the prevalence should be higher. A meta-analysis showed that the global prevalence of sleep problems reached 35% and is the highest among COVID-19 sufferers (Jahrami et al. 2021).

The patient usually presents with insomnia, hypersomnia, difficulty waking up in the morning, and early evening sleepiness according to the subtype of the disorder. Its diagnosis is mainly clinical; diagnostic criteria may help. Imaging, actigraphy, and overnight polysomnography may help to establish the diagnosis and exclude the differentials.

12.3 Post COVID-19 Condition

The CDC (USA) and WHO agreed on using the term "post-COVID condition" to define the various post-COVID sufferings. They defined post-COVID conditions as a wide range of new, returning, or ongoing health problems people can experience 4 or more weeks after first being infected with the virus that causes COVID-19 (CDC 2021). The estimated prevalence of post-covid conditions ranges from 10% to 50% in various literature (Carfì et al. 2020; Logue et al. 2021; Townsend et al. 2020; Mahmud et al. 2021).

There is an overlap between COVID-19 and post-COVID features; fatigue, cough, and dyspnea are important. In addition, sleep disturbances during COVID-19 also overlapped with the post-COVID state, like insomnia, circadian rhythm abnormalities, excessive daytime sleepiness, and restless leg syndrome (Fig. 12.1).

12.4 Post-COVID-19 condition and Sleep-Circadian Rhythm Disorders

12.4.1 Epidemiology

The impact of the COVID-19 pandemic on the life of the people of the whole world is enormous. The fear of COVID-19 and its consequences, enforced isolation, social insecurity, financial hardship, an atypical work schedule, and inappropriate use of electronic devices and social media are harming people's physical and mental health. The ultimate consequence is that people become more anxious and depressed. Sleep problem is also a consequence among the general population and professional groups. In a recent meta-analysis, about 40% of the general population and

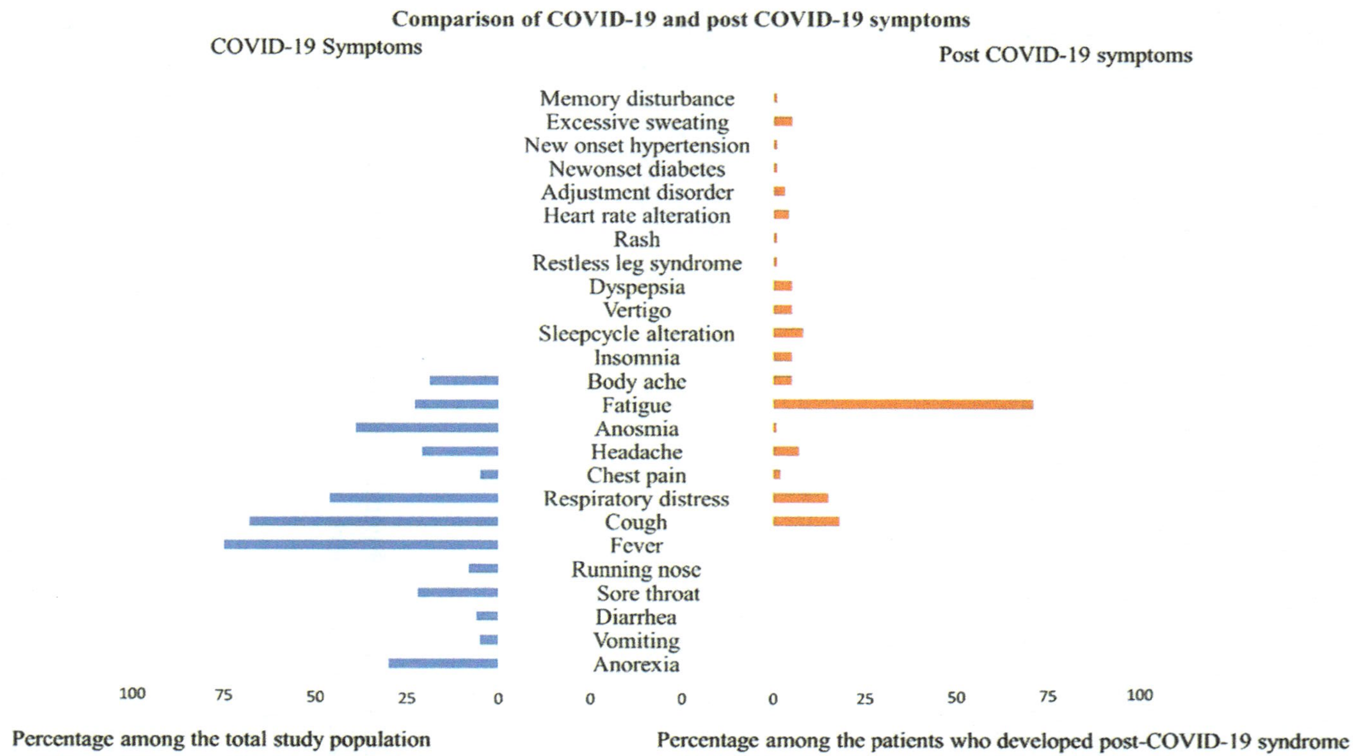

Fig. 12.1 Clinical features of COVID-19 and post-COVID conditions (Mahmud et al. 2021, PLoS One). Comparison of COVID-19 and post-COVID 19, most of the symptoms overlap in the COVID-19 and post-COVID conditions

healthcare workers were found to have sleep problems. The pooled prevalence of sleep problems among patients with COVID-19 was about 75% (Jahrami et al. 2021). There is a 20% increase in the consumption of sleeping pills. Circadian rhythm disturbances were highly prevalent sleep problems; about two-thirds of patients with sleep problems had some form of sleep cycle disturbances during this era. This problem also continued among the COVID-19-affected patient in the post-covid recovery. About 80% of the COVID-19-affected patients had circadian sleep–wake rhythm disturbances (Boiko et al. 2022). Moderate to severe sleep problem was found in 30% of the instances (El Sayed et al. 2021).

Sleep and circadian rhythm disturbances in the post-COVID state may be isolated or may be found as a comorbid condition along with other neuropsychiatric conditions. It was found as an isolated phenomenon in about 2–30% of the instances (Mahmud et al. 2021; Premraj et al. 2022).

12.4.2 Risk Factors for the Development of Sleep–Wake Cycle Disorder

Increasing age might be a potential risk factor. The circadian rhythm system is less efficient among the elderly than the young (Gulia and Kumar 2018). Melatonin, the hormonal regulator of the circadian rhythm, starts to secrete from 3 to 4 months of age, peaks at 8–10 years, and decreases after 40 years (Poza et al. 2018). It was found that patients aged >36 develop moderate to severe sleep problems in the post-COVID state (El Sayed et al. 2021).

Gender differences exist even in the duration and architecture of normal sleep (Mallampalli and Carter 2014); sleep latency is longer in women than men. Men remain more in NREM stage 1 and stage 2 sleep, and females remain more in slow wave sleep. Women have a 40% more risk of developing insomnia (Zhuang et al. 2017). Hormonal modulation of the circadian rhythm, psychosocial factors, and associated comorbid psychiatric disorders might play a role. Regarding post-COVID sleep-rhythm disturbance, females are 1.7 times more vulnerable to developing such a sleep disorder.

Sleep disparity also prevails in the race; poverty and the African-Latino race have increased poor sleep quality (Patel et al. 2010). Like other diseases, racial disparity exists in the outcome of COVID patients, both in terms of mortality and post-COVID conditions. Social disparity, environmental conditions, and genetic predisposition to a particular disease might play a role (Johnson 2020). Moderate to severe insomnia is highly prevalent among black.

Low-level functioning is associated with post-covid sleep cycle rhythm disturbances, increasing twice the risk of sleep disturbance in a post-covid state (El Sayed et al. 2021). In addition, a bidirectional relationship exists between physical functioning and sleep–wake cycle disturbance (Kline 2014).

In the pandemic era, many patients required ICU admission (Tan et al. 2012). Critical-care survival patients suffered physical, mental, and cognitive impairment even after 1 year of discharge (Heesakkers et al. 2022). Critically ill patients are susceptible to circadian rhythm dysrhythmia resulting in sleep disturbance and delirium (Daou et al. 2020). The Loss of the sensory cues to the master clock, abnormal melatonin secretion, the wrong thing at the wrong time, mitochondrial dysfunction due to inflammation, or drugs used in ICU like propofol, remdesivir, and steroids might play roles behind it. Remdesivir causes phase shift in Clock, Per1, and Per2 and has the potential to alter the sleep–wake cycle (Faltraco et al. 2021).

12.4.3 Comorbid Conditions

Anxiety, post-traumatic stress disorder, depression, and chronic fatigue syndrome are prevalent in the post-COVID state. There is a bidirectional relationship between these comorbid conditions and sleep cycle disturbances.

According to the WHO, the prevalence of anxiety and depression increased by 25% in the COVID era. About half of the COVID-19 hospital survivors developed anxiety, depression, or sleep-related problems (Fernández-de-Las-Peñas et al. 2021). An increase in anxiety state is associated with the delayed sleep phase, non-24 h rhythm, and irregular phase of sleep disorder (Daou et al. 2020).

Myalgic encephalomyelitis (ME) and chronic fatigue syndrome (CFS) is a chronic, disabling condition (Committee on the Diagnostic Criteria for Myalgic Encephalomyelitis 2015). The etiology is largely unknown. The symptoms might be triggered by an infection or other prodromal events, like vaccination, anesthesia, physical injury, or exposure to different environmental toxins (Carruthers et al. 2003). The disease has been renamed as systemic exertion intolerance disease (SEID) by the Institute of Medicine (IOM) (Committee on the Diagnostic Criteria for Myalgic Encephalomyelitis 2015).

Fatigue is the most common persistent symptom (Logue et al. 2021; Carfì et al. 2020; Townsend et al. 2020) in the post-COVID state. The estimated prevalence ranges from 10% to 50% in other literature. The etiology of this condition is still undetermined. The spread of the virus to the hypothalamus and the resulting formation of pro-inflammatory cytokines, interleukins, and interferon-gamma is one of the suggested mechanisms (Tenforde et al. 2020). Immunologic alteration (Qin et al. 2020; Chen et al. 2020; Zhou et al. 2020) due to SARS CoV-2 might also play a role in developing post-viral fatigue. Due to this fatigue, about one-third of the patient did not return to everyday life (Logue et al. 2021), and a similar number of patients did not return to their job (Townsend et al. 2020). Whether sleep and rhythm disturbances are due to chronic fatigue syndrome, or sleep and rhythm disturbance are the risk factor for developing chronic fatigue syndrome is a debate. The circadian rhythm disturbances desynchronize the body's physiology, disrupting metabolism and energy and giving rise to fatigue. Genetic alteration in the clock gene is identified in patients with chronic fatigue syndrome. SARS CoV-2 upregulates the transforming growth factor beta in the peripheral tissue. Other dysregulated cytokines may have a role in developing chronic fatigue syndrome.

12.4.4 The Pattern of Sleep Cycle Disturbance in the Post-COVID State

The duration of sleep increased among the general population during the pandemic era. A study showed that the average sleep duration before the pandemic was 6.9 h but after the pandemic it increased to 7.2 h (Mandelkorn et al. 2021). The frequency of the patient resting in the bed for less than 7 h decreased from 42% to 35%. Another study showed a higher mean PSQI score among post-covid patients (6.28 ± 2.11 vs. 3.22) (Choudhry et al. 2021).

The global PSQI score is about 16 among post-covid patients (El Sayed et al. 2021). In that study, among the sub-types of circadian rhythm disorders, delayed sleep phase disorder was found in half of the patients. Advanced sleep phase disorder was found in 25% of patients, irregular sleep phase disorders in 20%, and non-24-h rhythm in 10%. Another study reported that COVID-19 increased the risk of developing delayed sleep phase disorders by 1.5 times (Boiko et al. 2022). Another study showed an association between poor sleep quality, early morning awakening, and daytime sleepiness with COVID-19 (Partinen et al. 2021). The critically ill patient had increased stage 1 and 2 sleep, less slow wave sleep, and frequent arousal (Cooper et al. 2000).

12.4.5 Pathophysiology of Post-COVID-19 Circadian Rhythms Disturbances

Circadian rhythms are the natural body's timekeeping system with two components (Dibner et al. 2010), the *Central pacemaker* (the Suprachiasmatic nuclei) and the *Subsidiary peripheral clocks* (every cell of the body). The sleep-wake state is the prime regulator of the circadian rhythm but temperature, different hormones, nutrients, feeding status, and physical activity influence the circadian rhythm (Güldür and Otlu 2017). Borbely (1982) explained a two-process model of sleep. Process-S activate in wakefulness, and Process-C in sleep are two endogenous drivers of the sleep–wake cycle. The interaction between these two determines sleep onset, duration, and structure. The lateral hypothalamus and the ventrolateral preoptic nuclei (VLPO) interact to promote sleep or awake states. The orexin (also known as hypocretin) neurons in the lateral hypothalamus project to the mid-brain and hindbrain to excite the monoaminergic neurons, the cholinergic neurons, and the glutaminergic neurons. Inhibition or activation of monoaminergic or cholinergic neurons determines REM or NREM sleep (Foster 2020). Adrenergic fiber from the

SCN regulates the synthesis of melatonin by the pineal gland; receptor in the SCN (MT (1) and MT (2) melatonin receptors) also acts in a feedback fashion (Dubocovich 2007).

Several neuropathologic changes occur in the brain due to inflammation in COVID-19, like hippocampal (Solomon 2020) and cortical atrophy (Generoso et al. 2021), hypoxic ischemic changes (Radnis et al. 2020), and small vessel disease (Lowenstein and Solomon 2020). The blood–brain barrier does not protect the hypothalamus and other circumventricular organs. SARS CoV-2 activates the hypothalamic–pituitary–adrenocortical (HPA) axis with innate pro-inflammatory cytokines (TNF-α, IL-1, and IL-6) and interferons, and late acquired T cell cytokines (IL-2 and IFN-γ) (Mussa et al. 2021). The persistence of the virus and prolonged and dysregulated inflammation in the hypothalamus (Tate et al. 2022) disrupt sleep's neuronal and humoral mechanism. It also disrupts the interaction of process-C and process-S endogenous drivers. The circadian rhythm modulates gene expression (Hsu and Harmer 2014), body metabolism (Asher and Sassone-Corsi 2015), immune status (Labrecque and Cermakian 2015; Zhuang et al. 2017; Haspel et al. 2020), endocrine function (Kriegsfeld and Silver 2006), and body's physiology according to environmental changes (Rosbash 2009) as well as person's behavior (Roenneberg et al. 2015). The primary physiologic systems of the body, like the digestive system, blood pressure regulating system, immune system, autonomic system, and human behavior, need to synchronize with the sleep–wake cycle. The desynchronization leads to the disruption of different physiologic systems of the body. It explains sleep, circadian rhythm disturbance, and its consequences on health in the post-COVID-19 state.

The time-to-time adjustment (photoentrainment) in response to a change in the amount of environmental light is the process's primary time cue (zeitgeber) (Roenneberg and Foster 1997). A distinct pathway different from the image-forming visual pathway, the retinohypothalamic tract (RHT) conveys the photic stimulation to the suprachiasmatic nucleus (SCN) (Peirson et al. 2009). Due to quarantine, a shift of rhythm about 3 h ahead was observed, which reduces exposure to sunlight and physical activity (Daou et al. 2020). The quarantine-associated shift of sleep pattern may also be responsible for sleep cycle disorder in the post-COVID state.

Most of the body's cells contain corresponding circadian clock genes, which act autonomously to maintain the rhythm (Mohawk et al. 2012). They act through feedback and feed-forward methods to influence the transcription and translation of the proteins (Bailey et al. 2014). CLOCK and BMAL1 are the core clock genes. The core clock genes transactivate their negative regulator, the Per1, Per2, Cry1, and Cry2 genes (Yoshitane et al. 2019). Identifying the reason for the predisposition to certain groups needs further research. In addition, many genetic polymorphisms were found among the Clock genes. A study found that The *PER3* polymorphism (rs228697) may be associated with a diurnal preference for sleep (Hida et al. 2014). This field requires further exploration of the post-COVID-19 conditions.

The influence of shift work on the sleep cycle is noteworthy. Night-shift workers had a higher prevalence of COVID-19 infection as well (Loef et al. 2022).

The influence of some biological phenomena was observed in a study conducted to observe sleep patterns from continuous signals collected via wearable wristbands (Mekhael et al. 2022). They found a correlation between different heart rates, respiratory rate SPO_2 on sleep quality. In addition, they found that post-COVID patients had decreased light sleep phase and deep sleep time.

12.4.6 The Primary Mechanism of Circadian Sleep Disturbances in Post-COVID Conditions

The primary mechanism for the sleep cycle disturbances in post-COVID patients includes environmental LD (light-dark cycle) cycle disruption, changes in eating behavior, physiologic disruption, immune dysregulation, and genetic predisposition. These mechanisms are altered in the post-covid state by different conditions. All the mechanisms are co-related in a person on different scales resulting the sleep–wake cycle disturbances.

Lockdown or quarantine during COVID affected period disrupts the LD cycle and changes the eating and rest pattern (Cellini et al. 2020). In the strict lockdown, midsleep times were delayed and became less variable, reducing resting heart rate (Ong et al. 2021). A study on the 12-week confinement impact revealed a reduction in sunlight exposure, delay in taking the main meal, increment in sedentarism, delay in sleep onset, and decrease in sleep duration (Baquerizo-Sedano et al. 2021). Long-term hospitalization or remaining in ICU causes physiologic disruption and immune dysregulation along with these mechanisms. Increasing age causes physiologic and immunologic disruption, and gender additionally causes genetic predisposition causing cycle disruption. Night-shift work, prior sleep cycle disorder, and prior degenerative brain diseases increased the risk of such disorders in the post-COVID state. Financial suffering and confinement were also associated with all sleep-related problems in a post-COVID state (Partinen et al. 2021) (Table 12.1).

12.4.7 Possible Health Consequences of the Post-COVID Sleep Cycle Disturbances

Sleep cycle disturbance causes alteration of the body's physiology, giving rise to short- and long-term consequences. However, the long-term consequences of post-COVID sleep disturbances are yet to be studied.

Sleep deprivation activates the hypothalamic-pituitary-adrenal (HPA) axis and the sympathoadrenal-medullary (SAM) drive, increasing ACTH, cortisol, and adrenaline. Thus, it creates a stress response that causes hyperglycemia, tachycardia, elevated blood pressure, suppression of the immune response, and slowed digestion and cognition. Additionally, long-term sleep disturbances decrease leptin levels and

Table 12.1 The primary mechanism of circadian sleep disturbances in post-COVID conditions

	Mechanism						
	Environmental LD cycle disruption	Eating behavior	Rest behavior	Genetic disruption	Physiologic disruption	Immune dysregulation	Risk factor
Source							
Lockdown or quarantine during COVID affected period	Yes	Yes	Yes				
Long-term hospitalization or remaining in ICU	Yes	Yes	Yes		Yes	Yes	
Prior shift work		Yes	Yes				Yes
Unusual photoperiods	Yes						
Prior sleep cycle disorder							Yes
Elderly					Yes	Yes	
Gender				Yes	Yes	Yes	
Comorbid disease, like dementia				Yes			
Addiction to android/Facebook	Yes	Yes	Yes				
Increasing use of steroid and other immune modulation agents						Yes	

increase ghrelin levels, which may lead to obesity and type II diabetes. In the long run, it is a risk factor for certain malignancies (Foster 2020).

Circadian rhythm abnormalities are positively associated with depression and anxiety. Depression occurs more in multiple circadian abnormalities, and anxiety occurs more in single circadian rhythm abnormalities (Tao et al. 2021). The underlying mechanism is not precise; some studies revealed that circadian genes CRY1, CRY2, PER2, and NPAS2 polymorphism have a higher risk for major depressive disorder (Melhuish Beaupre et al. 2018), and a study in mice revealed a lack of CRY 1 and 2 proteins have an association with anxiety (Charrier et al. 2017).

A study showed that in post-COVID patients, insomnia severity was positively correlated with dysfunction in several domains of the quality-of-life scale SF 36, including physical functioning and role limitation due to emotional and physical problems (El Sayed et al. 2021).

12.4.8 Diagnosis of Circadian Rhythms in Post-COVID Patients

The diagnosis of circadian rhythm disorders is mainly clinical. According to types of circadian rhythm disorder, patients present with falling asleep, staying asleep, waking up too early, getting sleep but not feeling refreshed by it, and feeling alert during the day. Moreover, the patient might commonly have fatigue, cognitive disturbances, comorbid anxiety, and headaches.

In the delayed sleep phase disorder, patients sleep late and wake up late; in advanced sleep phase disorder, patients lie down early and wake up early; in irregular sleep–wake rhythm, patients take several naps in 24 h; in non-24-h sleep–wake syndrome, the internal clock is longer. In the post-COVID state, patients usually have delayed sleep phase disorder. The prevalence of the disorder is not associated with COVID-19; the prevalence is the same as the general population.

To diagnose correctly, the clinician should consider the sleep habits of the patient, including the usual bedtime, sleep duration, sleep latency, and frequency of wakening up.

12.4.9 Investigations to Diagnose Circadian Rhythm Disorders

In general, laboratory and sleep investigations are unnecessary in diagnosing circadian rhythm disorders. However, sometimes it is necessary to exclude narcolepsy, obstructive sleep apnea, hypersomnia, and other sleep disorder. Multiple sleep latency tests, polysomnograms, and actigraphy may help exclude sleep disorders that overlap or coexist with circadian rhythm disorders. Brain imaging is usually unnecessary and may be done in the case of comorbid dementia and other neurodegenerative diseases. In actigraphy, patients wear watches that record sleep onset time, wake time, and physiologic variables, like heart rate and respiratory rate. It helps to objectively document sleep patterns and sleep duration.

12.4.10 Countermeasures Against the Development of Rhythm Disturbances

The sleep cycle disturbances in the post-COVID state have an enormous impact on the well-being of the patients, and it delays the return to the job or reduces job performance. It has been shown that only 1.5 h of sleep reduction for one night reduces daytime alertness by 32% (Bonnet and Arand 1995). So, to get optimum productivity of the working population in the post-COVID era, proper intervention is necessary to overcome the sleep–wake cycle disturbances (Table 12.2).

Gradual extension in sleep time may reduce sleep latency, increase sleep duration, and improve visuospatial cognitive performances (Dewald-Kaufmann et al. 2013). In addition, a behavioral intervention like an appropriate bedtime recommendation may improve sleep initiation and maintenance by 53% (Priday et al. 2016).

Sleep hygiene is a behavior that can improve sleep quality and quantity. Peter Hauri first introduced this concept. Sleep hygiene intervention improves sleep quality (Tan et al. 2012). Sleep does not rely only on bedtime behavior but also the daytime activities as well. The daytime physical exercise, exposure to sunlight, stress, etc., affect sleep. We should have a consistent wake up time, bedtime, and wind-down time before going to sleep. It is advisable to take a short mid-day nap, not excessive naps in the day, and it should be less than 20 min and not within 6 h of night sleep. We should optimize the light and temperature during bedtime. The electronic devices emit blue light, which decreases the slow wave and REM sleep and increases the time lag between bedtime and sleep onset. So in the post-COVID-19 era, we should rationally use electronic devices. A way to relax is essential for sleep behavior. One should adopt relaxation techniques for himself, like deep, Calming music, quiet reading, yoga, etc. (Fig. 12.2).

Table 12.2 Countermeasure against the development of rhythm disturbances in post-COVID state

1. Sleep extension
2. Appropriate bedtime recommendation
3. Sleep hygiene
4. Reciprocity between physical activity and the circadian system
5. Time-of-day-restricted feeding (TRF)
6. Daytime exposure to natural light
7. Chronobiology
8. Drugs

Fig. 12.2 Sleep hygiene. Good quality sleep in post-COVID state requires maintenance of sleep hygiene during the day, before bed, bedroom environment, and in-bed behavior

Exercise induces some physiologic changes, including circadian rhythm (Buman and King 2010). For example, morning exercise enhances parasympathetic activity (Yamanaka et al. 2015), and evening exercise increases the heart rate, and delays sleep by 1 h. This change may be due to a phase shift of melatonin secretion (Buxton et al. 1997). So, the patient who recovered from COVID-19 should be advised to exercise regularly. Nevertheless, rehabilitation and exercise in recovering COVID-19 patients should be carefully planned, as some may still have hypoxia and breathing difficulty in recovery. WHO made a guideline for such a self-rehabilitation process (WHO 2020). Before recommending an exercise, the safety should be assessed. If one can speak the whole sentence without stopping and is not feeling breathless, then he can exercise harder. However, it should be graded, warming up exercise for 5 min, fitness exercise for 20–30 min, 5 days each week, and some strengthening exercise according to exercise capability and cool down exercise for 5 min.

In time-restricted feeding, feeding time is restricted to <12 h; it ensures >12 h of fasting and improves the circadian rhythm (Longo and Panda 2016). The circadian misalignment in feeding and sleep leads to increased glucose, decreased leptin, increased insulin secretion, reversed cortisol secretion rhythm, increased mean arterial pressure, and reduced sleep efficiency. The circadian rhythm fast 16:8 may be a suitable approach for the post-COVID patient to improve metabolic derangement. Several studies have found this approach to improve health-related biomarkers (Gabel et al. 2018; Moro et al. 2016).

The daytime behavior influences sleep at night. It is encouraged to have as much natural light as possible, promoting vitamin D synthesis. Vitamin D might have a pleiotropic effect regulating the sleep–wake cycle (Romano et al. 2020) as melatonin production. Vitamin D deficiency is a risk factor for developing restless leg syndrome, which also indirectly influences sleep quality at night. A study revealed that sunlight exposure correlated with COVID-19 recovery (Asyary and Veruswati 2020). Fair-skinned people of color need 20–30 min of mid-day sun exposure, and dark-skin-colored people require 30–40 min of sun exposure 3–4 times a week for optimum vitamin D synthesis (Sidiropoulou et al. 2021). So we can recommend that in a post-COVID state, people should have at least 30 min of sun exposure in mid-day at least thrice per week.

12.5 Chronobiology

The concept is to identify the body's circadian rhythm of various biologic responses and modify the treatment to gain optimum benefit. It is essential for the application of a drug to respect the circadian rhythm, its role in the management of the acute care setting is under investigation. Delirium is an essential consequence of circadian dysrhythmia in critical-care settings (McKenna et al. 2018). The sleep–awake inversion in critical-care settings may give rise to sleep cycle disturbance. Again, we need to respect the circadian rhythm of feeding time and immune response during an intervention. If not appropriately followed may give rise to metabolic derangement and impaired wound healing, respectively. So the concept of doing the right thing at a suitable time prevails (Table 12.3).

12.5.1 Melatonin

Melatonin is a chronobiologic agent; chemically, it is five methoxy-*N*-acetyltryptamine. It is mainly secreted from the pineal gland. However, sources like the retina, bone marrow cells, platelets, skin, lymphocytes, the Harderian gland, cerebellum, and gastrointestinal tract have been identified (Tordjman et al. 2017). Pinealocytes synthesize it from the amino acid tryptophan (Axelrod and Weissbach 1960).

Table 12.3 Chronobiologic approach to preventing dysrhythmia in ICU (McKenna et al. 2018)

Zeitgeber	Ideal ICU environment
Light	• Bright-light daylight hours (>1000 lx) • Minimize light pollution at night (<1 lx, consider) • Eye masks at night
Feeding	• Intermittent daytime feeding aligned with usual meal times • Avoid continuous and overnight feeding
Temperature	• Warmer environment during the day and cooler at night
Exercise rehabilitation	• Similar time each day • Take patient chronotype into consideration when planning schedule
Noise	• Minimize noise during the nighttime • Ear plugs at night
Medical and nursing interventions	• Minimize at night and cohort together • For non-sedated patients, allow patients to sleep according to their natural schedule, rather than a staff-based schedule
Sedative drugs	• Minimize use through regular review and avoid "sleeping tablets" • Consider melatonin (up to 5 mg) at 0900 h as circadian preserving sleep aide • When possible, give drugs at the time of day least disturbing to circadian rhythms

In the body, it is mainly synthesized in darkness, irrespective of sleep state, beginning to rise at 20.00, peaking at 00.00–03.00, and reaching baseline at 6.00 among those who sleep early and wake up early. The peak plasma concentration at night ranges from 100 to 200 pg/mL and 10 to 30 pg/mL during the day. It is mainly applied orally, reaches a peak within 30 min, and remains in the body for 4–8 h. Typical doses of melatonin are 2.5–3 mg in children and 5–10 mg in adolescents and adults. It acts through the MT1 and MT2 receptors. Melatonin resynchronizes the circadian rhythm and the sleep–wake cycle. It also modulates neuroendocrine and sexual behavior. It may have antioxidant properties and an oncostatin effect. It may be used in primary insomnia, comorbid insomnia with psychiatric disorders, or neurodegenerative diseases. It might be a potential treatment for COVID and post-COVID conditions (Wichniak et al. 2021).

12.6 Conclusion

Post-COVID sleep–wake rhythm disturbance is one of the morbidities in the post-COVID state. There is little research available to identify the condition's magnitude, pattern, and risk factors, and scarcity of research regarding the potential intervention of this problem. Extensive research is demanded regarding the epidemiology, extent of suffering, and potential intervention. Whether vaccination reduces the burden of this condition is unknown.

References

Alimoradi Z, Brostrom A, Tsang HWH et al (2021) Sleep problems during COVID-19 pandemic and its' association to psychological distress: a systematic review and meta-analysis. EClinicalMedicine 36:100916. https://doi.org/10.1016/j.eclinm.2021.100916

American Academy of Sleep Medicine (2014) International classification of sleep disorders, 3rd edn. American Academy of Sleep Medicine, Darien, IL

Asher G, Sassone-Corsi P (2015) Time for food: the intimate interplay between nutrition, metabolism, and the circadian clock. Cell 161:84–92

Asyary A, Veruswati M (2020) Sunlight exposure increased Covid-19 recovery rates: a study in the central pandemic area of Indonesia. Sci Total Environ 729:139016. https://doi.org/10.1016/j.scitotenv.2020.139016. Epub 2020 Apr 27. PMID: 32361458; PMCID: PMC7184988

Axelrod J, Weissbach H (1960) Enzymatic O-methylation of N-acetyl serotonin to melatonin. Science 131:1312

Bailey SM, Udoh US, Young ME (2014) Circadian regulation of metabolism. J Endocrinol 222(2): R75–R96

Baquerizo-Sedano L, Chaquila JA, Aguilar L, Ordovás JM et al (2021) Anti-COVID-19 measures threaten our healthy body weight: changes in sleep and external synchronizers of circadian clocks during confinement. Clin Nutr 25. https://doi.org/10.1016/j.clnu.2021.06.019. Epub ahead of print. PMID: 34246488

Boiko DI, Skrypnikov AM, Shkodina AD et al (2022) Circadian rhythm disorder and anxiety as mental health complications in post-COVID-19. Environ Sci Pollut Res Int 29(19): 28062–28069. https://doi.org/10.1007/s11356-021-18384-4. Epub 2022 Jan 5. PMID: 34988815; PMCID: PMC8730477

Bonnet MH, Arand DL (1995) We are chronically sleep deprived. Sleep 18(10):908–911. https://doi.org/10.1093/sleep/18.10.908. PMID: 8746400

Borbely AA (1982) A two process model of sleep regulation. Hum Neurobiol 1:195–204

Buman MP, King AC (2010) Exercise as a treatment to enhance sleep. Am J Lifestyle Med 4:500–514

Buxton OM, L'Hermite-Baleriaux M, Hirschfeld U et al (1997) Acute and delayed effects of exercise on human melatonin secretion. J Biol Rhythms 12:568–574. https://doi.org/10.1177/074873049701200611

Carfì A, Bernabei R, Landi F et al (2020) Persistent symptoms in patients after acute COVID-19. JAMA 324(6):603–605. https://doi.org/10.1001/jama.2020.12603. PMID: 32644129; PMCID: PMC7349096

Carruthers BM, Jain AK, De Meirleir KL et al (2003) Myalgic encephalomyelitis/chronic fatigue syndrome: clinical working case definition, diagnostic and treatment protocols (Canadian case definition). J Chronic Fatigue Syndr 11(1):7–115

Cellini N, Canale N, Mioni G et al (2020) Changes in sleep pattern, sense of time and digital media use during COVID-19 lockdown in Italy. J Sleep Res 29:e13074. https://doi.org/10.1111/jsr.13074

Centers for Disease Control and Prevention, USA (2021) Post-COVID conditions. https://www.cdc.gov/coronavirus/2019-ncov/long-term-effects/index.html. Accessed 19 Jul 2021

Charrier A, Olliac B, Roubertoux P et al (2017) Clock genes and altered sleep-wake rhythms: their role in the development of psychiatric disorders. Int J Mol Sci 18(5):E938

Chen G, Wu D, Guo W et al (2020) Clinical and immunological features of severe and moderate coronavirus disease 2019. J Clin Invest 130(5). https://doi.org/10.1172/JCI137244. PMID: 32217835

Choudhry AA, Shahzeen F, Choudhry SA et al (2021) Impact of COVID-19 infection on quality of sleep. Cureus 13(9):e18182. https://doi.org/10.7759/cureus.18182

Committee on the Diagnostic Criteria for Myalgic Encephalomyelitis/Chronic Fatigue Syndrome; Board on the Health of Select Populations; Institute of Medicine (2015. https://www.ncbi.nlm.nih.gov/books/NBK274235/) Beyond myalgic encephalomyelitis/chronic fatigue syndrome:

redefining an illness. National Academies Press (US), Washington, DC. https://doi.org/10.17226/19012

Cooper AB, Thornley KS, Young GB et al (2000) Sleep in critically ill patients requiring mechanical ventilation. Chest 117:809–818. https://doi.org/10.1378/chest.117.3.809

Daou M, Telias I, Younes M et al (2020) Abnormal sleep, circadian rhythm disruption, and delirium in the ICU: are they related? Front Neurol 18(11):549908. https://doi.org/10.3389/fneur.2020.549908. PMID: 33071941; PMCID: PMC7530631

Dewald-Kaufmann JF, Oort FJ, Meijer AM (2013) The effects of sleep extension on sleep and cognitive performance in adolescents with chronic sleep reduction: an experimental study. Sleep Med 14(6):510–517. https://doi.org/10.1016/j.sleep.2013.01.012. Epub 2013 Mar 21. PMID: 23523432

Dibner C, Schibler U, Albrecht U (2010) The mammalian circadian timing system: organization and coordination of central and peripheral clocks. Annu Rev Physiol 72:517–549. https://doi.org/10.1146/annurev-physiol-021909-135821. PMID: 20148687

Dubocovich ML (2007) Melatonin receptors: role on sleep and circadian rhythm regulation. Sleep Med 8(Suppl 3):34–42. https://doi.org/10.1016/j.sleep.2007.10.007. PMID: 18032103

El Sayed S, Gomaa S, Shokry D et al (2021) Sleep in post-COVID-19 recovery period and its impact on different domains of quality of life. Egypt J Neurol Psychiatr Neurosurg 57(1):172. https://doi.org/10.1186/s41983-021-00429-7. Epub 2021 Dec 14. PMID: 34924750; PMCID: PMC8669420

Faltraco F, Palm D, Coogan A et al (2021) Remdesivir shifts circadian rhythmicity to eveningness; similar to the most prevalent chronotype in ADHD. J Neural Transm 128(7):1159–1168. https://doi.org/10.1007/s00702-021-02375-3. Epub 2021 Jul 17. PMID: 34273024; PMCID: PMC8285716

Fernández-de-Las-Peñas C, Gómez-Mayordomo V, de la Llave-Rincón AI et al (2021) Anxiety, depression and poor sleep quality as long-term post-COVID sequelae in previously hospitalized patients: a multicenter study. J Infect 83(4):496–522. https://doi.org/10.1016/j.jinf.2021.06.022. Epub 2021 Jun 27. PMID: 34192525; PMCID: PMC8236125

Foster RG (2020) Sleep, circadian rhythms and health. Interface Focus 10(3):20190098. https://doi.org/10.1098/rsfs.2019.0098. Epub 2020 Apr 17. PMID: 32382406; PMCID: PMC7202392

Gabel K, Hoddy KK, Haggerty N et al (2018) Effects of 8-hour time restricted feeding on body weight and metabolic disease risk factors in obese adults: a pilot study. Nutr Healthy Aging 4(4):345–353. https://doi.org/10.3233/NHA-170036. PMID: 29951594; PMCID: PMC6004924

Generoso JS, Barichello De Quevedo JL, Cattani M et al (2021) Neurobiology of COVID-19: how can the virus affect the brain? Br J Psychiatry 43(6):650–664. https://doi.org/10.1590/1516-4446-2020-1488

Giri A, Srinivasan A, Sundar IK (2021) COVID-19: sleep, circadian rhythms and immunity - repurposing drugs and chronotherapeutics for SARS-CoV-2. Front Neurosci 18(15):674204. https://doi.org/10.3389/fnins.2021.674204. PMID: 34220430; PMCID: PMC8249936

Güldür T, Otlu HG (2017) Circadian rhythm in mammals: time to eat & time to sleep. Biol Rhythm Res 48(2):243–261

Gulia KK, Kumar VM (2018) Sleep disorders in the elderly: a growing challenge. Psychogeriatrics 18(3):155–165. https://doi.org/10.1111/psyg.12319. PMID: 29878472

Gupta R, Grover S, Basu A et al (2020) Changes in sleep pattern and sleep quality during COVID-19 lockdown. Indian J Psychiatry 62(4):370–378. https://doi.org/10.4103/psychiatry.IndianJPsychiatry_523_20. Epub 2020 Jul 27. PMID: 33165382; PMCID: PMC7597722

Haspel JA, Anafi R, Brown MK (2020) Perfect timing: circadian rhythms, sleep, and immunity - an NIH workshop summary. JCI Insight 5:e131487

Heesakkers H, van der Hoeven JG, Corsten S et al (2022) Clinical outcomes among patients with 1-year survival following intensive care unit treatment for COVID-19. JAMA 327(6):559–565. https://doi.org/10.1001/jama.2022.0040

Hida A, Kitamura S, Katayose Y et al (2014) Screening of clock gene polymorphisms demonstrates association of a PER3 polymorphism with morningness-eveningness preference and circadian rhythm sleep disorder. Sci Rep 9(4):6309. https://doi.org/10.1038/srep06309. PMID: 25201053; PMCID: PMC4158573

Hsu PY, Harmer SL (2014) Global profiling of the circadian transcriptome using microarrays. Methods Mol Biol 1158:45–56

Huang C, Huang L, Wang Y et al (2021) 6-month consequences of COVID-19 in patients discharged from hospital: a cohort study. Lancet 397(10270):220–232. https://doi.org/10.1016/S0140-6736(20)32656-8. Epub 2021 Jan 8. PMID: 33428867; PMCID: PMC7833295

Jahrami H, BaHammam AS, Bragazzi NL et al (2021) Sleep problems during the COVID-19 pandemic by population: a systematic review and meta-analysis. J Clin Sleep Med 17(2): 299–313

Jahrami HA, Alhaj OA, Humood AM et al (2022) Sleep disturbances during the COVID-19 pandemic: a systematic review, meta-analysis, and meta-regression. Sleep Med Rev 62: 101591. https://doi.org/10.1016/j.smrv.2022.101591. Epub 2022 Jan 22. PMID: 35131664; PMCID: PMC8782754

Johnson DA (2020) Sleep disparities in the era of the COVID-19 pandemic highlight the urgent need to address social determinants of health like the virus of racism. J Clin Sleep Med 16(8): 1401–1402. https://doi.org/10.5664/jcsm.8570. PMID: 32400388; PMCID: PMC7446086

Kline CE (2014) The bidirectional relationship between exercise and sleep: implications for exercise adherence and sleep improvement. Am J Lifestyle Med 8(6):375–379. https://doi.org/10.1177/1559827614544437. PMID: 25729341; PMCID: PMC4341978

Kriegsfeld LJ, Silver R (2006) The regulation of neuroendocrine function: timing is everything. Horm Behav 49:557–574

Labrecque N, Cermakian N (2015) Circadian clocks in the immune system. J Biol Rhythm 30:277–290

Lee EK (2020) Advanced sleep-wake rhythm disorder. In: Auger R (ed) Circadian rhythm sleep-wake disorders. Springer, Cham. https://doi.org/10.1007/978-3-030-43803-6_8

Loef B, Dollé MET, Proper KI et al (2022) Night-shift work is associated with increased susceptibility to SARS-CoV-2 infection. Chronobiol Int 39(8):1100–1109. https://doi.org/10.1080/07420528.2022.2069031. Epub 2022 May 2. PMID: 35502475

Logue JK, Franko NM, McCulloch DJ et al (2021) Sequelae in adults at 6 months after COVID-19 infection. JAMA Netw Open 4(2):e210830. https://doi.org/10.1001/jamanetworkopen.2021.0830. PMID: 33606031; PMCID: PMC7896197

Longo VD, Panda S (2016) Fasting, circadian rhythms, and time-restricted feeding in healthy lifespan. Cell Metab 23(6):1048–1059. https://doi.org/10.1016/j.cmet.2016.06.001. PMID: 27304506; PMCID: PMC5388543

Lowenstein CJ, Solomon SD (2020) Severe COVID-19 is a microvascular disease. Circulation 142(17):1609–1611

Mahmud R, Rahman MM, Rassel MA et al (2021) Post-COVID-19 syndrome among symptomatic COVID-19 patients: a prospective cohort study in a tertiary care center of Bangladesh. PLoS One 16(4):e0249644. https://doi.org/10.1371/journal.pone.0249644. PMID: 33831043; PMCID: PMC8031743

Mallampalli MP, Carter CL (2014) Exploring sex and gender differences in sleep health: a Society for Women's Health Research Report. J Womens Health (Larchmt) 23(7):553–562. https://doi.org/10.1089/jwh.2014.4816. Epub 2014 Jun 23. PMID: 24956068; PMCID: PMC4089020

Mandelkorn U, Genzer S, Choshen-Hillel S et al (2021) Escalation of sleep disturbances amid the COVID-19 pandemic: a cross-sectional international study. J Clin Sleep Med 17(1):45–53

McKenna H, van der Horst GTJ, Reiss I et al (2018) Clinical chronobiology: a timely consideration in critical care medicine. Crit Care 22:124. https://doi.org/10.1186/s13054-018-2041-x

Mekhael M, Lim CH, El Hajjar AH (2022) Studying the effect of long COVID-19 infection on sleep quality using wearable health devices: observational study. J Med Internet Res 24(7):e38000. https://doi.org/10.2196/38000. PMID: 35731968

Melhuish Beaupre L, Brown GM, Kennedy JL (2018) Circadian genes in major depressive disorder. World J Biol Psychiatry 26:1e11

Mohawk JA, Green CB, Takahashi JS (2012) Central and peripheral circadian clocks in mammals. Annu Rev Neurosci 35:445–462

Moro T, Tinsley G, Bianco A et al (2016) Effects of eight weeks of time-restricted feeding (16/8) on basal metabolism, maximal strength, body composition, inflammation, and cardiovascular risk factors in resistance-trained males. J Transl Med 14(1):290. https://doi.org/10.1186/s12967-016-1044-0. PMID: 27737674; PMCID: PMC5064803

Mussa BM, Srivastava A, Verberne AJM (2021) COVID-19 and neurological impairment: hypothalamic circuits and beyond. Viruses 13(3):498. https://doi.org/10.3390/v13030498. PMID: 33802995; PMCID: PMC8002703

Ong JL, Lau T, Karsikas M et al (2021) A longitudinal analysis of COVID-19 lockdown stringency on sleep and resting heart rate measures across 20 countries. Sci Rep 11(1):14413. https://doi.org/10.1038/s41598-021-93924-z. PMID: 34257380; PMCID: PMC8277902

Partinen M, Holzinger B, Morin CM et al (2021) Sleep and daytime problems during the COVID-19 pandemic and effects of coronavirus infection, confinement and financial suffering: a multinational survey using a harmonised questionnaire. BMJ Open 11(12):e050672. https://doi.org/10.1136/bmjopen-2021-050672. PMID: 34903540; PMCID: PMC8671846

Patel NP, Grandner MA, Xie D et al (2010) "Sleep disparity" in the population: poor sleep quality is strongly associated with poverty and ethnicity. BMC Public Health 10:475. https://doi.org/10.1186/1471-2458-10-475

Peirson SN, Halford S, Foster RG (2009) The evolution of irradiance detection: melanopsin and the non-visual opsins. Philos Trans R Soc Lond B Biol Sci. 364(1531):2849–2865. https://doi.org/10.1098/rstb.2009.0050. PMID: 19720649; PMCID: PMC2781857

Poza JJ, Pujol M, Ortega-Albás JJ, Romero O (2018) Melatonin in sleep disorders. Neurologia (Engl Ed). https://doi.org/10.1016/j.nrl.2018.08.002. Epub ahead of print. PMID: 30466801

Premraj L, Kannapadi NV, Briggs J et al (2022) Mid and long-term neurological and neuropsychiatric manifestations of post-COVID-19 syndrome: a meta-analysis. J Neurol Sci 434:120162. https://doi.org/10.1016/j.jns.2022.120162. Epub 2022 Jan 29. PMID: 35121209; PMCID: PMC8798975

Priday LJ, Byrne C, Totsika V (2016) Behavioural interventions for sleep problems in people with an intellectual disability: a systematic review and meta-analysis of single case and group studies. J Intellect Disabil Res 61(1):1–15. https://doi.org/10.1111/jir.12265. Epub 2016 Mar 7. PMID: 26952339

Qin C, Zhou L, Hu Z et al (2020) Dysregulation of immune response in patients with COVID-19 in Wuhan, China. Clin Infect Dis. https://doi.org/10.1093/cid/ciaa248. PMID: 32161940

Radnis C, Qiu S, Jhaveri M et al (2020) Radiographic and clinical neurologic manifestations of COVID-19 related hypoxemia. J Neurol Sci 418:117119

Roenneberg T, Foster RG (1997) Twilight times: light and the circadian system. Photochem Photobiol 66:549–561. https://doi.org/10.1111/j.1751-1097.1997.tb03188.x

Roenneberg T, Keller LK, Fischer D et al (2015) Human activity and rest in situ. Methods Enzymol 552:257–283. https://doi.org/10.1016/bs.mie.2014.11.028. Epub 2015 Jan 5. PMID: 25707281

Romano F, Muscogiuri G, Di Benedetto E et al (2020) Vitamin D and sleep regulation: is there a role for vitamin D? Curr Pharm Des 26(21):2492–2496. https://doi.org/10.2174/1381612826666200310145935. PMID: 32156230

Rosbash M (2009) The implications of multiple circadian clock origins. PLoS Biol 7:e62. https://doi.org/10.1371/journal.pbio.1000062

Sidiropoulou P, Docea AO, Nikolaou V et al (2021) Unraveling the roles of vitamin D status and melanin during Covid-19 (Review). Int J Mol Med 47(1):92–100. https://doi.org/10.3892/ijmm.2020.4802. Epub 2020 Nov 30. PMID: 33416113; PMCID: PMC7723676

Solomon IH, Normandin E, Bhattacharyya S et al (2020) Neuropathological features of Covid-19. N Engl J Med 383(10):989–992

Tan E, Healey D, Gray AR et al (2012) Sleep hygiene intervention for youth aged 10 to 18 years with problematic sleep: a before-after pilot study. BMC Pediatr 12:189. https://doi.org/10.1186/1471-2431-12-189

Tao S, Wu X, Li S et al (2021) Circadian rhythm abnormalities during the COVID-19 outbreak related to mental health in China: a nationwide university-based survey. Sleep Med 84:165–172. https://doi.org/10.1016/j.sleep.2021.05.028

Tate W, Walker M, Sweetman E et al (2022) Molecular mechanisms of neuroinflammation in ME/CFS and long COVID to sustain disease and promote relapses. Front Neurol 25(13): 877772. https://doi.org/10.3389/fneur.2022.877772. PMID: 35693009; PMCID: PMC9174654

Tenforde MW, Kim SS, Lindsell CJ (2020) Symptom duration and risk factors for delayed return to usual health among outpatients with COVID-19 in a multistate health care systems network: United States, March–June 2020. MMWR Morb Mortal Wkly Rep 69(30):993–998. https://doi.org/10.15585/mmwr.mm6930e1. PMID: 32730238

Tordjman S, Chokron S, Delorme R et al (2017) Melatonin: pharmacology, functions and therapeutic benefits. Curr Neuropharmacol 15(3):434–443. https://doi.org/10.2174/1570159X14666161228122115. PMID: 28503116; PMCID: PMC5405617

Townsend L, Dyer AH, Jones K et al (2020) Persistent fatigue following SARS-CoV-2 infection is common and independent of severity of initial infection. PLoS One 15(11):e0240784. https://doi.org/10.1371/journal.pone.0240784

Tsang S, Avery AR, Seto EYW et al (2021) Is COVID-19 keeping us up at night? Stress, anxiety, and sleep among adult twins. Front Neurosci 26(15):665777. https://doi.org/10.3389/fnins.2021.665777. PMID: 33981199; PMCID: PMC8107288

WHO (2020) Support for rehabilitation self-management after COVID-19-related illness. https://apps.who.int/iris/bitstream/handle/10665/333287/WHO-EURO-2020-855-40590-54571-eng.pdf. Accessed 1 Sept 2022

Wichniak A, Kania A, Siemiński M, Cubała WJ (2021) Melatonin as a potential adjuvant treatment for COVID-19 beyond sleep disorders. Int J Mol Sci 22(16):8623. https://doi.org/10.3390/ijms22168623

Yamanaka Y, Hashimoto S, Takasu NN et al (2015) Morning and evening physical exercise differentially regulate the autonomic nervous system during nocturnal sleep in humans. Am J Phys Regul Integr Comp Phys 309:R1112–R1121. https://doi.org/10.1152/ajpregu.00127.2015

Yoshitane H, Asano Y, Sagami A et al (2019) Functional D-box sequences reset the circadian clock and drive mRNA rhythms. Commun Biol 2:300. https://doi.org/10.1038/s42003-019-0522-3

Zhou Y, Fu B, Zheng X et al (2020) Pathogenic T cells and inflammatory monocytes incite inflammatory storm in severe COVID-19 patients. Natl Sci Rev 7(6):998–1002

Zhuang X, Rambhatla SB, Lai AG et al (2017) Interplay between circadian clock and viral infection. J Mol Med 95:1283–1289

Chapter 13
COVID-19 and Obstructive Sleep Apnoea

Laura M. Piggott, Cara M. Gill, and Brian D. Kent

Abstract Obstructive sleep apnoea (OSA) is the most common physical sleep disorder. Many factors placing patients at increased risk of severe COVID-19 illness are shared with OSA, including male sex, obesity, diabetes, and increasing age. OSA also shares several symptoms with post-acute or long COVID, such as poor quality, unrefreshing sleep, and daytime fatigue and tiredness. Hence, there has been significant interest in interactions between OSA and COVID-19. Several large, but predominantly retrospective, studies have suggested that OSA patients are at greater risk of severe COVID-19 illness. While CPAP usage ameliorates many factors associated with adverse outcomes in COVID-19, it remains unclear if this translates into CPAP therapy modifying the severity of SARS-CoV-2 infection. Indeed, several studies have identified treated OSA as an important predictor of COVID-19-related death but are likely significantly confounded by the presence of large numbers of undiagnosed OSA patients in their study populations. The COVID-19 has certainly had a marked impact on sleep services and their ability to diagnose and treat OSA, with pandemic-induced reductions in face-to-face care potentially being associated with reduced adherence to CPAP therapy in newly diagnosed patients. Finally, there is likely a high burden of OSA in long COVID cohorts; what effect identifying and treating this will have on their symptoms remains to be seen.

Keywords COVID-19 · OSA · CPAP · Adherence

L. M. Piggott · C. M. Gill
Department of Respiratory Medicine, St James' Hospital, Dublin, Ireland

B. D. Kent (✉)
Department of Respiratory Medicine, St James' Hospital, Dublin, Ireland

School of Medicine, Trinity College, Dublin, Ireland
e-mail: briankent@physicians.ie

A. S. BaHammam et al. (eds.), *COVID-19 and Sleep: A Global Outlook*, Progress in Sleep Research, https://doi.org/10.1007/978-981-99-0240-8_13

13.1 Introduction

Globally, there have been more than 600 million cases of coronavirus disease 2019 (COVID-19) caused by severe acute respiratory syndrome coronavirus 2 (SARS-CoV-2). This has resulted in millions of deaths worldwide, with significant long-term morbidity in countless others (WHO 2022). Age, obesity, hypertension, type 2 diabetes (T2D), cardiovascular disease, and ethnic minority status are risk factors for poor outcomes in COVID-19, most of which are also common to obstructive sleep apnoea (OSA) (Cappuccio and Siani 2020; Richardson et al. 2020). Mortality and morbidity in COVID-19 arise from hypoxemic respiratory failure, multi-organ failure, venous or arterial thromboembolism, and cardiovascular complications, such as myocarditis and myocardial infarction (Leisman et al. 2020; Elezkurtaj et al. 2021; Gacche et al. 2021). The World Health Organization (WHO) defines mild COVID-19 as ambulatory patients, moderate COVID-19 as hospitalised patients including those requiring oxygen therapy, and severe COVID-19 as hospitalised patients requiring high-flow nasal oxygen, non-invasive ventilation, or invasive ventilatory support (World Health Organisation 2020). Research has suggested that the rapid deterioration in patients with COVID-19 arises from a cytokine storm, coupled with coagulation dysfunction. These factors have also been demonstrated in less severe forms in OSA patients. The available evidence suggests that patients with OSA are at higher risk of morbidity and mortality from COVID-19 infection. There are numerous potential mechanisms for such an observation, though whether OSA is in itself an independent risk factor remains to be definitively proven. This chapter aims to explore the relationships between OSA and COVID-19 infection. We will discuss potential mechanistic pathways for increased susceptibility to COVID-19 infection, as well as factors related to worse outcomes. We will review studies and meta-analyses published to date that examine this hypothesis and will also review evidence suggesting that people with severe COVID-19 infection are at higher risk of subsequently developing OSA.

OSA is the most common sleep-related breathing disorder, affecting almost one billion people worldwide, a significant proportion of whom are undiagnosed (Benjafield et al. 2019). It is characterised by transient upper airway obstruction, resulting in disrupted breathing, oxygen desaturation, and arousal from sleep (Riha 2021). Based on the apnoea-hypopnoea index (AHI), the severity of sleep apnoea is categorised as mild (AHI 5–14), moderate (AHI 15–30), and severe (AHI >30) (Sleep-related breathing disorders in adults: recommendations for syndrome definition and measurement techniques in clinical research. The Report of an American Academy of Sleep Medicine Task Force 1999; Kheirandish-Gozal and Gozal 2019). Common daytime symptoms of OSA include excessive somnolence and cognitive dysfunction, and it is strongly associated with cardiovascular and metabolic co-morbidity (Gildeh et al. 2016; Kent et al. 2011). Untreated, it is associated with a two- to threefold increased risk of mortality (Punjabi et al. 2009). Treatment frequently involves devices to maintain upper airway structure (for example mandibular repositioning devices) or continuous positive airway pressure (CPAP) therapy.

13.2 Susceptibility to SARS-CoV-2 Infection

Hypoxemia, as a consequence of untreated OSA, results in dysregulation of the renin-angiotensin axis with over-expression of angiotensin-converting enzyme 2 (ACE2) (Jin and Wei 2016). ACE2 is a key entry receptor for SARS-CoV-2 in human cells (Hoffmann et al. 2020) and thus potentially represents a predisposition to contracting COVID-19. Several studies have suggested that this putative mechanism may translate into an increased risk of COVID-19 in OSA. A retrospective analysis of over 80,000 patients with OSA found that patients with more severe OSA were more likely to have contracted COVID-19 at that point of the pandemic (OR 1.27; 95% CI 1.07–1.51) (Shi et al. 2021). Furthermore, other authors have identified an even stronger correlation between the two, with Najafi et al. identifying that patients with severe OSA were twice as likely to contract COVID-19 than those with mild disease (Najafi et al. 2021), and Maas et al. found that people living with OSA had an eightfold increase in susceptibility to COVID-19 infection than those without (Maas et al. 2021). While other studies have not confirmed this relationship (Strausz et al. 2021), it will be difficult for future trials to clarify this particular question given the global prevalence of COVID-19 at this point.

13.3 Inflammation in OSA and COVID-19

OSA has been demonstrated to exacerbate systemic inflammation, oxidative stress, and endothelial dysfunction leading to an increased risk of cardiovascular morbidity and mortality (Kent et al. 2011; Lavie 2009; Bironneau et al. 2020). Tumour necrosis factor-α (TNFα), a classic pro-inflammatory cytokine, is upregulated during sleep deprivation. In turn, this leads to the activation of nuclear factor kβ pathways activating nitric oxide synthase, cyclooxygenase 2, and adenosine A1 receptors (Kheirandish-Gozal and Gozal 2019). Similarly, intermittent hypoxia and chronic sleep fragmentation induce adipose tissue inflammation and interleukin-6 (IL-6)-driven endothelial dysfunction (Kheirandish-Gozal and Gozal 2019). In addition, lung injury may be potentiated by repeated airway obstruction (Memtsoudis et al. 2020). Overall, the intermittent hypoxaemia associated with OSA, and sleep deprivation-driven effects, may perpetuate a pro-inflammatory state contributing to the cytokine storm that leads to the rapid clinical deterioration in COVID-19 (Mutti et al. 2020). This effect is not unique to COVID-19 infection and has been replicated for numerous other disease states long before this global pandemic. OSA is associated with worse outcomes in patients hospitalised with community-acquired pneumonia (Lindenauer et al. 2014), in addition to increasing the likelihood of hospitalisation with influenza infection (Mok et al. 2020).

13.4 OSA as a Risk Factor for Poor Outcomes in COVID-19

Several studies have suggested that OSA is associated with an increased risk of severe COVID-19 infection leading to hospitalisation, invasive ventilatory support, and death (Maas et al. 2021; Cariou et al. 2020; Tufik et al. 2020; McSharry and Malhotra 2020). While some have suggested that such associations are potentially related to various confounding interplays between other co-morbid conditions, others have hypothesised that OSA may be an independent risk factor for poor outcomes in COVID-19. One of the earliest studies suggesting that OSA may be an independent risk factor for poor outcomes in COVID-19 came from Cade et al. (2020). They assessed outcomes in patients with PCR-confirmed SARS-CoV-2 infection in a hospital group, stratified by the presence or absence of a diagnosis of OSA based on an administrative database. They observed an increased likelihood of death or intensive care unit (ICU) admission in patients with diagnosed OSA (OR 1.64; 95% CI 1.32–2.04). The patients were not stratified according to the severity of OSA, need for therapy, or adherence to therapy. Furthermore, this association was markedly attenuated in models adjusting for confounding variables such as age, sex, race, ethnicity, BMI, and medical co-morbidities (adjusted OR 1.04; 95% CI 0.81–1.34). Additionally, there was a relatively low observed prevalence of OSA at just 9.5% in the study population, likely suggesting a significant burden of undiagnosed OSA (Gill et al. 2021). These results have been replicated amongst other groups, including in a large cohort of French diabetics, amongst whom a diagnosis of OSA was one of the strongest predictors of death due to COVID-19 (Cariou et al. 2020; Voncken et al. 2022).

Many early studies evaluating the correlation between an OSA diagnosis and outcomes in COVID-19 have been included in a recently published meta-analysis. Incorporating 21 studies, the authors found that OSA was associated with severe COVID-19 (OR 1.70; 95% CI 1.18–2.45), ICU admission (OR 1.76; 95% CI 1.51–2.05), need for mechanical ventilation (OR 1.67; 95% CI 1.48–1.88), and mortality from COVID-19 (OR 1.74; 95% CI 1.39–2.19) (Hariyanto and Kurniawan 2021). While strengthening the evidence base for a link between OSA and COVID-19, associations were not corrected for confounding variables in this meta-analysis and thus may have been substantially driven by shared co-morbidities, such as obesity and diabetes.

Individual studies have attempted to stratify risk according to the severity of OSA. A single-centre retrospective cohort study of 461 patients with OSA and subsequent COVID-19 found that severe OSA was associated with a significantly greater likelihood of requiring hospital admission than mild OSA (OR 4.29; 95% CI 2.09–9.02; $p < 0.0001$). This association was also markedly attenuated and was no longer statistically significant, once adjusted for confounding variables (OR 1.83;

95% CI 0.72–4.64; p 0.20) (Kravitz et al. 2021). A similar retrospective cohort study did not identify an association between the severity of OSA and COVID-19 outcomes when using apnoea-hypopnoea index (AHI), oxygen desaturation index (ODI), or respiratory disturbance index (RDI). It did report an increased likelihood of hospitalisation for patients with severe OSA when using nadir oxyhaemoglobin desaturation (LSAT) as a measure of severe OSA (OR 1.048; 95% CI 1.003–1.096) (Ho et al. 2021). However, this may reflect impaired gas exchange due to ventilation-perfusion mismatch in severe COVID-19 pneumonia, in addition to hypoxaemia arising from upper airway obstruction in OSA.

There is still a scarcity of data prospectively examining the relationship between OSA severity and outcomes in COVID-19 patients. A single-centre Italian study performed sleep apnoea testing on 44 patients hospitalised with moderate-severe COVID pneumonia, finding that over 40% had at least moderately severe sleep-disordered breathing and that increasing AHI, along with BMI, predicted an increased risk of the need for ventilatory support (Perger et al. 2021). However, this study did not include patients with mild illness, or those requiring prolonged mechanical ventilation, so it is unclear if these findings can be applied to more general populations with COVID-19.

13.5 Does Treating OSA Reduce the Risk of Severe COVID-19?

Little is known regarding the influence of adequate CPAP therapy on overall COVID-19 outcomes. Treating OSA with CPAP has been shown to ameliorate many hyper-inflammatory pathways (Kent et al. 2011). CPAP reduces hypercoagulability through decreased platelet activation, haematocrit, and plasma viscosity (Mutti et al. 2020). Additionally, it increases functional residual capacity and in turn gas exchange (Lindner et al. 1987). By improving gas exchange, it reduces the expression of markers of tissue hypoxia and prevents stimulation of the inflammatory cascade culminating in lung injury (Gill et al. 2021). Previous studies have identified that CPAP use of at least 4 h per night for at least 70% of nights—with best results achieved when used for 6 h—has been shown to have a beneficial impact on arterial stiffness reducing pro-inflammatory cytokines including TNFα, IL-2, and IL-6 amongst others (Ning et al. 2019; Perrini et al. 2017; Sawyer et al. 2011). With this, benefits are largely dependent on patient adherence to treatment (Sawyer et al. 2011). Thus, one might hypothesise that treated patients with OSA who are adherent to therapy may have a lower risk of poor COVID-19 outcomes, as suggested by a prior systematic review by Miller and Cappuccio (2021).

As treating OSA has previously been shown to ameliorate systemic inflammation, it comes as a surprise that some investigators have identified patients with OSA treated with PAP therapy having worse outcomes than those not on treatment; one example of this is a retrospective analysis of Icelandic OSA patients, which identified a significantly greater risk of severe COVID-19 illness in patients prescribed CPAP (OR 2.4; 95% CI 1.3–4.5) (Rognvaldsson et al. 2022). Similarly, findings from the CORONADO group found that treated OSA had the highest risk of 7-day mortality of all variables assessed in a cohort of French diabetics (Cariou et al. 2020). However, these data were extracted from administrative database recordings and electronic patient records did not account for the underlying severity of OSA and did not adjust for levels of adherence to CPAP. A Spanish retrospective analysis has recently aimed to bridge this gap in knowledge by including adherence to CPAP in the 6 months prior to SARS-CoV-2 infection. In this OSA population adherent to CPAP therapy, they did not identify a statistically significant link between an OSA diagnosis and death or respiratory failure from COVID-19 (OR 1.02; 95% CI 0.616–1.662; p 0.948) (Sampol et al. 2022).

13.6 Cardiovascular Interactions Between COVID-19 and OSA

While hypoxaemic respiratory failure is the main contributor to mortality in COVID-19, death has also been reported from cardiac and thromboembolic complications of the disease. Myocarditis, myocardial infarction, and fatal arrhythmias have all been reported in COVID-19 (Bandyopadhyay et al. 2020). As OSA is also a known risk factor for coronary syndrome and cardiac failure (Kohli et al. 2011), there may be an additive or synergistic effect on cardiac morbidity and mortality in OSA patients with COVID-19. Similarly, hypoxaemia and oxidative stress in OSA can precipitate a tendency towards coagulopathy, accentuating COVID-19 thrombosis and embolism (Garcia Suquia et al. 2015). However, controlled, prospective data on this potential interaction are lacking.

13.7 Access to OSA Services and CPAP Usage During the Pandemic

Access to expert follow-up was drastically cut during the pandemic due to a variety of logistical factors. The ESADA group reported a significant reduction of nearly 80% in the induction and titration of PAP during the pandemic throughout Europe (Grote et al. 2020). Adherence rates to CPAP subsequently reduced with reflective figures of up to 11% ceasing treatment during the pandemic (Grote et al. 2020; Thorpy et al. 2020). This contradicts other studies, which suggest a minor improvement in overall compliance to CPAP therapy (Tepwimonpetkun et al. 2022), which may be attributed to more time spent at home and the publicity that COVID-19 is a disease of the lungs (Demirovic et al. 2021).

Perhaps the sleep patient cohort at the greatest disadvantage at this time were those just given a new diagnosis of OSA, or those undergoing induction of therapy at the beginning of the pandemic. Turnbull et al. suggest that improvised CPAP set-up pathways driven by infection control considerations led to significant reductions in adherence 30 days following the initiation of therapy (Turnbull et al. 2022). Of those who completed face-to-face set-up, 1.1% had CPAP turned on, 71.1% had set-up without CPAP turned on, and 27.9% had remote set-up. Overall, when comparing 2019 and 2020 data, there were lower rates of CPAP usage after the first wave of the COVID-19 pandemic for patients newly started on CPAP therapy (Turnbull et al. 2022). Potential reasons for this reduction may include remote consultation, delays in consultation, delays in treatment commencement, or indeed remote set-up—while remote telemonitoring of CPAP usage has been shown to improve adherence, there remains little evidence about outcomes with remote CPAP set-up (Hwang et al. 2018). Additionally, in the early days of the pandemic, international consortia published conflicting advice regarding the use of CPAP devices at home, given the concern for droplet spread (Barker et al. 2020; Baker and Sovani 2020; Craig and West n.d.). Factors influencing CPAP therapy titration and adherence will be discussed further in later chapters. However, it remains important to note that patients with moderate-to-severe OSA who required CPAP therapy may not have had access to appropriate supports during this time, and thus adherence, and potential benefits of therapy may have been lost for some.

13.8 Impact of COVID-19 on Established OSA Therapy

CPAP requirements in the treatment of established OSA patients may change post-COVID-19 infection, though data to date is limited. A small Turkish study assessed 19 patients who had been previously diagnosed with OSA established on CPAP therapy and had a COVID-19 history. Fifteen control subjects who were not infected with SARS-CoV-2 were also recruited. Prior to contracting COVID-19 infection, auto-CPAP settings showed average 95th percentile pressure (95thpp) 8.56 ± 0.17 cm H_2O compared to post-COVID-19 settings of 9.78 ± 0.21 cm H2O ($p < 0.01$). In addition, before COVID infection, median CPAP pressure was found to be 7.49 ± 0.16 cm H_2O versus post-acute infection pressure of 8.15 ± 0.19 cm H_2O ($p < 0.01$). AHI recorded by machine on auto-titrating of CPAP was also increased post-infection but was not statistically significant (Fidan et al. 2021). It remains conceivable that there may be a causal relationship between COVID-19 infection sequelae and incidence of OSA, beyond the realms of previously undiagnosed disease.

13.9 OSA and Post-COVID Complications

Fatigue is one of the most common symptoms that persist post-COVID-19 infection and is well recognised as a symptom of untreated OSA (Gandhi et al. 2020; Labarca et al. 2022a). A small study of 69 patients recruited in the weeks after contracting COVID-19 identified high rates of OSA. Of these, 33.3% had mild OSA, 20.8% had moderate, and 16.7% had severe disease. AHI results did not significantly correlate with sleepiness (*r* 0.154; *p* 0.302) but did significantly correlate with both BMI (*r* 0.464; *p* 0.001) and age (*r* 0.319; *p* 0.027) (Schwarzl et al. 2021). A multicentre prospective cohort study of OSA in patients with long COVID-19 is underway with a nearing primary completion date that is eagerly awaited. It remains physiologically plausible that OSA may contribute to COVID-19 fatigue symptoms and morbidity by mechanisms previously mentioned. The study will aim to identify the prevalence of OSA and its relationship with post-COVID fatigue symptoms in outpatient clinics at least 6 months post-infection (Algarve 2022). In addition to sharing many of the aforementioned risk factors, it has also been suggested that OSA is linked to ongoing pathological processes in the recovery phase of COVID-19 infection. Many pathological processes are recognised in the disease process of OSA including sleep fragmentation, intermittent hypoxaemia, and sympathetic activation—with studies suggesting multi-system impairment in those with OSA that is not treated (Labarca et al. 2020; Veasey and Rosen 2019). A prospective, observational cohort study by Labarca et al. suggests that in COVID-19 survivors, confirmed OSA diagnosis is linked to significant long-term symptoms and inflammatory cytokines. Those evaluated include the risk of incident insulin resistance, T2DM, uncontrolled hypertension (>140/90 mmHg), DLCO (<80%), and impairment of neurocognitive function reported at 1 year post-acute phase of COVID-19. Findings revealed increased baseline insulin resistance in those with OSA (despite similar BMI compared with the non-OSA group), a significant difference of circulating IL-6 levels in OSA subjects at 4 months, reduced DLCO, increased incidence of uncontrolled hypertension, and also a significant reduction in Montreal Cognitive Assessment (MOCA) scores in the OSA cohort (Labarca et al. 2022a). These differences are particularly notable in subsequent cognitive impairment in patients with OSA who developed acute respiratory distress syndrome during the course of COVID-19 disease (OR 20.4, 95% CI 1.04–504). This reflects prior studies that suggest that OSA patients are more likely to develop cognitive impairment than those without, hypothesised to be related to intermittent hypoxia and dysregulation of the sleep cycle (Leng et al. 2017).

It remains difficult to assess if COVID-19 infection does increase individual risk for developing sleep-disordered breathing versus ongoing underdiagnosis of the disease. Many studies evaluating sleep breathing in COVID-19 cohorts have done so after their viral illness, making it difficult to ascertain whether OSA has arisen post-COVID-19, or predated it. For example, the CONTAC group undertook a prospective observational study of 60 Chilean patients, including performing home sleep apnoea testing at 12 and 24 weeks post-COVID-19 infection. They found higher RDI in patients who had severe COVID-19 requiring ICU admission versus patients who did not require hospitalisation (13.5 ± 10.3 vs. 7.1 ± 8.6 events/h) (Labarca et al. 2022b). Similarly, a cross-sectional population-based web survey of 20,598 adults across 14 countries, which utilised the STOP questionnaire as a 'high risk of OSA' screening tool, suggested that respondents at high risk of OSA were more likely to have had COVID-19 and also twice as likely to have developed a severe disease, requiring hospitalisation or ICU admission (Chung et al. 2021). The direction of causality here—if any—is uncertain and will require large-scale prospective studies.

Aside from fatigue and impaired neurocognitive function, interstitial lung disease is another important potential consequence of COVID-19 illness. Patients with established ILD have been found to have a high prevalence of OSA (Lee et al. 2020), and OSA is recognised as an important co-morbidity that can affect patient survival (Lancaster et al. 2009). PAP therapy in these patients may lead to improvements in overall mortality and morbidity (Papadogiannis et al. 2021). The exact burden of pulmonary fibrosis post-COVID-19 infection has not yet been established, but rates of 2–6% after moderate illness have been estimated based on data from survivors of SARS-CoV-1 infection (Bazdyrev et al. 2021). A single-centre prospective cohort study found that at 4 months post-infection, 72% of mechanically ventilated and 20% of non-mechanically ventilated COVID-19 patients had fibrotic changes on lung imaging (McGroder et al. 2021), with some data suggesting OSA to be an independent risk factor for abnormal CT thorax finding at 4 months post-infection (Labarca et al. 2022b). Reduced diffusing capacity for carbon monoxide (DLCO) has also been shown to be significantly reduced in those with OSA post-COVID-19 infection than those without at 1-year follow-up (25% OSA vs. 3.6% non-OSA, $p = 0.02$) (Labarca et al. 2022a).

13.10 OSA and Vaccine Response

Some groups have suggested that increased COVID-19 morbidity and mortality in OSA are also a consequence of decreased immunisation efficacy. It has been postulated that sleep deprivation decreases immune response after vaccination (Mello et al. 2020; Benedict and Cedernaes 2021). However, a recent study did not identify any statistically significant associations between severe OSA and reduced vaccine efficacy, as determined by anti-SARS-CoV-2 immunoglobulin G antibody levels (Tufik et al. 2022). This may provide reassurance and further motivation for high-risk patients to pursue vaccination in line with local recommendations.

13.11 Conclusion

A complex web of confounding factors exist in OSA patients with COVID-19. OSA may increase the risk of acquiring SARS-CoV-2 infection, and there is a tentative suggestion that patients with OSA are at increased risk of severe COVID-19. This may be attenuated by adherence to CPAP therapy, although this too remains uncertain, and seems unlikely to be fully protective given the many factors influencing outcomes. Additionally, given the high burden of undiagnosed OSA, it is difficult to fully extrapolate any meaningful data from large cohort studies, unless all patients are fully evaluated prospectively for the presence of OSA. The COVID-19 pandemic hindered access to sleep medicine services, potentially impacting CPAP usage; newly diagnosed OSA patients may have been particularly affected by improvised diagnostic and therapeutic pathways. Finally, unidentified and untreated OSA may be a factor in patients with fatigue and sleep disruption post-COVID-19, and in those with ILD following severe COVID pneumonia.

References

Algarve CHUd (2022) Obstructive sleep apnea in patients with long COVID-19 fatigue: a multicentre prospective cohort study (PostCoV2OSA). ClinicalTrials.gov

Baker JG, Sovani M (2020) Case for continuing community NIV and CPAP during the COVID-19 epidemic. Thorax 75(5):368

Bandyopadhyay D et al (2020) COVID-19 pandemic: cardiovascular complications and future implications. Am J Cardiovasc Drugs 20(4):311–324

Barker J et al (2020) COVID-19: community CPAP and NIV should be stopped unless medically necessary to support life. Thorax 75(5):367

Bazdyrev E et al (2021) Lung fibrosis after COVID-19: treatment prospects. Pharmaceuticals (Basel) 14(8):807

Benedict C, Cedernaes J (2021) Could a good night's sleep improve COVID-19 vaccine efficacy? Lancet Respir Med 9(5):447–448

Benjafield AV et al (2019) Estimation of the global prevalence and burden of obstructive sleep apnoea: a literature-based analysis. Lancet Respir Med 7(8):687–698

Bironneau V et al (2020) Sleep apnoea and endothelial dysfunction: an individual patient data meta-analysis. Sleep Med Rev 52:101309

Cade BE et al (2020) Sleep apnea and COVID-19 mortality and hospitalization. Am J Respir Crit Care Med 202(10):1462–1464

Cappuccio FP, Siani A (2020) Covid-19 and cardiovascular risk: susceptibility to infection to SARS-CoV-2, severity and prognosis of Covid-19 and blockade of the renin-angiotensin-aldosterone system. An evidence-based viewpoint. Nutr Metab Cardiovasc Dis 30(8): 1227–1235

Cariou B et al (2020) Phenotypic characteristics and prognosis of inpatients with COVID-19 and diabetes: the CORONADO study. Diabetologia 63(8):1500–1515

Chung F et al (2021) The association between high risk of sleep apnea, comorbidities, and risk of COVID-19: a population-based international harmonized study. Sleep Breath 25(2):849–860

Craig S, West S (n.d.) Guidance regarding coronavirus (COVID-19) and Obstructive Sleep Apnoea (OSA). https://www.brit-thoracic.org.uk/media/455098/osa-alliance-cpap-covid-19-advice-20-3-20-v10.pdf

Demirovic S et al (2021) The COVID-19 lockdown and CPAP adherence: the more vulnerable ones less likely to improve adherence? Nat Sci Sleep 13:1097–1108
Elezkurtaj S et al (2021) Causes of death and comorbidities in hospitalized patients with COVID-19. Sci Rep 11(1):4263
Fidan V, Koyuncu H, Akin O (2021) Alteration of Auto-CPAP requirements in obstructive sleep apnea patients with COVID-19 history. Am J Otolaryngol 42(3):102919
Gacche RN et al (2021) Predictors of morbidity and mortality in COVID-19. Eur Rev Med Pharmacol Sci 25(3):1684–1707
Gandhi RT, Lynch JB, Del Rio C (2020) Mild or moderate Covid-19. N Engl J Med 383(18): 1757–1766
Garcia Suquia A et al (2015) High D-dimer levels after stopping anticoagulants in pulmonary embolism with sleep apnoea. Eur Respir J 46(6):1691–1700
Gildeh N et al (2016) Emerging co-morbidities of obstructive sleep apnea: cognition, kidney disease, and cancer. J Thorac Dis 8(9):E901–E917
Gill CM, Piggott LM, Kent BD (2021) Sleep medicine and coronavirus disease 2019. Curr Opin Pulm Med 27(6):529–534
Grote L, McNicholas WT, Hedner J (2020) Sleep apnoea management in Europe during the COVID-19 pandemic: data from the European Sleep Apnoea Database (ESADA). Eur Respir J 55(6):2001323
Hariyanto TI, Kurniawan A (2021) Obstructive sleep apnea (OSA) and outcomes from coronavirus disease 2019 (COVID-19) pneumonia: a systematic review and meta-analysis. Sleep Med 82: 47–53
Ho J et al (2021) Association between the degree of obstructive sleep apnea and the severity of COVID-19: an explorative retrospective cross-sectional study. PLoS One 16(9):e0257483
Hoffmann M et al (2020) SARS-CoV-2 cell entry depends on ACE2 and TMPRSS2 and is blocked by a clinically proven protease inhibitor. Cell 181(2):271–280.e8
Hwang D et al (2018) Effect of telemedicine education and telemonitoring on continuous positive airway pressure adherence. The Tele-OSA randomized trial. Am J Respir Crit Care Med 197(1): 117–126
Jin ZN, Wei YX (2016) Meta-analysis of effects of obstructive sleep apnea on the renin-angiotensin-aldosterone system. J Geriatr Cardiol 13(4):333–343
Kent BD, Ryan S, McNicholas WT (2011) Obstructive sleep apnea and inflammation: relationship to cardiovascular co-morbidity. Respir Physiol Neurobiol 178(3):475–481
Kheirandish-Gozal L, Gozal D (2019) Obstructive sleep apnea and inflammation: proof of concept based on two illustrative cytokines. Int J Mol Sci 20(3)
Kohli P, Balachandran JS, Malhotra A (2011) Obstructive sleep apnea and the risk for cardiovascular disease. Curr Atheroscler Rep 13(2):138–146
Kravitz MB et al (2021) Severity of sleep apnea and COVID-19 illness. OTO Open 5(2): 2473974X211016283
Labarca G et al (2020) Chronic intermittent hypoxia in obstructive sleep apnea: a narrative review from pathophysiological pathways to a precision clinical approach. Sleep Breath 24(2):751–760
Labarca G et al (2022a) Impact of obstructive sleep apnea (OSA) in COVID-19 survivors, symptoms changes between 4-months and 1 year after the COVID-19 infection. Front Med (Lausanne) 9:884218
Labarca G et al (2022b) Undiagnosed sleep disorder breathing as a risk factor for critical COVID-19 and pulmonary consequences at the midterm follow-up. Sleep Med 91:196–204
Lancaster LH et al (2009) Obstructive sleep apnea is common in idiopathic pulmonary fibrosis. Chest 136(3):772–778
Lavie L (2009) Oxidative stress—a unifying paradigm in obstructive sleep apnea and comorbidities. Prog Cardiovasc Dis 51(4):303–312
Lee JH, Park CS, Song JW (2020) Obstructive sleep apnea in patients with interstitial lung disease: prevalence and predictive factors. PLoS One 15(10):e0239963

Leisman DE et al (2020) Cytokine elevation in severe and critical COVID-19: a rapid systematic review, meta-analysis, and comparison with other inflammatory syndromes. Lancet Respir Med 8(12):1233–1244

Leng Y et al (2017) Association of sleep-disordered breathing with cognitive function and risk of cognitive impairment: a systematic review and meta-analysis. JAMA Neurol 74(10):1237–1245

Lindenauer PK et al (2014) Prevalence, treatment, and outcomes associated with OSA among patients hospitalized with pneumonia. Chest 145(5):1032–1038

Lindner KH, Lotz P, Ahnefeld FW (1987) Continuous positive airway pressure effect on functional residual capacity, vital capacity and its subdivisions. Chest 92(1):66–70

Maas, M.B., et al., Obstructive sleep apnea and risk of COVID-19 infection, hospitalization and respiratory failure. Sleep Breath, 2021 25(2): 1155–1157.

McGroder CF et al (2021) Pulmonary fibrosis 4 months after COVID-19 is associated with severity of illness and blood leucocyte telomere length. Thorax 76(12):1242

McSharry D, Malhotra A (2020) Potential influences of obstructive sleep apnea and obesity on COVID-19 severity. J Clin Sleep Med 16(9):1645

Mello MT et al (2020) Sleep and COVID-19: considerations about immunity, pathophysiology, and treatment. Sleep Sci 13(3):199–209

Memtsoudis SG et al (2020) Obesity as a risk factor for poor outcome in COVID-19-induced lung injury: the potential role of undiagnosed obstructive sleep apnoea. Br J Anaesth 125(2):e262–e263

Miller MA, Cappuccio FP (2021) A systematic review of COVID-19 and obstructive sleep apnoea. Sleep Med Rev 55:101382

Mok EM, Greenough G, Pollack CC (2020) Untreated obstructive sleep apnea is associated with increased hospitalization from influenza infection. J Clin Sleep Med 16(12):2003–2007

Mutti C et al (2020) Obstructive sleep apnea, cpap and COVID-19: a brief review. Acta Biomed 91(4):e2020196

Najafi A et al (2021) The effect of apnea management on novel coronavirus infection: a study on patients with obstructive sleep apnea. Sleep Health 7(1):14–18

Ning Y et al (2019) Effects of continuous positive airway pressure on cardiovascular biomarkers in patients with obstructive sleep apnea: a meta-analysis of randomized controlled trials. Sleep Breath 23(1):77–86

Papadogiannis G et al (2021) Patients with idiopathic pulmonary fibrosis with and without obstructive sleep apnea: differences in clinical characteristics, clinical outcomes, and the effect of PAP treatment. J Clin Sleep Med 17(3):533–544

Perger E et al (2021) Sleep-disordered breathing among hospitalized patients with COVID-19. Am J Respir Crit Care Med 203(2):239–241

Perrini S et al (2017) Correction of intermittent hypoxia reduces inflammation in obese subjects with obstructive sleep apnea. JCI Insight 2(17):e94379

Punjabi NM et al (2009) Sleep-disordered breathing and mortality: a prospective cohort study. PLoS Med 6(8):e1000132

Richardson S et al (2020) Presenting characteristics, comorbidities, and outcomes among 5700 patients hospitalized with COVID-19 in the New York City area. JAMA 323(20):2052–2059

Riha RL (2021) Defining obstructive sleep apnoea syndrome: a failure of semantic rules. Breathe (Sheff) 17(3):210082

Rognvaldsson KG et al (2022) Obstructive sleep apnea is an independent risk factor for severe COVID-19: a population-based study. Sleep 45(3)

Sampol J et al (2022) Impact of home CPAP-treated obstructive sleep apnea on COVID-19 outcomes in hospitalized patients. J Clin Sleep Med 18(7):1857–1864

Sawyer AM et al (2011) A systematic review of CPAP adherence across age groups: clinical and empiric insights for developing CPAP adherence interventions. Sleep Med Rev 15(6):343–356

Schwarzl G et al (2021) The prevalence of obstructive sleep apnea (OSA) in patients recovering from COVID-19. ERJ Open Res 7(Suppl 7):24

Shi J et al (2021) Impact of obstructive sleep apnea and positive airway pressure therapy on COVID-19 outcomes. Am J Respir Crit Care Med 203:A1108

Sleep-related breathing disorders in adults: recommendations for syndrome definition and measurement techniques in clinical research. The Report of an American Academy of Sleep Medicine Task Force (1999) Sleep 22(5):667–689

Strausz S et al (2021) Sleep apnoea is a risk factor for severe COVID-19. BMJ Open Respir Res 8(1):e000845

Tepwimonpetkun C et al (2022) Positive airway pressure therapy adherence for obstructive sleep apnea during the COVID-19 pandemic. Sleep Breath 26(3):1193–1199

Thorpy M et al (2020) Management of sleep apnea in New York City during the COVID-19 pandemic. Sleep Med 74:86–90

Tufik S et al (2020) Does obstructive sleep apnea lead to increased risk of COVID-19 infection and severity? J Clin Sleep Med 16(8):1425–1426

Tufik S et al (2022) Effects of obstructive sleep apnea on SARS-CoV-2 antibody response after vaccination against COVID-19 in older adults. Nat Sci Sleep 14:1203–1211

Turnbull CD et al (2022) COVID-19-related changes in outpatient CPAP setup pathways for OSA are linked with decreased 30-day CPAP usage. Thorax. https://doi.org/10.1136/thoraxjnl-2021-218635

Veasey SC, Rosen IM (2019) Obstructive sleep apnea in adults. N Engl J Med 380(15):1442–1449

Voncken SFJ et al (2022) Impact of obstructive sleep apnea on clinical outcomes in patients hospitalized with COVID-19. Sleep Breath 26(3):1399–1407

WHO (2022) WHO COVID-19 dashboard. World Health Organization, Geneva. https://covid19.who.int/

World Health Organisation (2020) A minimal common outcome measure set for COVID-19 clinical research. pp 1474–4457 (Electronic)

Chapter 14
The Correlation Between COVID-19 Infection and Restless Legs Syndrome

Andrea Maculano Esteves, Beatriz Franco, and Milca Abda de Morais

Abstract Restless legs syndrome (RLS) is considered a common sensorimotor disorder characterized by an urge to move the limbs, especially the legs. Studies demonstrate an association between RLS and mood disorders, and comorbidities such as hypertension, diabetes, and cardiovascular diseases. Exercise, alone or in combination with medication, is considered a good strategy to improve symptoms. Home confinement during COVID-19 affects the physiological health of both body and mind. Isolation at home brought about changes in people's mood, metabolism, and sleep, as well as an increasingly sedentary lifestyle. Based on this, the behavioral changes caused by COVID-19 could have impacted people with RLS or who are predisposed.

Keywords Sleep disorders · Exercise · Mood disorders · Comorbidities · Pandemic

14.1 Introduction

Restless legs syndrome (RLS), also known as Willis-Ekbom disease, was first described in 1672 by the English physician, Sir Thomas Willis (Coccagna et al. 2004), and is considered a common sensorimotor disorder characterized by an urge to move the limbs, especially the legs, to relieve uncomfortable sensations such as pain, burning, itching, and tension or discomfort between the knee and ankle.

A. M. Esteves (✉)
Faculdade de Ciências Aplicadas, Universidade Estadual de Campinas, Limeira, SP, Brazil

Faculdade de Educação Física, Universidade Estadual de Campinas, Campinas, SP, Brazil
e-mail: andrea.esteves@fca.unicamp.br

B. Franco
Faculdade de Educação Física, Universidade Estadual de Campinas, Campinas, SP, Brazil

M. A. de Morais
Faculdade de Ciências Aplicadas, Universidade Estadual de Campinas, Limeira, SP, Brazil

Faculdade de Medicina de Açailândia, FAMEAC/IDOMED, Açailândia, MA, Brazil

A. S. BaHammam et al. (eds.), *COVID-19 and Sleep: A Global Outlook*, Progress in Sleep Research, https://doi.org/10.1007/978-981-99-0240-8_14

Movement such as walking, stretching, or bending the legs relieves the discomfort (Trenkwalder et al. 2005). The syndrome presents within a circadian rhythm, with worsening symptoms at night (Kushida et al. 2007). Patients with RLS often have insomnia and periodic leg movements (PLMs) during sleep (PLMS) or resting wakefulness (PLMW) (Allen et al. 2014). Periodic limb movement disorder is defined as involuntary movements of the patient's limb or torso while awake or asleep, of which the patient is not aware (Trenkwalder et al. 2005).

Epidemiological studies find that RLS prevalence is between 5% and 15% of North American adults, 2–3% with severe symptoms. Restless legs syndrome affects twice as many women as men, with a mean age of onset between 30 and 40 years (Garcia-Borreguero and Cano-Pumarega 2017). Pregnancy and hormonal factors are risk factors for RLS, causing the increased prevalence in women (Hanson et al. 2004; Manconi et al. 2021), and it is argued that women are more aware of symptoms than men (Holzknecht et al. 2020). Studies also show an increased prevalence of RLS symptoms with increasing age (Ohayon et al. 2012). In addition, studies have also indicated that RLS is relatively common in children and adolescents, affecting 1–4% of this population (Yilmaz et al. 2011; Turkdogan et al. 2011).

RLS can be classified as idiopathic or primary and secondary (symptomatic) to comorbid conditions (Garcia-Borreguero and Cano-Pumarega 2017). Primary or idiopathic RLS occurs when the cause is truly unknown. Among the idiopathic RLS, 40.9–92% had a family history of RLS, indicating the important role of genetic factors (Winkelman et al. 1996; Winkelmann et al. 2002). On the other hand, secondary RLS is associated with a variety of neurological disorders, iron deficiency, pregnancy, or chronic renal failure (Çurgunlu et al. 2012; Srivanitchapoom et al. 2014; Winkelman et al. 1996). Some authors report that these classifications are being increasingly questioned because they might suggest an inappropriate causal relation. However in certain conditions (e.g., iron deficiency), treatment of the underlying disease may reduce or eliminate symptoms of RLS. The term "comorbid RLS" seems more appropriate for most other associations. Thus, RLS might be seen as a continuous spectrum with a major genetic contribution at one end and a major environmental or comorbid disease contribution at the other (Trenkwalder et al. 2016).

Studies in individuals with a high comorbidity burden have consistently observed a higher prevalence of RLS than studies of healthy individuals (Szentkirályi et al. 2014). A growing number of publications have reported associations between RLS and numerous diseases, including metabolic conditions, cardiovascular or renal disorders, arterial hypertension, autoimmune diseases, polyneuropathy, neurodegenerative disorders, and conditions associated with inflammation (Trenkwalder et al. 2016). It often affects the patient's sleep, and they might complain about having difficulty in their daily life, including their jobs and social activities (Kushida et al. 2007). A significantly higher probability of persons with attention deficit hyperactivity disorder (25%) was found in RLS patients compared to the general population (Pullen et al. 2011).

Furthermore, studies demonstrate an association between RLS and mood disorders (Lee et al. 2008, 2012), such as depression and anxiety (Becker and Sharon

2014). There may be an alteration in the sleep architecture by both RLS and mood disorders, and both can exacerbate psychiatric symptoms (Hornyak 2010; Lee et al. 2008). Many common antidepressants and antipsychotics can increase the sensory symptoms of RLS, worsening sleep disturbance as well as preexisting psychiatric disorders (Becker and Sharon 2014). Also, RLS and mood disorders may share similar pathophysiological networks in the central nervous system, including dopaminergic dysfunction (Connor et al. 2009; Lee et al. 2008; Muhle et al. 2008). Treatment of depression in RLS patients should be cautious because many antidepressants, such as selective serotonin reuptake inhibitors and tricyclic antidepressants, can worsen the symptoms of RLS (Mackie and Winkelman 2015).

Among the risk factors for RLS, it is known that genetic predisposition is an essential contributor (Manconi et al. 2021). Studies have found a positive family history of RLS in 20–60% of patients, and it is believed to vary by differences in ethnicity, study design, and sample size. Also, positive family history can be found in up to 60% of individuals with primary RLS (Allen et al. 2002; Ohayon et al. 2012).

Etiopathogenesis of RLS remains somewhat uncertain and is considered a complex condition in which genetic factors and environmental and gene-environment interactions predispose people to the disease and affect the expression of the full clinical phenotype (Trenkwalder et al. 2016). Several mechanisms, such as brain iron deficiency (Michaud et al. 2004), dopamine dysregulation (Galbiati et al. 2015), hyper-glutamatergic state (Allen et al. 2013), adenosine dysregulation (Ferré et al. 2018), and variations in genes such as MEIS1, BTBD9, TOX3, and PTPRD (Morais et al. 2021; Schormair et al. 2017), have been identified and proposed to play a major part in the pathophysiology. Some of these can be targets for therapeutic action (Manconi et al. 2021).

In many cases, RLS is still incurable, so treatments focus on alleviating disease symptoms, which can be effectively managed if treatment is timely and adequate (Manconi et al. 2021). Restless legs syndrome symptoms respond well to low doses of dopamine agonists, $\alpha 2\delta$ ligands, or opiates (Winkelmann et al. 2018). Dopamine agonists are among the most effective compounds in the treatment of RLS and are considered the first line of treatment (Garcia-Borreguero et al. 2013; Silber et al. 2013). However, long-term treatment is associated with augmentation, a severe, paradoxical, drug-related worsening of symptoms (Mackie and Winkelman 2015).

Considering this effect, some studies have demonstrated the positive effect of non-pharmacological treatments such as exercise (Esteves et al. 2009, 2016; Frank et al. 2012). Exercise can be effective in reducing symptoms and improving sleep patterns by increasing sleep efficiency and REM sleep time and reducing sleep latency (Aukerman et al. 2006; Esteves et al. 2009, 2011). Esteves et al. (2009) showed improvements in symptoms of leg movements during sleep in patients with RLS and mild, moderate, and severe PLM with engagement in aerobic exercise. The positive effects of exercise on RLS have been shown in studies with acute and chronic aerobic exercise at moderate to intense intensity and anaerobic exercise (Franco et al. 2019).

14.2 RLS and COVID-19

On the other hand, studies have shown that physical inactivity, and a lack of regular social interaction in home confinement caused by the COVID-19 pandemic, affects the physiological health of the body and mind (Brooks et al. 2020; Lee et al. 2012). Isolation at home brought about changes in mood, metabolism, and sleep, and an increasingly sedentary lifestyle in people. Based on this, the behavioral changes caused by COVID-19 can impact people with RLS or who are predisposed to RLS (Franco et al. 2020).

In February 2020, the World Health Organization (WHO) declared the severe acute respiratory syndrome coronavirus 2 (SARS-CoV-2) a threat to the entire world, and in March, it declared COVID-19 a global pandemic (Cucinotta and Vanelli 2020). From that moment on, the WHO suggested protective measures, and with the lack of effective treatments, lack of vaccines, and increase in cases and deaths, several countries decreed lockdown. The lockdown was the most effective measure to stop the contamination of the COVID-19 virus; however, it had consequences on people's health.

The lack of daily activities, leisure, and visiting friends and family due to confinement generated changes in mood in the population, such as anxiety, depression, stress, and feelings of loneliness (Hendriksen et al. 2021; Müller et al. 2021). Financial problems, insecurity, fear of infection, and inadequate information were also causes of changes in mood during the COVID-19 pandemic (Altena et al. 2020; Brooks et al. 2020).

Sleep also suffered during the COVID-19 confinement. Sleep dysfunctions such as insomnia, disrupted sleep continuity, changes in the sleep-wake cycle, feelings of non-restorative sleep, and decreased sleep quality arise either due to stresses related to fear of the virus itself or due to the psychosocial impact on daily living (Bhat and Chokroverty 2022). Thus, several additional stressors during the COVID-19 pandemic seem to have accentuated preexisting sleep disorders or started new ones.

Patients with RLS had worsening symptoms during the period of greater confinement (April and May 2020). This worsening may have been due to confinement, though it could also be attributed to other stressors such as emotional, social, and political, and the use of specific medication. Wipper et al. (2022), in a longitudinal observational study of patients using opioid medications, evaluated the treatment of RLS at three moments during the pandemic (outset of the pandemic, 6 months later, and approximately 1 year later). Symptom severity scores were still elevated on subsequent questionnaires completed over 6 months into the pandemic but had returned toward baseline 1 year later. Just as rates of anxiety and depression dropped after the easing of confinement during the pandemic (early 2021) (Jia et al. 2021), RLS symptoms re-emerged (Wipper et al. 2022).

Mood disorders such as depression and anxiety are related to RLS (Becker and Sharon 2014; Lee et al. 2008). This relationship may be due to the pathophysiological sharing of systems, such as the dopaminergic system (Lee et al. 2008), by the vicious cycle in which RLS worsens mood and mood worsens RLS, and by drugs used in both treatments (Hornyak 2010). Therefore, this relationship may have been exacerbated during the confinement of COVID-19, worsening the symptoms of RLS and mood disorders.

A reduction in physical activities is another factor that may impact the worsening of RLS symptoms. A study carried out during the period of greater confinement and social distancing showed that all types of physical activities carried out before the pandemic, such as vigorous or moderate exercise and walking, were reduced (Ammar et al. 2020). That is, people became less active. In addition, people reported increased time spent sitting during the day (Ammar et al. 2020). Exercise is considered a non-pharmacological treatment for RLS or even an aid in conjunction with drugs to improve symptoms (Esteves et al. 2009; Franco et al. 2019), so reducing physical activities can worsen symptoms in RLS patients and even in undiagnosed people.

Reduced exercise, increased stress, and changes in food consumption during social distancing caused comorbidities to worsen or appear during much of the COVID-19 pandemic (Ruiz-Roso et al. 2020). Hypertension, diabetes, and cardiovascular diseases were identified as risk factors for worsening COVID-19 symptoms, in addition to being related to RLS, with an increased prevalence of RLS in patients with these comorbidities (Akin et al. 2019; Walters and Rye 2009; Yoshihisa et al. 2019). Studies show an increased prevalence of RLS in people with cardiovascular disease, hypertension, and type 2 diabetes (Akin et al. 2019; Yoshihisa et al. 2019). The hypotheses of these relationships are based on dopaminergic alterations, sympathetic hyperactivity, and sleep alterations due to nocturnal excitability (Cubo et al. 2019; Trenkwalder et al. 2018). In addition, changes in sleep caused by RLS (e.g., sleep disruption and reduced total sleep time) favor the development of obesity, hypertension, heart disease, and type 2 diabetes (Cubo et al. 2019; Jin et al. 2020). Considering the data that physical inactivity can generate, and the relationship of RLS with comorbidities, the reduction of physical activities during the COVID-19 pandemic may have had short- and long-term consequences on health, sleep, and quality of life for people with RLS or who are predisposed.

Recent studies have also demonstrated the relationship of RLS with patients who experience long-COVID (LC), also known as long-haul COVID, post-COVID inflammatory syndrome, and post-acute sequelae of COVID-19 (Weinstock et al. 2022). In general, sleep disturbances occur in 34% of patients with COVID-19 (Deng et al. 2021). At 6 months after the infection, fatigue or muscle weakness and sleep difficulties were the main symptoms of patients who had recovered from COVID-19 (Huang et al. 2021). Weinstock et al. (2022) evaluated 136 participants with LC and 136 controls. Restless legs syndrome prevalence in females with LC was 5.7% pre-COVID-19 and 14.8% post-COVID-19 ($p < 0.01$) vs. 6.7% in control females. Logistic regression predicting post-COVID-19 RLS among females with LC failed to find significant effects of hospitalization, sleep apnea, neuropathic pain severity, or use of antihistamines and antidepressants. The authors hypothesize that postinfectious immunological mechanisms may be one potential way that RLS could be produced by COVID-19.

Although poor sleep quality is a common manifestation of COVID-19 infection, associations between changes in RLS severity, depression, anxiety, and sleep disturbance are important factors and require special attention.

The article by Franco et al. (2020) brought suggestions and recommendations to reduce RLS-related damage during confinement. However, the possible consequences caused or that remain in patients with RLS or predisposed people, who are not even aware of the existence of this movement disorder, are unknown. That brings us an important mission for studies that investigate and create strategies to reduce possible damage.

References

Akin S, Bölük C, Börü UT, Tasdemir M, Gezer T, Sahbaz FG et al (2019) Restless legs syndrome in type 2 diabetes mellitus. Prim Care Diabetes 13(1):87–91

Allen RP, La Buda MC, Becker P, Earley CJ (2002) Family history study of the restless legs syndrome. Sleep Med 3(Suppl):S3–S7

Allen RP, Barker PB, Horska A, Earley CJ (2013) Thalamic glutamate/glutamine in restless legs syndrome: increased and related to disturbed sleep. Neurology 80:2028–2034

Allen RP, Picchietti DL, Garcia-Borreguero D, Ondo WG, Walters AS, Winkelman JW, Zucconi M, Ferri R, Trenkwalder C, Lee HB (2014) International Restless Legs Syndrome Study Group. Restless legs syndrome/Willis-Ekbom disease diagnostic criteria: updated International Restless Legs Syndrome Study Group (IRLSSG) consensus criteria—history, rationale, description, and significance. Sleep Med 15(8):860–873

Altena E, Baglioni C, Espie CA, Ellis J, Gavriloff D, Holzinger B et al (2020) Dealing with sleep problems during home confinement due to the COVID-19 outbreak: practical recommendations from a task force of the European CBT-I Academy. J Sleep Res 29(4):e13052

Ammar A, Brach M, Trabelsi K, Chtourou H, Boukhris O, Masmoudi L et al (2020) Effects of COVID-19 home confinement on eating behaviour and physical activity: results of the ECLB-COVID19 International Online Survey. Nutrients 12(6):1583. https://doi.org/10.3390/nu12061583

Aukerman MM, Aukerman D, Bayard M, Tudiver F, Thorp L, Bailey B (2006) Exercise and restless legs syndrome: a randomized controlled trial. J Am Board Fam Med 19(5):487–493

Becker PM, Sharon D (2014) Mood disorders in restless legs syndrome (Willis-Ekbom disease). J Clin Psychiatry 75(7):e679–e694

Bhat S, Chokroverty S (2022) Sleep disorders and COVID-19. Sleep Med 91:253–261

Brooks SK, Webster RK, Smith LE, Woodland L, Wessely S, Greenberg N, Rubin GJ (2020) The psychological impact of quarantine and how to reduce it: rapid review of the evidence. Lancet 395:12–920

Coccagna G, Vetrugno R, Lombardi C, Provini F (2004) Restless legs syndrome: an historical note. Sleep Med 5(3):279–283

Connor JR, Wang XS, Allen RP, Beard JL, Wiesinger JA, Felt BT, Earley CJ (2009) Altered dopaminergic profile in the putamen and substantia nigra in restless leg syndrome. Brain 132 (Pt 9):2403–2412

Cubo E, Gallego-Nieto C, Elizari-Roncal M, Barroso-Perez T, Collazo C, Calvo S et al (2019) Is restless legs syndrome associated with an increased risk of mortality? A meta-analysis of cohort studies. Tremor Other Hyperkinet Mov 9. https://doi.org/10.7916/tohm.v0.650

Cucinotta D, Vanelli M (2020) WHO declares COVID-19 a pandemic. Acta Biomed 91(1):157

Çurgunlu A, Döventaş A, Karadeniz D, Erdinçler DS, Oztürk AK, Karter Y, Yaldiran A, Sipahioğlu F, Beğer T (2012) Prevalence and characteristics of restless legs syndrome (RLS) in the elderly and the relation of serum ferritin levels with disease severity: hospital-based study from Istanbul, Turkey. Arch Gerontol Geriatr 55(1):73–76. S0167494311001531. https://doi.org/10.1016/j.archger.2011.06.002

Deng J, Zhou F, Hou W, Silver Z, Wong CY, Chang O, Huang E, Zuo QK (2021) The prevalence of depression, anxiety, and sleep disturbances in COVID-19 patients: a meta-analysis. Ann N Y Acad Sci 1486(1):90–111

Esteves AM, de Mello MT, Pradella-Hallinan M, Tufik S (2009) Effect of acute and chronic physical exercise on patients with periodic leg movements. Med Sci Sports Exerc 41(1): 237–242

Esteves AM, Silva AAB, Tufik S, De Mello MT (2011) Impact of aerobic physical exercise on Restless Legs Syndrome. Sleep Sci 4:45–48

Esteves AM, Lopes C, Frank MK, Arida RM, Frussa-Filho R, Tufik S, de Mello MT (2016) Can physical exercise have a protective effect in an animal model of sleep-related movement disorder? Brain Res 1639:47–57

Ferré S, Quiroz C, Guitart X, Rea W, Seyedian A, Moreno E, Casadó-Anguera V, Díaz-Ríos M, Casadó V, Clemens S, Allen RP, Earley CJ, García-Borreguero D (2018) Pivotal role of adenosine neurotransmission in Restless Legs Syndrome. Front Neurosci 11:722

Franco B, Daubian-Nosé P, De-Mello MT, Esteves AM (2019) Exercise as a favorable non-pharmacologic treatment to Sleep-Related Movement Disorders: a review. Sleep Sci 12(2):116–121. https://doi.org/10.5935/1984-0063.20190064

Franco B, Morais MA, Holanda ASS, Manconi M, de Mello MT, Esteves AM (2020) Impact of Covid-19 on the restless legs syndrome. Sleep Sci 13(3):186–190

Frank MK, Esteves AM, Lopes C, Cavagnolli DA, Tufik S, de Mello MT (2012) The effects of physical exercise on the serum iron profile in spontaneously hypertensive rats. Biol Trace Elem Res 145(2):222–224

Galbiati A, Marelli S, Giora E, Zucconi M, Oldani A, Ferini-Strambi L (2015) Neurocognitive function in patients with idiopathic Restless Legs Syndrome before and after treatment with dopamine-agonist. Int J Psychophysiol 95(3):304–309

Garcia-Borreguero D, Cano-Pumarega I (2017) New concepts in the management of restless legs syndrome. BMJ 356:j104

Garcia-Borreguero D, Kohnen R, Silber MH, Winkelman JW, Earley CJ, Högl B, Manconi M, Montplaisir J, Inoue Y, Allen RP (2013) The long-term treatment of restless legs syndrome/ Willis-Ekbom disease: evidence-based guidelines and clinical consensus best practice guidance: a report from the International Restless Legs Syndrome Study Group. Sleep Med 14(7):675–684

Hanson M, Honour M, Singleton A, Crawley A, Hardy J, Gwinn-Hardy K (2004) Analysis of familial and sporadic restless legs syndrome in age of onset, gender, and severity features. J Neurol 251:1398–1401

Hendriksen PA, Garssen J, Bijlsma EY, Engels F, Bruce G, Verster JC (2021) COVID-19 lockdown-related changes in mood, health and academic functioning. Eur J Investig Health Psychol Educ 11(4):1440–1461. Published 2021 Nov 18

Holzknecht E, Hochleitner M, Wenning GK, Högl B, Stefani A (2020) Gender differences in clinical, laboratory and polysomnographic features of restless legs syndrome. J Sleep Res 29(3): e12875

Hornyak M (2010) Depressive disorders in restless legs syndrome: epidemiology, pathophysiology and management. CNS Drugs 24(2):89–98

Huang C, Huang L, Wang Y, Li X, Ren L, Gu X et al (2021) 6-month consequences of COVID-19 in patients discharged from hospital: a cohort study. Lancet 397(10270):220–232

Jia H, Guerin RJ, Barile JP, Okun AH, McKnight-Eily L, Blumberg SJ, Njai R, Thompson WW (2021) National and state trends in anxiety and depression severity scores among adults during the COVID-19 pandemic - United States, 2020–2021. MMWR Morb Mortal Wkly Rep 70(40): 1427–1432

Jin B, Wang A, Earley C, Allen R (2020) Moderate to severe but not mild RLS is associated with greater sleep-related sympathetic autonomic activation than healthy adults without RLS. Sleep Med 68:89–95

Kushida C, Martin M, Nikam P, Blaisdell B, Wallenstein G, Ferini-Strambi L, Ware JE Jr (2007) Burden of restless legs syndrome on health-related quality of life. Qual Life Res 16(4):617–624

Lee HB, Hening WA, Allen RP, Kalaydjian AE, Earley CJ, Eaton WW, Lyketsos CG (2008) Restless legs syndrome is associated with DSM-IV major depressive disorder and panic disorder in the community. J Neuropsychiatry Clin Neurosci 20(1):101–105

Lee IM, Shiroma EJ, Lobelo F, Puska P, Blair SN, Katzmarzyk PT, Lancet Physical Activity Series Working Group 2012 (2012) Effect of physical inactivity on major non-communicable diseases worldwide: an analysis of burden of disease and life expectancy. Lancet 380:219–229

Mackie S, Winkelman JW (2015) Long-term treatment of Restless Legs Syndrome (RLS): an approach to management of worsening symptoms, loss of efficacy, and augmentation. CNS Drugs 29(5):351–357

Manconi M, Garcia-Borreguero D, Schormair B, Videnovic A, Berger K, Ferri R, Dauvilliers Y (2021) Restless legs syndrome. Nat Rev Dis Primers 7:80

Michaud M, Dumont M, Selmaoui B, Paquet J, Fantini ML, Montplaisir J (2004) Circadian rhythm of restless legs syndrome: relationship with biological markers. Ann Neurol 55(3):372–380

Morais MA, Franco B, Holanda ASS, Simino LAP, Manconi M, Torsoni A, Esteves AM (2021) PTPRD as a candidate druggable target for therapies for restless legs syndrome? J Sleep Res 30(4):e13216

Muhle H, Neumann A, Lohmann-Hedrich K, Lohnau T, Lu Y, Winkler S, Waltz S, Fischenbeck A, Kramer PL, Klein C, Stephani U (2008) Childhood-onset restless legs syndrome: clinical and genetic features of 22 families. Mov Disord 23(8):1113–1121. quiz 1203

Müller F, Röhr S, Reininghaus U, Riedel-Heller SG (2021) Social isolation and loneliness during COVID-19 lockdown: associations with depressive symptoms in the German old-age population. Int J Environ Res Public Health 18(7):3615. https://doi.org/10.3390/ijerph18073615

Ohayon MM, O'Hara R, Vitiello MV (2012) Epidemiology of restless legs syndrome: a synthesis of the literature. Sleep Med Rev 16(4):283–295

Pullen SJ, Wall CA, Angstman ER, Munitz GE, Kotagal S (2011) Psychiatric comorbidity in children and adolescents with restless legs syndrome: a retrospective study. J Clin Sleep Med 7(6):587–596

Ruiz-Roso MB, Knott-Torcal C, Matilla-Escalante DC, Garcimartín A, Sampedro-Nuñez MA, Dávalos A, Marazuela M (2020) COVID-19 lockdown and changes of the dietary pattern and physical activity habits in a cohort of patients with type 2 diabetes mellitus. Nutrients 12(8): 2327

Schormair B, Zhao C, Bell S, Tilch E, Salminen AV, Pütz B et al (2017) Identification of novel risk loci for restless legs syndrome in genome-wide association studies in individuals of European ancestry: a meta-analysis. Lancet Neurol 16(11):898–907

Silber MH, Becker PM, Earley C, Garcia-Borreguero D, Ondo WG, Medical Advisory Board of the Willis-Ekbom Disease Foundation (2013) Willis-Ekbom Disease Foundation revised consensus statement on the management of restless legs syndrome. Mayo Clin Proc 88(9):977–986

Srivanitchapoom P, Pandey S, Hallett M (2014) Restless legs syndrome and pregnancy: a review. Parkinsonism Relat Disord 20(7):716–722

Szentkirályi A, Völzke H, Hoffmann W, Trenkwalder C, Berger K (2014) Multimorbidity and the risk of restless legs syndrome in 2 prospective cohort studies. Neurology 82(22):2026–2033

Trenkwalder C, Paulus W, Walters AS (2005) The restless legs syndrome. Lancet Neurol 4(8): 465–475

Trenkwalder C, Allen R, Högl B, Paulus W, Winkelmann J (2016) Restless legs syndrome associated with major diseases: a systematic review and new concept. Neurology 86(14): 1336–1343

Trenkwalder C, Allen R, Högl B, Clemens S, Patton S, Schormair B, Winkelmann J (2018) Comorbidities, treatment, and pathophysiology in restless legs syndrome. Lancet Neurol 17(11):994–1005

Turkdogan D, Bekiroglu N, Zaimoglu S (2011) A prevalence study of restless legs syndrome in Turkish children and adolescents. Sleep Med 12(4):315–321

Walters AS, Rye DB (2009) Review of the relationship of restless legs syndrome and periodic limb movements in sleep to hypertension heart disease and stroke. Sleep 32(5):589–597. https://doi.org/10.1093/sleep/32.5.589

Weinstock LB, Brook JB, Walters AS, Goris A, Afrin LB, Molderings GJ (2022) Restless legs syndrome is associated with long-COVID in women. J Clin Sleep Med 18(5):1413–1418

Winkelman JW, Chertow GM, Lazarus JM (1996) Restless legs syndrome in end-stage renal disease. Am J Kidney Dis 28:372–378

Winkelmann J, Muller-Myhsok B, Wittchen HU, Hock B, Prager M, Pfister H, Strohle A, Eisensehr I, Dichgans M, Gasser T, Trenkwalder C (2002) Complex segregation analysis of restless legs syndrome provides evidence for an autosomal dominant mode of inheritance in early age at onset families. Ann Neurol 52(3):297–302

Winkelmann J, Allen RP, Högl B, Inoue Y, Oertel W, Salminen AV, Winkelman JW, Trenkwalder C, Sampaio C (2018) Treatment of restless legs syndrome: evidence-based review and implications for clinical practice (Revised 2017). Mov Disord 33(7):1077–1091

Wipper B, Romero-Gutierrez C, Winkelman JW (2022) Restless legs syndrome severity in the National RLS Opioid Registry during the COVID-19 pandemic. Sleep Med 90:96–101

Yilmaz K, Kilincaslan A, Aydin N, Kor D (2011) Prevalence and correlates of restless legs syndrome in adolescents. Dev Med Child Neurol 53:40–47

Yoshihisa A, Misaka T, Kanno Y, Yokokawa T, Sato T, Oikawa M et al (2019) Prognostic impact of restless legs syndrome on patients with heart failure. Int Heart J 60(5):1098–1105

Chapter 15
What Happened to Central Disorders of Hypersomnolence During the COVID-19 Pandemic?

Giorgia Varallo, Christian Franceschini, and Giuseppe Plazzi

Abstract The COVID-19 pandemic posed a significant challenge to the global organization of health systems. The reallocation and concentration of available resources on the management of COVID-19 had a significant impact on all other health conditions due to the suspension or cancellation of screening and routine examinations, the reduction of referrals, and the redistribution of staff. The management of central disorders of hypersomnolence was also negatively affected, although some evidence has debated the role of the pandemic, which may have represented not only a challenge but also an opportunity for people with central disorders of hypersomnolence. On the one hand, the loss of zeitgebers as a result of changes in patients' routines during the lockdown period, such as working from home, spending extended periods indoors, and a reduction in social interactions, could have worsened the management of these disorders. On the other hand, prolonged home confinement during the pandemic may have afforded more opportunities to implement the recommended behavioral strategies for reducing excessive daytime sleepiness and maintaining a more flexible sleep schedule. Consequently, the purpose of this chapter is to describe the most significant findings from studies examining the impact of the COVID-19 pandemic on central sleep disorders, with a particular emphasis on narcolepsy.

Keywords COVID-19 · Central disorders of Hypersomnolence · Narcolepsy · Sleep Medicine · Excessive daytime sleepiness

G. Varallo (✉) · C. Franceschini
Department of Medicine and Surgery, University of Parma, Parma, Italy
e-mail: giorgia.varallo@unipr.it

G. Plazzi
IRCCS Istituto delle Scienze Neurologiche di Bologna, Bologna, Italy

Department of Biomedical, Metabolic, and Neural Sciences, Università degli Studi di Modena e Reggio Emilia, Modena, Italy

A. S. BaHammam et al. (eds.), *COVID-19 and Sleep: A Global Outlook*, Progress in Sleep Research, https://doi.org/10.1007/978-981-99-0240-8_15

15.1 Pitfalls in the Management of Central Disorders of Hypersomnolence During the COVID-19 Pandemic

The global COVID-19 pandemic has resulted in a catastrophic loss of human life and has impacted every facet of daily life, representing an unprecedented challenge to public health (Zhu et al. 2020). Confinement and social distancing were implemented in several countries as measures to prevent the spread of COVID-19 infection, and health resources were reallocated to prioritize the COVID-19 response. Authorities worldwide have advised postponing nonemergency procedures, such as diagnostic and follow-up evaluations of chronic diseases, to prevent unnecessary exposure and potential contamination of hospitals and safeguard both healthcare workers and patients (CMS 2020). Thus, the suspension or cancellation of screening and routine examinations, the reduction of referrals, and the redistribution of staff to provide COVID-19-specific care have disrupted care provided to patients with other conditions. It is not surprising that the management of sleep disorders has also been profoundly affected (American Academy of Sleep Medicine 2021).

Despite the widespread interest in the impact of the pandemic on the general and several clinical populations (Richards et al. 2020; Motolese et al. 2020; Al-Quteimat and Amer 2020; Izquierdo et al. 2021), few studies have examined patients with primary central disorders of hypersomnolence, which include narcolepsy types 1 (NT1) and 2 (NT2), idiopathic hypersomnia (IH), and Kleine-Levin syndrome (KLS). These disorders are characterized by excessive daytime sleepiness (EDS), defined as the inability to remain awake and alert during the major waking periods of the day, resulting in periods of irrepressible need for sleep or unintentional lapses in drowsiness or sleep (Sateia 2014). Although numerous treatment options are available for these disorders, the evidence supporting their use varies considerably depending on the intervention and the condition. Current therapeutic approaches are symptomatic and consist of non-pharmacological (e.g., scheduled naps and regular sleep schedule) and pharmacological interventions tailored to the patient's needs and daily routines (Franceschini et al. 2021). Therefore, follow-ups are required to adapt therapeutic interventions to changing needs and preferences. Due to the COVID-19 pandemic's restrictions on home confinement, public transport, and concern over cross-infection in hospitals, patients with narcolepsy may have neglected to perform follow-ups or obtain their prescriptions, resulting in pharmacological treatment discontinuation.

Furthermore, the circadian timing system relies on exogenous factors, also known as *zeitgebers*, which provide an external signal that synchronizes physiological 24-h periodicity, setting an organism's biological clock (Elmore et al. 1994; Honma et al. 2003). The loss of *zeitgebers* as a result of alterations in patients' routines due to the lockdown period, such as working from home, spending extended periods indoors, engaging in fewer outdoor activities, irregular mealtimes, and a reduction in social interactions, may have affected the circadian sleep-wake cycle. On the other hand, the extended period of home confinement during the pandemic may have provided more opportunities to implement the recommended behavioral strategies for reducing excessive daytime sleepiness and maintaining a more flexible sleep schedule. For this reason, the COVID-19 pandemic provided a unique opportunity to examine the effects of lockdown and its associated lifestyle modifications on sleep, quality of life, disease burden, and symptomatology in patients with central hypersomnolence disorders.

Thus, the goal of this chapter is to describe the most important results of studies that look at the effect of the COVID-19 pandemic on central sleep disorders, with a focus on narcolepsy.

15.2 Impact of COVID-19 Pandemic on Narcolepsy: Challenges or Opportunity?

Narcolepsy is a chronic and disabling central hypersomnolence disorder characterized by a pentad of symptoms, including EDS, cataplexy, sleep paralysis, hypnopompic and/or hypnagogic hallucinations, and disturbed nighttime sleep, all of which severely impair health-related quality of life (Plazzi et al. 2008; Bassetti et al. 2019). The third edition of the International Classification of Sleep Disorders classifies narcolepsy as type 1 (NT1) and type 2 (NT2) (Sateia 2014). NT1 is characterized by cataplexy and a deficiency of hypocretin-1 in cerebrospinal fluid due to orexinergic neuron loss in the lateral hypothalamus (Nishino et al. 2001). Cataplexy is absent in NT2, and the hypocretin-1 levels in cerebrospinal fluid are typically above 110 pg/mL.

The prolonged period of home confinement caused by COVID-19 may have altered sleep patterns and routines, allowing patients with narcolepsy to adhere to a more flexible sleep schedule due to fewer time constraints for work and daily activities; on the other hand, the stressful conditions were exacerbated by fewer opportunities for recreation and social interaction which serve as *zeitgebers* with a negative impact on the sleep-wake cycle (Cellini et al. 2020; Franceschini et al. 2020). Few studies on narcolepsy have produced contradictory findings, showing both improvement and deterioration of symptoms. The following is a summary of the primary studies and their results.

15.2.1 A Focus on the Self-Reported Sleep Schedule, Excessive Daytime Sleepiness, and Medication Intake

Most available studies on narcolepsy assess changes in self-reported sleep quality, sleep schedule, and habits, as well as changes in life quality and medication use and their potential effects on symptoms, especially excessive daytime sleepiness and cataplexy (for NT1). These findings may help determine how changes in lifestyle and habits affect narcolepsy-related symptoms and even adherence to nonbehavioral strategies such as scheduled naps and a consistent sleep schedule.

For example, Barateau et al. (2022) conducted a cross-sectional study on 102 patients (60.78% women; mean age 40.11 years) with both NT1 and NT2. Participants completed self-reported questionnaires twice, once referring to the study period during confinement and once recalling the time before confinement. During the lockdown, 13.83% of the participants had a regular working schedule, 28.72% were working/studying at home, 32.98% were retired/unemployed, and the remaining were off work or partially unemployed. Findings showed that patients during lockdown exhibited increased time spent in bed (*in hour:min*, 8:17 vs. 8:37), total sleep time both overnight (*in hour:min*, 7:28 vs. 7:47) and over 24 h (*in hour: min*, 7:53 vs. during lockdown 8:23), as well as higher number of naps on weekdays, compared to pre-lockdown. Patients reported significantly higher levels of EDS during lockdown (mean Epworth Sleepiness Scale score before lockdown: 14.12 vs. during lockdown: 14.86), and a decline in quality of life (mean European Quality of life five-dimension questionnaire score before lockdown: 73.61 vs. during lockdown: 67.02). The frequency of cataplexy attacks did not change significantly during the lockdown. Regarding medication adherence, 26.32% of patients adjusted their treatment, with the majority reducing their dosage without affecting symptom severity.

Postiglione and colleagues (2021) examined the effect of lockdown on sleep-wake patterns and severity of symptoms in 53 NT1 (40% males) with a mean age of 35.0 years at evaluation, receiving stable long-term pharmacotherapy patients. Participants were selected from patients previously followed by the Istituto delle Scienze Neurologiche di Bologna. Participants were offered a telephone follow-up during the lockdown period to ensure continuity of care. Self-reported data on work status, current sleep schedule, and narcolepsy symptoms were collected and compared to pre-lockdown data gathered during a 6-month prior assessment.

Individuals with NT1 who maintained their normal working schedule experienced increased disturbed nocturnal sleep (from 15% to 40%) with more nocturnal awakenings. Neither patients with unchanged work schedules nor those who lost their occupations due to the COVID-19 pandemic exhibited significant changes. In contrast, NT1 patients in smart working or studying at home appear to have improved adherence to behavioral strategies as evidenced by an increase in the duration of total sleep time (from 7.9 to 8.3 h) and the frequency of daytime napping. Specifically, the percentage of patients reporting multiple daytime scheduled naps increased from 23.7% to 54.5%, accompanied by a significant decrease in the severity of EDS (Epworth Sleepiness Scale scores before lockdown: 10.6 vs. during lockdown: 8.6). No differences were found with respect to cataplexy.

In line with this, Wu and colleagues conducted a cross-sectional retrospective study (Wu et al. 2021) in China on 40 patients with narcolepsy (70% male; mean age 20.15 years; range: 7–59 years) who were taking regular medication for the treatment of narcolepsy prior to COVID-19. According to results, during the COVID-19 pandemic, 62.5% of patients underwent irregular pharmacological treatment with 52.5% discontinuing medication due to a lack of prescription. Notably, only 15% of patients reported experiencing more daytime sleepiness during the pandemic even though pharmacological treatment was either inconsistent or stopped entirely. These findings might be explained by the fact that individuals with narcolepsy had longer total sleep time (533.55 min vs. 495.63 min) and later wake-up time (7:33 vs. 7:01) after the COVID-19 pandemic. Finally, during the COVID-19 pandemic, there were no significant differences in bedtime, number of naps, or duration of naps.

Rodrigues Aguilar et al. conducted a cross-sectional retrospective study that evaluated changes in work status, sleep habits, quality of life, and EDS during the quarantine period using an ad hoc self-report questionnaire developed by experts in sleep medicine. The total sample comprised 76 Brazilian patients diagnosed with narcolepsy (68.5% NT1; 68.4% female; mean age 36.9 years), 21.8% reported no change in EDS during the lockdown, while 78.2% experienced worsening despite 60.5% reporting an increase in the number of scheduled naps (Rodrigues Aguilar et al. 2021). While 42.1% of patients reported no change in dosage or discontinuation of medication during the lockdown, in the remaining sample, there was a significant increase in the percentage of patients using stimulants as monotherapy (25% before lockdown vs. 34.2% during lockdown) and a significant decrease in the use of antidepressants (7.9% before lockdown vs. 5.3% during lockdown) and stimulants associated with antidepressants (32.9% before lockdown vs. 5.3% during lockdown). Furthermore, 51.3% of the patients reported a decrease in their quality of life while 17.1% reported an improvement. Overall, patients reported a worsening of symptoms with 30.3% of the sample reporting an increase in hypnagogic and hypnopompic hallucinations, and 17% reporting an increase in sleep paralysis and cataplexy.

In contrast, Nigam and colleagues (2022) performed a retrospective cross-sectional study on a mixed sample of patients with NT1 and NT2 ($n = 131$). Participants were asked to complete the self-report survey recalling their pre-lockdown EDS and nighttime sleep, followed by an assessment of their current somnolence and sleep schedule to measure changes.

Both NT1 patients and NT2 patients reported a significant increase in nighttime sleep duration (NT1: 7.8 h vs. 8.5 h; NT2: 7.9 h vs. 8.5 h). In addition, only NT2 showed a significant decrease in daytime sleepiness during lockdown (mean EES: 14.3 vs. 12.9). This modest but significant change was achieved despite most participants maintaining (63.4%) or decreasing (29.6%) their stimulant dosage. Additionally, cataplexy improved significantly with 54.1% of the 61 participants with NT1 who typically experienced cataplexy prior to lockdown reporting a decrease or disappearance of cataplexy, while 14.8% reported an increase and 31.1% reported no change. Furthermore, 35.0% of NT1 participants experienced a decrease in sleep-related hallucinations, while 10.7% experienced an increase and 54.3% remained unchanged. Finally, 34.5% of NT2 participants reported a reduction in sleep-related hallucinations, 1.7% reported an increase, and 63.8% reported no change.

15.2.2 A Focus on Objective Sleep Parameters

However, all these studies rely solely on self-report instruments, frequently employed retrospectively. Although this method helps expand knowledge, it has significant limitations. When asked about the past, however, respondents tend to provide less precise responses than when asked about the present (Schnell 2012), placing high cognitive demands and introducing the possibility of recall bias (Schmier and Halpern 2004).

The only study that used objective measurements of sleep structure via actigraphy was conducted by Filardi and colleagues involving 18 children and adolescents with NT1 receiving stable pharmacological treatment (i.e., sodium oxybate).

Findings showed that during the lockdown, children and adolescents with NT1 both on weekdays and weekends went to bed later (bedtime during weekdays: 23: 06 vs. 1:20; bedtime during weekends: 23:46 vs. during lockdown 1:48) and awoke later (wake-up time during weekdays, 6:48 vs. 7:45; wake-up time during the weekend, 7:35 vs. during lockdown 8:34), with a significant reduction in social jet leg measured as the difference between midpoint of sleep of weekend and weekdays. In addition, patients napped more frequently (n° of naps before lockdown 2.89 vs. during lockdown 4.3), especially during weekdays (n° of naps before lockdown 3.61 vs. during lockdown 5.89). On the other hand, no differences were found in EDS and other sleep parameters such as total sleep time or sleep efficiency.

These findings suggested that self-managing daytime activities allowed patients with NT1 to receive stable pharmacological treatment to take more planned naps during the day. However, to date, few experimental data have demonstrated the effectiveness of napping in reducing daytime sleepiness in NT1 individuals.

15.3 Idiopathic Hypersomnia and the COVID-19 Pandemic

Idiopathic hypersomnia (IH) is characterized by prolonged undisrupted nocturnal sleep with long unrefreshing naps, and often sleep inertia. As with narcolepsy, a significant challenge for this population is reconciling their unique sleep requirements (e.g., increased total sleep time and/or napping) with their daytime obligations.

Despite the interest and the unique opportunity, even less evidence exists regarding the impact of the COVID-19 pandemic on IH patients. In a retrospective cross-sectional study on 81 patients with IH conducted by Barateau and colleagues, 31.65% of patients reported a regular working schedule, 22.78% were working/studying at home, 20.25% were retired, and the remaining were unemployed or off work. Furthermore, during the lockdown, patients reported increased bedtime (in hours:min, 8:30 vs. 9:09) and total sleep time overnight (8:17 vs. 8:55) and over 24 h (8:42 vs. 9:20) on weekdays.

On both weekends and weekdays, bedtimes and wake-up times were delayed, and nocturnal sleep latency (in min, before lockdown 11.06 vs. during lockdown 21.41) and number of awakenings per night (before lockdown 1.18 vs. during lockdown 1.85) increased during the lockdown. However, no changes were found in excessive daytime sleepiness, severity of symptoms, and quality of life.

The study conducted by Nigam showed that, as in narcolepsy, patients with IH reported an increase in nighttime sleep (in hours, 8.3 vs. 9.6), with both delayed bedtime and wake-up time. In addition, participants experienced a decrease in excessive daytime sleepiness (mean Epworth Sleepiness Score 13.4 vs. 12.5). In both studies, only self-report measures were used, and there is currently no study on IH patients that has performed an objective evaluation of sleep structure.

15.4 Overall Conclusion

It is difficult to draw a unified and uniform conclusion from these findings. For example, two out of five studies (Barateau et al. 2022; Rodrigues Aguilar et al. 2021) indicate a worsening of excessive daytime sleepiness in narcoleptics, despite an increase in bedtime and number of scheduled naps. On the other hand, three studies (Postiglione et al. 2021; Wu et al. 2021; Nigam et al. 2022) reported an improvement in excessive daytime sleepiness in conjunction with an increased number of naps and longer nighttime sleep duration. In contrast, not all studies have assessed changes in cataplexy; two studies reported substantial equality during lockdown relative to the previous period (Barateau et al. 2022; Postiglione et al. 2021), whereas one study (Nigam et al. 2022) reported a decrease/disappearance of cataplectic episodes, and another reported a worsening (Rodrigues Aguilar et al. 2021). Finally, the results indicate that medication intake remained mostly stable during the lockdown (Barateau et al. 2022; Wu et al. 2021). However, one study found that the majority of the sample discontinued their medication with a decrease in the use of antidepressants and increased use of stimulants (Rodrigues Aguilar et al. 2021).

Also concerning IH, the results are contradictory, with one study indicating a decrease in EDS during the lockdown and the other indicating no significant changes.

Numerous variables may have influenced the results of these studies, including national policies regarding COVID-19 containment, previous socioeconomic conditions, and impact of media communication on the population. In addition, these studies have not considered the potential effect of mood and the impact of stress on sleep architecture (Palagini et al. 2013), nor its potential impact on symptomatology and quality of life (Vignatelli et al. 2011). Numerous studies now emphasize mood changes in narcolepsy (Dauvilliers et al. 2009; Barateau et al. 2020) and the impact of anxious and depressive symptoms on the burden of illness (Fortuyn et al. 2010; Morse and Sanjeev 2018). Furthermore, during the pandemic, there was an increase in the prevalence of depressive and anxious symptoms (Morin et al. 2021).

Lockdown appears to have had a positive effect on the sleep schedule by making it more adaptable to the needs of the patients, as indicated by the fact that all studies indicate an increase in bedtime. However, these results are intriguing because they suggest that it may be useful to investigate how lifestyle modifications (i.e., teleworking) affect symptomatology and quality of life in this population. Ozaki and colleagues (2012) found that a flexible work schedule was associated with a higher quality of life in individuals with NT1, NT2, or IH. Despite this growing evidence, workplace and schedule adjustments may be inadequately provided. Providing greater access to telework or even just allowing adjustments to the work environment and schedule has the potential to improve sleepiness, work performance, and quality of life for those with central disorders of hypersomnia.

The lack of research on the effects of COVID on patients with Kleine-Levine syndrome is noteworthy, probably because it is an extremely rare disease. Only two case reports exist in which COVID infection was associated with a recurrence of symptoms (Nasrullah et al. 2021; Marčić et al. 2021).

15.5 Perspective on Telemedicine During (and After) COVID-19

The COVID-19 pandemic paved the way for implementing ICT-based solutions, altering current medical practice. The World Health Organization defines telemedicine as "the delivery of health services in which clients/patients and health workers are separated by distance" (World Health Organization 2019).

During the COVID-19 pandemic, telemedicine was a "forced" solution that successfully addressed several challenges in delivering healthcare services. Indeed, using telehealth technology has many benefits, especially for nonemergency and routine care; indeed, it reduces the use of health center resources and the risk of direct transmission of infectious agents. However, telemedicine's benefits extend well beyond its use in emergency healthcare settings, especially in improving access to specialized care and qualified specialists, and its use in sleep medicine is supported by the American Academy of Sleep Medicine (Singh et al. 2015).

In Italy, for instance, specialized sleep centers with multidisciplinary expertise to treat narcolepsy are scarce, forcing patients and their families to travel long distances with high associated costs, thereby increasing the disease-related burden. In light of this, the Narcolepsy Center of the IRCCS Istituto delle Scienze Neurologiche di Bologna launched the TENAR (TElemedicine for NARcolepsy) to evaluate the feasibility of a multidisciplinary approach via telemedicine procedures (i.e., televisits) for patients with narcolepsy during the COVID-19 pandemic emergency in Italy (Pizza et al. 2022). Overall, 39 participants were included (97.4% NT1; 56.4% male; median age of children: 16 years; median age of adults: 35 years). After an initial multidisciplinary televisit (i.e., neurological, metabolic, and psychosocial aspects), patients underwent follow-ups at +2, +4, + 6, and +12 months. The results showed a good level of acceptance from narcolepsy patients to participate in the TM follow-up and a substantial adherence to the proposed visits and procedures, confirming its feasibility. Moreover, TENAR supported patients in maintaining their clinical stability and improvement in EDS and body mass index at 6-month and 12-month follow-ups. It is interesting to note that telemedicine prevented the need for the entire period to be traveled by an estimated 50,000 km, which would have been necessary if the visits had taken place in person. Despite the preliminary nature of these findings, this study showed for the first time that a telemedicine intervention is feasible and safe for children and adults with narcolepsy and could reduce the cost of care for patients and their families.

References

Al-Quteimat OM, Amer AM (2020) The impact of the COVID-19 pandemic on cancer patients. Am J Clin Oncol Cancer Clin Trials 43(6):452–455

American Academy of Sleep Medicine (2021) COVID-19 mitigation strategies for sleep clinics & centers - REOPENING. https://aasm.org/covid-19-resources/covid-19-mitigation-strategies-sleep-clinics-labs/

Barateau L, Lopez R, Chenini S, Pesenti C, Rassu AL, Jaussent I et al (2020) Depression and suicidal thoughts in untreated and treated narcolepsy: systematic analysis. Neurology 95(20): E2755–E2768

Barateau L, Chenini S, Rassu AL, Denis C, Lorber Q, Dhalluin C, Lopez R, Jaussent I, Beziat S, Dauvilliers Y (2022) Changes in sleep pattern during the COVID-19 lockdown in patients with narcolepsy, idiopathic hypersomnia, and restless legs syndrome. Neurology 99(14):e1475–e1485

Bassetti CLA, Adamantidis A, Burdakov D, Han F, Gay S, Kallweit U et al (2019) Narcolepsy—clinical spectrum, aetiopathophysiology, diagnosis and treatment. Nat Rev Neurol 15(9): 519–539

Cellini N, Canale N, Mioni G, Costa S (2020) Changes in sleep pattern, sense of time and digital media use during COVID-19 lockdown in Italy. J Sleep Res 29(4):1–5

CMS (2020) CMS releases recommendations on adult elective surgeries, non-essential medical, surgical, and dental procedures during COVID-19 response [Internet]. https://www.cms.gov/newsroom/press-releases/cms-releasesrecommendations-adult-elective-surgeries-non-essential-medical-surgicaland-dental

Dauvilliers Y, Paquereau J, Bastuji H, Drouot X, Weil JS, Viot-Blanc V (2009) Psychological health in central hypersomnias: the French Harmony study. J Neurol Neurosurg Psychiatry 80(6):636–641

Elmore SK, Betrus PA, Burr R (1994) Light, social zeitgebers, and the sleep–wake cycle in the entrainment of human circadian rhythms. Res Nurs Health 17(6):471–478

Fortuyn HA, Lappenschaar MA, Furer JW, Hodiamont PP, Rijnders CA, Renier WO, Buitelaar JK, Overeem S (2010) Anxiety and mood disorders in narcolepsy: a case-control study. Gen Hosp Psychiatry 32(1):49–54

Franceschini C, Musetti A, Zenesini C, Palagini L, Scarpelli S, Quattropani MC et al (2020) Poor sleep quality and its consequences on mental health during the COVID-19 lockdown in Italy. Front Psychol 11(November):1–15

Franceschini C, Pizza F, Cavalli F, Plazzi G (2021) A practical guide to the pharmacological and behavioral therapy of Narcolepsy. Neurotherapeutics 18(1):6–19. https://doi.org/10.1007/s13311-021-01051-4

Honma KI, Hashimoto S, Nakao M, Honma S (2003) Period and phase adjustments of human circadian rhythms in the real world. J Biol Rhythm 18(3):261–270

Izquierdo JL, Almonacid C, González Y, Del Rio-Bermudez C, Ancochea J, Cárdenas R et al (2021) The impact of COVID-19 on patients with asthma. Eur Respir J 57(3):2003142

Marčić M, Marčić L, Marčić B (2021) SARS-CoV-2 infection causes relapse of Kleine-Levin syndrome: case report and review of literature. Neurol Int 13(3):328–334

Morin CM, Bjorvatn B, Chung F, Holzinger B, Partinen M, Penzel T et al (2021) Insomnia, anxiety, and depression during the COVID-19 pandemic: an international collaborative study. Sleep Med 87:38–45

Morse A, Sanjeev K (2018) Narcolepsy and psychiatric disorders: comorbidities or shared pathophysiology? Med Sci 6(1):16

Motolese F, Rossi M, Albergo G, Stelitano D, Villanova M, Di Lazzaro V et al (2020) The psychological impact of COVID-19 pandemic on people with multiple sclerosis. Front Neurol 11(October):1–8

Nasrullah A, Javed A, Ashraf O, Malik K (2021) Possible role of COVID-19 in the relapse of Klein-Levin Syndrome. Respir Med Case Rep 33(March):101445. https://doi.org/10.1016/j.rmcr.2021.101445

Nigam M, Hippolyte A, Dodet P, Gales A, Maranci JB, Al-Youssef S et al (2022) Sleeping through a pandemic: impact of COVID-19-related restrictions on narcolepsy and idiopathic hypersomnia. J Clin Sleep Med 18(1):255–263

Nishino S, Ripley B, Overeem S, Nevsimalova S, Lammers GJ, Vankova J et al (2001) Low cerebrospinal fluid hypocretin (orexin) and altered energy homeostasis in human narcolepsy. Ann Neurol 50(3):381–388

Ozaki A, Inoue Y, Hayashida K, Nakajima T, Honda M, Usui A et al (2012) Quality of life in patients with narcolepsy with cataplexy, narcolepsy without cataplexy, and idiopathic hypersomnia without long sleep time: comparison between patients on psychostimulants, drug-naïve patients and the general Japanese population. Sleep Med 13(2):200–206. https://doi.org/10.1016/j.sleep.2011.07.014

Palagini L, Baglioni C, Ciapparelli A, Gemignani A, Riemann D (2013) REM sleep dysregulation in depression: state of the art. Sleep Med Rev 17(5):377–390. https://doi.org/10.1016/j.smrv.2012.11.001

Pizza F, Vignatelli L, Oriolo C, Zenesini C, Mangiaruga A, Rossetti A, Moresco M, Vandi S, Citeroni F, Pagotto U, Ingravallo F, Plazzi G (2022) Multidisciplinary care of patients with narcolepsy during COVID-19 pandemic in Italy via televisit: the TElemedicine for NARcolepsy (TENAR) feasibility study. Sleep 45(12):zsac228

Plazzi G, Serra L, Ferri R (2008) Nocturnal aspects of narcolepsy with cataplexy. Sleep Med Rev 12(2):109–128

Postiglione E, Pizza F, Ingravallo F, Vignatelli L, Filardi M, Mangiaruga A et al (2021) Impact of COVID-19 pandemic lockdown on narcolepsy type 1 management. Brain Behav 11(1):1–8

Richards M, Anderson M, Carter P, Ebert BL, Mossialos E (2020) The impact of the COVID-19 pandemic on cancer care. Nat Cancer 1(6):565–567. https://doi.org/10.1038/s43018-020-0074-y

Rodrigues Aguilar AC, Frange C, Huebra L, Dias Gomes AC, Tufik S, Santos Coelho FM (2021) The effects of the COVID-19 pandemic on patients with narcolepsy. J Clin sleep Med 17(4): 621–627

Sateia MJ (2014) International classification of sleep disorders-third edition highlights and modifications. Chest 146(5):1387–1394. https://doi.org/10.1378/chest.14-0970

Schmier J, Halpern MT (2004) Patient recall and recall bias of health state and health status. Expert Rev Pharmacoecon Outcomes Res 4(2):159–163

Schnell R (2012) Survey-Interviews. Methoden standardisierter Befragungen, 2nd edn. Survey-Interviews

Singh J, Badr MS, Diebert W, Epstein L, Hwang D, Karres V, Khosla S, Mims KN, Shamim-Uzzaman A, Kirsch D, Heald JL, McCann K (2015) American Academy of Sleep Medicine

(AASM) position paper for the use of telemedicine for the diagnosis and treatment of sleep disorders. J Clin Sleep Med 11(10):1187–1198
Vignatelli L, Plazzi G, Peschechera F, Delaj L, D'Alessandro R (2011) A 5-year prospective cohort study on health-related quality of life in patients with narcolepsy. Sleep Med 12(1):19–23. https://doi.org/10.1016/j.sleep.2010.07.008
World Health Organization (2019) WHO guideline: recommendations on digital interventions for health system strengthening. WHO, Geneva
Wu M, Ren J, Li SX, Xue P, Su C, Zhou J (2021) Management of narcolepsy during COVID-19: a challenge or an opportunity? Sleep 44(2):1–3
Zhu N, Zhang D, Wang W, Li X, Yang B, Song J et al (2020) A novel coronavirus from patients with pneumonia in China, 2019. N Engl J Med 382(8):727–733

Chapter 16
Sleep, Physical Activity, and Dietary Patterns During COVID-19 Pandemic

Maha H. Alhussain, MoezAlIslam E. Faris, and Ahmed S. BaHammam

Abstract The COVID-19 outbreak has escalated into a major public health crisis. To prevent disease transmission and avoid overwhelming health systems, contingency measures involving home confinement, quarantine, and lockdown were implemented in most countries worldwide. The COVID-19 pandemic and its associated preventive measures had powerfully disrupted lifestyle behaviors, including sleep, physical activity, and dietary patterns. Current evidence provided insights into general patterns of these three behavioral modifications associated with the COVID-19 pandemic, including poor sleep quality, physical inactivity, and changes in dietary patterns.

Keywords Covid-19 · Lockdown · Lifestyle · Sleep · Sedentary behaviors · Diet

The COVID-19 outbreak has escalated into a major public health crisis. Its rapid spreading and high mortality rate have caused severe disruptions in different aspects of life, including economic, social, and health consequences. To prevent disease transmission and avoid overwhelming health systems, most countries implemented restrictive measures involving home confinement, quarantine, and lockdown of social, cultural, and part of economic and industrial activities. Although these

M. H. Alhussain
Department of Food Science and Nutrition, College of Food and Agriculture Sciences, King Saud University, Riyadh, Saudi Arabia
e-mail: mhussien@ksu.edu.sa

M. E. Faris
Department of Clinical Nutrition and Dietetics, College of Health Sciences/Research Institute of Medical and Health Sciences (RIMHS), University of Sharjah, Sharjah, UAE
e-mail: mfaris@sharjah.ac.ae

A. S. BaHammam (✉)
The University Sleep Disorders Center, Department of Medicine, College of Medicine, King Saud University, Riyadh, Saudi Arabia

Strategic Technologies Program of the National Plan for Sciences and Technology and Innovation in the Kingdom of Saudi Arabia, King Saud University, Riyadh, Saudi Arabia
e-mail: ashammam@ksu.edu.sa

A. S. BaHammam et al. (eds.), *COVID-19 and Sleep: A Global Outlook*, Progress in Sleep Research, https://doi.org/10.1007/978-981-99-0240-8_16

measures were the best option to prevent the dramatic spread of infections, they posed severe challenges for the population. Indeed, these preventive measures have brought far-reaching changes in all aspects of our lives and considerably disrupted the population's lifestyles. The importance of maintaining a healthy lifestyle through regular physical activity and adopting healthy sleep and dietary patterns in coping with such consequences has been emphasized (Natalucci et al. 2020).

Evidence that sleep, physical activity, and dietary patterns are key factors influencing health and well-being has given rise to multiple epidemiological and surveillance studies. Inadequate sleep duration, poor sleep quality, physical inactivity, and poor diet are major contributors to the global burden of disease, as they are risk factors associated with higher rates of chronic diseases, including hypertension, cardiovascular disease, diabetes, obesity, and depression. The interplay between sleep, physical activity, and dietary patterns is of great interest as it has been highlighted how their interaction with each other affects general health (Janssen et al. 2020).

Understanding the impact of a global health crisis such as COVID-19 and the lockdowns that ensued on sleep, physical activity, and dietary patterns is warranted because they are important aspects of health. This will guide psychological and behavioral measures aimed at populations, enhancing resilience, preventing diseases, and increasing the efficacy of health approaches to mitigate the effects of this contingency and others that may happen. This chapter discusses the changes in sleep, physical activity, and dietary patterns and their interaction in response to the COVID-19 pandemic and concomitant restrictions (i.e., lockdown) worldwide.

16.1 Sleep

Sleep has emerged as a critical lifestyle contributor to health and may represent a modifiable risk factor (along with physical activity and diet) for many adverse health conditions. Therefore, good sleep is integral to enhancing health and well-being (Itani et al. 2017; Alfawaz et al. 2021a; BaHammam et al. 2021). Numerous practice guidelines for age-specific sleep duration and physical activity for optimum health have been published, suggesting that individuals over the age of 18 years should sleep an average of 7–9 h per night (Alfawaz et al. 2021a; Hirshkowitz et al. 2015). Sleep quality is crucial to good health as sleep quantity and quality are more closely related to mental health and well-being (Wickham et al. 2020). Poor sleep quality has previously been documented in response to major stressful events such as natural disasters (Lavie 2001). The COVID-19 pandemic caused unprecedented changes in individuals' lives, which generated stress and anxiety for many people (Jahrami et al. 2022; Jahrami et al. 2021). Available data demonstrated a high prevalence of anxiety generated by confinement during the COVID-19 pandemic in many affected countries, including Spain, Greece (Papandreou et al. 2020), Saudi Arabia (Alfawaz et al. 2021b), and China (Cao et al. 2020).

The emergence of sleep disturbances in response to such a stressful life event is to be expected. Previous work performed during the pandemic consistently reported that sleep patterns were altered during the lockdowns, with people having poor sleep quality (Chouchou et al. 2021). Stress, anxiety, and fear of novelty and the high spread of COVID-19 might have contributed to the observed changes in sleep quality (Ingram et al. 2020; Morin and Carrier 2021). People with lower anxiety levels reported more satisfaction with their sleep quality during the pandemic (Frontini et al. 2021).

Beyond mental health, the restrictive measures accompanying the COVID-19 pandemic have upset daily routines that regulate sleep-wake rhythms to remain in synchrony with the 24-h cycle. Indeed, normal routines such as getting up in the morning, going to work, eating meals at a fixed time, and maintaining physical activities have all been disrupted during the pandemic as a result of stay-at-home restrictions. Irregular sleep-wake rhythms have an adverse effect on sleep quality. Sleep disturbances were linked to increased energy and fat intake as well as obesity risk (St-Onge et al. 2011; Lin et al. 2018). Moreover, individuals with adverse changes in sleep quality experienced weight gain (Micheletti Cremasco et al. 2021), and vice versa, individuals with obesity reported worsened sleep quality during the pandemic (Brown et al. 2021).

Individuals who develop sleep disturbances during COVID-19 may be at greater risk for long-term adverse outcomes because sleep disturbances such as insomnia and nightmares may persist long after the pandemic. More details about sleep and COVID-19 have been intensively discussed in the other chapters of the book.

16.2 Physical Activity

Physical activity has been defined by the World Health Organization (WHO) as "any bodily movement produced by skeletal muscles that require energy expenditure" (World Health Organization 2021). It is well documented that reduced physical activity and sedentary behavior are associated with adverse health outcomes and increased mortality (Alfawaz et al. 2021a; Jiménez-Pavón et al. 2020; Lavie et al. 2019; Stamatakis et al. 2019; Panahi and Tremblay 2018). To achieve substantial health benefits, current guidelines for physical activity encourage people to engage in a minimum of 150 min per week of moderate-intensity aerobic physical activity, 75 min per week of vigorous-intensity aerobic physical activity, or an equivalent combination (Alfawaz et al. 2021a; WHO 2020a). However, many people worldwide fail to achieve sufficient physical activity and still present with high levels of sedentary behavior (sitting, watching TV, playing video games, or using mobile devices) (Lavie et al. 2015; Fletcher et al. 2018; Wisloff and Lavie 2017). It would be assumed that this situation may further worsen as a result of the restrictions during the COVID-19 pandemic period. Home confinement involves online learning and working, limited outdoor activities, and in-gym exercises.

The immediate effects of the COVID-19 pandemic on physical activity have been assessed in many countries around the world, and the available data demonstrated a substantial rise in sedentary behavior and a decrease in physical activity in different age groups (Xiang et al. 2020; Jalal et al. 2021; Bağcı et al. 2021; Al Sabbah et al. 2022; Rawat et al. 2021).

The environment in which COVID-19 is prevalent has an adverse effect on physical activity (Park et al. 2021). An upward trend in smartphone applications, internet-connected devices, and TV-watching was noticed during the pandemic in many affected countries, including the USA, Spain, and Italy (Bhutani and Cooper 2020). Indeed, screen time soared during the pandemic period and is linked to weight gain, likely because of the dual conditions of sedentary behavior and positive correlation between screen time and snacking (Marsh et al. 2013).

The consequences of COVID-19-induced alteration in physical activity resulted in an increased risk of poor health-related quality of life and shortened life expectancy (Mattioli et al. 2020; Lee et al. 2012). Furthermore, changes in physical activity may exacerbate the deleterious changes in mental health reported during the pandemic. Unpleasant emotions like anger, sadness, or irritation might be experienced as a result of the reduction in physical activity. People who engaged in less sedentary behavior and did more physical activity reported better mental health outcomes during the COVID-19 pandemic (Ingram et al. 2020; Schuch et al. 2020).

Increasing physical activity is assumed to be an effective approach to maintaining a healthy lifestyle and mitigating the adverse effect of lockdown on well-being. Performing high physical activity was associated with lower anxiety values than moderate physical activity (Frontini et al. 2021). Physical activity can also enhance the immune system function and help cope with viral infections. Depending on the intensity of the exercise, the immune system adapts by improving its function (Laddu et al. 2021). It appears that increasing physical activity levels could diminish the severity of COVID-19 complications. Affected patients with lower physical activity levels reported more severe complications of the COVID-19 disease (Tavakol et al. 2021).

Becoming physically active during the pandemic time is of utmost importance and has fortunately been shown in some studies. For example, studies in Spain (López-Bueno et al. 2020) and Italy (Di Renzo et al. 2020) reported that many people increased their physical activity levels during the lockdown.

16.3 Dietary Pattern

As mentioned earlier, the COVID-19 pandemic and the accompanying lockdown have produced sudden and radical changes in population's daily routines. A deviation from usual routines would be anticipated to have broad-ranging effects on the dietary pattern. Such circumstances might lead to irregular dietary patterns and frequent snacking, both of which have deleterious impacts on energy expenditure, glucose responses, and, consequently, weight gain (Alhussain et al. 2016; Alhussain et al. 2022; Cheikh Ismail et al. 2020). Alterations in meal distribution, higher consumption of snacks and small meals, and reduced consumption of main meals (breakfast, lunch, and dinner) were documented during the COVID-19 confinement (Enriquez-Martinez et al. 2021). Changes in work status due to the COVID-19 pandemic were associated with an unhealthier diet (Ingram et al. 2020). The interruption of the work routine due to the lockdown may cause boredom, which in turn is linked to a greater energy intake (Moynihan et al. 2015). Boredom feelings are often associated with overeating as a means to escape monotony (Havermans et al. 2015). It has been observed that the most prevalent unhealthy change in dietary behavior at the lockdown time was increased energy intake (Radwan et al. 2021). It was estimated that during the pandemic lockdown, daily energy intake increased by 6% among the Spanish population (Batlle-Bayer et al. 2020). An increase in energy intake could be driven by fear, stress, and anxiety (Radwan et al. 2021).

Reading or hearing continuously about the pandemic from the media can be stressful. Quarantine also induces stress, and links between stress and unhealthy dietary patterns are consistently reported (Al-Musharaf 2020; Haddad et al. 2020; Barcin-Guzeldere and Devrim-Lanpir 2022). Some people cope with stress by eating in an attempt to feel better, even overriding other signals of satiety and hunger (Singh 2014). Emotional eating triggered by stress during the pandemic time was associated with choosing foods based on convenience and sensory appeal (Shen et al. 2020). Stressed people are more likely to crave food high in energy, sugars, and fats (Wardle et al. 2000). An inverse relationship between stress and higher consumption of vegetables, fruits, and dairy products has been reported (Sadeghi et al. 2021).

An increase in the consumption of unhealthy snacking, such as sweet and savory snacks, was also demonstrated in normal and overweight adults in many countries, including Italy, Poland, Ireland, and the USA (Shen et al. 2020; Pellegrini et al. 2020; Sidor and Rzymski 2020; Spyreli et al. 2021). Sugar-sweetened energy-dense foods were frequently consumed as a way to avoid boredom and calm stress and other negative feelings triggered by COVID-19 (Spyreli et al. 2021). When people are experiencing negative emotions or feel the need to reward themselves, they are more likely to turn to comfort foods (Verhoeven et al. 2015). Consuming more high-sugar and high-fat energy-dense foods (considering that high energy intake is the triggering factor for inflammation and the subsequent elevation in the inflammatory cytokines that trigger the mental stress) would be expected to increase energy intake with low intake of essential nutrients leading to long-term health and weight implications (Larson and Story 2013).

Accordingly, it was predictable that COVID-19 period restrictions on daily life and increased stress could contribute to less healthy diet patterns. A previous study concluded that people who reported feeling anxious and those with a confirmed COVID-19 diagnosis were more likely to adopt less healthy dietary patterns (Enriquez-Martinez et al. 2021).

An unhealthy dietary pattern is associated with the increased risk of developing obesity and its related diseases beyond a chronic state of inflammation, which has been proven to increase the risk for more severe COVID-19 complications. In contrast, a healthy dietary pattern that involves more consumption of fruits correlated significantly with less severe complications of the COVID-19 disease (Tavakol et al. 2021).

On the contrary, some positive consequences of the COVID-19 pandemic and lockdown have been documented. As evident from previous research in Ireland, England, and Italy, individuals cooked and consumed more homemade meals during the lockdown period than before (Di Renzo et al. 2020; Murphy et al. 2020; Thompson et al. 2020). The additional time available throughout the lockdown in conjunction with the closure of restaurants encouraged the intention to prepare homemade meals and use fresh ingredients (e.g., fruit, vegetables, and fresh meat) (Spyreli et al. 2021). Moreover, the desire of parents to provide their families with a healthy diet during the pandemic further contributed to consuming homemade meals (Spyreli et al. 2021). The WHO published a document that supports healthy dietary intake during quarantine time and encourages the consumption of fresh fruits and vegetables and cooking recipes at home (WHO 2020b).

People who cook regularly are more likely to adopt healthy dietary patterns compared with those who frequently consume ready-made foods (Larson et al. 2006). Undoubtedly, homemade meals are linked to better nutritional outcomes than ready-made food. On the other hand, frequent consumption of fast food and ready-made meals seems to increase caloric intake, which in turn, in the long term, contributes to positive energy balance and obesity and is more likely to contribute to developing mental health distress (Poti and Popkin 2011).

Although preparation and consumption of homemade meals were motivated, limited access to daily grocery shopping caused by restricted store-opening hours during the pandemic time might have led to lower consumption of fresh foods. A considerable number of studies confirmed that the consumption of fresh products, including fruits and vegetables, decreased during the pandemic time (Di Renzo et al. 2020; Deschasaux-Tanguy et al. 2021; Litton and Beavers 2021).

A micronutrient deficit often accompanies a reduction in the intake of fresh foods. Vitamin and mineral deficiency is associated with impaired immune responses, thus making individuals more susceptible to viral infections (Childs et al. 2019). In addition, limited access to fresh food also leads to an increase in the consumption of energy-dense ultra-processed foods, such as snacks, convenience foods, junk foods, and ready-to-eat cereals, which tend to be high in sugars, fat, and salt (Gupta et al. 2019).

16.4 Interaction Between Sleep, Physical Activity, and Dietary Pattern

It is well established that a healthy lifestyle benefits people in different age groups. Therefore, maintaining a healthy lifestyle is of great interest, particularly in a period when the immune system might need to fight back.

A healthy lifestyle needs more than one healthy behavior to improve health and well-being. For example, getting adequate sleep, being physically active, and sitting less have been shown to be associated with better health (Tan et al. 2018). The restrictions put in place to inhibit the spread of COVID-19 put more challenges in the adoption of multiple health behaviors. Engaging in negative lifestyle behaviors was closely related to lower health and quality of life (Duncan et al. 2014).

Emerging evidence suggests that different health behaviors, including sleep, physical activity, and diet, interrelate (Lippke 2014). When individuals lack sleep or have poor sleep quality, they tend to have a higher tendency to crave an unhealthy diet; they are also likely to feel sleepy and lack motivation to perform daily activities, which may, in turn, result in poor work performance and social participation, as well as limited life aspirations (Kyle et al. 2010). A study by Tan et al. reported that the duration and quality of sleep, daytime functioning, and consuming a low-fat diet are significantly interrelated with each other (Tan et al. 2018). Another study found that sleep problems have been associated with a tendency to consume high-fat food and weight gain (Beccuti and Pannain 2011). Irregular eating patterns, frequent snacking between meals, and consuming less vegetables have also been linked to insufficient sleep duration (Imaki et al. 2002). Moreover, poor sleepers had lower adherence to the Mediterranean diet and higher body mass index compared with good sleepers (Muscogiuri et al. 2020). Poor sleep quality adversely impacts both psychological and physiological health by causing poor dietary habits (Chaput 2014).

Recent studies carried out in the context of lockdown caused by the COVID-19 pandemic highlighted the interaction between lifestyle behaviors. A study conducted among college students in the USA to ascertain the effects of the COVID-19 pandemic and concomitant restrictions on sleep and dietary patterns found that students who did not change their sleep duration during the lockdown were more likely not to have changes in their food intake or dietary patterns (Rotvold et al. 2021). Another study conducted among the general public in Greece showed that people who ate a well-balanced diet and engaged in adequate physical activity maintained good sleep quality (Papazisis et al. 2021).

Numerous studies confirmed the positive correlation between changes in diet and perceived sleep quality, which could be partly ascribed to the dietary inflammatory potential and the possible role of inflammation in worsening sleep quality (Ingram et al. 2020; Faris et al. 2022; Masaad et al. 2021). Home confinement has favored flexible work hours at home, ensuing in late sleeping, which has been associated with skipping breakfast, an unhealthy pattern linked to obesity and cardiovascular diseases (Rong et al. 2019).

A positive correlation between changes in diet and changes in physical activity, as well as changes in sleep quality and physical activity, was also reported (Ingram et al. 2020).

Lack of physical activity as a result of home confinement has been acknowledged as a potential risk factor that negatively affects sleep quality (Cellini et al. 2020). A reduction in walking time was also related to a worsened sleep quality during the lockdown period (Papazisis et al. 2021). Sleep quality was positively correlated with the intensity of physical activity (Papazisis et al. 2021).

A multinational study of female adults showed negative perceived effects on sleep, fitness, and diet during the pandemic, and these effects were intercorrelated (Choi et al. 2022).

A study conducted among UAE residents showed that the COVID-19 restrictions caused a high prevalence of unhealthy lifestyle behaviors, such as consuming more calories, performing less physical activity, insufficient sleep, and consequently weight gain (Radwan et al. 2021). Another study from the UK described undesirable changes in diet and physical activity behavior (most participants reported snacking more frequently) and experiencing barriers to body weight management compared to before the COVID-19 pandemic lockdown (Robinson et al. 2021). This study also showed that a higher body mass index during the lockdown was related to lower diet quality, a more reported frequency of overeating, and reduced physical activity levels (Robinson et al. 2021). In Italy, a study of youths with obesity revealed that, compared to data collected in 2019, 21 days into lockdown, participants spent less time in sports activities and consumed more unhealthy foods (Pietrobelli et al. 2020). Results from a longitudinal study demonstrated that a lockdown period due to COVID-19 had an inverse effect on physical activity levels, sleep quality, and well-being among physically active Spanish adults (Martínez-de-Quel et al. 2021).

Multiple health behavior change studies are lacking. Further research is needed to assess whether such multiple behavior change approaches yield better health results and are more cost effective than single health behavior change approaches. In addition, multiple health behavior change studies would offer insights into how best to support an individual's health and well-being at such a vulnerable time.

References

Al Sabbah H, Taha Z, Qasrawi R, Assaf EA, Cheikh Ismail L, Al Dhaheri AS, Hoteit M, Al-Jawaldeh A, Tayyem R, Bawadi H, AlKhalaf M, Bookari K, Kamel I, Dashti S, Allehdan S, Alalwan TA, Hammouh F, Waly MI, Al-Halawa DA, Mansour R, Abu Farha A, The Regional Corona Cooking Survey Group (2022) The impact of COVID-19 on physical (in) activity behavior in 10 Arab countries. Int J Environ Res Public Health 19(17):10832. https://doi.org/10.3390/ijerph191710832

Alfawaz RA, Aljuraiban GS, AlMarzooqi MA, Alghannam AF, BaHammam AS, Dobia AM, Alothman SA, Aljuhani O, Aljaloud KS (2021a) The recommended amount of physical activity, sedentary behavior, and sleep duration for healthy Saudis: a joint consensus statement of the Saudi public health authority. Ann Thorac Med 16(3):239–244. https://doi.org/10.4103/atm.atm_33_21

Alfawaz H, Yakout SM, Wani K, Aljumah GA, Ansari MGA, Khattak MNK, Hussain SD, Al-Daghri NM (2021b) Dietary intake and mental health among Saudi adults during COVID-19 lockdown. Int J Environ Res Public Health 18(4):1653

Alhussain MH, Macdonald IA, Taylor MA (2016) Irregular meal-pattern effects on energy expenditure, metabolism, and appetite regulation: a randomized controlled trial in healthy normal-weight women. Am J Clin Nutr 104(1):21–32

Alhussain MH, Macdonald IA, Taylor MA (2022) Impact of isoenergetic intake of irregular meal patterns on thermogenesis, glucose metabolism, and appetite: a randomized controlled trial. Am J Clin Nutr 115(1):284–297

Al-Musharaf S (2020) Prevalence and predictors of emotional eating among healthy young Saudi women during the COVID-19 pandemic. Nutrients 12(10):2923. https://doi.org/10.3390/nu12102923

Bağcı TAB, Kanadıkırık A, Somyürek E, Gerçek G, Tanrıkulu HB, Öntaş E, Uzun S (2021) Impact of COVID-19 on eating habits, sleeping behaviour and physical activity status of final-year medical students in Ankara, Turkey. Public Health Nutr 24(18):6369–6376

BaHammam AS, Alghannam AF, Aljaloud KS, Aljuraiban GS, AlMarzooqi MA, Dobia AM, Alothman SA, Aljuhani O, Alfawaz RA (2021) Joint consensus statement of the Saudi public health authority on the recommended amount of physical activity, sedentary behavior, and sleep duration for healthy Saudis: background, methodology, and discussion. Ann Thorac Med 16(3): 225–238. https://doi.org/10.4103/atm.atm_32_21

Barcin-Guzeldere HK, Devrim-Lanpir A (2022) The association between body mass index, emotional eating and perceived stress during COVID-19 partial quarantine in healthy adults. Public Health Nutr 25(1):43–50. https://doi.org/10.1017/S1368980021002974

Batlle-Bayer L, Aldaco R, Bala A, Puig R, Laso J, Margallo M, Vázquez-Rowe I, Antó JM, Fullana-i-Palmer P (2020) Environmental and nutritional impacts of dietary changes in Spain during the COVID-19 lockdown. Sci Total Environ 748:141410

Beccuti G, Pannain S (2011) Sleep and obesity. Curr Opin Clin Nutr Metab Care 14(4):402

Bhutani S, Cooper JA (2020) COVID-19–related home confinement in adults: weight gain risks and opportunities. Obesity 28(9):1576–1577. https://doi.org/10.1002/oby.22904

Brown A, Flint SW, Kalea AZ, O'Kane M, Williams S, Batterham RL (2021) Negative impact of the first COVID-19 lockdown upon health-related behaviours and psychological wellbeing in people living with severe and complex obesity in the UK. EClinicalMedicine 34:100796

Cao W, Fang Z, Hou G, Han M, Xu X, Dong J, Zheng J (2020) The psychological impact of the COVID-19 epidemic on college students in China. Psychiatry Res 287:112934. https://doi.org/10.1016/j.psychres.2020.112934

Cellini N, Canale N, Mioni G, Costa S (2020) Changes in sleep pattern, sense of time and digital media use during COVID-19 lockdown in Italy. J Sleep Res 29(4):e13074. https://doi.org/10.1111/jsr.13074

Chaput JP (2014) Sleep patterns, diet quality and energy balance. Physiol Behav 134:86–91

Cheikh Ismail L, Osaili TM, Mohamad MN, Al Marzouqi A, Jarrar AH, Abu Jamous DO, Magriplis E, Ali HI, Al Sabbah H, Hasan H, AlMarzooqi LMR, Stojanovska L, Hashim M, Shaker Obaid RR, Saleh ST, Al Dhaheri AS (2020) Eating habits and lifestyle during COVID-19 lockdown in The United Arab Emirates: a cross-sectional study. Nutrients 12(11):3314. https://doi.org/10.3390/nu12113314

Childs CE, Calder PC, Miles EA (2019) Diet and immune function. Nutrients 11(8):1933

Choi KW, Kim HH, Basu A, Kwong AS, Hernandez-Diaz S, Wyszynski DF, Koenen KC (2022) COVID-19 perceived impacts on sleep, fitness, and diet and associations with mental health during pregnancy: a cross-national study. J Affect Disord Rep 7:100288

Chouchou F, Augustini M, Caderby T, Caron N, Turpin NA, Dalleau G (2021) The importance of sleep and physical activity on well-being during COVID-19 lockdown: Reunion Island as a case study. Sleep Med 77:297–301

Deschasaux-Tanguy M, Druesne-Pecollo N, Esseddik Y, De Edelenyi FS, Allès B, Andreeva VA, Baudry J, Charreire H, Deschamps V, Egnell M (2021) Diet and physical activity during the coronavirus disease 2019 (COVID-19) lockdown (March–May 2020): results from the French NutriNet-Santé cohort study. Am J Clin Nutr 113(4):924–938

Di Renzo L, Gualtieri P, Pivari F, Soldati L, Attinà A, Cinelli G, Leggeri C, Caparello G, Barrea L, Scerbo F (2020) Eating habits and lifestyle changes during COVID-19 lockdown: an Italian survey. J Transl Med 18(1):1–15

Duncan MJ, Kline CE, Vandelanotte C, Sargent C, Rogers NL, Di Milia L (2014) Cross-sectional associations between multiple lifestyle behaviors and health-related quality of life in the 10,000 steps cohort. PLoS One 9(4):e94184

Enriquez-Martinez OG, Martins MCT, Pereira TSS, Pacheco SOS, Pacheco FJ, Lopez KV, Huancahuire-Vega S, Silva DA, Mora-Urda AI, Rodriguez-Vásquez M, Montero López MP, Molina MCB (2021) Diet and lifestyle changes during the COVID-19 pandemic in Ibero-

American countries: Argentina, Brazil, Mexico, Peru, and Spain. Front Nutr 8:671004. https://doi.org/10.3389/fnut.2021.671004

Faris ME, Vitiello MV, Abdelrahim DN, Cheikh Ismail L, Jahrami HA, Khaleel S, Khan MS, Shakir AZ, Yusuf AM, Masaad AA, Bahammam AS (2022) Eating habits are associated with subjective sleep quality outcomes among university students: findings of a cross-sectional study. Sleep Breath 26(3):1365–1376. https://doi.org/10.1007/s11325-021-02506-w

Fletcher GF, Landolfo C, Niebauer J, Ozemek C, Arena R, Lavie CJ (2018) Promoting physical activity and exercise: JACC health promotion series. J Am Coll Cardiol 72(14):1622–1639

Frontini R, Rebelo-Gonçalves R, Amaro N, Salvador R, Matos R, Morouço P, Antunes R (2021) The relationship between anxiety levels, sleep, and physical activity during COVID-19 lockdown: an exploratory study. Front Psychol 12:659599

Gupta S, Hawk T, Aggarwal A, Drewnowski A (2019) Characterizing ultra-processed foods by energy density, nutrient density, and cost. Front Nutr 6:70. https://doi.org/10.3389/fnut.2019.00070

Haddad C, Zakhour M, Bou Kheir M, Haddad R, Al Hachach M, Sacre H, Salameh P (2020) Association between eating behavior and quarantine/confinement stressors during the coronavirus disease 2019 outbreak. J Eat Disord 8:40. https://doi.org/10.1186/s40337-020-00317-0

Havermans RC, Vancleef L, Kalamatianos A, Nederkoorn C (2015) Eating and inflicting pain out of boredom. Appetite 85:52–57

Hirshkowitz M, Whiton K, Albert SM, Alessi C, Bruni O, DonCarlos L, Hazen N, Herman J, Katz ES, Kheirandish-Gozal L, Neubauer DN, O'Donnell AE, Ohayon M, Peever J, Rawding R, Sachdeva RC, Setters B, Vitiello MV, Ware JC, Adams Hillard PJ (2015) National Sleep Foundation's sleep time duration recommendations: methodology and results summary. Sleep Health 1(1):40–43. https://doi.org/10.1016/j.sleh.2014.12.010

Imaki M, Hatanaka Y, Ogawa Y, Yoshida Y, Tanada S (2002) An epidemiological study on relationship between the hours of sleep and life style factors in Japanese factory workers. J Physiol Anthropol Appl Hum Sci 21(2):115–120

Ingram J, Maciejewski G, Hand CJ (2020) Changes in diet, sleep, and physical activity are associated with differences in negative mood during COVID-19 lockdown. Front Psychol 11:588604

Itani O, Jike M, Watanabe N, Kaneita Y (2017) Short sleep duration and health outcomes: a systematic review, meta-analysis, and meta-regression. Sleep Med 32:246–256. https://doi.org/10.1016/j.sleep.2016.08.006

Jahrami H, BaHammam AS, AlGahtani H, Ebrahim A, Faris M, AlEid K, Saif Z, Haji E, Dhahi A, Marzooq H, Hubail S, Hasan Z (2021) The examination of sleep quality for frontline healthcare workers during the outbreak of COVID-19. Sleep Breath 25(1):503–511. https://doi.org/10.1007/s11325-020-02135-9

Jahrami H, Haji EA, Saif ZQ, Aljeeran NO, Aljawder AI, Shehabdin FN, Fekih-Romdhane F, Trabelsi K, BaHammam AS, Vitiello MV (2022) Sleep quality worsens while perceived stress improves in healthcare workers over two years during the COVID-19 pandemic: results of a longitudinal study. Healthcare (Basel) 10(8):1588. https://doi.org/10.3390/healthcare10081588

Jalal SM, Beth MRM, Al-Hassan HJM, Alshealah NMJ (2021) Body mass index, practice of physical activity and lifestyle of students during COVID-19 lockdown. J Multidiscip Healthc 14:1901

Janssen I, Clarke AE, Carson V, Chaput J-P, Giangregorio LM, Kho ME, Poitras VJ, Ross R, Saunders TJ, Ross-White A, Chastin SFM (2020) A systematic review of compositional data analysis studies examining associations between sleep, sedentary behaviour, and physical activity with health outcomes in adults. Appl Physiol Nutr Metab 45(10 (Suppl. 2)):S248–S257. https://doi.org/10.1139/apnm-2020-0160

Jiménez-Pavón D, Carbonell-Baeza A, Lavie CJ (2020) Physical exercise as therapy to fight against the mental and physical consequences of COVID-19 quarantine: special focus in older people. Prog Cardiovasc Dis 63(3):386

Kyle SD, Espie CA, Morgan K (2010) "... Not just a minor thing, it is something major, which stops you from functioning daily": quality of life and daytime functioning in insomnia. Behav Sleep Med 8(3):123–140

Laddu DR, Lavie CJ, Phillips SA, Arena R (2021) Physical activity for immunity protection: inoculating populations with healthy living medicine in preparation for the next pandemic. Prog Cardiovasc Dis 64:102

Larson N, Story M (2013) A review of snacking patterns among children and adolescents: what are the implications of snacking for weight status? Child Obes 9(2):104–115

Larson NI, Perry CL, Story M, Neumark-Sztainer D (2006) Food preparation by young adults is associated with better diet quality. J Am Diet Assoc 106(12):2001–2007

Lavie P (2001) Sleep disturbances in the wake of traumatic events. N Engl J Med 345(25): 1825–1832

Lavie CJ, Arena R, Swift DL, Johannsen NM, Sui X, Lee D-c, Earnest CP, Church TS, O'Keefe JH, Milani RV (2015) Exercise and the cardiovascular system: clinical science and cardiovascular outcomes. Circ Res 117(2):207–219

Lavie CJ, Ozemek C, Carbone S, Katzmarzyk PT, Blair SN (2019) Sedentary behavior, exercise, and cardiovascular health. Circ Res 124(5):799–815

Lee IM, Shiroma EJ, Lobelo F, Puska P, Blair SN, Katzmarzyk PT, Lancet Physical Activity Series Working Group (2012) Effect of physical inactivity on major non-communicable diseases worldwide: an analysis of burden of disease and life expectancy. Lancet 380(9838):219–229

Lin CL, Lin CP, Chen SW, Wu HC, Tsai YH (2018) The association between sleep duration and overweight or obesity in Taiwanese adults: a cross-sectional study. Obes Res Clin Pract 12(4): 384–388. https://doi.org/10.1016/j.orcp.2016.07.005

Lippke S (2014) Modelling and supporting complex behavior change related to obesity and diabetes prevention and management with the compensatory carry-over action model. J Diab Obes 1:1–5

Litton MM, Beavers AW (2021) The relationship between food security status and fruit and vegetable intake during the COVID-19 pandemic. Nutrients 13(3):712

López-Bueno R, Calatayud J, Casaña J, Casajús JA, Smith L, Tully MA, Andersen LL, López-Sánchez GF (2020) COVID-19 confinement and health risk behaviors in Spain. Front Psychol 11:1426

Marsh S, Mhurchu CN, Maddison R (2013) The non-advertising effects of screen-based sedentary activities on acute eating behaviours in children, adolescents, and young adults. A systematic review. Appetite 71:259–273

Martínez-de-Quel Ó, Suárez-Iglesias D, López-Flores M, Pérez CA (2021) Physical activity, dietary habits and sleep quality before and during COVID-19 lockdown: a longitudinal study. Appetite 158:105019

Masaad AA, Yusuf AM, Shakir AZ, Khan MS, Khaleel S, Cheikh Ismail L, Faris MAE, Jahrami HA, Shivappa N, Hebert JR, Bahammam AS (2021) Sleep quality and dietary inflammatory index among university students: a cross-sectional study. Sleep Breath 25(4):2221–2229. https://doi.org/10.1007/s11325-020-02169-z

Mattioli AV, Ballerini Puviani M, Nasi M, Farinetti A (2020) COVID-19 pandemic: the effects of quarantine on cardiovascular risk. Eur J Clin Nutr 74(6):852–855. https://doi.org/10.1038/s41430-020-0646-z

Micheletti Cremasco M, Mulasso A, Moroni A, Testa A, Degan R, Rainoldi A, Rabaglietti E (2021) Relation among perceived weight change, sedentary activities and sleep quality during COVID-19 lockdown: a study in an academic Community in Northern Italy. Int J Environ Res Public Health 18(6):2943

Morin CM, Carrier J (2021) The acute effects of the COVID-19 pandemic on insomnia and psychological symptoms. Sleep Med 77:346–347. https://doi.org/10.1016/j.sleep.2020.06.005

Moynihan AB, WAV T, Igou ER, Wisman A, Donnelly AE, Mulcaire JB (2015) Eaten up by boredom: consuming food to escape awareness of the bored self. Front Psychol 6:369

Murphy B, Benson T, McCloat A, Mooney E, Elliott C, Dean M, Lavelle F (2020) Changes in consumers' food practices during the COVID-19 lockdown, implications for diet quality and the food system: a cross-continental comparison. Nutrients 13(1):20

Muscogiuri G, Barrea L, Aprano S, Framondi L, Di Matteo R, Laudisio D, Pugliese G, Savastano S, Colao A (2020) Sleep quality in obesity: does adherence to the Mediterranean diet matter? Nutrients 12(5):1364

Natalucci V, Carnevale Pellino V, Barbieri E, Vandoni M (2020) Is it important to perform physical activity during coronavirus pandemic (COVID-19)? Driving action for a correct exercise plan. Front Public Health 8:602020. https://doi.org/10.3389/fpubh.2020.602020

Panahi S, Tremblay A (2018) Sedentariness and health: is sedentary behavior more than just physical inactivity? Front Public Health 6:258

Papandreou C, Arija V, Aretouli E, Tsilidis KK, Bulló M (2020) Comparing eating behaviours, and symptoms of depression and anxiety between Spain and Greece during the COVID-19 outbreak: cross-sectional analysis of two different confinement strategies. Eur Eat Disord Rev 28(6): 836–846. https://doi.org/10.1002/erv.2772

Papazisis Z, Nikolaidis PT, Trakada G (2021) Sleep, physical activity, and diet of adults during the second lockdown of the COVID-19 pandemic in Greece. Int J Environ Res Public Health 18(14):7292

Park JH, Yoo E, Kim Y, Lee JM (2021) What happened pre-and during COVID-19 in South Korea? Comparing physical activity, sleep time, and body weight status. Int J Environ Res Public Health 18(11):5863

Pellegrini M, Ponzo V, Rosato R, Scumaci E, Goitre I, Benso A, Belcastro S, Crespi C, De Michieli F, Ghigo E, Broglio F, Bo S (2020) Changes in weight and nutritional habits in adults with obesity during the "lockdown" period caused by the COVID-19 virus emergency. Nutrients 12(7):2016

Pietrobelli A, Pecoraro L, Ferruzzi A, Heo M, Faith M, Zoller T, Antoniazzi F, Piacentini G, Fearnbach SN, Heymsfield SB (2020) Effects of COVID-19 lockdown on lifestyle behaviors in children with obesity living in Verona, Italy: a longitudinal study. Obesity 28(8):1382–1385. https://doi.org/10.1002/oby.22861

Poti JM, Popkin BM (2011) Trends in energy intake among US children by eating location and food source, 1977-2006. J Am Diet Assoc 111(8):1156–1164

Radwan H, Al Kitbi M, Hasan H, Al Hilali M, Abbas N, Hamadeh R, Saif ER, Naja F (2021) Indirect health effects of COVID-19: unhealthy lifestyle behaviors during the lockdown in The United Arab Emirates. Int J Environ Res Public Health 18(4):1964

Rawat D, Dixit V, Gulati S, Gulati S, Gulati A (2021) Impact of COVID-19 outbreak on lifestyle behaviour: a review of studies published in India. Diabetes Metab Syndr Clin Res Rev 15(1): 331–336. https://doi.org/10.1016/j.dsx.2020.12.038

Robinson E, Boyland E, Chisholm A, Harrold J, Maloney NG, Marty L, Mead BR, Noonan R, Hardman CA (2021) Obesity, eating behavior and physical activity during COVID-19 lockdown: a study of UK adults. Appetite 156:104853

Rong S, Snetselaar LG, Xu G, Sun Y, Liu B, Wallace RB, Bao W (2019) Association of skipping breakfast with cardiovascular and all-cause mortality. J Am Coll Cardiol 73(16):2025–2032. https://doi.org/10.1016/j.jacc.2019.01.065

Rotvold A, Parker K, Honrath K, Rhee Y (2021) Sleep and diet patterns during COVID-19 pandemic compared to pre-pandemic patterns. Current developments. Nutrition 5(Supplement_2):449–449. https://doi.org/10.1093/cdn/nzab038_061

Sadeghi O, Keshteli AH, Afshar H, Esmaillzadeh A, Adibi P (2021) Adherence to Mediterranean dietary pattern is inversely associated with depression, anxiety and psychological distress. Nutr Neurosci 24(4):248–259. https://doi.org/10.1080/1028415X.2019.1620425

Schuch FB, Bulzing RA, Meyer J, Vancampfort D, Firth J, Stubbs B, Grabovac I, Willeit P, Tavares VDO, Calegaro VC, Deenik J, López-Sánchez GF, Veronese N, Caperchione CM, Sadarangani KP, Abufaraj M, Tully MA, Smith L (2020) Associations of moderate to vigorous physical activity and sedentary behavior with depressive and anxiety symptoms in self-isolating people during the COVID-19 pandemic: a cross-sectional survey in Brazil. Psychiatry Res 292:113339. https://doi.org/10.1016/j.psychres.2020.113339

Shen W, Long LM, Shih C-H, Ludy M-J (2020) A humanities-based explanation for the effects of emotional eating and perceived stress on food choice motives during the COVID-19 pandemic. Nutrients 12(9):2712

Sidor A, Rzymski P (2020) Dietary choices and habits during COVID-19 lockdown: experience from Poland. Nutrients 12(6):1657

Singh M (2014) Mood, food, and obesity. Front Psychol 5:925

Spyreli E, McKinley MC, Woodside JV, Kelly C (2021) A qualitative exploration of the impact of COVID-19 on food decisions of economically disadvantaged families in Northern Ireland. BMC Public Health 21(1):1–16

Stamatakis E, Gale J, Bauman A, Ekelund U, Hamer M, Ding D (2019) Sitting time, physical activity, and risk of mortality in adults. J Am Coll Cardiol 73(16):2062–2072

St-Onge MP, Roberts AL, Chen J, Kelleman M, O'Keeffe M, RoyChoudhury A, Jones PJ (2011) Short sleep duration increases energy intakes but does not change energy expenditure in normal-weight individuals. Am J Clin Nutr 94(2):410–416. https://doi.org/10.3945/ajcn.111.013904

Tan SL, Whittal A, Lippke S (2018) Associations among sleep, diet, quality of life, and subjective health. Health Behav Policy Rev 5(2):46–58

Tavakol Z, Ghannadi S, Tabesh MR, Halabchi F, Noormohammadpour P, Akbarpour S, Alizadeh Z, Nezhad MH, Reyhan SK (2021) Relationship between physical activity, healthy lifestyle and COVID-19 disease severity; a cross-sectional study. J Public Health 31:267. https://doi.org/10.1007/s10389-020-01468-9

Thompson C, Hamilton LK, Dickinson A, Fallaize R, Mathie E, Rogers S, Wills W (2020) The impact of Covid-19 and the resulting mitigation measures on food and eating in the East of England: interim report. University of Hertfordshire, Hatfield

Verhoeven AA, Adriaanse MA, de Vet E, Fennis BM, de Ridder DT (2015) It's my party and I eat if I want to. Reasons for unhealthy snacking. Appetite 84:20–27

Wardle J, Steptoe A, Oliver G, Lipsey Z (2000) Stress, dietary restraint and food intake. J Psychosom Res 48(2):195–202

WHO (2020a) WHO guidelines on physical activity and sedentary behaviour: at a glance. Geneva. World Health Organization. https://apps.who.int/iris/handle/10665/337001. Accessed 20 Oct 2022

WHO (2020b) WHO guideline. stay physically active during self–quarantine. World Health Organization. http://www.euro.who.int/en/health-topics/healthemergencies/coronavirus-covid-19/novel-coronavirus-2019-ncov-technicalguidance/stay-physically-active-during-self-quarant. Accessed 2 Nov 2022

Wickham SR, Amarasekara NA, Bartonicek A, Conner TS (2020) The big three health behaviors and mental health and Well-being among young adults: a cross-sectional investigation of sleep, exercise, and diet. Front Psychol 11:579205. https://doi.org/10.3389/fpsyg.2020.579205

Wisloff U, Lavie CJ (2017) Taking physical activity, exercise, and fitness to a higher level. Prog Cardiovasc Dis 60(1):1–21

World Health Organization (2021) Physical activity fact sheet. https://www.who.int/publications/i/item/WHO-HEP-HPR-RUN-2021.2. Accessed 5 Oct 2022

Xiang M, Zhang Z, Kuwahara K (2020) Impact of COVID-19 pandemic on children and adolescents' lifestyle behavior larger than expected. Prog Cardiovasc Dis 63(4):531–532. https://doi.org/10.1016/j.pcad.2020.04.013

Chapter 17
Impact of Screen Time During the Pandemic of COVID-19 on Sleep Habits

Omar Boukhris, Haitham Jahrami, Khaled Trabelsi, Jordan M. Glenn, and Nicola Luigi Bragazzi

Abstract This chapter focuses on the impact of the COVID-19 pandemic on screen time and the effects of increased screen time on sleep patterns. Additionally, some strategies to cope with excessive screen time and sleep disruption will be outlined. To date, available data indicate restrictions imposed during the COVID-19 pandemic increased screen time, independent of age. This excessive use of digital devices, especially before bedtime during the pandemic, resulted in sleep perturbations. Sleep is pivotal for health, well-being, and overall health-related quality of life and is thus essential for optimal immune function. As a result, implementing strategies aimed at reducing the negative effects of increased screen time to improve sleep may help prevent the deleterious effects of the virus infection.

O. Boukhris (✉)
SIESTA Research Group, School of Allied Health, Human Services and Sport, La Trobe University, Melbourne, Australia

Sport, Performance, and Nutrition Research Group, School of Allied Health, Human Services and Sport, La Trobe University, Melbourne, Australia
e-mail: O.Boukhris@latrobe.edu.au

H. Jahrami
Department of Psychiatry, Ministry of Health, Manama, Bahrain

Department of Psychiatry, College of Medicine and Medical Sciences, Arabian Gulf University, Manama, Bahrain

K. Trabelsi
Higher Institute of Sport and Physical Education of Sfax, University of Sfax, Sfax, Tunisia

Research Laboratory: Education, Motricity, Sport and Health, EM2S, LR19JS01, University of Sfax, Sfax, Tunisia

J. M. Glenn
Department of Health, Exercise Science Research Center Human Performance and Recreation, University of Arkansas, Fayetteville, AR, USA

N. L. Bragazzi
Laboratory for Industrial and Applied Mathematics, Department of Mathematics and Statistics, York University, Toronto, ON, Canada

A. S. BaHammam et al. (eds.), *COVID-19 and Sleep: A Global Outlook*, Progress in Sleep Research, https://doi.org/10.1007/978-981-99-0240-8_17

Keywords Coronavirus · Digital technology · Sleep-wake behavior · Health

17.1 Introduction

During the pandemic of COVID-19, most countries worldwide implemented severe restrictions such as quarantine and lockdowns to limit the spread of the virus. Due to these restrictions, several activities were postponed, including the closure of schools, universities, workplaces, gyms, and sports centers. These restrictions, and subsequent home confinement, generated reductions in social interaction and life satisfaction, as well as an impairment in mental health (Ammar et al. 2020). To decrease this psychosocial strain during the lockdown, Ammar et al. (2020) recommended the use of digital technology to promote social inclusion. However, the use of digital technology also resulted in an increase in screen time during this period (Drumheller and Fan 2022; Runacres et al. 2021). Screen time is defined as the amount of time spent using digital devices for work aims, education, or entertainment (Pandya and Lodha 2021). Previous studies indicated that screen time (e.g., TV, social media, and internet) increased significantly compared to before the COVID-19 pandemic (Drumheller and Fan 2022; Guo et al. 2021; Meyer et al. 2020; Bruni et al. 2022). Although digital technologies have limited feelings of loneliness and social withdrawal due to social distancing during home confinement, this increase in screen time also leads to health concerns (Pandya and Lodha 2021). Several factors including, but not limited to, sleep disturbances, markedly reduced physical activity, stress, anxiety, and obesity are all reported to relate to this increase in screen time, especially in children and adolescents (Drumheller and Fan 2022; Hammoudi et al. 2021).

Given the above considerations, this chapter aims to review the current state of knowledge about the effects of the COVID-19 pandemic on screen time and the impact of increased screen time on health and specifically sleep habits. Additionally, some strategies to cope with challenges imposed by the COVID-19 pandemic will be suggested.

17.2 Increasing Screen Time During the COVID-19 Pandemic

Characteristics of the included studies examining the pandemic's effects on screen time in children and adolescents are summarized in Table 17.1. The studies were published between the years 2020 and 2022. The highest number of participants was 1115 (Eyimaya and Irmak 2021), whilst the remaining studies ranged between 57 and 860 participants. The mean participant age ranged from 6 to 17 years.

Almost all studies conducted during the pandemic of COVID-19 reported that screen time increased significantly compared to before the pandemic independently of age and population. Additionally, excessive screen time (>1 h per day) was often

Table 17.1 Effects of COVID-19 pandemic on screen time in children and adolescents

Study	Country	Sample size	Age (years)	Restrictions	Method of screen time measurement	Results
Carroll et al. (2020)	Canada	310	5.7 ± 2.0	Closure of parks, school closures, and social distancing	Author developed	↑ to 2.4 ± 1.6 h/day
López-Bueno et al. (2020)	Spain	860	9.6 ± 3.9	Quarantine, social distancing, and lockdown requirements	Author developed	↑ by 2.9 ± 2.3 h/day
Munasinghe et al. (2020)	Australia	582	17.0 ± 1.0	School closures, social distancing, and lockdown	Adolescent sedentary activities questions	≥4 h/day on social media and ≥1 h/day on TV
Palladino et al. (2020)	Italy	57	8.0 ± 1.6	Lockdown	A brief questionnaire about the use of mobile media devices	↓ from 2.5 h during pre-lockdown to 5.8 h during lockdown
Abid et al. (2021)	Tunisia	100	8.66 ± 3.3	Lockdown	SergeTisseron's questionnaire	↑ from 1.53 ± 0.79 h/day to 4.45 ± 1.41 h/day
Eyimaya and Irmak (2021)	Turkey	1115	9.0 ± 2.0	Lockdown	Author developed	↑ to 6.42 ± 3.07 h/day
Burkart et al. (2022)	USA	231	7–12	Lockdown	Author developed	↑ by 97 min/day
Dutta et al. (2022)	India	153	8–16	Lockdown	Author developed	↑ from 4–8 h/day

(continued)

Table 17.1 (continued)

Study	Country	Sample size	Age (years)	Restrictions	Method of screen time measurement	Results
So et al. (2022)	China	577 (during the pandemic), 146 (during school closures), and 293 (when schools partially reopened)	12.85 ± 2.61 (during the pandemic), 12.14 ± 2.90 (during school closures), and 11.93 ± 2.11 (when schools partially reopened)	School closures and lockdown	Author developed	↑ from 5.38 h/day before the pandemic to 6.23 h/day during school closures and to 8.52 h/day when schools partially reopened

↑ increased, *NM* not mentioned

reported in children, indicating that this population exceeded the recommendation thresholds related to permissible screen time (Gupta et al. 2022).

Characteristics of the included studies examining the effect of the pandemic of COVID-19 on screen time in adults and older adults are summarized in Table 17.2. The studies were published between the years 2020 and 2022. The highest number of participants was 45,161 (Malta et al. 2020), whilst the remaining studies ranged between 117 and 13,754 participants. The mean ages of participants ranged from 18 to 80 years.

17.3 Impact of Increased Screen Time on Overall Health

Mental and physical health are the factors most adversely impacted by increased screen time (Pandya and Lodha 2021). Specific to children and adolescents, a high risk of myopia is considered a negative result of these increases (Singh and Balhara 2021). There is evidence that excessive digital use also has negative long-term physical impacts, including sleep perturbation, eye strain, and neck pain among them (Pandya and Lodha 2021). Excessive exposure to screens before bedtime may lead to melatonin suppression, which negatively influences sleep quality, making children susceptible to the negative effects of artificial light at night (Abid et al. 2021). Concerning mental health, excessive screen time adversely affects attention, social intelligence, and concentration, as well as engenders isolation, emotional dysregulation, anxiety, depression, and technology addiction (Pandya and Lodha 2021).

For adults, sleep routines and regular physical activity tend to be replaced by sedentary behaviors, characterized by excessive screen time, provoking negative effects such as decreased sleep duration or sleep-wake inversion, neck pain, headache, myopia, high blood pressure, obesity, insulin resistance, and digital eye syndrome (World Health Organization 2020). Excessive digital technology use can also result in collateral damage to visual health, sleep patterns, and food habits (Pandya and Lodha 2021). Furthermore, the relationship between poor mental health and increased screen time has been previously reported (Pandya and Lodha 2021). More importantly, as sleep and the immune system are related to each other, the negative effects of excessive screen time before bed, during the COVID-19 pandemic, on sleep quality and duration may negatively impact the immune system. Bryant et al. (2004) reported that sleep loss or perturbation negatively influences immune function. In this context, greater odds to contract COVID-19 infection are correlated with sleep loss or perturbation, which is not the case for people who have a greater duration of sleep (Ragnoli et al. 2022).

Table 17.2 Effects of COVID-19 pandemic on screen time in adults and older adults

Study	Country	Sample size	Age (years)	Restrictions	Method of screen time measurement	Results
Carroll et al. (2020)	Canada	351	38.5 ± 5.2	Closure of parks, school closures, and social distancing	Author developed	↑ to 2.8 ± 1.7 h/day
Cheikh Ismail et al. (2020)	United Arab Emirates	1012	NM	Quarantine and social distancing	IPAQ short form	36.2% of participants spent >5 h/day for entertainment during COVID-19 compared to 12.9% pre-COVID-19
Colivicchi et al. (2020)	France	124	71.0 ± 14.0	Lockdown	Telephone interviews	↑ in 50% of participants
Górnicka et al. (2020)	Poland	3241	46.2 ± 15.3	Social distancing and lockdown	Canadian Health Measures Survey	↑ to ≥8 h/day during COVID-19 pandemic
Hu et al. (2020)	China	1033	18–60	Lockdown	Author developed	↑ in 70% of participants
Husain and Ashkanani (2020)	Kuwait	415	38.5 ± 12.7	Lockdown	Author developed	Participants watching >6 h/day ↑ by 27.5%
Malta et al. (2020)	Brazil	45,161	18–29	Social distancing and lockdown	Internally validated questionnaire	↑ from 1.5 ± 0.1 h/day to 5.3 ± 0.1 h/day for computer or tablet uses, and from 1.5 ± 0.1 h/day to 3.3. ± 0.1 h/day for TV
Mon-López et al. (2020)	Spain	120	30–39	Quarantine, social distancing, and lockdown	Author developed	↑ from 403.0 ± 203.4 min/day to 615.6 ± 331.6 min/day
Qin et al. (2020)	China	12,107	18–80	Lockdown	IPAQ-short form	↑ to 261.3 ± 189.8 min/day during lockdown
Khare et al. (2021)	India	NM	NM	Lockdown	Author developed	61.4% of participants ↑ their screen time during lockdown

(continued)

Table 17.2 (continued)

Study	Country	Sample size	Age (years)	Restrictions	Method of screen time measurement	Results
Richardson et al. (2021)	United Kingdom	117	75.0 ± 4.0	Social distancing and lockdown	Author developed	↑ from 426.0 ± 27.0 min/day pre-COVID-19 to 490.0 ± 25.0 min/day during COVID-19
Rodríguez-Larrad et al. (2021)	Spain	13,754	22.8 ± 5.3	Lockdown	Combination of IPAQ and modified Sedentary Behaviour Questionnaires	↑ from 217 min/day pre-COVID-19 to 373 min/day during COVID-19
Rolland et al. (2020)	France	11,391	22.8 ± 5.3	Lockdown	Author developed	↑ in 64.6% of participants

↑ increased, *NM* not mentioned, *IPAQ* International Physical Activity Questionnaire

17.4 Impact of Increased Screen Time on Sleep Parameters

17.4.1 Children and Adolescents

Sleep is known to be important for child and adolescent health and well-being, and there is an increased possibility of sleep problems developing or worsening during home confinement (Bruni et al. 2022). Sleep disturbances are linked to increased levels of psychological and emotional distress, as a result of changes in family financial circumstances, uncertainty about the future, and health concerns (Becker and Gregory 2020). Additionally, social distancing, advice to stay indoors, and remote learning tend to reduce sunlight exposure, create more flexibility in wake and sleep times, increase the possibility of taking longer daytime naps, and favor the use of technology for learning during the day. In this context, the results of the review by Paterson et al. (2021) revealed an increase in sleep duration during the first year of the COVID-19 pandemic in children and adolescents. Bedtime and wake time were also delayed during the pandemic (Paterson et al. 2021). Furthermore, a systematic review and meta-analysis concluded that around half of healthy children did not accumulate the recommended total sleep time (i.e., 9–11 h for school children; Hirshkowitz et al. 2015) (Sharma et al. 2021).

Unfortunately, there are limited studies (with inconsistent findings) examining the relationship between screen time and sleep parameters in children or adolescents during the COVID-19 pandemic. In this context, Zhang et al. (2021) reported that sleep disturbances were not associated with screen time, whilst Moitra and Madan (2022) found a significant correlation between sleep problems and increased screen time. Therefore, further studies are needed to determine the relationship between sleep parameters and screen time in children and adolescents during future pandemics.

17.4.2 Adults

Contrary to children and adolescents, several studies have investigated the relationship between screen time and sleep parameters in adults.

17.4.2.1 Relationship Between Sleep Latency and Screen Time

It has been shown that sleep latency increased significantly when screen time was higher during the pandemic of COVID-19 (Drumheller and Fan 2022). In this context, Salfi et al. (2021) reported that excessive digital use before bedtime is related to increases in sleep latency. In addition, Facer-Childs et al. (2021) reported that Australian athletes who increased their screen time during the pandemic had a higher sleep latency (i.e., 37 min), compared to a sleep latency of 22 min for athletes who reduced their screen time. Furthermore, a significant correlation existed between excessive screen time and sleep latency during the COVID-19 pandemic (Cellini et al. 2020).

17.4.2.2 Relationship Between Delayed Bedtime and Screen Time

It has been reported that delayed bedtimes were due to excessive screen time before sleep during the pandemic (Drumheller and Fan 2022). It has been shown that excessive screen time during the pandemic leads to a significant delay in bedtime, whilst a decrease or a lack of change in screen use did not lead to a delayed bedtime (Salfi et al. 2021). Moreover, Cellini et al. (2020) reported that delayed bedtime was probably due to increased screen time.

17.4.2.3 Relationship Between Sleep Duration and Screen Time

Several studies have examined the effect of screen time on sleep duration during the pandemic of COVID-19. A sleep duration of ≤ 7 h per day was attributed to the excessive use of digital devices (Souza et al. 2022). Likewise, Salfi et al. (2021) found that an increase in screen time caused a decrease in sleep duration, whereas a decrease or a lack of change in screen time did not produce a reduction in sleep duration. In addition, a significant correlation has been observed between excessive screen time and decreased sleep duration during the pandemic in individuals studying or working from home (Majumdar et al. 2020). Furthermore, Akulwar-Tajane et al. (2020) reported that 52% of students linked their reduction in sleep duration to excessive screen time. However, Ali et al. (2022) reported a significant correlation between increased screen time and longer sleep duration during the pandemic.

17.4.2.4 Relationship Between Wake Time and Screen Time

Two studies have examined the relationship between screen time and wake time during the COVID-19 pandemic. Cellini et al. (2020) found a significant correlation between wake time and excessive screen time. Similarly, Salfi et al. (2021) found that an increase in screen time before bedtime resulted in later wake times during the COVID-19 pandemic, whereas a decrease or a lack of change in screen time did not cause later wake times.

17.4.2.5 Relationship Between Sleep Quality and Screen Time

Five studies examined the relationship between screen time and sleep quality during the COVID-19 pandemic. Of these, four reported sleep quality to be significantly impaired by increased digital use (Barrea et al. 2020; Dai et al. 2021; Salfi et al. 2021; Werneck et al. 2020), whilst one study reported that sleep quality was not affected by increased digital use (Cellini et al. 2020). Salfi et al. (2021) found that an increase in screen time before bedtime resulted in decreased sleep quality, as measured via the Pittsburgh Sleep Quality Index (PSQI) questionnaire, whereas a decrease or no change in screen time did not cause sleep quality impairment. Additionally, Barrea et al. (2020) found that sleep quality measured by PSQI decreased significantly in online workers; however, these results were not observed in offline workers. Moreover, Werneck et al. (2020) and Dai et al. (2021) reported significant declines in sleep quality due to excessive screen time before bedtime. In contrast, Cellini et al. (2020) found that sleep quality was not influenced by excessive screen time.

Although there are a high number of studies investigating the relationship between sleep and screen time in adults, a subset of studies only examined the impact of the COVID-19 pandemic on sleep, without examining the relationship between screen time and sleep. A systematic review and meta-analysis conducted by Jahrami et al. (2021) found that nearly 40% of general people have sleep problems during the COVID-19 pandemic. Another systematic review conducted in 2021 found that sleep perturbations were very common during the COVID-19 pandemic among the general population (Lin et al. 2021). Recently, Jahrami et al. (2022) concluded through a systematic review and meta-analysis that sleep problems were reported among four in every ten individuals and that children, adolescents, and patients infected with COVID-19 appeared to be the most affected groups during the lockdown.

It should be acknowledged that studies examining the relationship between sleep and screen time in athletes during the COVID-19 pandemic are lacking. Therefore, impairment in sleep quality and inadequate sleep duration, reported in athletes during the COVID-19 pandemic (da Silva Santos et al. 2021; Romdhani et al. 2022a, b; Kurniarobbi et al. 2022), could not be attributed to increased screen time. Nevertheless, this relationship should be investigated in future studies.

17.5 Practical Recommendations

To limit or reduce the negative effects of screen use and sleep disruption during the COVID-19 pandemic and especially during home confinement, the following recommendations are proposed:

- For adults, spend less than 4 h per day for digital use with breaks every 20 min. Technology may be used to schedule this time (Agarwal et al. 2022). However, children (24–59 months) should not exceed 1 h per day of screen time use (with each session not more than 20–30 min); the less the better (Gupta et al. 2022). The maximum duration of 2 h per day including recreational screen time, and time spent on screen at home to complete educational and extracurricular assignments, was recommended for children aged 5–10 years (Gupta et al. 2022). For adolescents (10–18 years), screen time should be balanced with other activities that are required for overall development (e.g., physical activity, hobbies, and family time) (Gupta et al. 2022).
- Use audio calls instead of video calls and typing messages to minimize screen fatigue (Pandya and Lodha 2021).
- Do not bring digital devices into the bedroom or switch them off before bed to decrease sleep perturbation owing to light exposure (Altena et al. 2020).
- Use strategies that promote sleep instead of screen use (e.g., slow breathing techniques) (Borges et al. 2021).
- Avoid late and long napping during the day and caffeine consumption 4–5 h before sleep (Romdhani et al. 2022a, b).
- Heavy meals are not recommended and consume snacks rich in tryptophan 1 h before sleep (Romdhani et al. 2022a, b).
- Some activities should be implemented like reading books, indoor games, and physical exercises such as meditation and Yoga for better overall individual health (Agarwal et al. 2022).
- People should be made aware of the detrimental effects of increased screen time and the value of sleep for overall health by health departments and policymakers.

17.6 Conclusion

In conclusion, although the use of digital devices helped people to reduce social distancing during the pandemic of COVID-19, excessive screen time may lead to negative effects on physical and mental health. Sleep problems are considered among the most negative effects of increased screen time, given that adequate sleep is essential for proper immune functioning which, in turn, helps to combat viral infections.

References

Abid R, Ammar A, Maaloul R, Souissi N, Hammouda O (2021) Effect of COVID-19-related home confinement on sleep quality, screen time and physical activity in Tunisian boys and girls: a survey. Int J Environ Res Public Health 18(6):3065

Agarwal R, Tripathi A, Khan IA, Agarwal M (2022) Effect of increased screen time on eyes during COVID-19 pandemic. J Family Med Prim Care 11(7):3462–3467

Akulwar-Tajane I, Parmar KK, Naik PH, Shah AV (2020) Rethinking screen time during COVID-19: impact on psychological well-being in physiotherapy students. Int J Clin Exp Med Res 4(4): 201–216

Ali A, Siddiqui AA, Arshad MS, Iqbal F, Arif TB (2022). Effects of COVID-19 pandemic and lockdown on lifestyle and mental health of students: a retrospective study from Karachi, Pakistan. In Annales Médico-psychologiques, Revue Psychiatrique Elsevier Masson, Paris. 180(6): S29–S37

Altena E, Baglioni C, Espie CA, Ellis J, Gavriloff D, Holzinger B et al (2020) Dealing with sleep problems during home confinement due to the COVID-19 outbreak: practical recommendations from a task force of the European CBT-I Academy. J Sleep Res 29(4):e13052

Ammar A, Chtourou H, Boukhris O, Trabelsi K, Masmoudi L, Brach M, ECLB-COVID19 Consortium et al (2020) COVID-19 home confinement negatively impacts social participation and life satisfaction: a worldwide multicenter study. Int J Environ Res Public Health 17(17): 6237

Barrea L, Pugliese G, Framondi L, Di Matteo R, Laudisio D, Savastano S et al (2020) Does Sars-Cov-2 threaten our dreams? Effect of quarantine on sleep quality and body mass index. J Transl Med 18(1):1–11

Becker SP, Gregory AM (2020) Editorial perspective: perils and promise for child and adolescent sleep and associated psychopathology during the COVID-19 pandemic. J Child Psychol Psychiatry 61:757–759. https://doi.org/10.1111/jcpp.13278

Borges U, Lobinger B, Javelle F, Watson M, Mosley E, Laborde S (2021) Using slow-paced breathing to foster endurance, well-being, and sleep quality in athletes during the COVID-19 pandemic. Front Psychol 12:624655

Bruni O, Malorgio E, Doria M, Finotti E, Spruyt K, Melegari MG, Villa MP, Ferri R (2022) Changes in sleep patterns and disturbances in children and adolescents in Italy during the Covid-19 outbreak. Sleep Med 91:166. (in press)

Bryant PA, Trinder J, Curtis N (2004) Sick and tired: does sleep have a vital role in the immune system? Nat Rev Immunol 4(6):457–467

Burkart S, Parker H, Weaver RG, Beets MW, Jones A, Adams EL et al (2022) Impact of the COVID-19 pandemic on elementary schoolers' physical activity, sleep, screen time and diet: a quasi-experimental interrupted time series study. Pediatr Obes 17(1):e12846

Carroll N, Sadowski A, Laila A, Hruska V, Nixon M, Ma DW, Haines J (2020) The impact of COVID-19 on health behavior, stress, financial and food security among middle to high income Canadian families with young children. Nutrients 12(8):2352

Cellini N, Canale N, Mioni G, Costa S (2020) Changes in sleep pattern, sense of time and digital media use during COVID-19 lockdown in Italy. J Sleep Res 29(4):e13074

Cheikh Ismail L, Osaili TM, Mohamad MN, Al Marzouqi A, Jarrar AH, Abu Jamous DO et al (2020) Eating habits and lifestyle during COVID-19 lockdown in The United Arab Emirates: a cross-sectional study. Nutrients 12(11):3314

Colivicchi F, Di Fusco SA, Magnanti M, Cipriani M, Imperoli G (2020) The impact of the coronavirus disease-2019 pandemic and Italian lockdown measures on clinical presentation and management of acute heart failure. J Card Fail 26(6):464–465

da Silva Santos AM, Rossi FE, Nunes DS, de Moura H, de Sousa Junior AVM, Machado DC, Neves LM et al (2021) COVID-19 pandemic impacts physical activity levels and sedentary time but not sleep quality in young badminton athletes. Sport Sci Health 17(4):969–977

Dai W, Zhou J, Li G, Zhang B, Ma N (2021) Maintaining normal sleep patterns, lifestyles and emotion during the COVID-19 pandemic: the stabilizing effect of daytime napping. J Sleep Res 30(4):e13259

Drumheller K, Fan CW (2022) Unprecedented times and uncertain connections: a systematic review examining sleep problems and screentime during the COVID-19 pandemic. Sleep Epidemiol 2:100029

Dutta K, Mukherjee R, Sen D, Sahu S (2022) Effect of COVID-19 lockdown on sleep behavior and screen exposure time: an observational study among Indian school children. Biol Rhythm Res 53(4):628–639

Eyimaya AO, Irmak AY (2021) Relationship between parenting practices and children's screen time during the COVID-19 pandemic in Turkey. J Pediatr Nurs 56:24–29

Facer-Childs ER, Hoffman D, Tran JN, Drummond SP, Rajaratnam SM (2021) Sleep and mental health in athletes during COVID-19 lockdown. Sleep 44(5):zsaa261

Górnicka M, Drywień ME, Zielinska MA, Hamułka J (2020) Dietary and lifestyle changes during COVID-19 and the subsequent lockdowns among Polish adults: a cross-sectional online survey PLifeCOVID-19 study. Nutrients 12(8):2324

Guo YF, Liao MQ, Cai WL, Yu XX, Li SN, Ke XY et al (2021) Physical activity, screen exposure and sleep among students during the pandemic of COVID-19. Sci Rep 11(1):1–11

Gupta P, Shah D, Bedi N, Galagali P, Dalwai S, Agrawal S et al (2022) Indian Academy of Pediatrics guidelines on screen time and digital wellness in infants, children and adolescents. Indian Pediatr 59(3):235–244

Hammoudi SF, Mreydem HW, Abou Ali BT, Saleh NO, Chung S, Hallit S, Salameh P (2021) Smartphone screen time among university students in Lebanon and its association with insomnia, bedtime procrastination, and body mass index during the COVID-19 pandemic: a cross-sectional study. Psychiatry Investig 18(9):871

Hirshkowitz M, Whiton K, Albert SM, Alessi C, Bruni O, DonCarlos L et al (2015) National sleep foundation's updated sleep duration recommendations. Sleep Health 1(4):233–243

Hu Z, Lin X, Kaminga AC, Xu H (2020) Impact of the COVID-19 epidemic on lifestyle behaviors and their association with subjective well-being among the general population in mainland China: cross-sectional study. J Med Internet Res 22(8):e21176

Husain W, Ashkanani F (2020) Does COVID-19 change dietary habits and lifestyle behaviours in Kuwait: a community-based cross-sectional study. Environ Health Prev Med 25(1):1–13

Jahrami H, BaHammam AS, Bragazzi NL, Saif Z, Faris M, Vitiello MV (2021) Sleep problems during the COVID-19 pandemic by population: a systematic review and meta-analysis. J Clin Sleep Med 17(2):299–313

Jahrami HA, Alhaj OA, Humood AM, Alenezi AF, Fekih-Romdhane F, AlRasheed MM et al (2022) Sleep disturbances during the COVID-19 pandemic: a systematic review, meta-analysis, and meta-regression. Sleep Med Rev 101591:101591

Khare R, Mahour J, Ohary R, Kumar S (2021) Impact of online classes, screen time, naps on sleep, and assessment of sleep-related problems in medical college students during lockdown due to coronavirus disease-19 pandemic. Natl J Physiol Pharm Pharmacol 11(1):56–56

Kurniarobbi J, Chikih C, Andeansah M, Lestari R, Sukendar I (2022) Athletes sleep duration during COVID-19 pandemic and its relationship with health condition. Int J Public Health 11(1):61–68

Lin YN, Liu ZR, Li SQ, Li CX, Zhang L, Li N et al (2021) Burden of sleep disturbance during COVID-19 pandemic: a systematic review. Nat Sci Sleep 13:933

López-Bueno R, López-Sánchez GF, Casajús JA, Calatayud J, Gil-Salmerón A, Grabovac I et al (2020) Health-related behaviors among school-aged children and adolescents during the Spanish Covid-19 confinement. Front Pediatr 8:573

Majumdar P, Biswas A, Sahu S (2020) COVID-19 pandemic and lockdown: cause of sleep disruption, depression, somatic pain, and increased screen exposure of office workers and students of India. Chronobiol Int 37(8):1191–1200

Malta DC, Szwarcwald CL, de Azevedo Barros MB, Gomes CS, Machado ÍE, de Souza Júnior PRB et al (2020) The COVID-19 pandemic and changes in adult Brazilian lifestyles: a cross-sectional study, 2020. Epidemiol Serv Saúde 29(4):e2020407

Meyer J, McDowell C, Lansing J, Brower C, Smith L, Tully M, Herring M (2020) Changes in physical activity and sedentary behavior in response to COVID-19 and their associations with mental health in 3052 US adults. Int J Environ Res Public Health 17(18):6469

Moitra P, Madan J (2022) Impact of screen time during COVID-19 on eating habits, physical activity, sleep, and depression symptoms: a cross-sectional study in Indian adolescents. PLoS One 17(3):e0264951

Mon-López D, Bernardez-Vilaboa R, Fernandez-Balbuena AA, Sillero-Quintana M (2020) The influence of COVID-19 isolation on physical activity habits and its relationship with convergence insufficiency. Int J Environ Res Public Health 17(20):7406

Munasinghe S, Sperandei S, Freebairn L, Conroy E, Jani H, Marjanovic S, Page A (2020) The impact of physical distancing policies during the COVID-19 pandemic on health and well-being among Australian adolescents. J Adolesc Health 67(5):653–661

Palladino F, Merolla E, Solimeno M, de Leva MF, Lenta S, Di Mita O et al (2020) Is Covid-19 lockdown related to an increase of accesses for seizures in the emergency department? An observational analysis of a paediatric cohort in the southern Italy. Neurol Sci 41(12):3475–3483

Pandya A, Lodha P (2021) Social connectedness, excessive screen time during COVID-19 and mental health: a review of current evidence. Front Hum Dyn 3

Paterson DC, Ramage K, Moore SA, Riazi N, Tremblay MS, Faulkner G (2021) Exploring the impact of COVID-19 on the movement behaviors of children and youth: a scoping review of evidence after the first year. J Sport Health Sci 10(6):675–689

Qin F, Song Y, Nassis GP, Zhao L, Dong Y, Zhao C et al (2020) Physical activity, screen time, and emotional well-being during the 2019 novel coronavirus outbreak in China. Int J Environ Res Public Health 17(14):5170

Ragnoli B, Pochetti P, Pignatti P, Barbieri M, Mondini L, Ruggero L et al (2022) Sleep deprivation, immune suppression and SARS-CoV-2 infection. Int J Environ Res Public Health 19(2):904

Richardson DL, Duncan MJ, Clarke ND, Myers TD, Tallis J (2021) The influence of COVID-19 measures in the United Kingdom on physical activity levels, perceived physical function and mood in older adults: a survey-based observational study. J Sports Sci 39(8):887–899

Rodríguez-Larrad A, Mañas A, Labayen I, González-Gross M, Espin A, Aznar S et al (2021) Impact of COVID-19 confinement on physical activity and sedentary behaviour in Spanish university students: role of gender. Int J Environ Res Public Health 18(2):369

Rolland B, Haesebaert F, Zante E, Benyamina A, Haesebaert J, Franck N (2020) Global changes and factors of increase in caloric/salty food intake, screen use, and substance use during the early COVID-19 containment phase in the general population in France: survey study. JMIR Public Health Surveill 6(3):e19630

Romdhani M, Ammar A, Trabelsi K, Chtourou H, Vitale J, Masmoudi L et al (2022a) Ramadan observance exacerbated the negative effects of COVID-19 lockdown on sleep and training behaviors: a international survey on 1,681 Muslim athletes. Front Nutr 9:925092

Romdhani M, Rae DE, Nédélec M, Ammar A, Chtourou H, Al Horani R et al (2022b) COVID-19 lockdowns: a worldwide survey of circadian rhythms and sleep quality in 3911 athletes from 49 countries, with data-driven recommendations. Sports Med 52(6):1433–1448

Runacres A, Mackintosh KA, Knight RL, Sheeran L, Thatcher R, Shelley J, McNarry MA (2021) Impact of the COVID-19 pandemic on sedentary time and behaviour in children and adults: a systematic review and meta-analysis. Int J Environ Res Public Health 18(21):11286

Salfi F, Amicucci G, Corigliano D, D'Atri A, Viselli L, Tempesta D, Ferrara M (2021) Changes of evening exposure to electronic devices during the COVID-19 lockdown affect the time course of sleep disturbances. Sleep 44(9):zsab080

Sharma M, Aggarwal S, Madaan P, Saini L, Bhutani M (2021) Impact of COVID-19 pandemic on sleep in children and adolescents: a systematic review and meta-analysis. Sleep Med 84:259–267

Singh S, Balhara YPS (2021) "Screen-time" for children and adolescents in COVID-19 times: need to have the contextually informed perspective. Indian J Psychiatry 63(2):192

So HK, Chua GT, Yip KM, Tung KT, Wong RS, Louie LH et al (2022) Impact of COVID-19 pandemic on school-aged Children's physical activity, screen time, and sleep in Hong Kong: a cross-sectional repeated measures study. Int J Environ Res Public Health 19(17):10539

Souza TC, Oliveira LA, Daniel MM, Ferreira LG, Della Lucia CM, Liboredo JC, Anastácio LR (2022) Lifestyle and eating habits before and during COVID-19 quarantine in Brazil. Public Health Nutr 25(1):65–75

Werneck AO, Silva DR, Malta DC, Lima MG, Souza-Júnior PR, Azevedo LO et al (2020) The mediation role of sleep quality in the association between the incidence of unhealthy movement behaviors during the COVID-19 quarantine and mental health. Sleep Med 76:10–15

World Health Organization (2020). Regional office for the eastern Mediterranean. Excessive screen use and gaming considerations during COVID-19. https://apps.who.int/iris/handle/10665/333467

Zhang X, Dimitriou D, Halstead EJ (2021) Sleep, anxiety, and academic performance: a study of adolescents from public high schools in China. Front Psychol 12:2567

Chapter 18
Dreams and Nightmares During the COVID-19 Pandemic

Deirdre Barrett

Abstract This chapter reviews studies from different countries about the influences of the COVID-19 pandemic on dreaming. These find that direct content about the pandemic is reported by many dreamers and that general dreams have been more anxious throughout this period. The chapter then focuses in detail on a survey of 12,000+ pandemic dreams by the current author and three analyses of it: (1) a qualitative reading for literal and metaphoric themes, (2) a comparison with normal-era dreams utilizing the Linguistic Inquiry and Word Count, and (3) a comparison with waking conversation about COVID-19 as differentiated by a deep learning algorithm.

Keywords Dreams · Nightmares · COVID-19 · Post-traumatic stress disorder · *Continuity hypothesis*

Crises prior to the COVID-19 pandemic have influenced dream content toward more anxious dreams, many of which are about those crises. This has been documented for the September 11, 2001, terrorist attacks (Barrett 2002; Bulkeley and Kahan 2008; Hartmann and Basile 2003), the Iraqi invasion of Kuwait (Barrett and Behbehani 2003), the Oakland/Berkeley Firestorm (Siegel 1996), and detention of Allied troops in Nazi POW camps during WWII (Barrett et al. 2013). Some of these studies have also reported metaphors standing in for the actual crisis in some dreams (Barrett 2002; Barrett and Behbehani 2003; Hartmann and Basile 2003); especially common are natural disasters such as tidal waves, hurricanes, and wildfires (Hartmann and Basile 2003).

In a survey in the USA conducted in early May 2020, 8.15% of respondents reported having dreamt about COVID-19 (Schredl and Bulkeley 2020). In Brazil, during May and June 2020, 33% reported dreams about COVID-19 (Musse et al. 2020). In a Canadian survey during April and May 2020, dreams about the pandemic

---

D. Barrett (✉)
Department of Psychiatry, Harvard Medical School, Cambridge, MA, USA
e-mail: deirdre_barrett@hms.harvard.edu

A. S. BaHammam et al. (eds.), *COVID-19 and Sleep: A Global Outlook*, Progress in Sleep Research, https://doi.org/10.1007/978-981-99-0240-8_18

were reported by almost 37% of the general population (Robillard et al. 2021). In the US study during the fall of 2020, the proportion of COVID-19 dream reporters increased to 45–65% (Gallagher and Incelli 2021). During the pandemic, dream imagery was characterized by a high presence of virus-related themes, particularly anxieties and concerns related to COVID-19 contagion (e.g., Giovanardi et al. 2022; Le Bel and DeCicco 2022; MacKay and DeCicco 2020; Margherita et al. 2021; Monaco et al. 2022; Pesonen et al. 2020; Šćepanović et al. 2022; Sommantico et al. 2021a, b, 2022; Wang et al. 2021).

Furthermore, findings in the literature of longitudinal dream studies during the different waves of COVID-19 (e.g., Alfonsi et al. 2021; Conte et al. 2021; Goncalves et al. 2022; Gorgoni et al. 2022a, b; Scarpelli et al. 2021, 2022) highlighted that its impact persisted on dream features, such as frequency, length, vividness, and negative emotionality, despite a sort of adaptation to the pandemic situation (e.g., Conte et al. 2021; Meaklim et al. 2022).

Studies on populations subject to the COVID-19 lockdown measures (e.g., Borghi et al. 2021; Cong et al. 2022; Gallagher and Incelli 2021; Guo and Shen 2021; Kilius et al. 2021; Mariani, Mariani, Monaco et al. 2022; Margherita et al. 2021; Marogna et al. 2021; Mota et al. 2020; 2022b) have shown that the isolation-quarantine situation specifically has a strong impact on dreaming.

Several studies indicated that aside from content about the pandemic, the COVID-19 pandemic has increased dream and nightmare recall frequency (e.g., Alghamdi et al. 2022; Conte et al. 2021; Fränkl et al. 2021; Giovanardi et al. 2022; Gorgoni et al. 2021; Gorgoni, Scarpelli, Alfonsi, and De Gennaro 2022a; Guerrero-Gomez et al. 2021; Guo and Shen 2021; Kennedy et al. 2022; Margherita et al. 2021; Musse et al. 2020; Sommantico et al. 2021a; Scarpelli et al. 2021; Schredl et al. 2022; Schredl and Bulkeley 2020; Solomonova et al. 2021; Sommantico et al. 2021b), as well as vividness and emotionality in dreams. In particular, participants who experienced COVID-19 more directly reported higher emotional intensity in their dreams, characterized by a higher presence of sensory impressions (e.g., Barrett 2020; Schredl and Bulkeley 2020; Sommantico et al. 2021a, b).

Reporting dreams about COVID-19 correlated with sleep problems reported in other chapters of this volume. In a Columbian sample, people who recalled having dreams about the COVID-19 pandemic averaged a worse quality of sleep than those who did not, as assessed by items such as sleep latency, sleep duration, reported sleep disturbances, and use of sleeping medication (Cabeza et al. 2022).

The survey reported in the remainder of this chapter is an ongoing one by the author of 12,000+ dreams reported by 45,000 dreamers from 86 countries. To date, it has been used in three very different analyses: (1) categorizations of literal and metaphoric themes based on a qualitative close read (Barrett 2020a); (2) a comparison with dreams from normal times using the Linguistic Inquiry and Word Count (LIWC) categories for emotions, illness, death, and body references (Barrett 2020); and (3) a deep learning algorithm identification of symptoms and emotions in pandemic survey dreams vs. waking conversation about COVID-19.

18.1 Method

18.1.1 Participants

Respondent's ages ranged from 18 to 91 years, with a mean of 40.08 and a standard deviation of 16.89. There were 1998 women and 890 men. Sixty-eight subjects were identified as either gender-neutral or transgender; this subset was deemed not large enough or consistent enough to analyze in the present study, but it may be included in future analyses as the survey N grows. Nationalities, in decreasing frequency, were US 2011, British 249, Italian 212, Canadian 173, Spanish 91, Indian 54, Peruvian 54, German 46, Mexican 42, Australian 34, Brazilian 33, French 22, Polish 22, and 73 other nationalities with less than 20 respondents each.

18.1.2 Materials

A survey was posted on March 23rd, and responses were downloaded for analysis on July 15th. The survey asked respondents to submit "any dreams you have had related to the COVID-19 coronavirus." The survey also inquired about age, gender, and nationality. The survey was announced on 11 Facebook groups—3 smaller ones (611–7461 members) focused on dreams and 8 larger ones (27,961–41,351 members) focused on the pandemic. The survey has also been linked to articles in major media in the USA, Europe, South America, Australia, New Zealand, and India. The survey is ongoing, but the responses as of July 15, 2020, were utilized for the analyses reported here.

18.2 Study 1: A Qualitative Close Read for Content Categories

18.2.1 Procedure

18.2.1.1 Results

A common category of dreams in my survey is simply catching the virus. Dreamers have trouble breathing or spike a fever. Other symptoms are more dreamlike: one woman sees a dark aura from a person she passes on the street who touches her body and knows that it has infected her. Another looks down and notices bright blue stripes on her stomach and remembers that it is the first sign of infection. Variations include one's children or elderly parents coming down with the virus.

Not all dreams about the virus are literal. After 9/11, I saw some metaphoric dreams. However, due to the dramatic images associated with that event, a majority dreamed of buildings falling, planes smashing into things, and/or hijackers with knives. Our dreaming mind is intensely visual, so when it feels fear, it searches for an image to match that feeling. Bugs express what many are feeling about COVID-19. Swarms of flying insects—bees, hornets, wasps, gnats, and horseflies—attack. Masses of toxic worms writhe in front of dreamers. Armies of cockroaches race toward them. Bedbugs, stink bugs: One woman dreamed of giant grasshoppers with vampire fangs. Bugs are the definitive metaphor now probably partly because of our slang use of the word "bug" to mean a virus or other illness, as in "I've got a bug," but they also make a very good representation for lots of tiny things that cumulatively could harm or kill one.

Another metaphoric creature unique to this epidemic is the invisible monster. Some dreamers must cross exposed outdoor areas and know that some monsters could kill but which they cannot spot. Others wander through building complexes and hear steps behind them or spot subtle shadows moving when they cannot directly see the monsters.

The standard metaphors for dreams about any crisis also showed up in the pandemic survey. Tsunamis, tornados, hurricanes, earthquakes, wildfires, and mass shooters are some of the common metaphors in the responses.

Metaphoric dreams may also be made with direct reference to some detail of this pandemic—interspersing the scary visual metaphor with actual guidance. New York's Governor Cuomo tells people that they have to shelter in place because of the swarms of bugs or shooters in the streets. President Trump announced that there was no tsunami, calling it "fake news."

I saw similar traumatic nightmares in the pandemic survey from healthcare workers as those on the front line of other crises. Six hundred healthcare workers have responded and many of their dreams reenact the worst moments of their days. They are trying to intubate a patient whose airway is too constricted. They get a tube into someone, and it slips back out. The ventilator malfunctions. Sometimes, there is a dreamlike element as with the Italian doctor falling out of the window with his patient.

As the lockdown/stay-at-home/shelter-in-place/quarantine orders have continued, there has been an increase in dreams which focus on this rather than the virus itself. Those who are home alone dream scenarios that exaggerate their loneliness: they are in all sorts of prisons, or the person I quoted in the introduction to the book got stranded alone on Saturn and another woman had a very similar dream in which she had been selected to be the first colonist on Mars—a job she had *not* applied for.

At the other end of the spectrum, people were sheltering with even more extended family than those with whom they usually reside. They dreamed of their house being converted into a homeless shelter, a testing site, or neighbors barging in and "borrowing" all their toilet paper.

Parents, mostly mothers, who were homeschooling their children early in the pandemic had exaggerated dreams about the frustrations of this. One mother of a 10-year-old dreamed that her child's school texted her to say that they were sending the entire class to her home for her to teach for the rest of the pandemic. Another dreamed that homeschooling consisted of the parents taking the children's exams and her son was going to have a year of Fs in middle-school math.

Dreams about masking were a category that evolved over time. Early in the pandemic, only respondents in Asia dreamed about masks—usually "whoops, I forgot my mask" or more dreamlike versions where the mask turns into a small animal clinging to their face which jumps off and runs away. By the summer of 2020, these dreams were common in North America and Europe also. In 2020, all dreams about not having a proper mask on were fearful ones about exposure to disease. By 2021, and increasingly through 2022, when dreamers realized that they were not wearing a mask, the response was often more one of social shame and hoping that others had not noticed their omission.

A few of the pandemic survey dreams were positive with the main two happy themes being the following: (1) the dreamer is cured of the virus or they or someone else discovers a cure for all mankind or (2) the post-COVID world is greatly improved by psychological lessons learned and/or reduced pollution and return to natural conditions.

18.3 Study 2: Computer Text Analysis Comparison with Pre-pandemic Dreams

18.3.1 Procedure

It was hypothesized that dreams since the outset of the COVID-19 pandemic, when compared to ones from a normative, noncrisis time, would demonstrate a shift from positive to negative emotions, especially anxiety, and a higher rate of concern with disease and death. The measure selected for testing the hypotheses was Linguistic Inquiry and Word Count (Pennebaker et al. 2015). This computer text analysis program is easily utilized on large samples of dream accounts and to contain emotion and other content categories that correspond closely to equivalent ones of the Hall and van de Castle rating scales (Barrett 2015; Bulkeley and Graves 2018), which are the most standard human-scored scales for a dream content (Domhoff 2003). The survey dreams were analyzed using nine scales of the LIWC: positive emotions, negative emotions, anxiety, anger, sadness, biological processes, body, health, and death. The results were compared to the 981 dreams utilized for norming the Hall and van de Castle rating scales and which have been used as a comparison sample in more dream research than any others (Domhoff 2003). Unequal variance t-tests were conducted with Benjamini-Hochberg corrections applied.

18.3.2 Results

All hypotheses were confirmed at a minimum of a corrected level of .05 with most reaching much higher significance (see Table 18.1).

The predicted increase in the variable health was the only one significant at as high a level as for women. Positive emotions, negative emotions, anxiety, and death changed in the predicted direction at lower significance levels than the effects of women's pandemic dreams. The variables anger, sadness, and body did not differ between the pandemic dreams and the normative sample (see Table 18.2).

Table 18.1 Female pandemic survey dreams vs. Hall and van de Castle female normative dreams

LIWC category	Pandemic mean	SD	Normative mean	SD	$t=$	$p=$
Positive emotions	1.11	1.82	1.48	1.52	4.64	<.0001[a]
Negative emotions	2.31	3.32	1.40	1.47	9.14	<.0001[a]
Anxiety	.76	2.20	.46	.74	5.05	<.0001[a]
Anger	.42	1.32	.31	.64	2.66	.0078
Sadness	.46	1.37	.27	.63	4.55	<.0001[a]
Biological processes	2.43	2.87	1.80	1.97	5.75	<.0001[a]
Body	1.07	1.83	.83	1.32	3.81	.0009[b]
Health	.91	1.69	.48	.90	7.76	<.0001[a]
Death	.51	1.75	.15	.52	7.89	<.0001[a]

[a] Benjamini-Hochberg corrected $p < .001$
[b] Benjamini-Hochberg corrected $p < .01$

Table 18.2 Male pandemic survey dreams vs. Hall and van de Castle male normative dream

LIWC category	Pandemic male mean	SD	Normative male mean	SD	$t=$	$p=$
Positive emotions	.96	3.81	1.48	1.46	3.47	.0009[a]
Negative emotions	1.91	2.95	1.50	1.52	3.28	.0011[a]
Anxiety	.48	1.06	.33	.64	3.17	.0015[a]
Anger	.45	1.27	.47	.93	.33	.7506
Sadness	.34	1.37	.27	.53	1.32	.1953
Biological processes	1.99	3.21	1.67	1.92	2.24	.0252[b]
Body	.73	1.44	.78	1.21	.67	.5032
Health	.76	1.69	.38	.76	5.52	<.0001[c]
Death	.74	5.33	.23	.67	2.67	.0077[a]

[a] Benjamini-Hochberg corrected $p < .01$
[b] Benjamini-Hochberg corrected $p < .05$
[c] Benjamini-Hochberg corrected $p < .001$

18.4 Study 3: Comparison of COVID-19 Pandemic Dreams with Tweets About the COVID-19 Pandemic

18.4.1 Procedure

This study was performed in collaboration with computer scientists Sanja Šćepanović, Luca Maria Aiello, and Daniele Quercia. All computation in this section was performed by them, and more on the deep learning AI algorithm utilized is reported elsewhere (Šćepanović et al. 2022). The dataset used for this study is one of Twitter's "tweets" containing words related to COVID-19, which is openly shared via GitHub (Chen et al. 2020). From this initial dataset, 57,287,490 English tweets posted in the year 2020 from February 1st to April 30th by 11,318,634 unique users were extracted. This timeframe includes the period of the initial spread of the virus until the peak of the number of deaths worldwide during the first wave of infections. Given that English tweets were selected, most of the users in the dataset came from the USA and the UK. In the USA, 22% of the population uses Twitter, and notably, Twitter users are gender-balanced though (50% women).

The procedure consisted of three main steps. First, on both of the datasets (i.e., pandemic discussions and dream reports), a deep learning natural language processing (NLP) method called MedDL (Scepanovic et al. 2020) was applied to extract mentions of medical conditions. Second, the extracted conditions were classified based on their relative prevalence in the two pandemic datasets into three groups: conditions typical of waking discussions, those equally prevalent, and those types of dream reports. Third, a co-occurrence network of dream conditions was constructed and analyzed using graph modeling to find important conditions (i.e., conditions linked to many other conditions), and groups of conditions that co-occur frequently.

As is detailed further in the results, conditions that are more prevalent in dream reports are often characterized by unreal imagery. To discover how those special types of conditions are related to concrete medical conditions, a co-occurrence graph of the 1732 unique medical conditions in dream reports was separately applied. This captured the semantic relatedness of medical conditions: symptoms or diseases that were often mentioned together are likely to describe the same condition, and served to discover groups of related conditions as, in the network, they form densely connected clusters of nodes. Out of the 1732 unique medical conditions in dream reports, 313 did not co-occur with any other condition (i.e., they resulted in singleton nodes in the graph), so they were discarded. The graph containing the remaining 1419 nodes had 4084 edges.

18.4.2 Results

The MedDL analyses found that common COVID-19 symptoms were present in both waking discussions during the pandemic (and not in those pre-pandemic) and dream reports. Yet, it also found sleep-deformed real-life experiences, in that dream reports tended to contain metaphorical embodiments of actual symptoms.

MedDL found that those commonly mentioned in both datasets included *coronavirus*, *anxiety*, *cancer*, *coughing*, and *stress*. Conditions such as *infectious disease* and *Ebola* were highly ranked among the top mentions in waking discussions, but not in dreams, whereas mentions of *seasick* and *bleeding* were more frequent in dream reports. Some of the rare conditions found in tweets included *hypergammaglobulinemia, hyperreactivity pulmonary destruction,* or *vaping-related lung illnesses*, while rare conditions in dreams included *feeling like water fills my lungs*, *extreme déjà vu*, or *grabbing at my throat and swollen tongue*. In summary, in the group of equally prevalent conditions in waking discussions and dream reports, we found different mentions of COVID-19 itself (e.g., *coronavirus, coronavirus, COVID,* or *virus*), or of its common symptoms (e.g., *fever, pain, sore throat, migraine,* and *cough*).

In addition to the conditions common to the two sets of pandemic data, some conditions differentiated the two. Symptoms found more in waking discussions were realistic symptoms directly linked to COVID-19 (e.g., *body aches, abnormal heart rate, bronchitis*, *pneumonia*, and *nasal pain*), or related to similar conditions (e.g., *influenza, SARS, H1N1, bird flu, allergy, flu-like symptoms*), showing that people were discussing other infectious diseases as well. On the other hand, conditions mentioned mostly in *dream reports* included those that did not occur with the virus (*maggots, deformities, red virus, and snakebites*) or surreal ones (*teeth suddenly falling out, body crumbling into the sand,* and *rodents moving around in my periphery*), likely reflecting an exaggerated visual depiction of something wrong with the body or of dramatic scenarios resulting from the virus (see Table 18.3).

The network of symptom co-occurrence sheds light on how the waking and dreaming associations to aspects of COVID-19 diverge. Anxiety and choking link to each other and have equally dense links to other symptoms for both waking and dreaming data. However, for the waking discussions, *anxiety* and *choking* also link to other potential realistic symptoms such as *diarrhea* and *throwing up*. The dreams' *anxiety* and *choking* references link with symptoms such as *balding* and *teeth falling out*. Interestingly, the symptom *nightmare* is equally densely linked in both waking and dream networks. In the waking discussions, however, it links to potential causes of nightmares: *trauma* and *PTSD*. The dream reports link to the imagery of nightmares: *dark clouds, a buzzing sound, and ice running through veins*.

In the deep learning analysis, a method using contextual embeddings trained on social media data is applied to extract symptom references. Symptoms seen more in waking discussions about COVID-19 were realistic potential symptoms ("body aches," "abnormal heart rate," and "nasal pain") while ones mentioned mostly or only in dreams included ones not actually occurring with the virus ("maggots,"

Table 18.3 Medical conditions ranked from most to least prevalent for combined reports versus waking discussion versus dream reports

Equally prevalent	Typical of waking discussions	Typical of dream reports
Corona virus	Infectious disease	Sleep paralysis
Sick	Influenza	Trouble breathing
Cough	HIV	Coughing up blood
Virus	AIDS	Gasping for air
Bronchitis	Common cold	Spitting out teeth
Fever	Heart disease	Thickness or pressure in my chest
COVID-19	Bird flu	Maggot
Cancer	Mental illness	Seizure disorder
Allergy	Lupus	Social anxiety
Cold	Immunocompromise	Overdose
Stress	Measles	Overwhelmingly large eyes
Infection	Malaria	Teeth started falling out
Anxiety	Mental health	Heart was beating out of my chest
Pneumonia	Lyme disease	Teeth breaking off and coming out
Plague	Nasal pain	Alien invasion

"deformities," and "snakebites") or surreal, impossible ones ("teeth suddenly falling out" and "body crumbling into sand"). Symptoms mentioned more in pandemic dreams by men than women included many around the core symptom of respiratory difficulty ("respiratory distress," "trouble breathing," and "lungs stopped working") while ones mentioned more in women's pandemic dreams than men's were often psychological ("loneliness," "fatigue," and "PTSD").

18.5 Discussion

The two main takeaways from the close read of the survey dreams are that they show very similar patterns in how crises manifest in literal content and metaphors while also making dreams in general much more anxious than in normal times, but also that they show some distinctive dream metaphors such as bug attacks and invisible monsters representing this more insidiously menacing crisis with fewer direct visual images of its own.

In study 2, while hypotheses were supported for women, and more than half the variables reached at least modest significance in the predicted direction for men, the other striking result was the gender difference in how much more pandemic dreams varied from the norms for women compared to the variance for men.

Upon examination of the literature on waking gender differences of effects from the pandemic, these results are not surprising and represent further confirmation of the continuity hypothesis. Women average performing three times as much unpaid care work as men (United Nations Women 2020a). They are more likely to be caregivers for the sick individuals in the family, making them more vulnerable to infection. Globally, women make up 70% of healthcare workers. They occupy the lower salary/lower authority of these jobs on average and are less well supplied with personal protective equipment (Wenham et al. 2020a, b). In many countries, COVID-19 infections among female health workers are twice that of their male counterparts (United Nations Women 2020b). Women are underrepresented in clinical trials for vaccines and drugs, and they are underrepresented in pandemic decision-making bodies (United Nations Women 2020b).

Disease outbreaks increase existing inequalities between the genders (Wenham et al. 2020a, b). During the lockdown, women are at greater risk of domestic violence (United Nations Population Fund 2020) and are disproportionately disadvantaged by reduced access to sexual and reproductive health services. Because women are more likely than men to have fewer hours of employed work and be on insecure contracts, they are more affected by job losses, and women lost disproportionately more jobs since the onset of the pandemic (United Nations Women 2020b). Women are reported to be suffering a greater increase in depression and anxiety since the beginning of the pandemic than the increase for men (Özdin and Özdin 2020).

The women's dreams reflect both this general increase in stress and many of their specific challenges. Female dreams included a nanny who dreamed of the parents terminating their work-from-home arrangements to focus on the children and turning her out in the streets. One mother who was homeschooling her 10-year-old dreamed that the school contacted her to say that the child's whole class was being sent to her condominium where she must homeschool all of them for the duration of the pandemic. The changes in the body and biological processes for women only probably reflect how much dreaming about unwanted sexualization and lack of privacy there is and dreams where the dreamer is pregnant and bizarre things are going wrong with the pregnancy related to the pandemic.

The finding of study 3 that waking discussions and dream reports contain proportionally equal mentions of COVID-19 itself and its major symptoms is consistent with the continuity hypothesis of dreaming (Domhoff 2003, 2017; Schredl 2017). It indicates that the waking and dreaming minds are equally focused on the threat of the COVID-19 pandemic—it is the resulting associations and style of thinking about it that are so divergent. The differences in frequency for other categories support the idea that the mind is thinking about this concern in two very different states consistent with what we know about brain activation in waking vs. sleep states (Dang-Vu et al. 2010). That waking discussions contain more references to similar diseases by name and realistic symptoms of other disorders reflects what people use in referencing known facts to try to figure out more about the threat posed by COVID-19 via a linear, logical process. The phrases more frequent in the dream reports about bizarre body dysfunctions represent a metaphoric manner of thinking about COVID-19. This is consistent with the observation that dreams are generally concerned with the same topics but filtered through distinctive brain states during sleep (Domhoff 2003, 2017; Schredl 2017). This is a further indication of the waking rational and verbal vs. the dreams' visual, emotional, and metaphoric approach to dealing with the same issue.

The survey is ongoing, and it is anticipated that more cross-nationality analyses can be done as the sample grows and that comparisons can be made between trends at the start, middle, and (hopefully!) the end of the pandemic.

References

Alfonsi V, Gorgoni M, Scarpelli S, Zivi P, Sdoia S, Mari E et al (2021) Changes in sleep pattern and dream activity across and after the COVID-19 lockdown in Italy: a longitudinal observational study. J Sleep Res 31:e13500. https://doi.org/10.1111/jsr.13500

Alghamdi AA, Alafif NO, BaHammam AS, Almuammar M, Alharbi NS, Alhefdhi NA et al (2022) Nightmares' pattern and predictors among a Saudi university students during COVID-19 pandemic. Int J Environ Res Public Health 19:1776. https://doi.org/10.3390/ijerph19031776

Barrett D (2002) Night wars. Presentation at the International Association for the Study of dreams international conference, Medford, MA, May

Barrett D (2015) The LIWC: a comparison of ITS psychological categories to those of the hall and van de castle dream rating system. Presentation at the International Association for the Study of dreams international conference, Virginia Beach, VA, June

Barrett D, Behbehani J (2003) Posttraumatic nightmares in Kuwait following the Iraqi invasion. In: Krippner S, McIntyre T (eds) Psychological impact of war on civilians: an international perspective. Baywood, Amityville, NY, pp 135–144

Barrett D (2020) Dreams about COVID-19 versus normative dreams: trends by gender. Dreaming 30:216–221. https://doi.org/10.1037/drm0000149

Barrett DL (2020a) Pandemic dreams. Oneiroi Press

Barrett D, Sogolow Z, Oh A, Panton J, Grayson M, Justiniano M (2013) Content of dreams from WWII POWs. Imagin Cogn Pers 33(1):193–204

Borghi L, Bonazza F, Lamiani G, Musetti A, Manari T, Filosa M et al (2021) Dreaming during lockdown: a quali-quantitative analysis of the Italian population dreams during the first COVID-19 pandemic wave. Res Psychother 24:200–211. https://doi.org/10.4081/ripppo.2021.547

Bulkeley K, Graves M (2018) Using the LIWC program to study dreams. Dreaming 28(1):43

Bulkeley K, Kahan TL (2008) The impact of September 11 on dreaming. Conscious Cogn 17 (4):1248–1256

Cabeza G, Caballero H, Castiblanco F, Cabrera D, Martinez F, Lara G (2022) Impact of the COVID-19 pandemic on the quality of sleep of mental health workers in Colombia. Dreaming 32(2):124

Chen E, Lerman K, Ferrara E (2020) COVID-19-TweetIDs. https://github.com/echen102/COVID-19-TweetIDs. Accessed 10 Sep 2020

Cong B, Xue M, Ye Y (2022) Relationships between attitudes toward dreams, typical dreams, and emotions among Chinese college students. Dreaming 32:1–14. https://doi.org/10.1037/drm0000191

Conte F, Rescott ML, De Rosa O, Cellini N, Coppola A, Cerasuolo M et al (2021) Changes in dream features across the first and second waves of the Covid-19 pandemic. J Sleep Res 31: e13425. https://doi.org/10.1111/jsr.13425

Dang-Vu TT, Schabus M, Desseilles M, Sterpenich V, Bonjean M, Maquet P (2010) Functional neuroimaging insights into the physiology of human sleep. Sleep 33(12):1589–1603

Domhoff GW (2003) The scientific study of dreams: neural networks, cognitive development, and content analysis. American Psychological Association, Washington, DC

Domhoff GW (2017) The invasion of the concept snatchers: the origins, distortions, and future of the continuity hypothesis. Dreaming 27(1):14–39

Fränkl E, Scarpelli S, Nadorff MR, Bjorvatn B, Bolstad CJ, Chan NY et al (2021) How our dreams changed during the COVID-19 pandemic: effects and correlates of dream recall frequency – a multinational study on 19,355 adults. Nat Sci Sleep 13:1573–1591. https://doi.org/10.2147/NSS.S324142

Gallagher SP, Incelli NA (2021) Pandemic dreams are susceptible to priming and unrelated to COVID-19 exposure. J Evid Based Psychother 21:189–206. https://doi.org/10.24193/jebp.2021.2.18

Giovanardi G, Fiorini Bincoletto A, Baiocco R, Ferrari M, Gentile D, Siri M et al (2022) Lockdown dreams: dream content and emotions during the COVID-19 pandemic in an Italian sample. Psychoanal Psychol 39:111–126. https://doi.org/10.1037/pap0000385

Goncalves M, Henriques A, Costa AR, Correia D, Severo M, Lucas R et al (2022) Insomnia and nightmare profiles during the COVID-19 pandemic in Portugal: characterization and associated factors. Sleep Med 90:44–52. https://doi.org/10.1016/j.sleep.2021.12.018

Gorgoni M, Scarpelli S, Alfonsi V, Annarumma L, Cordone S, Stravolo S, De Gennaro L (2021) Pandemic dreams: quantitative and qualitative features of the oneiric activity during the lockdown due to COVID-19 in Italy. Sleep Med 81:20–32. https://doi.org/10.1016/j.sleep.2021.02.006

Gorgoni M, Scarpelli S, Alfonsi V, Annarumma L, Pellegrini E, Fasiello E et al (2022a) The oneiric activity during and after the COVID-19 Total lockdown in Italy: a longitudinal study. Int J Environ Res Public Health 19:3857. https://doi.org/10.3390/ijerph19073857

Gorgoni M, Scarpelli S, Alfonsi, & De Gennaro, L. (2022b) Dreaming during the COVID-19 pandemic: a narrative review. Neurosci Biobehav Rev 138:104710. https://doi.org/10.1016/j.neubiorev.2022.104710

Guerrero-Gomez A, Nöthen-Garunja I, Schredl M, Homberg A, Vulcan M, Brusić A et al (2021) Dreaming in adolescents during the COVID-19 health crisis: survey among a sample of European school students. Front Psychol 12:652627. https://doi.org/10.3389/fpsyg.2021.652627

Guo H, Shen H (2021) Media exposure to COVID-19 epidemic and threatening dream frequency: a moderated mediation model of anxiety and coping efficacy. Dreaming 31:1–19. https://doi.org/10.1037/drm0000159

Hartmann E, Basile R (2003) Dream imagery becomes more intense after 9/11/01. Dreaming 13(2):61–66

Kennedy KER, Bastien CH, Ruby PM, Killgore WDS, Wills CCA, Grandner MA (2022) Nightmare content during the COVID-19 pandemic: influence of COVID-related stress and sleep disruption in the United States. J Sleep Res 31:e13439. https://doi.org/10.1111/jsr.13439

Kilius E, Abbas NH, McKinnon L, Samson DR (2021) Pandemic nightmares: COVID-19 lockdown associated with increased aggression in female university students' dreams. Front Psychol 12:644636. https://doi.org/10.3389/fpsyg.2021.644636

Le Bel S, DeCicco TL (2022) What dreams reveal about people's emotional state due to the impacts of COVID-19. Int J Dream Res 15:18–26. https://doi.org/10.11588/ijodr.2022.1.82110

MacKay C, DeCicco TL (2020) Pandemic dreaming: the effect of COVID-19 on dream imagery, a pilot study. Dreaming 30:222–234. https://doi.org/10.1037/drm0000148

Margherita G, Gargiulo A, Lemmo D, Fante C, Filosa M, Manari T et al (2021) Are we dreaming or are we awake? A quali–quantitative analysis of dream narratives and dreaming process during the COVID-19 pandemic. Dreaming 31:373–387. https://doi.org/10.1037/drm0000180

Marogna C, Montanari E, Contiero S, Lleshi K (2021) Dreaming during COVID-19: the effects of a world trauma. Res Psychother Psychopathol Process Outcome 24(2)

Meaklim H, Burge M, Le F, Bains SK, Saunders W, Ghosh SK et al (2022) Strange themes in pandemic dreams: insomnia was associated with more negative, anxious and death-related dreams during the COVID-19 pandemic. J Sleep Res, First Online June 14:e13655. https://doi.org/10.1111/jsr.13655

Monaco S, Citti T, Gaudio B, Mariani R, Di Trani M (2022) "Dreams Drawer": analysis of dreams during lockdown in the Italian population. Dreaming 32:111–123. https://doi.org/10.1037/drm0000205

Mota NB, Weissheimer J, Ribeiro M, de Paiva M, Avilla-Souza J, Simabucuru G et al (2020) Dreaming during the Covid-19 pandemic: computational assessment of dream reports reveals mental suffering related to fear of contagion. PLoS One 15:e0242903. https://doi.org/10.1371/journal.pone.0242903

Musse FCC, Castro LDS, Sousa KMM, Mestre TF, Teixeira CDM, Pelloso SM et al (2020) Mental violence: the COVID-19 nightmare. Front Psychiatry 11:579289. https://doi.org/10.3389/fpsyt.2020.579289

Özdin S, Özdin ŞB (2020) Levels and predictors of anxiety, depression and health anxiety during COVID-19 pandemic in Turkish society: the importance of gender. Int J Soc Psychiatry 66(5): 504–511

Pennebaker JW, Boyd RL, Jordan K, Blackburn K (2015) The development and psychometric properties of LIWC2015. University of Texas, Austin, TX

Pesonen A, Lipsanen J, Halonen R, Elovainio M, Sandman N, Mäkelä J et al (2020) Pandemic dreams: network analysis of dream content during the COVID-19 lockdown. Front Psychol 11: 573961. https://doi.org/10.3389/fpsyg.2020.573961

Robillard R, Dion K, Pennestri MH, Solomonova E, Lee E, Saad M et al (2021) Profiles of sleep changes during the COVID 19 pandemic: demographic, behavioral and psychological factors. J Sleep Res 30(1):e13231. https://doi.org/10.1111/jsr.13231

Scarpelli S, Alfonsi V, Gorgoni M, Musetti A, Filosa M, Quattropani MC et al (2021) Dreams and nightmares during the first and second wave of the COVID-19 infection: a longitudinal study. Brain Sci 11:1375. https://doi.org/10.3390/brainsci11111375

Scarpelli S, Gorgoni M, Alfonsi V, Annarumma L, Di Natale V, Pezza E, De Gennaro L (2022) The impact of the end of COVID confinement on pandemic dreams, as assessed by a weekly sleep diary: a longitudinal investigation in Italy. J Sleep Res 31:e13429. https://doi.org/10.1111/jsr.13429

Scepanovic S, Martin-Lopez E, Quercia D, Baykaner K (2020) Extracting medical entities from social media. In Proceedings of the ACM conference on health, inference, and learning. pp. 170–181

Šćepanović S, Aiello LM, Barrett D, Quercia D (2022) Epidemic dreams: dreaming about health during the COVID-19 pandemic. R Soc Open Sci 9(1):211080

Schredl M (2017) Theorizing about the continuity between waking and dreaming: comment on Domhoff. Dreaming 27(4):351–359

Schredl M, Bulkeley K (2020) Dreaming and the COVID-19 pandemic: a survey in a U.S. sample. Dreaming 30:189–198. https://doi.org/10.1037/drm0000146

Schredl M, Remedios A, Marin-Dragu S, Sheikh S, Forbes A, Iyer RS et al (2022) Dream recall frequency, Lucid Dream frequency, and personality during the Covid-19 pandemic. Imagin Cogn Pers 42(2):113–133. https://doi.org/10.1177/02762366221104214

Siegel A (1996) Dreams of firestorm survivors. In: Barrett D (ed) Trauma and dreams. Harvard University Press, Cambridge, MA, pp 159–176

Solomonova E, Picard-Deland C, Rapoport IL, Pennestri MH, Saad M, Kendzerska T et al (2021) Stuck in a lockdown: dreams, bad dreams, nightmares, and their relationship to stress, depression, and anxiety during the COVID-19 pandemic. PLoS One 16:e0259040. https://doi.org/10.1371/journal.pone.02590

Sommantico M, Iorio I, Lacatena M, Parrello S (2021a) Dreaming during COVID-19 lockdown: a comparison between Italian adolescents and adults. Res Psychother 24:212–223. https://doi.org/10.4081/ripppo.2021.53

Sommantico M, Iorio I, Parrello S (2021b) Mood, sleep quality, and dreaming during the third wave of the COVID-19 pandemic in Italy. Int J Dream Res 14:309–319. https://doi.org/10.11588/ijodr.2021.2.82109

Sommantico M, DeCicco TL, Osorio Guzmán M, Prado Romero C, Le Bel S, Parrello S (2022) Illness attitudes, mood, and dreams during the second wave of the COVID-19 pandemic: an international study. Int J Dream Res 15:104–117. https://doi.org/10.11588/ijodr.2022.1.84712

United Nations Population Fund (2020) Gender-based violence spikes amid pandemic, shelters need support. https://www.unfpa.org/news/gender-based-violence-spikes-amid-pandemic-shelters-need-support

United Nations Women (2020a) "Unpaid care work: your load and why it matters"

United Nations Women (2020b) "UN Secretary-General's policy brief: the impact of COVID-19 on women". https://www.unwomen.org/en/digital-library/publications/2020/04/policy-brief-the-impact-of-covid-19-on-women#view

Wang J, Zemmelman SE, Hong D, Feng X, Shen H (2021) Does COVID-19 impact the frequency of threatening events in dreams? An exploration of pandemic dreaming in light of contemporary dream theories. Conscious Cogn 87:103051. https://doi.org/10.1016/j.concog.2020.103051

Wenham C, Smith J, Davies SE, Feng H, Grépin KA, Harman S, Herten-Crabb A, Morgan R (2020a) Women are most affected by pandemics-lessons from past outbreaks. Nature 583(7815):194–198

Wenham C, Smith J, Morgan R (2020b) COVID-19: the gendered impacts of the outbreak. Lancet 395(10227):846–848

Chapter 19
The Impact of COVID-19 Pandemic on Sleep Medicine Services

Jennifer Y. So and Shannon S. Sullivan

Abstract COVID-19 pandemic has led to significant disruption in medical service delivery, leading to adaptation of innovative ways to provide sleep medicine care. In the acute phase, the focus was on navigation of infection control ventilation guidance, and procedures for clinic spaces and staff areas, as well as assimilating new scientific data, balancing staff allocation demands, and evolving public health and institutional policies. Sleep laboratory services were impacted substantially, and novel changes to protocols and processes were made to meet clinical needs. Alternative pathways for care, including telemedicine and remote testing, became overnight mainstays of practice. Payment models were adjusted as well, enabled by temporary changes in payor policy. Subsequently, focus had shifted to include vaccination policy and increasing attention to workforce-related issues, such as scarcity and burnout, as well as the impact healthcare access and outcomes disparities. Though more research and integration are needed, sleep medicine as a field has proven to be flexible, dynamic, and creative in finding ways to meet the evolving needs of patients despite the challenges presented by the pandemic.

Keywords COVID-19 pandemic · Sleep medicine · Sleep laboratory · Pandemic response

---

J. Y. So (✉)
University of Maryland Sleep Laboratory, Division of Pulmonary, Critical Care and Sleep Medicine, University of Maryland School of Medicine, Baltimore, MD, USA
e-mail: jennifer.so@som.umaryland.edu

S. S. Sullivan
Division of Pediatric Pulmonary, Asthma, and Sleep Medicine, Department of Pediatrics; and by courtesy Division of Sleep Medicine, Department of Psychiatry, Stanford University School of Medicine, Palo Alto, CA, USA

A. S. BaHammam et al. (eds.), *COVID-19 and Sleep: A Global Outlook*, Progress in Sleep Research, https://doi.org/10.1007/978-981-99-0240-8_19

19.1 Introduction

COVID-19 pandemic has led to significant disruption and re-imagining of medical services delivery. Since its onset in early 2020, many changes have been made to the way medical care and sleep medicine services are provided in the United States and around the globe. Given its easy and rapid mode of transmission and the potential consequences of hospitalization, respiratory failure, multisystem dysfunction, and death, healthcare systems around the world were strained with medical resources, especially in healthcare personnel, personal protective equipment (PPE), and ventilators (Zhu et al. 2020; Paules et al. 2020). Healthcare providers had to respond rapidly to the dynamic and uncertain situation and face many challenges driven by the shortage of healthcare resources, infection prevention mitigations, changes in patient behaviors, and evolving and expanding knowledge about the virus. Virus specific vaccines were introduced in late 2020 to early 2021, which helped to ease healthcare shortage and rates of COVID-19 infection (Oliver et al. 2020; Pilishvili et al. 2021; Self et al. 2021). However, communities around the world have faced intermittent surges of the virus, stemming from the rise of virus variants, relaxation of social distancing and masking mandates, and waning of vaccine immunity (Lemey et al. 2021). With this, there have been many adjustments to the initial COVID-19 pandemic response and many lessons learned (Khosla et al. 2022). In this chapter, we discuss the various challenges faced and adaptations made by sleep medicine services during the COVID-19 pandemic.

19.2 Alterations in Healthcare Resources in Sleep Medicine

Specific to the practice of sleep medicine in the U.S., a number of key workstreams and decisions became paramount from the earliest days of the pandemic. These included the navigation of infection control procedures for clinic spaces, staff areas, as well as laboratory spaces and equipment, all the while integrating emerging scientific data about the virus and balancing staff allocation demands and evolving public health and institutional policies. The goals for sleep medicine, like for many practice areas, focused on keeping patients and staff safe while delivering necessary care. As the pandemic has worn on, this "acute phase" focus has shifted to include vaccination policy with increasing attention to workforce-related issues, such as scarcity and burnout, as well as the impact of disparities on healthcare access and outcomes.

In the initial phase of the pandemic, there was significant acute shortage of healthcare resources. Medical personnel and equipment had to be consolidated and re-allocated to care for sudden rise of COVID-19 patients in various parts of the U.S. and around the world (Emanuel et al. 2020). Many clinics were closed, and physicians in different fields were brought in to care for the high-acuity and rising number of COVID-19 patients. To meet the demand, many places halted non-urgent outpatient clinic services. Patients also stopped going to the doctors due to the fear of COVID-19 spread, and non-COVID-19 related medical care decreased significantly (Jeffery et al. 2020; Grote et al. 2020). This abrupt cessation of services was later mitigated in the pandemic with evolving alternative pathways to care, such as wide adoption of telemedicine as a mainstay of clinical practice (see Chap. 22) (Bokolo 2021; Hollander and Carr 2020).

In addition, many practice areas have faced a shortage of not just personal protective equipment (PPE) for staff, but also respiratory equipment and ventilators for patients due to the rising number of patients with respiratory failure and requiring ventilatory support. This has led to repurposing of outpatient non-invasive ventilators (NIV) and positive airway pressure (PAP) devices to those who need ventilation (Dar et al. 2021). Treatment of urgent respiratory failure was prioritized over chronic sleep disordered breathing, and some patients paused use of their devices. This was further impacted by overall supply shortage during the pandemic due to factory closures, as well as the June, 2021 Philips Respironics voluntary recall in the U.S. of certain models of continuous positive airway pressure (CPAP), bilevel positive airway pressure (BiPAP), and mechanical ventilator devices, estimated to impact over five million devices worldwide (Phillips 2022).

During the early and rapidly evolving days of the pandemic, in light of the resource limitations outlined above, as well as infection control considerations, a number of professional organizations provided up-to-date information and guidance for sleep medicine professionals. Examples of these are the American Academy of Sleep Medicine (AASM), which established a task force and maintained a website providing regularly updated infection control and reopening guidance and links to the latest guidance from the Centers for Disease Control and Prevention (CDC), the U.S. Food and Drug Administration (FDA), Centers of Medicare and Medicaid Services (CMS) and the National Institutes of Health (NIH) (AASM 2022). American Thoracic Society (ATS) published guidance on restoring sleep medicine services and on pediatric sleep (Wilson et al. 2020; Taylor et al. 2021), and the European Respiratory Society (ERS) (Schiza et al. 2021) also published a perspective on an algorithm to assist in reopening of sleep laboratories. These efforts, led by professional societies with substantial sleep medicine membership, tended to allow for alignment across practice locations, though substantial variation in practice existed due to differences in local and regional public health guidance, disease transmission, and later, vaccination related policies and local staffing concerns. These various guidance resources also tended to provide a context for discussion among professionals, something highlighted as potentially important as health care workers varied significantly in their opinions and beliefs regarding sleep medicine services as the pandemic persisted (Johnson et al. 2021a).

19.3 Clinic Services

At the start of the pandemic, all non-emergent medical care was halted, particularly during the stay-at-home orders in late Spring, 2020. Clinics were closed in order to accommodate for hospital personnel and minimize viral spread (Grote et al. 2020). Even after a process of reopening, clinic capacity became more limited in order to adhere to social distancing measures (Wilson et al. 2020; Taylor et al. 2021; Schiza et al. 2021). In the face of these rapid developments, temporary changes to reimbursement, and rapid adaptation of technology across many practice environments allowed transition of in-person visits to telemedicine visits.

Importantly, the US Department of Health and Human Services temporarily waived CMS reimbursement requirements for face-to-face office visits following initiation of PAP therapy in light of COVID-19 pandemic (Center for Medicare and Medicaid Services (CMS) 2021; Shamim-Uzzaman et al. 2021). This allowed for platforms, which could be used at out of center locations for both care providers and patients including at home, to provide some relief to overburdened in-person clinic models, many of whom had been challenged by reduced appointment availability and census due to staff shortages, illness, ventilation requirements, and social distancing.

While the use of telemedicine in clinical sleep practice predated the onset of the pandemic, the uptake was overall spotty and relatively uncommon. The pandemic served to launch a new phase of sleep medicine care for adults and pediatrics alike (Paruthi 2020; Ramar 2020). In 2021, the AASM released an update on its 2015 position paper on the use of telemedicine in clinical sleep medicine, largely in response to the advent of the pandemic (Shamim-Uzzaman et al. 2021). It highlighted the importance of needed research on the efficacy of telemedicine compared to traditional in-person models of care. Additionally, the update pointed out that the bulk of discussion regarding the use of telemedicine in the practice of sleep medicine has centered on the diagnosis and treatment of sleep apnea; much less is known about how telemedicine can play an effective role in the diagnosis and treatment of other common sleep disorders such as insomnia, narcolepsy, restless legs syndrome, parasomnias, and circadian rhythm sleep-wake disorders, though some pathways have been suggested (Irfan et al. 2020). Only after telemedicine has been evaluated in these contexts, evidence-based workflows can fully evolve and be used to diagnose and manage the broader sleep medicine clinical patient populations.

An additional observation regarding sleep telemedicine is that it may allow for broader outreach to those with difficulty physically presenting to clinics for visits. Examples of these include caregivers and patients living in rural areas or with substantial travel requirements to reach clinic, those unable to miss work during the workday, and those with physical and/or medical reasons limiting access in-person visits. However, outside the scope of this chapter, this is not a well-studied area in sleep medicine contexts. It should also be noted that telemedicine may increase barriers for some individuals (Chang et al. 2021) with socio-economic disadvantage, language barrier, physical or cognitive impairment, or those who have difficulty navigating device applications, many of which are not straightforward or user friendly (Sharma et al. 2022).

19.4 Sleep Laboratory Services

Sleep lab operations underwent significant change through the different phases of the pandemic. First and foremost, all non-emergent and non-essential testing were halted during the COVID-19 pandemic, driven by lack of healthcare resources and attempts to mitigate COVID-19 spread in hospital setting (Centers for Disease Control and Prevention 2020; Voulgaris et al. 2020). This especially affected sleep labs, where patients spend prolonged period of time receiving testing and have close contact with sleep technicians during study setup. In fact, one study shows that 93.6% of sleep centers stopped at least one type of sleep study and 90.4% stopped or reduced in-lab studies by at least 90% (Johnson et al. 2021b). Titration studies were also stopped due to concern regarding risk of viral aerosolization (Grote et al. 2020). Various societies in America and Europe (AASM, ERJ, ATS) also recommended postponement of non-urgent studies, especially in areas with high COVID-19 spread (Wilson et al. 2020; Schiza et al. 2021; American Academy of Sleep Medicine 2021a, b; Spicuzza and Sanna 2021).

To minimize the amount of time patients might spend in the lab, technicians utilized telephone visits prior to patient's arrival with the patient to go over study instruction and information. Patients were also screened for COVID-19 symptoms and exposures prior to presenting to the lab and upon arrival to the lab (Ayas et al. 2020a). One study showed that 77.1% of the sleep labs screened for symptoms and about 17.8% labs required COVID-19 polymerase chain reaction (PCR) testing prior to studies (Johnson et al. 2021b).

Arrangements had to be made at the testing site to allow for social distancing in the waiting room, exam room, and technologist rooms. Admission of caregivers during the study also had to be limited, which made it difficult especially for those with disabilities or mobility limitations requiring caregivers (World Health Organization 2020a, b).

Sleep labs also had to adhere to recommended infection prevention procedures for PPE. Initially, full airborne precautions, including protective gown, facemask (N95 or equivalent in filtration), and goggles were recommended. As the prevalence of the virus decreased and scientific field learned more about the mode of COVID-19 viral transmission, some of these requirements were slowly reduced to just facemasks in many places.

Many sleep laboratories also instituted additional protective measures to equipment use, including placement of viral filters at exhalation vents, non-vented facemasks for titrations, and transitioning to disposable parts, such as belts and masks (Thorpy et al. 2020). All equipment was to be stored with protective covers and not be reused, if able. Adequate time in between testing and addition of air filtration system were suggested to allow for adequate ventilation of the rooms (Cilea et al. 2021). Reusable equipment were recommended to undergo thorough disinfection and isolation period (>72 h), and more thorough disinfection and documentation of cleaning procedures were recommended (American Academy of Sleep Medicine 2021a).

Many societies recommended use of home sleep apnea testing (HSAT) or disposable HSAT equipment rather than in-lab studies to minimize disease spread, in appropriate patients (Ayas et al. 2020a; Zhang and Xiao 2020a, b). Use of HSAT for diagnosis of OSA, followed by auto titrating CPAP instead of in-lab CPAP titration, was more widely utilized during the pandemic, as it minimized in-person contact and potential viral exposure (American Academy of Sleep Medicine 2021b; Spicuzza and Sanna 2021). In addition, this was a way for sleep laboratory to continue serving community needs with laboratories and clinics being closed. Many saw advantages of this and continued to utilize this model even after reopening of sleep labs. Disposable HSAT devices, where cleaning and viral transmission can be limited, significantly grew in popularity, leading to increased development and marketing by manufactures.

Interestingly, the pandemic has prompted a fresh look at the role of polysomnography in sleep medicine (Patel and Donovan 2020). This was especially notable in vulnerable populations such as those with immunosuppression, multiple chronic diseases, and in pediatrics where in-lab polysomnography is the gold standard for diagnosing OSA. During the lockdown, some centers have experimented with using HSATs for pediatric studies (Schiza et al. 2021; Jones et al. 2022) and other centers utilized flexibilities available during the public health emergency to provide care in the absence of PSG (Gupta et al. 2021).

19.5 Positive Airway Pressure

The shortage in ventilators and repurposing of PAP devices to meet high demand for ventilators and the risks associated with aerosolization from PAP use also significantly affected the way OSA is treated (Simonds et al. 2010). In certain parts of the world, PAP was distributed based on the severity of the disease, with priority given to those with more severe disease. Many providers also turned to non-PAP treatment, such as oral mandibular devices or positional therapy, during device shortage.

Persistent concern regarding the potential for viral aerosolization and spread using PAP also limited the use of PAP devices at home and in hospital settings, with a variety of mitigations suggested depending upon circumstances and necessity (Kryger and Thomas 2020; Voulgaris et al. 2020; Ayas et al. 2020b).

Both the providers and patients increased the use of remote cloud systems to assess PAP adherence and usage (Spicuzza and Sanna 2021). These monitoring systems provide information regarding compliance, pressure settings, leaks, and residual apnea hypopnea index, and can help troubleshoot any problems patients may face, especially in setting of self-driven CPAP setup and mask fitting. These can also help improve overall compliance and treatment effectiveness (Spicuzza and Sanna 2021; Chumpangern et al. 2021; Schutte-Rodin 2020).

19.6 Durable Medical Equipment (DME) Companies

In the United States, DME companies are the main provider of PAP, NIV, and all related supply distribution. Prior to the pandemic, patients were often setup for their initial PAP device, mask, and supplies at DME offices or in-house visits. Ideally, this in-person contact allowed for some training on usage and maintenance of the PAP device, as well as mask fitting. This practice was replaced by virtual (phone or video call) setup during the pandemic to minimize in-person contact. Packages containing multiple mask sizes to self-fit masks and various software that virtually estimate best mask fit for patients were also developed and distributed by various companies (Ayas et al. 2020a; Tran et al. 2021; Ma et al. 2021). With increase in device demand, especially for autoCPAP after OSA diagnosis, and conversion to virtual care, DME companies also saw increased call volumes, and some experienced delays in patient care coordination (Pusalavidyasagar et al. 2020). Unlike telemedicine services, recent data shows that the CPAP adherence rate was lower in those who received virtual setup compared to in-center setup (Stanchina et al. 2022). More research is needed to further optimize utilization of non-in-person equipment distribution and fitting.

19.7 Disparity, Sleep Medicine, and Overall Impact of the Pandemic

The population health consequences of the Covid-19 pandemic are varied, complex, and still being understood (Bann et al. 2021). Access to care, health-impacting behaviors and responses, job security, environment, and other drivers of health inequalities may all conspire to interact with pandemic impacts to drive sleep health disparity (Johnson et al. 2018, 2022; Alhasan et al. 2022). While incompletely understood, it is emerging that some patient populations with sleep disorders have been disproportionally affected by the changes wrought by the pandemic (Billings et al. 2020). Despite optimistic observations about the nimble response of the field of sleep medicine to meet the challenges presented by the pandemic, there is growing evidence that the overall impact of the pandemic has reduced availability of sleep medicine services (Grote et al. 2021), as well as treatments (Sunkonkit et al. 2022), a concerning observation for a field already stretched to fill gaps and meet patient needs. Sleep medicine workforce challenges also exist, including pandemic-augmented concerns regarding healthcare worker burnout, which itself has complex systemic drivers and which has been termed a crisis by the U.S. Surgeon General in May 2022 (U.S. Department of Health and Human Services 2022).

19.8 Summary and Future Directions

The Covid-19 pandemic fundamentally challenged Sleep Medicine—upending usual practice, human, physical, laboratory and testing resources, balance sheets, logistics, and referring partners and equipment services. The novelty and severity of the virus initially caused significant disruption healthcare system. Now, with changes in viral prevalence and virulence, the advent of effective vaccination, use of masking and appropriate PPE, ventilation and other mitigations, and updated payer rules allowing for remote care, the field of sleep medicine has continued to adapt.

That said, much remains to be learned and more work is needed to understand the outcomes of the sleep field's adaptations to best prepare for uncertainty in the future. For example, investigation and research regarding all aspects of novel models of care introduced in the pandemic are needed. Importantly, outcomes for all those with sleep disorders—including those who may be vulnerable—require research and inquiry, to understand the ways in which the recent flurry of pandemic-induces adaptations may have impacted patient outcomes. Additionally, updated payor requirements, delivery models, hospital engagement, regulatory policy, and workforce efforts required to back up such systemic change are all needed.

More research is needed regarding models of care, including group appointments, remote testing in certain nontraditional populations, the impact of telehealth on access and outcomes, resource mobilization, and identification of and efforts to reduce barriers to care. In addition, the roles of industry and consumer technology are important and need further research in how consumer sleep technologies, data generated by medical devices, wearables and nearables, may be incorporated into algorithms for remote patient monitoring.

Sleep medicine as a field has proven to be flexible, dynamic, and creative in finding ways to meet the evolving needs of patients despite the challenges presented over the past 2 years. Work is far from over, and perhaps just begun, in understanding the influence of the pandemic on sleep medicine and sleep health at the populations level.

References

AASM (2022). https://aasm.org/covid-19-resources/ https://aasm.org/covid-19-resources/. Accessed 31 July 2022

Alhasan DM, Gaston SA, Jackson CL (2022) Investigate the complexities of environmental determinants of sleep health disparities. Sleep 23:zsac145

American Academy of Sleep Medicine (2021a) Summary of CDC recommendations relevant for sleep practices during COVID-19. https://aasm.org/covid-19-resources/covid-19-mitigation-strategies-sleep-clinics-labs. Accessed 15 July 2020

American Academy of Sleep Medicine (2021b). https://aasm.org/covid-19-resources/considerations-practice-sleep-medicine. Accessed 15 July 2020

Ayas NT, Fraser KL, Giannouli E, Hanly PJ, Kendzerska T, Katz SL, Lachmann BN, Lajoie A, Minville C, Morrison D, Narang I, Povitz M, Skomro R, Spurr KF (2020a) Key highlights from the Canadian Thoracic Society's position statement on optimizing the management of sleep disordered breathing during the coronavirus disease 2019 pandemic. Chest 158(3):899–900

Ayas NT, Fraser KL, Giannouli E et al (2020b) Helping Canadian health care providers to optimize Sleep Disordered Breathing management for their patients during the COVID-19 pandemic. Can J Resp Crit Care Sleep Med. https://doi.org/10.1080/24745332.2020.1758442

Bann D, Villadsen A, Maddock J, Hughes A, Ploubidis GB, Silverwood R, Patalay P (2021) Changes in the behavioural determinants of health during the COVID-19 pandemic: gender, socioeconomic and ethnic inequalities in five British cohort studies. J Epidemiol Community Health 75(12):1136–1142

Billings ME, Hale L, Johnson DA (2020) Physical and social environment relationship with sleep health and disorders. Chest 157(5):1304–1312

Bokolo AJ (2021) Application of telemedicine and eHealth technology for clinical services in response to COVID-19 pandemic. Health Technol (Berl) 11(2):359–366. https://doi.org/10.1007/s12553-020-00516-4

Center for Medicare and Medicaid Services (CMS) (2021) COVID-19 emergency declaration blanket waivers for health care providers. https://www.cms.gov/files/document/summary-covid-19-emergency-declaration-waivers.pdf. Accessed 7 July 2022

Centers for Disease Control and Prevention (2020) CDC prevention control recommendations 2019. Social distancing: keep a safe distance to slow the spread. https://www.cdc.gov/coronavirus/2019-ncov/preventgetting-sick/social-distancing.html

Chang JE, Lai AY, Gupta A, Nguyen AM, Berry CA, Shelley DR (2021) Rapid transition to telehealth and the digital divide: implications for primary care access and equity in a post-COVID era. Milbank Q 99(2):340–368

Chumpangern W, Muntham D, Chirakalwasan N (2021) Efficacy of a telemonitoring system in continuous positive airway pressure therapy in Asian obstructive sleep apnea. J Clin Sleep Med 17(1):23–29

Cilea R, Guaraldi P, Barletta G, Lucchi M, Baschieri F, Baldelli L, Loddo G, Rocca A, Cortelli P, Calandra-Buonaura G, Provini F (2021) Performing sleep studies after the COVID-19 outbreak: practical suggestions from Bologna's sleep unit. Sleep Med 77:45–50. https://doi.org/10.1016/j.sleep.2020.11.005. Epub 2020 Nov 7. PMID: 33310113; PMCID: PMC7837108

Dar M, Swamy L, Gavin D, Theodore A (2021) Mechanical-ventilation supply and options for the COVID-19 pandemic. Leveraging all available resources for a limited resource in a crisis. Ann Am Thorac Soc 18(3):408–416. https://doi.org/10.1513/AnnalsATS.202004-317CME. PMID: 33202144; PMCID: PMC7919160

Emanuel EJ, Persad G, Upshur R, Thome B, Parker M, Glickman A, Zhang C, Boyle C, Smith M, Phillips JP (2020) Fair allocation of scarce medical resources in the time of Covid-19. N Engl J Med 382(21):2049–2055

Grote L, McNicholas WT, Hedner J, ESADA collaborators (2020) Sleep apnoea management in Europe during the COVID-19 pandemic: data from the European Sleep Apnoea Database (ESADA). Eur Respir J 55(6):2001323

Grote L, Theorell-Haglöw J, Ulander M, Hedner J (2021) Prolonged effects of the COVID-19 pandemic on sleep medicine services-longitudinal data from the Swedish Sleep Apnea Registry. Sleep Med Clin 16(3):409–416

Gupta M, Ish P, Chakrabarti S, Sen MK, Mishra P, Suri JC (2021) Diagnostic accuracy and feasibility of portable sleep monitoring in patients with obstructive sleep apnea: re-exploring the utility in the current COVID-19 pandemic. Monaldi Arch Chest Dis 92(1)

Hollander JE, Carr BG (2020) Virtually perfect? Telemedicine for Covid-19. N Engl J Med 382(18):1679–1681

Irfan M, Khalil W, Iber C (2020) Practicing sleep medicine amidst a pandemic: a paradigm shift. J Clin Sleep Med 16(8):1405–1408

Jeffery MM, D'Onofrio G, Paek H et al (2020) Trends in emergency department visits and hospital admissions in health care systems in 5 states in the first months of the COVID-19 pandemic in the US. JAMA Intern Med 180(10):1328–1333

Johnson DA, Billings ME, Hale L (2018) Environmental determinants of insufficient sleep and sleep disorders: implications for population health. Curr Epidemiol Rep 5(2):61–69

Johnson KG, Sullivan SS, Rastegar V, Gurubhagavatula I (2021a) Sleep medicine health-care worker concerns about COVID-19: an early pandemic survey. Respir Care 66(11):1729–1738

Johnson KG, Sullivan SS, Nti A, Rastegar V, Gurubhagavatula I (2021b) The impact of the COVID-19 pandemic on sleep medicine practices. J Clin Sleep Med 17(1):79–87

Johnson DA, Ohanele C, Alcántara C, Jackson CL (2022) The need for social and environmental determinants of health research to understand and intervene on racial/ethnic disparities in obstructive sleep apnea. Clin Chest Med 43(2):199–216

Jones S, Hanwell R, Chowdhury T, Orgill J, van den Eshof K, Farquhar M, Joseph D, Gringras P, Trucco F (2022) Feasibility and parental perception of home sleep studies during COVID-19: a tertiary sleep centre experience. Arch Dis Child 107(2):189–191. https://doi.org/10.1136/archdischild-2021-322184. Epub 2021 Sep 22. PMID: 34551900

Khosla S, Beam E, Berneking M, Cheung J, Epstein LJ, Meyer BJ, Ramar K, So JY, Sullivan SS, Wolfe LF, Gurubhagavatula I (2022) The COVID-19 pandemic and sleep medicine: a look back and a look ahead. J Clin Sleep Med 18(8):2045

Kryger MH, Thomas R (2020) Home PAP devices in patients infected with COVID-19. J Clin Sleep Med 16(7):1217–1219

Lemey P, Ruktanonchai N, Hong SL, Colizza V, Poletto C, Van den Broeck F, Gill MS, Ji X, Levasseur A, Oude Munnink BB, Koopmans M, Sadilek A, Lai S, Tatem AJ, Baele G, Suchard MA, Dellicour S (2021) Untangling introductions and persistence in COVID-19 resurgence in Europe. Nature 595(7869):713–717

Ma Z, Hyde P, Drinnan M, Munguia J (2021) Development of a smart-fit system for CPAP interface selection. Proc Inst Mech Eng H 235(1):44–53. https://doi.org/10.1177/0954411920959879. Epub 2020 Sep 28. PMID: 32988316; PMCID: PMC7780270

Oliver S, Gargano J, Marin M et al (2020) The Advisory Committee on Immunization Practices' interim recommendation for use of Pfizer-BioNTech COVID-19 vaccine—United States, December 2020. MMWR Morb Mortal Wkly Rep 69:1922–1924. https://doi.org/10.15585/mmwr.mm6950e2

Paruthi S (2020) Telemedicine in pediatric sleep. Sleep Med Clin 15(3S):e1–e7

Patel SR, Donovan LM (2020) The COVID-19 pandemic presents an opportunity to reassess the value of polysomnography. Am J Respir Crit Care Med 202(3):309–310

Paules CI, Marston HD, Fauci AS (2020) Coronavirus infections-more than just the common cold. JAMA 323(8):707–708

Phillips (2022). https://www.usa.philips.com/healthcare/e/sleep/communications/src-update. Accessed 31 July 2022

Pilishvili T, Fleming-Dutra KE, Farrar JL et al (2021) Interim Estimates of Vaccine Effectiveness of Pfizer-BioNTech and Moderna COVID-19 Vaccines Among Health Care Personnel — 33 U.-S. Sites, January–March 2021. MMWR Morb Mortal Wkly Rep 70:753–758. https://doi.org/10.15585/mmwr.mm7020e2

Pusalavidyasagar S, Silbernick E, Stachel E, Xiong M, Iber C (2020) Sleep medicine durable medical equipment management during the COVID-19 pandemic: our center's initial experience. J Clin Sleep Med 16(9):1637–1638. https://doi.org/10.5664/jcsm.8638. PMID: 32620191; PMCID: PMC7970592

Ramar K (2020) AASM takes the pulse of the sleep field and responds to COVID-19. J Clin Sleep Med 16(11):1939–1942

Schiza S, Simonds A, Randerath W et al (2021) Sleep laboratories reopening and COVID-19: a European perspective. Eur Respir J 57:2002722

Schutte-Rodin S (2020) Telehealth, telemedicine and obstructive sleep apnea. Sleep Med Clin 15(3):359–375

Self WH, Tenforde MW, Rhoads JP et al (2021) Comparative effectiveness of Moderna, Pfizer-BioNTech, and Janssen (Johnson & Johnson) vaccines in preventing COVID-19 hospitalizations among adults without immunocompromising conditions—United States, March–August 2021. MMWR Morb Mortal Wkly Rep 70:1337–1343. https://doi.org/10.15585/mmwr.mm7038e1externalicon

Shamim-Uzzaman QA, Bae CJ, Ehsan Z, Setty AR, Devine M, Dhankikar S, Donskoy I, Fields B, Hearn H, Hwang D, Jain V, Kelley D, Kirsch DB, Martin W, Troester M, Trotti LM, Won CH, Epstein LJ (2021) The use of telemedicine for the diagnosis and treatment of sleep disorders: an American Academy of Sleep Medicine update. J Clin Sleep Med 17(5):1103–1107

Sharma AE, Khoong EC, Sierra M, Rivadeneira NA, Nijagal MA, Su G, Lyles CR, DeFries T, Tuot DS, Sarkar U (2022) System-level factors associated with telephone and video visit use: survey of safety-net clinicians during the early phase of the COVID-19 pandemic. JMIR Form Res 6(3):e34088

Simonds AK, Hanak A, Chatwin M, Morrell M, Hall A, Parker KH, Siggers JH, Dickinson RJ (2010) Evaluation of droplet dispersion during non-invasive ventilation, oxygen therapy, nebuliser treatment and chest physiotherapy in clinical practice: implications for management of pandemic influenza and other airborne infections. Health Technol Assess 14(46):131–172. https://doi.org/10.3310/hta14460-02. PMID: 20923611

Spicuzza L, Sanna A (2021) Continuum of care for patients with obstructive sleep apnea after one year from the COVID-19 pandemic onset: no time for further delays: practical issues for a safe and effective management. Sleep Med 84:98–106. https://doi.org/10.1016/j.sleep.2021.05.026. Epub 2021 May 31. PMID: 34144451; PMCID: PMC8166158

Stanchina M, Lincoln J, Prenda S, Holt M, Leon I, Donat W, Corrao W, Jabbour E, Koenig S, Malhotra A (2022) The impact of different CPAP delivery approaches on nightly adherence and discontinuation rate in patients with obstructive sleep apnea. J Clin Sleep Med 18(8):2023–2027. https://doi.org/10.5664/jcsm.10038. PMID: 35499279

Sunkonkit K, Selvadurai S, Voutsas G, Benzon D, Baker A, Trinh M, Narang I (2022) To wear or not wear the mask: decline in positive airway pressure usage in children with sleep disordered breathing during the COVID-19 pandemic. Nat Sci Sleep 14:621–633

Taylor JB, Oermann CM, Deterding RR, Redding G, Davis SD, Piccione J, Moore PE, Kupfer O, Santiago MT, Rosenfeld M, Ingram DG, Ross K, DeBoer EM (2021) Innovating and adapting in pediatric pulmonology and sleep medicine during the COVID-19 pandemic: ATS pediatric assembly web committee consensus statement for initial COVID-19 virtual response. Pediatr Pulmonol 56(2):539–550

Thorpy M, Figuera-Losada M, Ahmed I, Monderer R, Petrisko M, Martin C, Akhtar J, Thorpy J, Haines C (2020) Management of sleep apnea in New York City during the COVID-19 pandemic. Sleep Med 74:86–90. https://doi.org/10.1016/j.sleep.2020.07.013. Epub 2020 Jul 18. PMID: 32841850; PMCID: PMC7368649

Tran T, Nonoyama M, Cithiravel N, Syed F, Janevski J, Chiang J, Amin R (2021) Virtual mask fitting in pediatric patients during COVID-19: a case series. Can J Respir Ther 57:93–98. https://doi.org/10.29390/cjrt-2021-023. PMID: 34345656; PMCID: PMC8302071

U.S. Department of Health and Human Services (2022). https://www.hhs.gov/surgeongeneral/priorities/health-worker-burnout/index.html. Accessed 31 July 2022

Voulgaris A, Ferini-Strambi L, Steiropoulos P (2020) Sleep medicine and COVID-19. Has a new era begun? Sleep Med 73:170–176

Wilson KC, Kaminsky DA, Michaud G, Sharma S, Nici L, Folz RJ, Barjaktarevic I, Bhakta NR, Cheng G, Chupp GL, Cole A, Dixon AE, Finigan JH, Graham B, Hallstrand TS, Haynes J, Hankinson J, MacIntyre N, Mandel J, McCarthy K, McCormack M, Patil SP, Rosenfeld M, Senitko M, Sethi S, Swenson ER, Stanojevic S, Teodorescu M, Weiner DJ, Wiener RS, Powell

CA (2020) Restoring pulmonary and sleep services as the COVID-19 pandemic lessens. From an Association of Pulmonary, Critical Care, and Sleep Division Directors and American Thoracic Society-coordinated Task Force. Ann Am Thorac Soc 17(11):1343–1351

World Health Organization (2020a) Operational considerations for case management of COVID-19 in health facility and community: interim guidance, 19 March 2020. World Health Organization. https://apps.who.int/iris/handle/10665/331492. Accessed 20 July 2022

World Health Organization (2020b) Clinical management of COVID-19: interim guidance, 27 May 2020. World Health Organization. https://apps.who.int/iris/handle/10665/332196. Accessed 20 July 2022

Zhang XL, Xiao Y (2020a) Sleep health service in China during the COVID-19 outbreak. J Clin Sleep Med. https://doi.org/10.5664/jcsm.8472

Zhang XL, Xiao Y (2020b) Sleep health service in China during the coronavirus disease outbreak. J Clin Sleep Med 16(7):1221–1222. https://doi.org/10.5664/jcsm.8472. PMID: 32248894; PMCID: PMC7954045

Zhu N, Zhang D, Wang W, Li X, Yang B, Song J, Zhao X, Huang B, Shi W, Lu R, Niu P, Zhan F, Ma X, Wang D, Xu W, Wu G, Gao GF, Tan W, China Novel Coronavirus Investigating and Research Team (2020) A novel coronavirus from patients with pneumonia in China, 2019. N Engl J Med 382(8):727–733

Chapter 20
Protocols for Sleep Medicine Services During COVID-19 and Pandemics

Rouf Pirzada and Ahmed S. BaHammam

Abstract Coronavirus disease-2019 (COVID-19) brought almost all healthcare systems worldwide to a breakdown or near breakdown, prompting the suspension of elective services like sleep medicine services and reflecting unpreparedness for providing the service during pandemics. However, over time, COVID-19 proved to be a steadfast settler, running a protracted epidemiological course with crests and troughs. Eventually, the suspension of elective services, like sleep medicine, added to overall COVID-19 mortality and morbidity due to additive effects. Therefore, the focus shifted from suspension to resumption, with a concept of "living with COVID-19" and working despite COVID-19; however, it warrants a guarded approach and robust preparation. This chapter discusses proposed protocols for continuing sleep medicine services during pandemics relying on the experience gained during the COVID-19 pandemic.

Keywords COVID-19 · Sleep study · Polysomnography · Home sleep apnea test

20.1 Introduction

The COVID-19 pandemic brought almost everything around us, including healthcare systems, to near breakdown due to rapid spread and novel characteristics (Ferrara and Albano 2020; Geyman 2021). The magnitude of morbidity and mortality puts the healthcare system under immense stress. It demonstrated crests and

---

R. Pirzada
North Cumbria Integrated Care (NCIC), NHS, Carlisle, UK
e-mail: Pirzada.baba@ncic.nhs.uk

A. S. BaHammam (✉)
The University Sleep Disorders Center, Department of Medicine, College of Medicine, King Saud University, Riyadh, Saudi Arabia

Strategic Technologies Program of the National Plan for Sciences and Technology and Innovation in the Kingdom of Saudi Arabia, King Saud University, Riyadh, Saudi Arabia
e-mail: ashammam@ksu.edu.sa

A. S. BaHammam et al. (eds.), *COVID-19 and Sleep: A Global Outlook*, Progress in Sleep Research, https://doi.org/10.1007/978-981-99-0240-8_20

troughs resulting in healthcare collapse or near collapse during crests in many countries. Healthcare systems struggled, but nevertheless responded appropriately by suspending nearly all elective services during earlier phases of the pandemic, concentrating mainly on acute services, and infusing practical contingency plans at all levels (Geyman 2021; Moynihan et al. 2021; Pagel 2022; Alfadda et al. 2021). The novel nature of the virus, in terms of pathogenesis, rapid transmission, and high mortality and morbidity compounded by the global patient volume, created disturbances in the services provided worldwide (Moynihan et al. 2021; Johnson et al. 2021). At epidemiological peaks, it was a virtual overhaul of healthcare services in the form of staff rationing, staff relocations, staff sickness breaks, and winding up of outpatient services (later taken over by virtual and telephonic clinics) (Garfan et al. 2021; Vas et al. 2022; Webster 2020). However, all exigent plans provided temporary and partial respite.

During the course of the current pandemic, COVID-19 demonstrated a protracted course with epidemiological rises and falls, primarily due to mutations, thus stipulating dynamic planning to correspond to epidemiological demands (Pagel 2022). During this time, the elective backlog increased exponentially, and the association of the backlog with COVID-19 proved to be additive to overall COVID-19 morbidity and mortality (Johnson et al. 2021; Bagenal 2022; Strausz et al. 2021; Center disease prevention and control (CDC) 2019; Sahni and Cao 2021; Pena Orbea et al. 2021; Clark et al. 2020). Thus, there was organization and reorganization to re-establish full or partial elective care despite COVID-19. Throughout the course, a need to devise a strategy and plan to offer optimal or near-optimal elective healthcare services like sleep medicine services despite COVID-19 came forth. These strategies evolved with a gradual understanding of COVID-19 pathogenesis, identifying high-risk cohorts (Center disease prevention and control (CDC) 2019). In addition, rolling out multiple vaccines markedly changed the scenario by mitigating morbidity and mortality, particularly in the high-risk population (Holstiege et al. 2021).

Formulating a plan to offer elective services like sleep medicine services despite pandemics involves a multipronged approach at the hospital and community levels as well as taking epidemiological input from local and international health authorities (Pirzada et al. 2020a). Hospitals need to prepare logistics and staffing; health authorities must closely watch the epidemiological dynamics of the pandemic (Pirzada et al. 2020a; Schiza et al. 2021). It is imperative to mention that guidelines and advisories cannot be generalized as different countries, different sleep societies, and infection control bodies develop, recommend, and implement different, although fundamentally the same, recommendations according to the local needs.

Elective services in sleep medicine are outpatient and inpatient services for evaluation, diagnostics, therapeutics, and follow-up. Planning to initiate these services needs stepwise planning. This chapter discusses the proposed protocols for sleep medicine services during pandemics taking the current COVID-19 pandemic as a model to build on for possible future pandemics. The first step can be defined as center preparedness followed by an epidemiological allowance assessment; these two work together in tandem, taking the lead from the epidemiological situation and assessing center preparedness to commence or suspend sleep medicine services

(Pirzada et al. 2020a; Schiza et al. 2021). Nevertheless, uncompromising diligence in following the policies and protocols is the goal for sleep center mitigation. Figure 20.1 shows a proposed algorithm for sleep center mitigation strategies during pandemics.

20.2 Center Preparedness

The center preparedness encompasses both staffing and logistics with contingency planning. Center preparedness must be optimal and target defined. The basics of center preparedness are to follow infection control guidelines accordingly, train the staff, and procure the logistics (Pirzada et al. 2020a; Bielicki et al. 2020; American Academy of Sleep Medicine 2022a).

Sleep disorders' center preparedness encompasses logistics (healthcare resources) and staff education and training to provide optimal services without compromising infection control guidelines whenever planning sleep diagnostics and therapeutics (Pirzada et al. 2020a; American Academy of Sleep Medicine 2022a).

20.2.1 Staff Training, Logistics, and Infection Control

Collaboration/coordination is the key; thus, each center should designate a staff member, usually from the nonclinical workforce, to coordinate center preparedness. The designated person will be responsible for regularly monitoring the pandemic updates from the state, local health departments, and hospital managerial body to update and coordinate demands; nonetheless, report and respond to all the deficiencies observed. A designated staff member would thus act as a key as well as a link in a designated hub (Pirzada et al. 2020a; American Academy of Sleep Medicine 2022a). In addition, the designated person will keep a record of preparedness.

20.2.1.1 Staff Training, Mitigation, and Welfare

The staff in the center should have training regarding infection control measures and protocols, with regular updates about them. Staff well-being must be ensured; it can be achieved by monitoring staff workloads that might reflect the impact of increased duties due to the pandemic and reduced staffing levels due to illness, quarantine, or redeployment to other frontline services. Additional shift burden on the staff must be minimized by curtailing the services (Ferrara and Albano 2020; Bielicki et al. 2020; Gross et al. 2021; Sanghera et al. 2020; Centre for disease control and prevention 2022; Centre for disease prevention and control 2022). For example, identifying vulnerable staff to get COVID-19 and severe COVID-19, as a part of the mitigation strategy to run the services smoothly, helps in staff allocations and diversion.

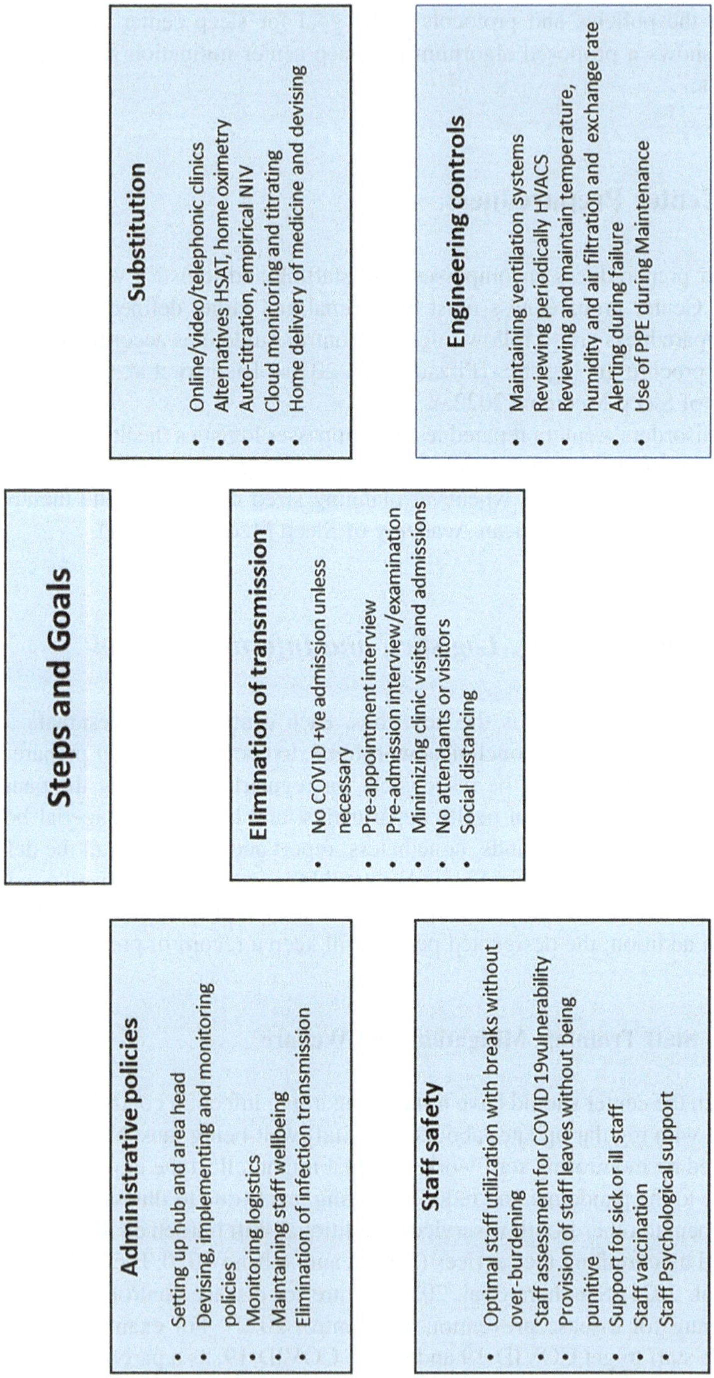

Fig. 20.1 A proposed algorithm for sleep center mitigation strategies during pandemics

Hospital policies meant for routine hospital functioning need to be revisited to encourage staff to report sickness without being punitive (Gross et al. 2021; Nguyen et al. 2020). Vaccination of the staff needs to be encouraged, incentivized, and implemented to attain goals. Vaccination information, benefits, and center locations should be advertised through all forms of media in hospitals and outside. Vaccination of staff taking high-risk procedures like aerosol-generating procedures should be prioritized. Additionally, all the staff must be provided with adequate workplace support, including psychological well-being (Pirzada et al. 2020a; Bielicki et al. 2020; American Academy of Sleep Medicine 2022a; Gross et al. 2021; Sanghera et al. 2020; Australian Sleep Association 2020; WHO 2022).

All the staff should be trained to follow infection control protocols, like donning and doffing of personal protective equipment (PPE). In addition to practicing judicious use, it should be ensured that the staff is educated about the indications of use, i.e., when and where PPE use is indicated, like handling aerosol-generating procedures or cleaning used sleep study equipment (Schiza et al. 2021; WHO 2022). Furthermore, it is essential to actively monitor and secure PPE supplies (Pirzada et al. 2020a; American Academy of Sleep Medicine 2022a; Australian Sleep Association 2020). Finally, it is important to understand that at epidemiological peaks, logistic diversion occurs, which might impair the supply chain for the sleep disorder center, with an exigent diversion from infection control norms that is discouraged (Toomey et al. 2021).

Staff should be well educated and trained about screening procedures and protocols to ensure optimal physical screening as well as screening of documentary/electronic proofs of well-being and vaccination status. In addition, telephonic/remote screening, which is usually done 24–48 h before an appointment, can be challenging at times; thus, staff needs to be trained to get information optimally without being coercive or intimidating (Pirzada et al. 2020a; American Academy of Sleep Medicine 2022a,b; Center for disease prevention and control 2022; Gupta et al. 2020).

Entry point screening for staff is also recommended. Staff should be encouraged to reveal health issues, particularly respiratory symptoms. Staff should be encouraged to communicate about symptoms through the telephone to avoid contact; besides, advisories need to be practical without being punitive (Gross et al. 2021; Center for disease prevention and control 2022).

20.2.1.2 Logistics: Supply, Monitoring, and Optimal Use

The staff in a sleep disorder center should be provided with all the necessary PPE, including a surgical gown, apron, gloves, and face mask, to manage all patients as potentially infected (American Academy of Sleep Medicine 2022a; Center for disease prevention and control 2022). It is more critical while taking a potential aerosol-generating procedure, where a face mask FFP2 (filtering facepiece) or higher or N95 mask and face shield or goggles are recommended; a face mask needs to be prescribed according to the "fit test" (Pirzada et al. 2020a; American Academy of Sleep Medicine 2022a; Smith et al. 2016). Wearing and removing PPE need to be performed following specific procedures with the help of an assistant to avoid contamination (Australian Sleep Association 2020; Cilea et al. 2021; Bartoszko et al. 2020). All healthcare staff must be adequately trained on PPE and educated to follow hospital disease control and prevention protocols, including ambient and medical device sanitization (Ong et al. 2020; Ferioli et al. 2020; Grote et al. 2020; Verbeek et al. 2020).

20.2.1.3 Infection Control: From Equipment Handling to Screening

Staff assigned to use and clean medical equipment, whether reusable or nonreusable, should be trained in handling and cleaning (Pirzada et al. 2020a). In addition, they should be trained to identify any breach in packaging or cleaning protocols. Generally, local, national, or international infection control protocols and procedures are followed for cleaning and inspecting all patient-related equipment; however, for cleaning reusable equipment, the manufacturer guidelines should be followed to maintain the integrity of the equipment (Pirzada et al. 2020a; American Academy of Sleep Medicine 2022a; Center for disease prevention and control 2022). A common method of cleaning reusable equipment is to use mild household detergent; for disinfection, 70–90% isopropyl alcohol, 8.25–10% chlorine bleach, or 2% glutaraldehyde besides less popular options like heat sterilization and hydrogen peroxide gas plasma can be used (Center for disease prevention and control 2022; Philips 2022; ResMed 2022a). As a principle to break the chain, all reusable equipment or its components should be removed from the services for 72 h before the next use, even after the disinfection procedure.

The conclusion has been drawn from the hypothesis that, via the nasal cannula, the virus may reach the nasal pressure transducer, which cannot be sterilized. To avoid the risk, the American Academy of Sleep Medicine (AASM) recommends no use of the polygraph for 72 h between tests on different patients (American Academy of Sleep Medicine 2022a; Technologies 2011; Nonin 2022). This can pose a considerable limitation, especially considering the high demand after the initial suspension of sleep studies. Low-cost ultraviolet C-machines have been proposed to sterilize non-disposable equipment as an alternative to standard hospital sanitization procedures (Australian Sleep Association 2020). Nevertheless, single-use, disposable equipment is commercially available for sleep recordings, but the cost-effectiveness is a limitation. Other strategies to improve safety are based on substituting nasal cannula/thermistor with the use of respiratory inductive plethysmography (RIP), but this limits the diagnostic information (Australian Sleep Association 2020; Cilea et al. 2021).

Once indicated and planned, taking up a titration study is the most challenging step because of droplet/aerosol dispersion generated during positive airway pressure (PAP) therapy (Pirzada et al. 2020b; Schweller 2021). In-hospital titration should only be taken in negative-pressure rooms with full PPE. Thus, using automatic continuous positive airway pressure (CPAP) devices with remote monitoring is preferred in OSA patients. It is preferred that the same PAP device used for automatic titration should be prescribed to the patient to avoid the use of a single ventilator for titration in different patients. In many sleep units, a fixed-pressure device is provided for titration, and pressure adjustment is performed remotely by telemonitoring (Australian Sleep Association 2020; Schweller 2021). When vented masks are used, droplet generation and dispersion might represent a real, preventable risk. To circumvent the risk, it is proposed to use a circuit with non-vented masks, with two antibacterial/antiviral filters and a safety valve, especially in cases of relevant epidemiological risk (Australian Sleep Association 2020; Gupta et al. 2020; Pirzada et al. 2020b; Schweller 2021; Wilson et al. 2020).

Other common strategies to avoid the risk of aerosolization are to show the patient how to put on the mask first without turning on the machine, and not use a humidifier for in-laboratory titration, besides constant use of full or extended personal protective equipment (PPE), including N95 masks, by health professionals. Home care personnel should follow the same principles (Australian Sleep Association 2020; Cilea et al. 2021; Pirzada et al. 2020b).

Although empirical noninvasive ventilation (NIV) initiation is advised, remote titration is preferred for NIV titration. Nevertheless, for therapeutically exigent situations, the same precautions as with in-hospital CPAP titration would apply. In addition to that, transcutaneous capnography should be obtained using disposable probes or transcutaneous measurement via ear clip to minimize exposure (American Academy of Sleep Medicine 2022a,b; Australian Sleep Association 2020; Cilea et al. 2021).

20.2.1.4 Sleep Center Screening and Hygiene Protocols

To ensure clean and hygienic patient areas and workstations, we need to follow local infection control guidelines; nevertheless, general principles for cleaning contact surfaces like bedposts and tables after cleaning for any visible contamination with soap water/soap-soaked cloth, disinfection with ethanol 70–90%, chorine-based products (0.1–0.5%), or 0.5% hydrogen peroxide are recommended (WHO 2022; Center for disease prevention and control 2022).

Patient screening protocols need to be implemented in letter and spirit. Pre-appointment screening, usually telephonic, is done 24–48 h before admission or arrival to ascertain fitness to attend; this is the backbone of the screening procedure. In many centers, patients are asked to get COVID-19 PCR 24 h before admission, particularly for aerosol-generating procedures. In an almost similar fashion, many centers ask for COVID-19 vaccination status prior to admission. Patients are given online information about guidelines and procedures, like physical and time gaps. All this pre-admission exercise is no shortcut for admission-time screening. Admission or point-of-care screening is to be arranged at entry points. Patient appointments should be arranged to allow adequate time and space gaps. Attendants or caregivers are discouraged, and if translation services are needed, remote services are preferred (Pirzada et al. 2020a; American Academy of Sleep Medicine 2022a; Australian Sleep Association 2020; CDC 2022; Serra 2021).

Hospital engineering departments have an important role to play in maintaining workplace hygiene. It is achieved by active monitoring of HVACs (heating, ventilation, and air-conditioning) and other ventilation systems. The aim is to maintain air exchange rates, ambient temperature, and humidity. It is done by maintaining airflow and filtration systems. Nonetheless, engineering staff assigned to this job also should be trained in PPE use and indication since ventilation systems are also the source of potential transmission (Sodiq et al. 2021).

20.3 Epidemiological Inputs and Sleep Medicine Inpatient Services

After ensuring center preparedness, the second step is to take epidemiological leads to plan sleep medicine services. The first principle remains to postpone and reschedule all sleep studies for the patients who are COVID-19 positive, or COVID-19 suspects. Moreover, it is imperative that no penalties are imposed on patients for cancellations and missed appointments related to respiratory illness, and rescheduling should be readily available to encourage patients to declare symptoms (Pirzada et al. 2020a; American Academy of Sleep Medicine 2022a; Australian Sleep Association 2020; Gupta et al. 2020; Wilson et al. 2020). Figure 20.2 shows a proposed epidemiology-based algorithm for a stepwise approach to sleep studies during pandemics.

The epidemiological status of any city/province/country is always a dynamic situation and thus warrants day-to-day assessment and subsequent planning and decision-making. In the scenario of a high epidemic situation, there is large-scale community transmission, healthcare staffing is significantly impacted, and multiple cases within the community are reported daily. For the high-epidemic areas, sleep studies are to be suspended, except in case of emergency, after weighing the risk-benefit ratio by the treating physician; it is always a difficult decision. Emergency sleep studies are mostly performed for complex cases where the therapeutic requirement is overwhelming (Pirzada et al. 2020a; Schiza et al. 2021; American Academy of Sleep Medicine 2022a; Australian Sleep Association 2020). Whenever commenced, diagnostic and therapeutic studies for any critically ill patient with sleep-disordered breathing should be commenced after a careful COVID-19 screening and thorough consideration of the potential benefits and risks. Patients need to be tested for COVID-19 in a reasonable time frame to ensure result validity (CDC 2022).

Nevertheless, home sleep apnea testing and initiating remote titration or empirical noninvasive ventilation are preferred whenever possible (Schiza et al. 2021; American Academy of Sleep Medicine 2022a; Australian Sleep Association 2020). It is necessary to mention that due to the high risk of aerosolization, titration studies and split-night studies should be postponed except in highly emergent conditions. In the latter setting, PAP administration should be taken in a negative-pressure room (when feasible) and ensure that the technologist uses the appropriate PPE. Due to the high

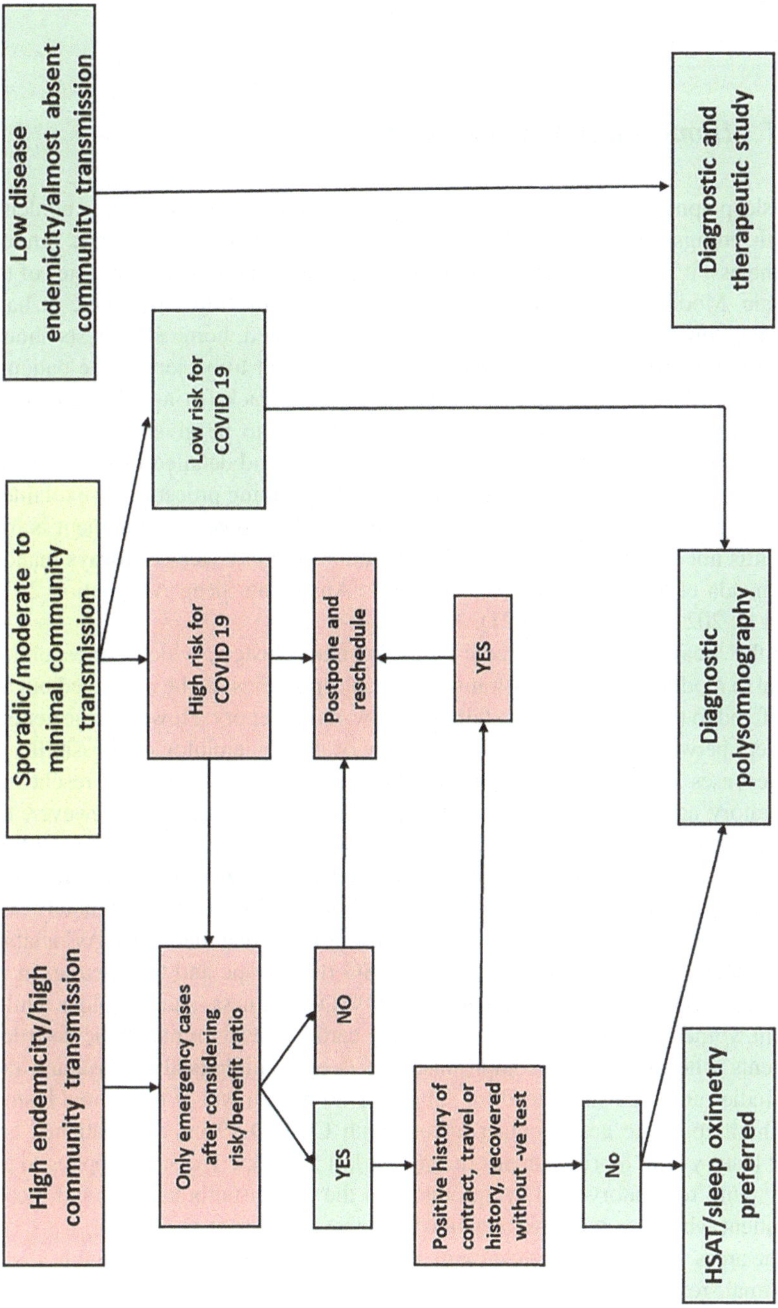

Fig. 20.2 A proposed epidemiology-based algorithm for a stepwise approach to sleep studies during pandemics

risk of aerosolization, PAP devices should not be operated in the clinical rooms even for demonstration (Pirzada et al. 2020a; Australian Sleep Association 2020; Cilea et al. 2021).

20.3.1 Home Sleep Apnea Testing

Home sleep apnea testing (HSAT) can provide diagnostic leeway in these epidemiological settings, but its role is limited since the diagnostic information is limited. Nevertheless, it is practical and helpful during the epidemiological exigencies of the pandemic. Models which can be delivered home and are fully disposable or have disposable components are preferred. Wherever indicated, home sleep tests should be preferred. Recommendations focus on the equipment to be sent to the patient's home, either through courier or by post in a sealed pack (to avoid medicolegal implications), and instructions regarding montage are to be given through video conferencing or online videos, if taken by the patient, and detailed instructions can still be given remotely. In a similar fashion, the demounting procedure is explained, besides repackaging and drop-off methods. Although disposable equipment is preferred, it is not cost effective. External sanitization of the devices is always mandatory (Pirzada et al. 2020a; Schiza et al. 2021; Australian Sleep Association 2020; Cilea et al. 2021; Ayas et al. 2021).

For the areas with only sporadic cases or small clusters, which correspond to minimal to moderate community transmission, sleep studies can be continued but are only offered to those patients who fall in the low-risk category. However, the general agreement between different societies, in view of the epidemiological possibility of multiple cases of COVID-19 in the community, is to postpone and reschedule in-laboratory administration of PAP therapy except in emergencies; however, the same can be duly considered if the center can spare a negative-pressure room for the procedure, in case of emergency (Pirzada et al. 2020a; Schiza et al. 2021; Australian Sleep Association 2020). It is preferred to initiate PAP therapy remotely and noninvasive ventilation empirically (Schiza et al. 2021; Australian Sleep Association 2020). Generally, we need to postpone all PSG (diagnostic and therapeutic) in all patients at a higher risk of getting severe COVID-19 illness, such as older adults, pregnant women, and patients with underlying cardiac, renal, or metabolic disorders or patients who are immunocompromised. It is epidemiologically warranted that sleep studies are to be suspended for subjects who have travel or residential history in the high-epidemic area, contact history with COVID-19-infected patients, and contact history with febrile patients during the last 2 weeks. The same applies to the subjects with respiratory tract symptoms from the cluster outbreak area and recovered patient who does not have negative PCR test (Pirzada et al. 2020a).

In the areas where disease endemicity is low or community transmission is absent or minimal, resumption of routine in-laboratory diagnostic as well as therapeutic studies for both adults and children can be done, in addition to PAP use in the clinic setting. In addition, in-person appointments can be restarted as needed, including clinic visits and PAP setup demonstrations. Additionally, HSAT can be resumed following the usual protocols and instructions (restriction is no longer warranted as previously described) (Pirzada et al. 2020a; Schiza et al. 2021; Australian Sleep Association 2020).

Nevertheless, continuous monitoring of state and local public health communication for warnings of any increase in community transmission is mandatory. Moreover, practical emphasis on patient screening and reviewing local epidemiological dynamics is highly recommended; any patient with confirmed or suspected COVID-19 infection must be postponed. Under emergency conditions, remote control multimodal ventilator, wherever available, is suggested to ensure healthcare worker safety. For patients who have recovered from COVID-19 infection, there should be strong evidence of negative results of nucleic acid testing before undergoing sleep studies, diagnostic as well as therapeutic; otherwise, it is best to be postponed (Pirzada et al. 2020a).

20.3.2 Outpatient Consultation and Follow-Up Services

Like sleep diagnostics for outpatient services, follow the epidemiological leads about disease burden in the community. However, outpatient services are much easier and more practical to be managed remotely compared to sleep diagnostics and mechanical therapeutics; it is recommended to utilize the service more liberally (Moynihan et al. 2021; Johnson et al. 2021; Garfan et al. 2021; Webster 2020; American Academy of Sleep Medicine 2022a; Gupta et al. 2020; Wilson et al. 2020; Valentino et al. 2020; Voulgaris et al. 2021). Nonetheless, patients who need to attend the clinic physically should follow the same protocol and preparation as applies to inpatient services regarding infection control. It is imperative to mention here that outpatient in-person services need to be suspended in areas with moderate to high disease burdens (Pirzada et al. 2020a). In emergent situations during these epidemiological burdens, if a patient needs to visit the clinic, PAP devices should be kept at home; if brought to the facility, it is never to be operated in the facility (Pirzada et al. 2020a; Schiza et al. 2021).

Almost all patients with sleep disorders can be evaluated, monitored, and followed up remotely through different modalities, nevertheless with some limitations. Remote consultations can be in real-time one-is-to-one consultations, also known as synchronous interaction through zoom calls or other video calling methods. On the other hand, it can be asynchronous, like messaging and data transmission through messaging services in the form of patient details, assessment charts, questionnaires, PSG, or actigraphy results (Irfan et al. 2020). Both can be used efficiently in sleep medicine clinics for screening, assessments, interpretations, diagnostics, and therapeutics (Gupta et al. 2020; Irfan et al. 2020). Patients with sleep disorders often warrant a collateral history, such as sleep-disordered breathing or parasomnias. Hence, it is best to have a bed partner present during the evaluation interviews in such clinical scenarios, even when done remotely.

Physical assessment is not possible during remote consultation, although there are remote examination programs that can be utilized with some limitations (Benziger et al. 2021; Russell and Artandi 2022). Nevertheless, primary care physicians can provide relevant information for physical examination in some countries; however, this poses a challenge during any diagnostic ambiguity (Australian Sleep Association 2020; Benziger et al. 2021; Russell and Artandi 2022).

The scope of telemedicine or remote monitoring and therapeutic services in sleep medicine is broad. Since sleep medicine therapeutics is mainly based on device prescriptions and data collection and interpretation, it can be managed remotely without calling patients to the clinic. Devices can be prescribed, and mask fit can be assessed remotely through one-is-to-one consultations. Follow-up data can be downloaded from the cloud services, and device settings can be changed remotely through e-services; however, it is not available in all countries due to patients' data confidentiality issues and the need for working internet services with fair data transmission speed. In addition to that, it is not possible on older devices. Although it seems practical and easy, it has limitations, particularly in patients who are naïve to the device, prone to complications, and pose compliance issues (ResMed 2022b; Keenan and Schwab 2021).

Pharmaceutical prescriptions can follow remote consultations. Medications can be dispensed and delivered to the doorstep. However, it has practical limitations since many medications used in sleep medicine fall under the restricted category, so that delivery can be a huge medicolegal challenge (Pirzada et al. 2020a). Besides prescriptions and medication, another category of interest is providing online cognitive behavioral therapy (CBT), which has proved to be promising in sleep disorders, particularly insomnia, and thus can be used efficiently during the COVID-19 pandemic and other pandemics; in fact, it has proven not to be inferior to in-person CBT (Hasan et al. 2022). Another advantage is that the patient can use it when he or she has time to devote. There are now exclusive electronic/mobile applications available that have eased out the therapy (Hsieh et al. 2020; Kallestad et al. 2021; Okujava et al. 2019).

For other medicolegal issues, keeping all the data collected through remote consultations safe and secure with clear documentation about time, mode, and other related issues is recommended. Moreover, it is essential to maintain call log data, preferably with electronic evidence. Figure 20.3 demonstrates a proposed algorithmic summary of guidelines for sleep centers during pandemics.

20.4 Conclusion

The resumption of sleep medicine services despite COVID-19 is needed to optimize healthcare outcomes for sleep disorders as well as COVID-19. However, it needs preparation and watchful monitoring. Although a challenging task, it is achievable

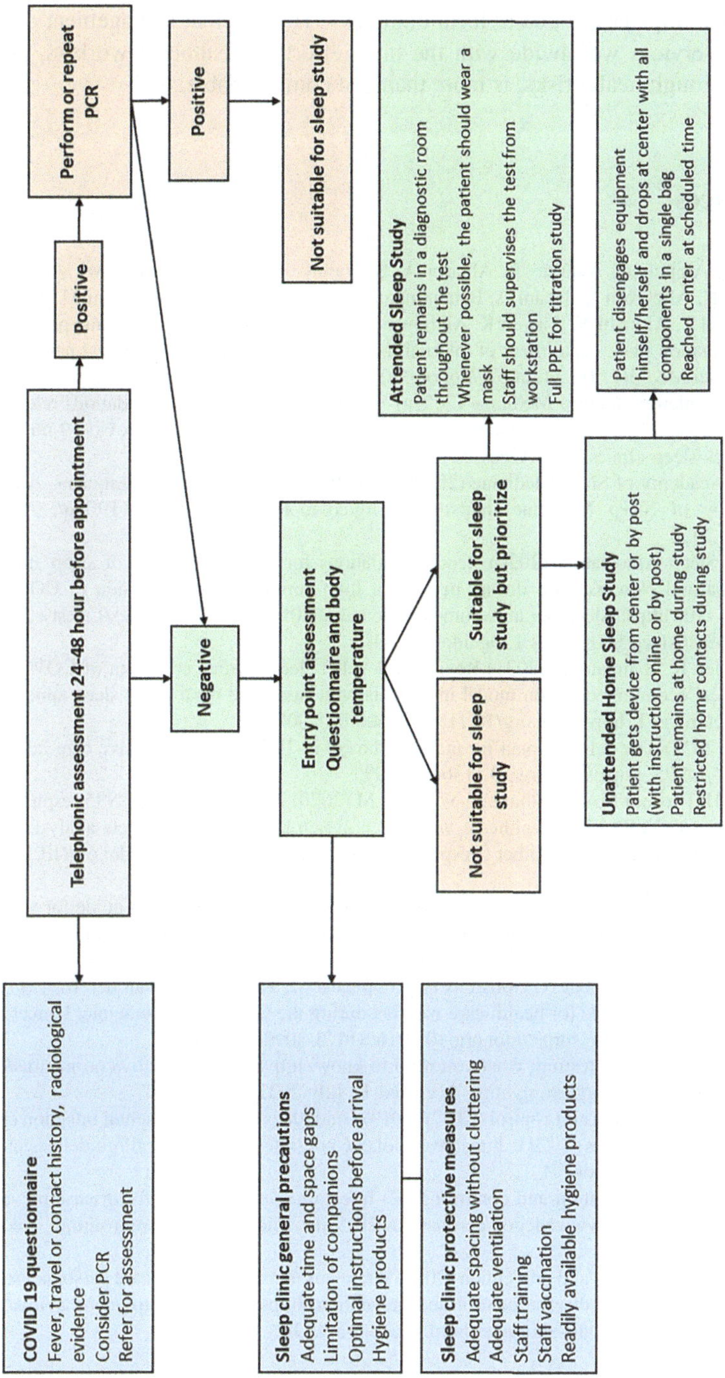

Fig. 20.3 A proposed algorithmic summary of guidelines for sleep centers during pandemics

through coordination and optimal use of resources. Nevertheless, it often gets hurdles and impedances due to local outbreaks. The efficient management of sleep medicine services worldwide with the huge effort of healthcare workers, despite running through health risks, is more than just commendable.

References

Alfadda A, Albilali A, Alqurtas E, Alharbi A, Ekhzaimy A, Muayqil T, Bedaiwi M, Barry M, Azzam N, Aljebreen A, Isnani A, Bahammam A, Nasser A, Nouh T, Alzahrani M, Alsaleh K, Alfaadhel T, Alotaibi N, Alayed K, Alkhowaiter M (2021) COVID-19 pandemic preparedness and mitigation plan: department of internal medicine experience from a clinical perspective. J Nature Sci Med 4(1):16–24. https://doi.org/10.4103/jnsm.jnsm_109_20

American Academy of Sleep Medicine (2022a) Summary of CDC recommendations relevant for sleep practices during COVID-19. https://aasm.org/covid-19-resources/covid-19-mitigation-strategies-sleep-clinics-labs. Accessed 18 July 2022

American Academy of Sleep Medicine (2022b) COVID-19: FAQs for sleep clinicians. American Academy of Sleep Medicine. https://aasm.org/covid-19-resources/covid-19-faq/. Accessed 17 July 2022

Australian Sleep Association (2020) Recommendations for the performance of sleep studies in Australia and New Zealand during periods of high community transmission of COVID-19. 2022: 1–10. https://sleep.org.au/common/Uploaded%20files/Public%20Files/Latest%20News/COVID%20sleep%20studies%20guidelines.pdf

Ayas NT, Jen R, Baumann B (2021) Revisiting level II sleep studies in the era of COVID-19: a theoretical economic decision model in patients with suspected obstructive sleep apnea. Sleep Sci Pract 5(1):11. https://doi.org/10.1186/s41606-021-00063-5

Bagenal J (2022) The delivery plan for tackling the covid-19 backlog of elective care falls short. BMJ 377:o995. https://doi.org/10.1136/bmj.o995

Bartoszko JJ, Farooqi MAM, Alhazzani W, Loeb M (2020) Medical masks vs N95 respirators for preventing COVID-19 in healthcare workers: a systematic review and meta-analysis of randomized trials. Influenza Other Respir Viruses 14(4):365–373. https://doi.org/10.1111/irv.12745

Benziger CP, Huffman MD, Sweis RN, Stone NJ (2021) The telehealth ten: a guide for a patient-assisted virtual physical examination. Am J Med 134(1):48–51. https://doi.org/10.1016/j.amjmed.2020.06.015

Bielicki JA, Duval X, Gobat N, Goossens H, Koopmans M, Tacconelli E, van der Werf S (2020) Monitoring approaches for health-care workers during the COVID-19 pandemic. Lancet Infect Dis 20(10):e261–e267. https://doi.org/10.1016/S1473-3099(20)30458-8

CDC (2022) COVID-19 testing: what you need to know. https://www.cdc.gov/coronavirus/2019-ncov/symptoms-testing/testing.html. Accessed 12 July 2022

Center disease prevention and control (CDC) (2019) Guidelines for environmental infection control in health-care facilities. 1–241. https://espanol.cdc.gov/infectioncontrol/pdf/guidelines/environmental-guidelines-P.pdf

Center for disease prevention and control (2022) Infection control. Center for disease prevention and control. https://wwwcdcgov/coronavirus/2019-ncov/hcp/infection-controlhtml. Accessed 15 July 2022

Centre for disease control and prevention (2022) Strategies to mitigate healthcare personnel staffing shortages. Centre for disease control and prevention. https://www.cdc.gov/coronavirus/2019-ncov/hcp/mitigating-staff-shortages.html. Accessed 15 July 2022

Centre for disease prevention and control (2022) Potential exposure at work. Centre for disease prevention and control. https://www.cdc.gov/coronavirus/2019-ncov/hcp/guidance-risk-assesment-hcp.html. Accessed 15 July 2022

Cilea R, Guaraldi P, Barletta G, Lucchi M, Baschieri F, Baldelli L, Loddo G, Rocca A, Cortelli P, Calandra-Buonaura G, Provini F (2021) Performing sleep studies after the COVID-19 outbreak: practical suggestions from Bologna's sleep unit. Sleep Med 77:45–50. https://doi.org/10.1016/j.sleep.2020.11.005

Clark A, Jit M, Warren-Gash C, Guthrie B, Wang HHX, Mercer SW, Sanderson C, McKee M, Troeger C, Ong KL, Checchi F, Perel P, Joseph S, Gibbs HP, Banerjee A, Eggo RM, Centre for the Mathematical Modelling of Infectious Diseases COVID-19 working group (2020) Global, regional, and national estimates of the population at increased risk of severe COVID-19 due to underlying health conditions in 2020: a modelling study. Lancet Glob Health 8(8):e1003–e1017. https://doi.org/10.1016/S2214-109X(20)30264-3

Ferioli M, Cisternino C, Leo V, Pisani L, Palange P, Nava S (2020) Protecting healthcare workers from SARS-CoV-2 infection: practical indications. Eur Respir Rev 29(155):200068. https://doi.org/10.1183/16000617.0068-2020

Ferrara P, Albano L (2020) COVID-19 and healthcare systems: what should we do next? Public Health 185:1–2. https://doi.org/10.1016/j.puhe.2020.05.014

Garfan S, Alamoodi AH, Zaidan BB, Al-Zobbi M, Hamid RA, Alwan JK, Ahmaro IYY, Khalid ET, Jumaah FM, Albahri OS, Zaidan AA, Albahri AS, Al-Qaysi ZT, Ahmed MA, Shuwandy ML, Salih MM, Zughoul O, Mohammed KI, Momani F (2021) Telehealth utilization during the Covid-19 pandemic: a systematic review. Comput Biol Med 138:104878. https://doi.org/10.1016/j.compbiomed.2021.104878

Geyman J (2021) COVID-19 has revealed America's broken health care system: what can we learn? Int J Health Serv 51(2):188–194. https://doi.org/10.1177/0020731420985640

Gross JV, Mohren J, Erren TC (2021) COVID-19 and healthcare workers: a rapid systematic review into risks and preventive measures. BMJ Open 11(1):e042270. https://doi.org/10.1136/bmjopen-2020-042270

Grote L, McNicholas WT, Hedner J, ESADA collaborators (2020) Sleep apnoea management in Europe during the COVID-19 pandemic: data from the European sleep apnoea database (ESADA). Eur Respir J 55(6):2001323. https://doi.org/10.1183/13993003.01323-2020

Gupta R, Kumar VM, Tripathi M, Datta K, Narayana M, Ranjan Sarmah K, Bhatia M, Devnani P, Das S, Shrivastava D, Gourineni RD, Singh TD, Jindal A, Mallick HN (2020) Guidelines of the Indian Society for Sleep Research (ISSR) for practice of sleep medicine during COVID-19. Sleep Vigil 4(2):61–72. https://doi.org/10.1007/s41782-020-00097-2

Hasan F, Tu YK, Yang CM, James Gordon C, Wu D, Lee HC, Yuliana LT, Herawati L, Chen TJ, Chiu HY (2022) Comparative efficacy of digital cognitive behavioral therapy for insomnia: a systematic review and network meta-analysis. Sleep Med Rev 61:101567. https://doi.org/10.1016/j.smrv.2021.101567

Holstiege J, Akmatov MK, Kohring C, Dammertz L, Ng F, Czihal T, von Stillfried D, Batzing J (2021) Patients at high risk for a severe clinical course of COVID-19–small-area data in support of vaccination and other population-based interventions in Germany. BMC Public Health 21(1):1769. https://doi.org/10.1186/s12889-021-11735-3

Hsieh C, Rezayat T, Zeidler MR (2020) Telemedicine and the management of insomnia. Sleep Med Clin 15(3):383–390. https://doi.org/10.1016/j.jsmc.2020.05.004

Irfan M, Khalil W, Iber C (2020) Practicing sleep medicine amidst a pandemic: a paradigm shift. J Clin Sleep Med 16(8):1405–1408. https://doi.org/10.5664/jcsm.8590

Johnson KG, Sullivan SS, Nti A, Rastegar V, Gurubhagavatula I (2021) The impact of the COVID-19 pandemic on sleep medicine practices. J Clin Sleep Med 17(1):79–87. https://doi.org/10.5664/jcsm.8830

Kallestad H, Scott J, Vedaa O, Lydersen S, Vethe D, Morken G, Stiles TC, Sivertsen B, Langsrud K (2021) Mode of delivery of cognitive behavioral therapy for insomnia: a randomized controlled

non-inferiority trial of digital and face-to-face therapy. Sleep 44(12):zsab185. https://doi.org/10.1093/sleep/zsab185
Keenan BT, Schwab RJ (2021) Using the remote monitoring framework to promote adherence to continuous positive airway pressure. Sleep Med Clin 16(1):85–99. https://doi.org/10.1016/j.jsmc.2020.11.001
Moynihan R, Sanders S, Michaleff ZA, Scott AM, Clark J, To EJ, Jones M, Kitchener E, Fox M, Johansson M, Lang E, Duggan A, Scott I, Albarqouni L (2021) Impact of COVID-19 pandemic on utilisation of healthcare services: a systematic review. BMJ Open 11(3):e045343. https://doi.org/10.1136/bmjopen-2020-045343
Nguyen LH, Drew DA, Graham MS, Joshi AD, Guo CG, Ma W, Mehta RS, Warner ET, Sikavi DR, Lo CH, Kwon S, Song M, Mucci LA, Stampfer MJ, Willett WC, Eliassen AH, Hart JE, Chavarro JE, Rich-Edwards JW, Davies R, Capdevila J, Lee KA, Lochlainn MN, Varsavsky T, Sudre CH, Cardoso MJ, Wolf J, Spector TD, Ourselin S, Steves CJ, Chan AT, Consortium COPE (2020) Risk of COVID-19 among front-line health-care workers and the general community: a prospective cohort study. Lancet Public Health 5(9):e475–e483. https://doi.org/10.1016/S2468-2667(20)30164-X
Nonin (2022) Reliable Nonin Solutions. https://www.nonin.com/products/. Accessed 29 July 2022
Okujava N, Malashkhia N, Shagidze S, Tsereteli A, Arevadze B, Chikhladze N, de Weerd A, Van Straten A (2019) Digital cognitive behavioral therapy for insomnia–the first Georgian version. Can we use it in practice? Internet Interv 17:100244. https://doi.org/10.1016/j.invent.2019.100244
Ong SWX, Tan YK, Chia PY, Lee TH, Ng OT, Wong MSY, Marimuthu K (2020) Air, surface environmental, and personal protective equipment contamination by severe acute respiratory syndrome coronavirus 2 (SARS-CoV-2) from a symptomatic patient. JAMA 323(16):1610–1612. https://doi.org/10.1001/jama.2020.3227
Pagel C (2022) The covid waves continue to come. BMJ 377:o1504. https://doi.org/10.1136/bmj.o1504
Pena Orbea C, Wang L, Shah V, Jehi L, Milinovich A, Foldvary-Schaefer N, Chung MK, Mashaqi S, Aboussouan L, Seidel K, Mehra R (2021) Association of sleep-related hypoxia with risk of COVID-19 hospitalizations and mortality in a large integrated health system. JAMA Netw Open 4(11):e2134241. https://doi.org/10.1001/jamanetworkopen.2021.34241
Philips (2022) Ventilation system support. Philips. https://www.philips.sa/en/healthcare/medical-specialties/covid-19/sleep-and-respiratory-care-covid-19/clinical-resource-and-training-page#triggername%3Dmenu_resources=&triggername=menu_one. Accessed 12 July 2022
Pirzada A, Awadh AA, Aleissi SA, Almeneessier AS, BaHammam AS (2020a) Reopening sleep medicine services in the conundrum of an ongoing COVID-19 pandemic: a global view. Sleep Vigil 4(2):73–80. https://doi.org/10.1007/s41782-020-00100-w
Pirzada AR, Aleissi SA, Almeneessier AS, BaHammam AS (2020b) Management of aerosol during noninvasive ventilation for patients with sleep-disordered breathing: important messages during the COVID-19 pandemic. Sleep Vigil 4(2):89–94. https://doi.org/10.1007/s41782-020-00092-7
ResMed (2022a) Re-opening Sleep Laboratories in the United States following COVID-19. ResMed. https://document.resmed.com/en-us/documents/covid-19/reopening-sleep-labs-following-covid19-white-paper_amer_eng.pdf?utm_source=rmd&utm_medium=web&utm_content=hcp-covid. Accessed 1 Aug 2022
ResMed (2022b) AirView key benefits. https://www.resmed.co.uk/healthcare-professional/sleep-apnoea/treatment/monitoring-treatment/. Accessed 28 July 2022
Russell SW, Artandi MK (2022) Approach to the telemedicine physical examination: partnering with patients. Med J Aust 216(3):131–134. https://doi.org/10.5694/mja2.51398
Sahni A, Cao M (2021) Obstructive sleep apnea and severe COVID-19 infection: is there a plausible link? J Clin Sleep Med 17(11):2145–2146. https://doi.org/10.5664/jcsm.9644
Sanghera J, Pattani N, Hashmi Y, Varley KF, Cheruvu MS, Bradley A, Burke JR (2020) The impact of SARS-CoV-2 on the mental health of healthcare workers in a hospital setting-a systematic review. J Occup Health 62(1):e12175. https://doi.org/10.1002/1348-9585.12175

Schiza S, Simonds A, Randerath W, Fanfulla F, Testelmans D, Grote L, Montserrat JM, Pepin JL, Verbraecken J, Ersu R, Bonsignore MR (2021) Sleep laboratories reopening and COVID-19: a European perspective. Eur Respir J 57(3):2002722. https://doi.org/10.1183/13993003.02722-2020

Schweller J (2021) Sleep labs in the pandemic era. American Association for Respiratory Care. https://www.aarc.org/an21-sleep-labs-in-the-pandemic-era/. Accessed 14 July 2022

Serra L (2021) Rationale for reverse-transcription polymerase chain reaction for SARS-CoV-2 screening in patients undergoing in-laboratory sleep studies. J Clin Sleep Med 17(6):1333. https://doi.org/10.5664/jcsm.9232

Smith JD, MacDougall CC, Johnstone J, Copes RA, Schwartz B, Garber GE (2016) Effectiveness of N95 respirators versus surgical masks in protecting health care workers from acute respiratory infection: a systematic review and meta-analysis. CMAJ 188(8):567–574. https://doi.org/10.1503/cmaj.150835

Sodiq A, Khan MA, Naas M, Amhamed A (2021) Addressing COVID-19 contagion through the HVAC systems by reviewing indoor airborne nature of infectious microbes: will an innovative air recirculation concept provide a practical solution? Environ Res 199:111329. https://doi.org/10.1016/j.envres.2021.111329

Strausz S, Kiiskinen T, Broberg M, Ruotsalainen S, Koskela J, Bachour A, Gen F, Palotie A, Palotie T, Ripatti S, Ollila HM (2021) Sleep apnoea is a risk factor for severe COVID-19. BMJ Open Respir Res 8(1):e000845. https://doi.org/10.1136/bmjresp-2020-000845

Grass Technologies (2011) Genuine grass gold disc electrode care. https://bio-medical.com/media/support/genuine_grass_gold_disc_electrode_care.pdf. Accessed 25 July 2022

Toomey EC, Conway Y, Burton C, Smith S, Smalle M, Chan XS, Adisesh A, Tanveer S, Ross L, Thomson I, Devane D, Greenhalgh T (2021) Extended use or reuse of single-use surgical masks and filtering face-piece respirators during the coronavirus disease 2019 (COVID-19) pandemic: a rapid systematic review. Infect Control Hosp Epidemiol 42(1):75–83. https://doi.org/10.1017/ice.2020.1243

Valentino LA, Skinner MW, Pipe SW (2020) The role of telemedicine in the delivery of health care in the COVID-19 pandemic. Haemophilia 26(5):e230–e231. https://doi.org/10.1111/hae.14044

Vas V, North S, Rua T, Chilton D, Cashman M, Malhotra B, Garrood T (2022) Delivering outpatient virtual clinics during the COVID-19 pandemic: early evaluation of clinicians' experiences. BMJ Open Qual 11(1):e001313. https://doi.org/10.1136/bmjoq-2020-001313

Verbeek JH, Rajamaki B, Ijaz S, Sauni R, Toomey E, Blackwood B, Tikka C, Ruotsalainen JH, Kilinc Balci FS (2020) Personal protective equipment for preventing highly infectious diseases due to exposure to contaminated body fluids in healthcare staff. Cochrane Database Syst Rev 4:CD011621. https://doi.org/10.1002/14651858.CD011621.pub4

Voulgaris A, Ferini-Strambi L, Economou NT, Nena E, Steiropoulos P (2021) Sleep telemedicine practices: checkpoint list and practical considerations in the COVID-19 era. Front Neurol 12:664274. https://doi.org/10.3389/fneur.2021.664274

Webster P (2020) Virtual health care in the era of COVID-19. Lancet 395(10231):1180–1181. https://doi.org/10.1016/S0140-6736(20)30818-7

WHO (2022) Infection prevention and control during health care when coronavirus disease (COVID-19) is suspected or confirmed. https://apps.who.int/iris/bitstream/handle/10665/342620/WHO-2019-nCoV-IPC-2021.1-eng.pdf?sequence=1&isAllowed=y. Accessed 12 July 2022

Wilson KC, Kaminsky DA, Michaud G, Sharma S, Nici L, Folz RJ, Barjaktarevic I, Bhakta NR, Cheng G, Chupp GL, Cole A, Dixon AE, Finigan JH, Graham B, Hallstrand TS, Haynes J, Hankinson J, MacIntyre N, Mandel J, McCarthy K, McCormack M, Patil SP, Rosenfeld M, Senitko M, Sethi S, Swenson ER, Stanojevic S, Teodorescu M, Weiner DJ, Wiener RS, Powell CA (2020) Restoring pulmonary and sleep services as the COVID-19 pandemic lessens. From an association of pulmonary, critical care, and sleep division directors and American thoracic society-coordinated task force. Ann Am Thorac Soc 17(11):1343–1351. https://doi.org/10.1513/AnnalsATS.202005-514ST

Chapter 21
Management of Hospital and Home Positive Airway Pressure Machines During Pandemics

Rouf Pirzada and Ahmed S. BaHammam

Abstract Noninvasive ventilation (NIV) has found its intensive use in respiratory medicine, at home, and in hospitals. Besides, NIV was pivotal in managing respiratory failure during the coronavirus disease 2019 (COVID-19) pandemic. However, due to aerosolization, the initial reaction at the advent of COVID-19 was adverse to NIV use in COVID-positive patients, and there were advisories to suspend all NIV at home wherever possible due to aerosolization. Aerosolization poses an immediate threat to all who work around it, like healthcare workers and in domiciliary-use households and caregivers. Nonetheless, COVID-19 mortality and morbidity are augmented by comorbid respiratory diseases, including sleep-disordered breathing. Thus, we need to continue NIV use during COVID-19 but with precautions. We need to plan the therapy and forge guidelines and recommendations to keep NIV safe during infections. Here, we have reviewed the available literature and applied our experience to formulate guidelines and recommendations. However, updates and appraisals are evolving rapidly, and we need to keep our eyes open to tailor our approach.

Keywords Noninvasive mechanical ventilation · Mask interface · Aerosol therapy · Respiratory failure · COVID-19

R. Pirzada
North Cumbria Integrated Care (NCIC), NHS, Carlisle, UK
e-mail: Pirzada.baba@ncic.nhs.uk

A. S. BaHammam (✉)
The University Sleep Disorders Center, Department of Medicine, College of Medicine, King Saud University, Riyadh, Saudi Arabia

Strategic Technologies Program of the National Plan for Sciences and Technology and Innovation in the Kingdom of Saudi Arabia, King Saud University, Riyadh, Saudi Arabia
e-mail: ashammam@ksu.edu.sa

A. S. BaHammam et al. (eds.), *COVID-19 and Sleep: A Global Outlook*, Progress in Sleep Research, https://doi.org/10.1007/978-981-99-0240-8_21

21.1 Introduction

Noninvasive ventilators (NIV) and their close cohorts, positive airway pressure (PAP) devices, opened novel options for treating respiratory failures and sleep-disordered breathing, respectively. In clinical practice, for arbitrary convenience, it is generally accepted to use NIV for both prototype NIV and PAP devices; therefore, in this chapter, we use the term/name NIV for both prototype NIV and PAP devices (Rochwerg et al. 2017). The use of NIV in clinical practice ushered operational ease in hospitals and a new era in domiciliary treatment (Ambrosino and Guarracino 2011). Nevertheless, NIV use is also fraught with complications for patients, caregivers, and households. One of the major issues with NIV use is the aerosol generation, which is a health hazard for hospital staff, caregivers, and households due to the potential source of infection transmission (Pirzada et al. 2020). Practically, NIV management during COVID-19 means aerosol management during NIV use to prevent infection transmission since the rest of management remains the same.

The initial response to NIV use during COVID-19 was overwhelmingly cautious, advising to suspend NIV use during infection or suspicion of infection (Barker et al. 2020). However, with time, NIV proved to be a boon during COVID-19: bypassing and delaying intubations, avoiding reintubation, cutting down on ICU admissions, and mitigating logistic burnouts (Arulkumaran et al. 2020; Menzella et al. 2021). Thus, advisories evolved into more practical handouts. Over time, different bodies and organizations developed guidelines and advisories to render NIV use safe during the COVID-19 pandemic. This chapter discusses potential risks and their management with NIV use during COVID-19. Thus, it covers the aerosol generation, hazards of aerosol generation, risks of infectious disease transmission, and precautions needed to minimize infection (COVID-19) transmission during NIV use at home and in healthcare facilities.

Aerosol, or suspension of fine solid/liquid particles/droplets, has been a well-recognized source for many pathogens, bacteria, viruses, and even fungi for decades. However, recent epidemics and pandemics, SARS, MERS-CoV, and COVID-19, brought it under enhanced limelight; thus, multiple bodies showed concern about aerosol management (Esquinas et al. 2014). Aerosol generation and transmission have been studied mainly using air-sampling techniques and culture and molecular detection methods, mainly on viruses and fungi. A landmark review by Beggs emphasized the underestimation of airborne transmission, with subsequent better understanding over the years of aerosol generation, dispersion, and transmission opening better preventive strategies (Beggs 2003). Since aerosol generation, dispersion, and transmission determine the pathogenicity of the phenomenon, it is essential to handle aerosol-generating procedures carefully and abide by the guidelines diligently (Hemmes et al. 1960; Yu et al. 2004, 2005; Pshenichnaya and Nenadskaya 2015; Jackson et al. 2020; Pirzada et al. 2020).

The recent experience with SARS, followed by MERS-CoV and then COVID-19, highlighted aerosol transmission as an important mode of infection. Aerosol proved hazardous not only for healthcare workers but also to the home care workers and households where NIV devices are used. However, it was COVID-19, due to its pandemic epidemiology, enormous mortality, and morbidity, which pushed for more stringent but practical aerosol management (Davies et al. 2009; Arulkumaran et al. 2020; Pirzada et al. 2020). As a result, multiple societies and organizations came forward to build a safety protocol, from device modifications to environmental management, for NIV use.

21.2 Aerosols

Aerosols are tiny particles suspended in the air that can contain a variety of pathogens, including viruses, bacteria, and even fungi. Aerosols are generally poly-dispersed, with variable sizes and, hence, have debatable classifications and definitions. "Airborne Aerosol Hygiene Research" described aerosol as droplets of respiratory secretions, generated into the air due to any physiological, diagnostic, or therapeutic exercise, which then evaporate to become "droplet nuclei," which remain suspended in air currents, and may drift away to considerable distances (>1 m). Arbitrarily, aerosols are categorized as small droplets (usually called aerosols exclusively), which have the potential to desiccate and form droplet nuclei that travel long distances. In contrast, large droplets do not evaporate before settling on surfaces. This aerosol classification has clinical relevance in relation to its dispersal properties and patterns, and hence infection transmission. Another clinically relevant classification considers the deposit sites in the upper or lower respiratory tract because it has a bearing on pathogenesis. Most often, 5 μm in diameter is taken as a cutoff for small droplets, while another possible cutoff between aerosol types is 20 μm since aerosols less than or equal to 20 μm in diameter can desiccate to form droplet nuclei. Particles and droplets with aerodynamic diameters <5 μm can readily travel and penetrate deep into the alveolar region of the lungs of a bystander (Tran et al. 2012; Buonanno et al. 2020). In contrast, relatively large droplets are thought to arise from the upper respiratory tract, usually settling quickly and relatively close to their source, diminishing the infectivity. However, these large droplets also tend to aerosolize by human handling.

Aerodynamic studies of respiratory droplets have shown that the 100 μm droplets take about 10 s, whereas the 10 μm droplets take 17 min to fall to the floor. The 5 μm droplets originating from an average height (160 cm) of speaking or coughing take 9 min to reach the ground (Knight 1980; Thomas 2013; Somsen et al. 2020). Droplet size, thus, determines aerodynamic properties and penetration ability. Although arbitrary, from this droplet size aerodynamic behavior, we have deduced 1-m safe spatial separation for social distancing during the COVID-19 pandemic (Hui et al. 2019; Judson and Munster 2019). However, aerosol generation and dispersion are also affected by the position of the patient, whether supine, reclined, or erect, and air direction drifts like air-conditioning (AC), fan, or open currents, which have a bearing on transmission, which we will examine further in this chapter (Guo et al. 2021).

Respiratory therapeutics, like NIV, are known to stimulate coughing and promote the generation of aerosols; however, their risk of transmission of infection is difficult to estimate; hence, it cannot be numerically quantified with accuracy. Thus, these procedures must be handled meticulously to minimize the spread among healthcare workers and, at times, in the household (Ferioli et al. 2020). General principles to follow are good hygiene practices in maintaining devices, minimizing exposure time, maximizing the distance, allowing adequate air exchanges, and using negative-pressure facilities (in hospitals) (Ahn et al. 2020; Robles-Romero et al. 2022).

Inferences about the dispersion of aerosols during aerosol-generating procedures come from scientific studies conducted in a negative-pressure room on a high-fidelity human patient simulator (HPS) representing a 70 kg adult male sitting on a 45° inclined hospital bed. These estimates are not always practical since the settings are ideal and do not include standard hospital rooms or normal homerooms, and exhaled air dispersion distance from the simulator has been evaluated using a laser smoke visualization method and calculated on the median sagittal plane (Ferioli et al. 2020; van Doremalen et al. 2020).

Deductions and hence application of these scientific studies in NIV can be summarized as follows:

- **CPAP via oronasal mask** at pressures 5, 10, 15, or 20 cm H_2O, exhaled air dispersed evenly in all directions through the mask vent holes at a very low normalized smoke concentration irrespective of the severity of lung injury, amounting to negligible.
- **CPAP via nasal cannula (nasal pillows)** is related to increased air dispersion with increasing CPAP and reduced air dispersion with worsening lung injury. Using two types of nasal pillows, Nuance Pro Gel and Swift FX, at a maximum CPAP of 20 cm H_2O and with a normal lung, maximum air dispersion distances were 26.4 cm and 33.2 cm, respectively.
- **NIV via a full-face mask** in the bilevel setting (inspiratory positive airway pressure (IPAP) 10 cm H_2O and expiratory positive airway pressure (EPAP) 5 cm H_2O) using a single-limb circuit; the exhaled air jet spread through the mask's holes up to 69.3 cm, 61.8 cm, and 58 cm in the normal lung, mild lung injury, and severe lung injury setting, respectively. When IPAP was increased, exhaled air dispersion distance increased; for example, with IPAP, 18 cm H_2O exhaled air jet reaches 91.6 cm, the average being 90 cm.
- **NIV via helmets** with IPAP 12 cm H_2O and EPAP 10 cm H_2O; the exhaled air dispersion distance is 17 cm in normal lung and 15 cm in mild or severe lung injury. With IPAP 20 cm H_2O, the air dispersion distances in three different settings (normal lung, mild lung injury, and severe lung injury) were found to be 27 cm, 23 cm, and 18 cm, respectively. In a double-limb circuit, the air cushion around the neck–helmet interface has negligible air dispersion during NIV application. Dispersion (maximum) distances during various NIV protocols are summarized in Table 21.1 (Hui et al. 2019; Ferioli et al. 2020; Hamilton et al. 2022).

Table 21.1 Dispersion (maximum) distances during various NIV protocols

NIV type and pressures	Mask type	Maximum dispersion in cm
CPAP		
CPAP at 5, 10, 15, 20 cm H_2O	Oronasal	Negligible
CPAP 20 cm H_2O	Nasal pillow	33
BPAP		
BPAP I/E = 18/5 cm H_2O	Full face	92
BPAP I/E = 20/10 cm H_2O	A helmet without a tight cushion	27
BPAP I/E = 20/10 cm H_2O	Helmet with tight cushion	Negligible

Thus, estimations are variable depending on multiple factors like mask type, mask brand, pressures, pressure types (bilevel or continuous), and circuitry. Inferences from all these estimations can guide choosing device type, circuitry, pressures, and interface besides training physicians and other therapists about anticipation and precautions. Under logistic burnout or when resources and infrastructure are pushed beyond limits during a pandemic, following the principles becomes a real challenge in hospitals and homes.

For practical considerations, we divide our discussion about NIV management and hence aerosol management into two sections, at home, and in hospitals, since more and more NIV use is taking place in the domiciliary setting for multiple respiratory disorders.

21.2.1 At Home

The home NIV systems, either CPAP or BPAP, pose a threat to household caregivers. The WHO currently lists NIV, BPAP, as well as CPAP as high-risk aerosol-generating procedures. NIV use at home is common particularly due to high prevalence of sleep-disordered breathing, chronic obstructive airway diseases, and hypoventilation syndromes. NIV accounts for high-dose viral transmission risk to family and caregivers during the treatment of chronic respiratory diseases, and sleep-disordered breathing, especially obstructive sleep apnea (Singh and Sterk 2008; Esquinas et al. 2014; Arulkumaran et al. 2020).

Various studies have demonstrated that the viral load is comparable in nasal and throat swabs of symptomatic and asymptomatic patients with COVID-19 (Baker and Sovani 2020), suggesting that NIV-induced aerosolization of nasopharyngeal secretions from asymptomatic patients may pose similar risks for high-dose viral transmission in households. Therefore, it indirectly means that all the patients who are even COVID-19 suspects should follow the COVID-19 protocol during NIV use unless proven otherwise by PCR testing. Furthermore, aerosolized severe acute respiratory syndrome coronavirus-2 (SARS-CoV-2) has a half-life of 1.1 h in the air; therefore, a period of prolonged isolation may be required following NIV usage, which is difficult to carry on at home particularly for the patients who are dependent on caregivers (Ferioli et al. 2020; van Doremalen et al. 2020; Zou et al. 2020). For this reason, the initial response to handle the hazards was to stop NIV use at home, even in COVID-19 suspects. However, patients with hypoventilation syndromes, or COPD, are at risk of respiratory failure, and the pandemic has taken a protracted and unpredictable course, besides the fact that CPAP use can have a bearing on overall mortality and morbidity in these patients if infected with SARS-CoV-2; therefore, we need to work on safe NIV use rather than suspending NIV (Pazarli et al. 2021; Perger et al. 2021; Cardoso et al. 2022).

Based on the literature review and personal experience for home NIV management (Baker and Sovani 2020; Barker et al. 2020; Pirzada et al. 2020; Wang et al. 2021a; Craig and West 2022), the following is suggested:

- Universal isolation for NIV at home is impractical since many require specialized home assistance, and many cannot afford it; hence, it must be individualized according to the needs and available resources. The attending physician should decide about all the "pros and cons" to minimize the transmission without compromising therapeutic benefits.
- A temporary cessation is an unplausible option, considering the risks of withholding NIV in patients with hypercapnia; however, this needs to be discussed with the treating physician. Moreover, consent and decision should be formally discussed and individualized. In patients with whom NIV therapy cannot be temporarily stopped, such as patients with severe hypoventilation syndromes, complete home isolation is preferred to minimize aerosol risk in the household but can be a formidable challenge.
- Other plausible options are applying mask and circuitry modifications to minimize aerosol dispersion. Modifications of the device and interface of NIV may reduce the risk of aerosol dispersion, following the experimental outcomes as already discussed (the modifications will be discussed below).
- Applying NIV in a supine or semi-inclined position to augment the settling down of aerosols and minimize the spread area, as discussed above, is from the inference from experimental studies.
- Following routine hygiene advisories is highly recommended: changing machine filters routinely, cleaning surfaces, cleaning masks and tubing with hand-hot soap water (washing-up liquid), and washing hands regularly.
- Oxygen systems, though not quantified, are highly infective. Thus, the same precautions are to be followed in the case of supplemental oxygen during NIV. In addition, it is better to integrate O_2 tubing into the circuitry (further discussion below), although it decreases FiO_2.
- It is pertinent to mention for home titration studies using NIV, and it is advisable to use auto-titration devices or use remote titration system (Kryger and Thomas 2020; Rapoport 2020).

Home NIV also warrants environmental measures to keep the space clean during and after NIV use by diluting air through good ventilation, not only in the patients' room/bedroom but also in the washroom and anteroom. Since the WHO recommends a ventilation rate of 288/h/person, which is not possible by natural ventilation, it can be bypassed using portable air purifiers in emergency situations (Pirzada et al. 2020; Guo et al. 2021; Duval et al. 2022). However, these devices need maintenance and cleaning under infection control guidelines. These situations emerge if the COVID-19 patient is managed at home, which is not uncommon during the pandemic.

The bottom line is that NIV use during the infection or suspicion of infection is to be individualized since the risk of infection transmission is high and protocol for prevention at home is clearly tedious and, at times, impractical.

21.2.2 In Hospitals

NIV emerged as a therapeutic respite during the COVID-19 pandemic by its judicious use as an alternative to intubation and prevented re-intubation in patients with respiratory failure. Thus, it helped to maintain logistic support during a pandemic when the healthcare structure was stretched beyond manageable limits. In addition, it provided a practical leeway to manage respiratory failures outside the intensive care units (Rochwerg et al. 2017; Menzella et al. 2021; Wang et al. 2021b).

For COVID-19 respiratory failures, both CPAP and BPAP are extensively used; in fact, there were governmental directives to companies to enhance the supply to meet the demand. However, NIV in these clinical situations carries a considerable risk for healthcare workers and other patients. Considering all the advantages NIV offers over invasive ventilation during COVID-19 treatment, aerosolization is a huge disadvantage since aerosolization in invasive ventilation (closed circuit) is almost negligible compared to NIV.

For in-hospital NIV use, dispersion of the virus is common if robust preventive measures are not applied. NIV use in hospital, in a setting of suspected or confirmed cases of COVID-19, warrants infectious disease protocol of isolation in a negative-pressure room, preferably with an anteroom and attached washroom (well ventilated) and personal protective equipment for attending staff; however, it is not possible in high-volume patient inflow. Even under such unavoidable conditions, PPE for the attending staff is mandatory (Bartoszko et al. 2020; Ferioli et al. 2020; McEnery et al. 2020; Bazant and Bush 2021).

A practical approach is to modify the NIV device, mostly interface and tubing, to minimize the dispersion and protect healthcare workers and domiciliary NIV household members or caregivers. CPAP or NIV/BPAP ventilators for patients with sleep-disordered breathing (in fact, most used NIV devices at home and high patient volume centers) are built to work with a single tube and a "relatively constant intentional leak"; the same holds true when hospitals must use out-of-ICU ventilators. The leak occurs in three parts: mouth leak if a nose mask is used, leak around the mask/face contact, and leak through the intentional leak port. Interventions to reduce leak (unintentional) include the use of a full-face mask (discarding nasal masks) with the optimal fit. To bypass dispersion through intentional leak, the mask must be non-vented with an exhalation port in the circuit to avoid carbon dioxide

(CO_2) rebreathing, and then fix a filter before the exhalation port. Thus, a guide to choosing a mask to minimize viral dispersion would be the following (Aarrestad et al. 2020; Bartoszko et al. 2020; Pirzada et al. 2020):

- **The first choice** is a full-face non-vented mask with an expiratory viral filter and exhalation port in the circuit beyond the filter so that air that is breathed out goes through the filter to ensure the viral trap. A good mask seal for face masks is important to minimize droplet dispersion and maximize effectiveness.
- **The second choice** is a helmet with an air cushion for CPAP, following the same filter and exhalation port protocol. But usually, it has low acceptability by the patients, and many find it claustrophobic.
- **The third choice** is a helmet without an air cushion for CPAP, following the same filter and exhalation port protocol.
- **The fourth choice** is a standard face mask with the same filter and exhalation port modifications.

From the recommendations, we infer that the choice of mask depends on minimizing aerosol exposure and ensuring patient compliance and acceptability.

For practical considerations, the preferential setting at the initiation of therapy to minimize aerosol exposure is to follow the default setting:

- For CPAP, the initial default setting is 10 cm H_2O with humidity (higher pressures will increase the exposure of healthcare staff to exhaled (infected) air). In fact, most patients find it difficult to accept pressures more than 10 cm H_2O, and most centers do not escalate pressures to more than 10 cm H_2O when treating acute respiratory failures with CPAP.
- For BBPAP (NIV), we should start with IPAP 15/EPAP 5 for mild cases and IPAP 20 30/EPAP 5–10 for severe cases. First, use a minimum-pressure CPAP alone for oxygenation, and then adjust the machine to add IPAP for respiratory support. It is more practical and result oriented.
- Oxygen can be blended into the system, typically at relatively high rates, as oxygen flow does not translate mathematically from nasal oxygen cannula settings. High air pressure in the conduit dilutes it; thus, delivered O_2 concentration will fall when we escalate the pressures, even while the patient has benefits. Monitoring oxygen saturation helps but must be interpreted with this in mind (Aarrestad et al. 2020; Pirzada et al. 2020).

Since it is essential while using a mask system to humidify the gas flow, a heat and moisture exchanger (HME)/viral filter should be fitted in the exhaust systems to reduce droplet spread. HME filters can result in waterlogging and, thus, must be watched carefully and changed to prevent high and unidirectional flow. It must be ensured that the ventilator mode employed supports the use of non-vented masks and exhalation ports. Likewise, the sequence of events protocol for hooking and unhooking the ventilator should be followed in prescribed steps to prevent spread; the steps are to put the mask on, tie it, and then put the ventilator on, and to disconnect, first put the ventilator off, untie it, and then take off the mask (James et al. 2020; Pirzada et al. 2020).

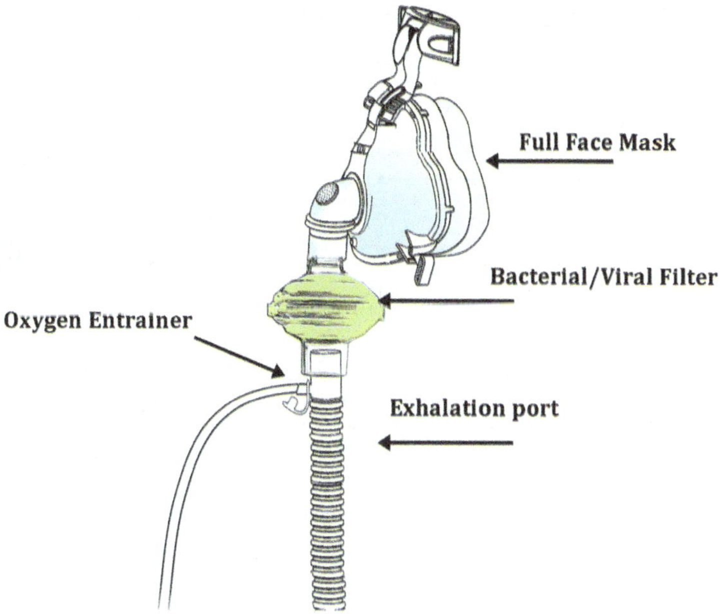

Fig. 21.1 A non-vented mask is attached to a bacterial filter, a connection to a safety valve and an oxygen entrainer port, and an exhalation port with a valve

To prevent accidental escape of breaths via a dislodged oxygen tubing while maintaining intentional leak, a viral/bacterial filter can be placed in the circuit between the mask and the oxygen and exhalation ports (Fig. 21.1) (Aarrestad et al. 2020; Kryger and Thomas 2020; Pirzada et al. 2020). This viral/bacterial filter can replace any filter at the machine end of the circuit and ideally needs to be changed every 24 h to prevent any resistance to flow due to moisture absorption, which can mimic clinical deterioration. It is important to mention that external humidification should not be used in such systems. Additionally, by adding a filter in a single-lumen NIV, PAP must negotiate the filter as it is fitted in the same lumen in the absence of a T-piece. Thus, the pressure gets reduced; it is to be considered and might require pressure escalation to maintain efficacy (Kryger and Thomas 2020; Pirzada et al. 2020; Rapoport 2020; Delorme et al. 2021). Figure 21.1 illustrates a non-vented mask attached to a bacterial filter, a connection to a safety valve and an oxygen entrainer port, and an exhalation port with a valve.

Commonly used NIV bilevel devices are single-lumen circuits and do not have a separate exhalation port. To act during the exigencies of patient volume and meager resources, devices are to be modified to prevent virus transmission and maintain circuitry. It is suggested to follow "jury rig," which is a system tailored to circumvent the lack of a commercial bilevel circuit with an exhalation port. This can be done by connecting a T- or Y-connector to the mask and then fitting a filter in the new limb, besides introducing an exhalation port (or resistor containing leak), which can be inserted before (or after) the filter (Pirzada et al. 2020; Rapoport 2020).

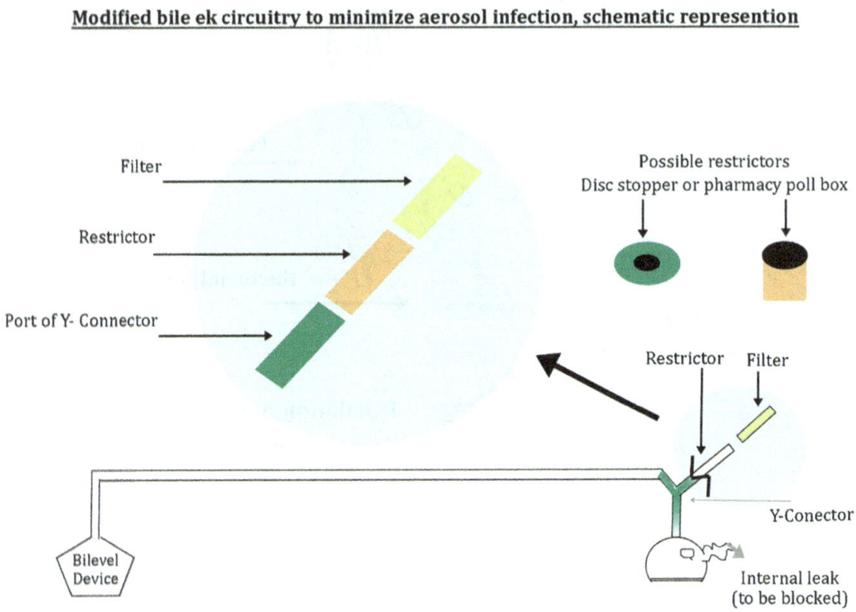

Fig. 21.2 An example of modified bilevel circuitry with a Y-connector added to reduce aerosol dispersion, with one limb connected to the device and the other to a filter through a restrictor

This addition of a "T" or "Y" tube piece with an "overflow" port allows filtering of all but the unintentional face leak (which can be managed using an appropriate mask). However, it creates dead space, which can be minimized by connecting the Y or T piece as close as feasible to the mask (Fig. 21.2) (Pirzada et al. 2020). However, merely adding a filter on the end of the "overflow" tube will dissipate much of the delivered pressure from the ventilator, thus compromising therapeutic benefits. An additional modification overcomes this in the circuit by adding a resistor. Thus, the combination of a filter and a resistor/restrictor in the right range will correct this issue. If a resistor is not commercially available, it can be tailored using a plug with a single 3.5 mm (1/8 or 5/32 inch) hole drilled in it or twenty 0.8 mm holes drilled in a standard medication pillbox that fits the tubing; holes are punctured/melted into the bottom with a hot paper clip or 18-gauge needle (Pirzada et al. 2020, Rapoport 2020). This arrangement should provide a leak flow of 12 l/min at 3 cm H_20 and 35 l/min at 25 cm H_20, providing adequate ventilation at most bilevel settings without significant CO2 rebreathing (Rapoport 2020).

The setup would replicate the typical leak port flows of existing masks or expiratory port circuits directed to a filter (Fig. 21.2) (Pirzada et al. 2020). The sterility of this part of the circuit is not a significant issue as the flow is unidirectional away from the patient (Rapoport 2020).

21.2.3 Environmental Control in Hospitals (Heating, Ventilation, and Air-Conditioning)

During the COVID-19 surge, hospital facilities were used for COVID-19 management, often bypassing area restrictions for NIV use. High-risk areas in the hospital are the isolation beds and ICU. Both these areas are used for NIV, particularly during an epidemic or pandemic like COVID-19. However, during phases of a sudden surge, almost all hospital facilities are used for COVID-19 patients, and NIV is used ubiquitously in the hospitals, and many patients are connected to NIV on arrival only in the emergency reception area. Thus, aerosol management in hospitals needs a comprehensive plan and program. The workforce must be prepared for the challenges associated with this infection control. The cornerstone is streamlining rapid diagnosis and isolation, clinical management, and infection prevention. Healthcare workers are at very high risk of acquiring infection and transmitting it to coworkers and other patients.

Standard infection control measures, as already discussed, are to be followed with professional conviction; however, aerosol management demands much more than that, to dispose of any contaminated usable and to clean the atmosphere. In addition, for aerosol management, the control of airborne bacteria and viruses should be guaranteed to minimize cross infection in hospital wards.

21.3 A Summary of Management Strategies According to Various Organizations

A. Ventilation: Wherever possible, maintain natural ventilation as much as possible; air dilution is an easy, although inadequate, way to dispose of infected aerosol. However, this is not possible in hospital settings as the environment is controlled, most often maintained by "heating, ventilation, and air-conditioning (HVAC) systems." The HVAC systems control the exchange and regulate cooling/heating and filtering. There are guidelines to maintain an optimal environment; thus, for general wards, 6 of total air and 2 of fresh air "air change per hour" (ACH) is recommended by the US and UK standards (ASHRAE Standard 170–200,831) and the UK standard (HTM 03–01). The European standard (CEN/TS 16244:2018) specifies the air quality requirements, air ventilation rate, control of airflow direction, and reduction of bio-contamination for hospitals which are practically the same. The WHO guidelines for natural ventilation for infection control in healthcare settings recommend a ventilation rate of 288 m^3/h (nearly 12 air change rate per hour) per person, at least to effectively eliminate the risk of infection. However, air exchanges for the ICU should be 12–20 ACH, 12 for the negative-pressure room, and 20 for the positive-pressure room.

For HVAC systems, to increase the amount of outdoor air in ventilation systems, open the minimum outdoor air dampers as high as 100% if possible. Disable demand-controlled ventilation to get the maximum. Central air filters should be improved as much as possible, at least to the grade of MERV-13 (Minimum Efficiency Reporting Value), which can filter all particles, including microorganisms as well as viruses, and seal the edges of the filter to limit bypass. Air circulation should be maintained from safe to unsafe, not from unsafe to safe area/room/ward, meaning the air from the area where NIV is used with potential infectivity should never go to the area of no infectivity.

Making judicious use of HEPA filters can theoretically remove at least 99.97% of dust, pollen, molds, bacteria, and any airborne particles of 0.3 mm in size. Keep the system running for longer hours; if possible, keep the system running 24 h a day, 7 days a week (7/24). For pandemic exigencies, "portable room air cleaners" with HEPA filters can be used in the areas where HVAC systems are not available particularly in makeshift areas during heavy patient burden.

B. Humidification and temperature: The indoor relative humidity between 50 and 60% is suggested to reduce the risk of spreading airborne infectious diseases as it prevents desiccation, dispersion, and human mucus membrane vulnerability. Thus, humidification must be optimal to minimize aerosol production and transmission. Since viruses can usually survive in low temp (4 C°) for longer hours, maintaining a higher temperature is recommended. Recommendations are not "comfort oriented" but "infection control oriented" to maintain higher temperatures during the pandemic.

C. UV light: UV lights can be installed in an air-handling unit or directly in the ventilation enclosure without impacting airflow pattern circulation. It works by breaking down certain chemical bonds and disrupting the structure of DNA, RNA, and proteins, causing a microorganism to become unable to multiply. However, if used in open spaces, like rooms or wards, it will require eye protection; thus, such use is discouraged.

It is relevant here to mention that exhaust from rooms should be low-lying and high air entry to minimize the aerosol movement along air currents. HVAC exhaust should be installed at the top of the hospital building, and the staff handling and maintaining the HVAC should use PPE during the handling protocols.

Thus, we conclude that aerosol management during NIV needs interventions, modification, and planning at multiple levels. We need to modify devices, circuitry, and interface, besides following environmental control by ensuring ventilation, natural and mechanical. General infection disease protocols must be adhered to with conviction (Memarzadeh 2011; Ding et al. 2020; Saran et al. 2020; Guo et al. 2021).

21.4 Practical Key Messages

From our discussion above, we can deduce the following:

- The use of noninvasive ventilation should not be absolutely discouraged at home and in the hospital during infections but should be weighed and considered in a tailored clinical context with precautions.
- Infection control guidelines are the backbone of aerosol management not only in the hospital but also at home.
- Interfaces and devices used during everyday NIV use might require to be changed at the time of infection.
- During healthcare exigencies like the COVID-19 pandemic, devices can be modified without replacing the whole assembly, at times with household tools and appendages.
- Environmental cleaning is essential at home and in healthcare facilities for better infection control during aerosol therapy.
- All the measures followed for NIV use during COVID-19 do not bypass standard NIV guidelines and precautions.

References

Aarrestad S, Wold M, Rasch-Halvorsen O, Indrekvam S and Fondenes O (2020) "NIV to patients in hospital with uncertain or detected Covid-19. Recommendations on practical implementation (non-peer-review article)." https://www.researchgate.net/publication/340285488_NIV_to_patients_in_hospital_with_uncertain_or_detected_Covid-340285419_Recommendations_on_practical_implementation_NIV_til_pasienter_innlagt_pa_sykehus_ved_usikker_eller_pavist_Covid-340285419_-_NKH_anbefalinger

Ahn JY, An S, Sohn Y, Cho Y, Hyun JH, Baek YJ, Kim MH, Jeong SJ, Kim JH, Ku NS, Yeom JS, Smith DM, Lee H, Yong D, Lee YJ, Kim JW, Kim HR, Hwang J, Choi JY (2020) Environmental contamination in the isolation rooms of COVID-19 patients with severe pneumonia requiring mechanical ventilation or high-flow oxygen therapy. J Hosp Infect 106(3):570–576

Ambrosino N, Guarracino F (2011) Unusual applications of noninvasive ventilation. Eur Respir J 38(2):440

Arulkumaran N, Brealey D, Howell D, Singer M (2020) Use of non-invasive ventilation for patients with COVID-19: a cause for concern? Lancet Respir Med 8(6):e45

Baker JG, Sovani M (2020) Case for continuing community NIV and CPAP during the COVID-19 epidemic. Thorax 75(5):368

Barker J, Oyefeso O, Koeckerling D, Mudalige NL, Pan D (2020) COVID-19: community CPAP and NIV should be stopped unless medically necessary to support life. Thorax 75(5):367

Bartoszko JJ, Farooqi MAM, Alhazzani W, Loeb M (2020) Medical masks vs N95 respirators for preventing COVID-19 in healthcare workers: a systematic review and meta-analysis of randomized trials. Influenza Other Respir Viruses 14(4):365–373

Bazant MZ, Bush JWM (2021) A guideline to limit indoor airborne transmission of COVID-19. Proc Natl Acad Sci U S A 118(17):e2018995118

Beggs CB (2003) The airborne transmission of infection in hospital buildings: fact or fiction? Indoor Built Environ 12(1–2):9–18

Buonanno G, Stabile L, Morawska L (2020) Estimation of airborne viral emission: quanta emission rate of SARS-CoV-2 for infection risk assessment. Environ Int 141:105794

Cardoso E, Herrmann MJ, Grize L, Hostettler KE, Bassetti S, Siegemund M, Khanna N, Sava M, Sommer G, Tamm M, Stolz D (2022) Is sleep disordered breathing a risk factor for COVID-19 or vice versa? ERJ Open Res 8(2):00034–02022

Craig S and West S (2022) "Guidance regarding coronavirus (COVID-19) and Obstructive Sleep Apnoea (OSA): for people who routinely use continuous positive airway pressure (CPAP), their families and health care workers." 2022, 1. https://www.brit-thoracic.org.uk/media/455098/osa-alliance-cpap-covid-455019-advice-455020-455093-455020-v455010.pdf

Davies A, Thomson G, Walker J, Bennett A (2009) A review of the risks and disease transmission associated with aerosol generating medical procedures. J Infect Prev 10(4):122–126

Delorme M, Leroux K, Boussaid G, Lebret M, Prigent H, Leotard A, Louis B, Lofaso F (2021) Protective recommendations for non-invasive ventilation during COVID-19 pandemic: a bench evaluation of the effects of instrumental dead space on alveolar ventilation. Arch Bronconeumol 57:28–33

Ding J, Yu CW, Cao SJ (2020) HVAC systems for environmental control to minimize the COVID-19 infection. Indoor Built Environ 29(9):1195–1201

Duval D, Palmer JC, Tudge I, Pearce-Smith N, O'Connell E, Bennett A, Clark R (2022) Long distance airborne transmission of SARS-CoV-2: rapid systematic review. BMJ 377:e068743

Esquinas AM, Egbert Pravinkumar S, Scala R, Gay P, Soroksky A, Girault C, Han F, Hui DS, Papadakos PJ, Ambrosino N, International NIV Network (2014) Noninvasive mechanical ventilation in high-risk pulmonary infections: a clinical review. Eur Respir Rev 23(134): 427–438

Ferioli M, Cisternino C, Leo V, Pisani L, Palange P, Nava S (2020) Protecting healthcare workers from SARS-CoV-2 infection: practical indications. Eur Respir Rev 29(155):200068

Guo M, Xu P, Xiao T, He R, Dai M, Miller SL (2021) Review and comparison of HVAC operation guidelines in different countries during the COVID-19 pandemic. Build Environ 187:107368–107368

Hamilton FW, Gregson FKA, Arnold DT, Sheikh S, Ward K, Brown J, Moran E, White C, Morley AJ, Bzdek BR, Reid JP, Maskell NA, Dodd JW (2022) Aerosol emission from the respiratory tract: an analysis of aerosol generation from oxygen delivery systems. Thorax 77(3):276

Hemmes JH, Winkler KC, Kool SM (1960) Virus survival as a seasonal factor in influenza and poliomyelitis. Nature 188:430–431

Hui DS, Chow BK, Lo T, Tsang OTY, Ko FW, Ng SS, Gin T, Chan MTV (2019) Exhaled air dispersion during high-flow nasal cannula therapy versus CPAP via different masks. Eur Respir J 53(4):1802339

Jackson T, Deibert D, Wyatt G, Durand-Moreau Q, Adisesh A, Khunti K, Khunti S, Smith S, Chan XHS, Ross L, Roberts N, Toomey E, Greenhalgh T, Arora I, Black SM, Drake J, Syam N, Temple R, Straube S (2020) Classification of aerosol-generating procedures: a rapid systematic review. BMJ Open Respir Res 7(1):e000730

James A, Verdonk F, Bougle A, Constantin JM (2020) Non-invasive ventilation for acute respiratory failure (in COVID-19 patients): the non-ending story? Anaesth Crit Care Pain Med 39(5): 549–550

Judson SD, Munster VJ (2019) Nosocomial transmission of emerging viruses via aerosol-generating medical procedures. Viruses 11(10):940

Knight V (1980) Viruses as agents of airborne contagion. Ann N Y Acad Sci 353:147–156

Kryger MH, Thomas R (2020) Home PAP devices in patients infected with COVID-19. J Clin Sleep Med 16(7):1217–1219

McEnery T, Gough C, Costello RW (2020) COVID-19: respiratory support outside the intensive care unit. Lancet Respir Med 8(6):538–539

Memarzadeh F (2011) Improved strategy to control aerosol-transmitted infections in a hospital suite. ASHRAE Conference, Las Vegas, NV

Menzella F, Barbieri C, Fontana M, Scelfo C, Castagnetti C, Ghidoni G, Ruggiero P, Livrieri F, Piro R, Ghidorsi L, Montanari G, Gibellini G, Casalini E, Falco F, Catellani C, Facciolongo N (2021) Effectiveness of noninvasive ventilation in COVID-19 related-acute respiratory distress syndrome. Clin Respir J 15(7):779–787

Pazarli AC, Ekiz T, Ilik F (2021) Coronavirus disease 2019 and obstructive sleep apnea syndrome. Sleep Breath 25(1):371

Perger E, Soranna D, Pengo M, Meriggi P, Lombardi C, Parati G (2021) Sleep-disordered breathing among hospitalized patients with COVID-19. Am J Respir Crit Care Med 203(2):239–241
Pirzada AR, Aleissi SA, Almeneessier AS, BaHammam AS (2020) Management of aerosol during noninvasive ventilation for patients with sleep-disordered breathing: important messages during the COVID-19 pandemic. Sleep Vigil 4(2):89–94
Pshenichnaya NY, Nenadskaya SA (2015) Probable Crimean-Congo hemorrhagic fever virus transmission occurred after aerosol-generating medical procedures in Russia: nosocomial cluster. Int J Infect Dis 33:120–122
Rapoport, DM (2020) "CPAP, Non Invasive Ventilation (NIV) by mask and use of NIV systems for intubated patients in the time of corona patient surge." www.thoracic.org. Accessed 13 Jan 2021
Robles-Romero JM, Conde-Guillen G, Safont-Montes JC, Garcia-Padilla FM, Romero-Martin M (2022) Behaviour of aerosols and their role in the transmission of SARS-CoV-2; a scoping review. Rev Med Virol 32(3):e2297
Rochwerg B, Brochard L, Elliott MW, Hess D, Hill NS, Nava S, Navalesi P, Antonelli M, Brozek J, Conti G, Ferrer M, Guntupalli K, Jaber S, Keenan S, Mancebo J, Mehta S, Raoof S (2017) Official ERS/ATS clinical practice guidelines: noninvasive ventilation for acute respiratory failure. Eur Respir J 50(2):1602426
Saran S, Gurjar M, Baronia A, Sivapurapu V, Ghosh PS, Raju GM, Maurya I (2020) Heating, ventilation and air conditioning (HVAC) in intensive care unit. Crit Care 24(1):194
Singh A, Sterk PJ (2008) Noninvasive ventilation and the potential risk of transmission of infection. Eur Respir J 32(3):816
Somsen GA, van Rijn C, Kooij S, Bem RA, Bonn D (2020) Small droplet aerosols in poorly ventilated spaces and SARS-CoV-2 transmission. Lancet Respir Med 8(7):658–659
Thomas RJ (2013) Particle size and pathogenicity in the respiratory tract. Virulence 4(8):847–858
Tran K, Cimon K, Severn M, Pessoa-Silva CL, Conly J (2012) Aerosol generating procedures and risk of transmission of acute respiratory infections to healthcare workers: a systematic review. PLoS One 7(4):e35797
van Doremalen N, Bushmaker T, Morris DH, Holbrook MG, Gamble A, Williamson BN, Tamin A, Harcourt JL, Thornburg NJ, Gerber SI, Lloyd-Smith JO, de Wit E, Munster VJ (2020) Aerosol and surface stability of SARS-CoV-2 as compared with SARS-CoV-1. N Engl J Med 382(16): 1564–1567
Wang CC, Prather KA, Sznitman J, Jimenez JL, Lakdawala SS, Tufekci Z, Marr LC (2021a) Airborne transmission of respiratory viruses. Science 373(6558):eabd9149
Wang YC, Lu MC, Yang SF, Bien MY, Chen YF, Li YT (2021b) Respiratory care for the critical patients with 2019 novel coronavirus. Respir Med 186:106516
Yu IT, Li Y, Wong TW, Tam W, Chan AT, Lee JH, Leung DY, Ho T (2004) Evidence of airborne transmission of the severe acute respiratory syndrome virus. N Engl J Med 350(17):1731–1739
Yu IT, Wong TW, Chiu YL, Lee N, Li Y (2005) Temporal-spatial analysis of severe acute respiratory syndrome among hospital inpatients. Clin Infect Dis 40(9):1237–1243
Zou L, Ruan F, Huang M, Liang L, Huang H, Hong Z, Yu J, Kang M, Song Y, Xia J, Guo Q, Song T, He J, Yen HL, Peiris M, Wu J (2020) SARS-CoV-2 viral load in upper respiratory specimens of infected patients. N Engl J Med 382(12):1177–1179

Chapter 22
Telemedicine and Telemonitoring of Sleep Disorders During the COVID-19 Pandemic

Amir Sharafkhaneh, Max Hirshkowitz, and Habibollah Khazaie

Abstract COVID-19 pandemic took the whole world by surprise. The widespread and devastating effects of the pandemic diverted the resources in taking care of patients infected with the virus while resulting in near-complete shutdown of sleep services. With the better control of the pandemic, sleep services gradually expanded their services, while in many localities, still they are below the pre-pandemic levels. The pandemic also pushed the use of telehealth services to the forefront of medical care. Sleep medicine services and workflows can easily adopt telehealth. Although the use of telehealth for sleep medicine grew during the pandemic, lack of infrastructure and of familiarity with telehealth are the main issues. In this chapter, we review the effect of pandemic on telehealth services as it relates to the application of telehealth for sleep services.

Keywords Telemedicine · Sleep-disordered breathing · Behavioral health · CBT-I · Insomnia

22.1 Introduction

COVID-19 pandemic took the world by surprise. It not only resulted in significant mortality and morbidity, but also created complexities in access to healthcare services and resulted in organizational changes in healthcare systems worldwide.

A. Sharafkhaneh (✉)
Baylor College of Medicine, Houston, TX, USA
e-mail: amirs@bcm.edu

M. Hirshkowitz
Baylor College of Medicine, Houston, TX, USA

Division of Public Mental Health and Population Sciences, Stanford University School of Medicine, Stanford, CA, USA

H. Khazaie
Sleep Disorders and Research Center, Kermanshah University of Medical Sciences, Kermanshah, Iran

A. S. BaHammam et al. (eds.), *COVID-19 and Sleep: A Global Outlook*, Progress in Sleep Research, https://doi.org/10.1007/978-981-99-0240-8_22

At the beginning of the pandemic and in many societies up to now, lockdowns created physical limitations to medical care. Further, majority of medical capabilities and resources were diverted to take care of the pandemic medical needs. The pandemic is affecting the sleep services even to this date (Hwang and Pépin 2020).

Access to sleep services is provided traditionally through face-to-face patient-provider interactions. The limitations of the current model of delivering medical care include timely access to a subspecialist, need for travel to long distances, and issue of space for clinical care and related expenses. The new models of care overcome these limitations through the use of new and upcoming telecommunication technologies.

Telehealth broadly is defined as access to healthcare providers and services anywhere and at any time. Telehealth services include real-time services (like an office visit but through telehealth) or remote review of the patient-related data. Telehealth covers a variety of services including inpatient and outpatient care, population health, home healthcare, second medical opinion, provider-to-provider consultation, group/individual education and counseling, group/individual physical rehabilitation, group/individual anti-addiction counseling and care, diagnostic interpretative services, remote patient monitoring, and quality control of various medical services (Kvedar et al. 2014).

Telehealth workflow can easily be adapted to accommodate various sleep services. These include diagnosis, review of the data, patient consultation, and initiation and follow-up of treatment. The diagnosis of sleep disorders heavily relies on the history and use of various questionnaires. The physician examination is mostly audiovisual rather than relying on palpation. Further, the relevant diagnostic tests are entirely in digital format and, thus, easily transferrable. Various device therapy options for sleep apnea provide remote access to the use of data. Thus, telehealth in the era of COVID-19 can be an excellent alternative to traditional models of sleep care (Verbraecken 2016; Bruyneel 2016).

22.2 COVID-19 Pandemic and Sleep Services

SARS-CoV-2, the pathogen causing the COVID-19 pandemic, is transmitted through droplets, airborne particles, and close contact with people. Due to concerns related to the viral transmission, the pandemic of COVID-19 brought the sleep services to a complete halt (Hwang and Pépin 2020). At the earlier stages of the pandemic, health authorities around the world recommended that nonemergency procedures, diagnostic evaluations, and laboratory tests in chronic patients, especially in cases related to the respiratory system, should be postponed or, if necessary, performed in a safer environment in the future (Voulgaris et al. 2020; Johnson et al. 2021).

Data from various countries show somewhere around 50–90% decline in sleep services due to the pandemic, and to this date, they have not recovered to pre-pandemic level. Polysomnography requires close contacts with patients (Grote et al. 2021). Attended PAP titration causes aerosolization and increases the possibility of infection. The devices and equipment used for sleep studies may harbor the virus and thus transmit the infection to others. Accordingly, all planned laboratory sleep studies, home sleep studies, and follow-up examinations of patients who had previously received sleep treatments were postponed at the earlier phase of this pandemic.

During the pandemic, not only sleep services diminished, but also incidences of sleep disorders like insomnia increased significantly (Kumar et al. 2022). Since the occurrence of sleep disorders during the epidemic is directly related to the occurrence of mood disorders, not paying attention to the treatment of sleep disorders caused a greater risk to public health during pandemics. Despite the dangers of ignoring sleep problems and the lack of sufficient evidence on how sleep laboratories work, efforts to provide remote medical care through telehealth increased to reduce patients' and providers' contact with pathogenic agents and conditions (Johnson et al. 2021).

22.2.1 Sleep Services at the Beginning and During the Pandemic Phase

At the beginning of the pandemic, there were no clear instructions on providing medical services in sleep laboratories. There were many challenges in conducting sleep studies at the beginning of the pandemic. Investigations show that in situations similar to the COVID-19 disease, sleep laboratories are known as one of the centers prone to spreading infection among patients and health professionals. Patients on nocturnal ventilatory support experience an inherently higher risk for respiratory infections. The droplets created by the ventilators due to the use of interfaces are spread up to a distance of 1 m, which increases the potential risk of infection to the treatment staff.

Sleep tests including level I and II polysomnography, and level III polygraphy, also known as a home cardiorespiratory study, carry risks of spreading the COVID-19 infection. The equipment of the polysomnography device that was connected to the patients during the study is made of fibers that get damaged by disinfectants. Also, there is a possibility of damage to polysomnography devices due to the use of disinfectants. The SDB group of the Canadian Thoracic Society, in relation to the management of patients with SDB during the epidemic, stated that laboratory sleep studies, home sleep apnea tests, and PAP treatment should not be performed during the COVID-19 epidemic. The society stated that in cases of patients with critical conditions, such as unstable cardiopulmonary diseases, in which SDB plays a fundamental role, disposable equipment should be used. Alternatively, if the reusable devices are used, after completing the treatment, the used equipment, based on its manufacturer specifications, should be placed out of use for a certain period of time to assure that the devices and equipment do not harbor the SARS-CoV-2 (Drummond 2020). Also, alternative treatment with auto-titrating PAP devices was recommended to prevent patients from going to hospitals (Voulgaris et al. 2020).

The Australian and British sleep associations also emphasized the safety of patients and healthcare workers. Their statements emphasized sleep centers to prioritize patients and to provide services based on the needs and severity of the diseases, as well as conducting screening tests for the infection with SARS-CoV-2 in patients. The American Academy of Sleep Medicine presented its recommendations regarding telephone screenings before home sleep apnea testing. Based on these recommendations, patients are evaluated for various symptoms of COVID-19, including fever, cough, and shortness of breath. Also, daily screening was done for the medical staff, and the staff was required to use personal protective equipment. French regional health agencies also emphasized the importance of gradually resuming screening and diagnostic activities to start timely treatment, especially in patients with SDB. This document is a practical framework for the relevant experts, and based on it, permission to treat patients with SDB was issued after the end of the quarantine. The statement recommended prioritization of patients in terms of the severity of the disease and the importance of protecting patients and treatment staff from the possible transmission and contracting of viral infections (Voulgaris et al. 2020; Bastier et al. 2020; Kole 2020). The European Sleep Association and the British Sleep Association provided guidelines for the provision of cognitive behavioral therapy in the treatment of insomnia and the management of obstructive sleep apnea at the beginning of the COVID-19 pandemic (Kanchan et al. 2021).

With the significant reduction of face-to-face visits for patients with sleep disorders, the role of telehealth in providing appropriate care became more apparent (Hwang and Pépin 2020). Healthcare facilities with already established telehealth services transitioned sleep-related services to telehealth even before recommendations from various sleep and respiratory societies emerged. In our center at the Veterans Administration, all face-to-face visits were moved to video visits, in-lab titrations were temporarily suspended, and most of the patients with SDB were managed by automatic PAP and telehealth visit. Figure 22.1 shows numerous applications of telehealth in the effective diagnosis and management of patients with various sleep disorders (Sharafkhaneh and Kuna 2021). Subsequent to explosion in the use of telehealth for sleep services, the American Academy of Sleep Medicine updated its 2015 guideline for the use of telehealth for sleep services (Shamim-Uzzaman et al. 2021; Singh et al. 2015). The reader is encouraged to review these two publications. Interestingly, growth of telemedicine in sleep also resulted in a chapter in the seventh edition of Principles and Practice of Sleep Medicine (Sharafkhaneh and Kuna 2021).

22.2.2 *Reopening of Sleep Services*

The 2019 coronavirus (COVID-19) pandemic caused disruptions in nonurgent health services to reduce the risk of infection and limit the spread of the virus,

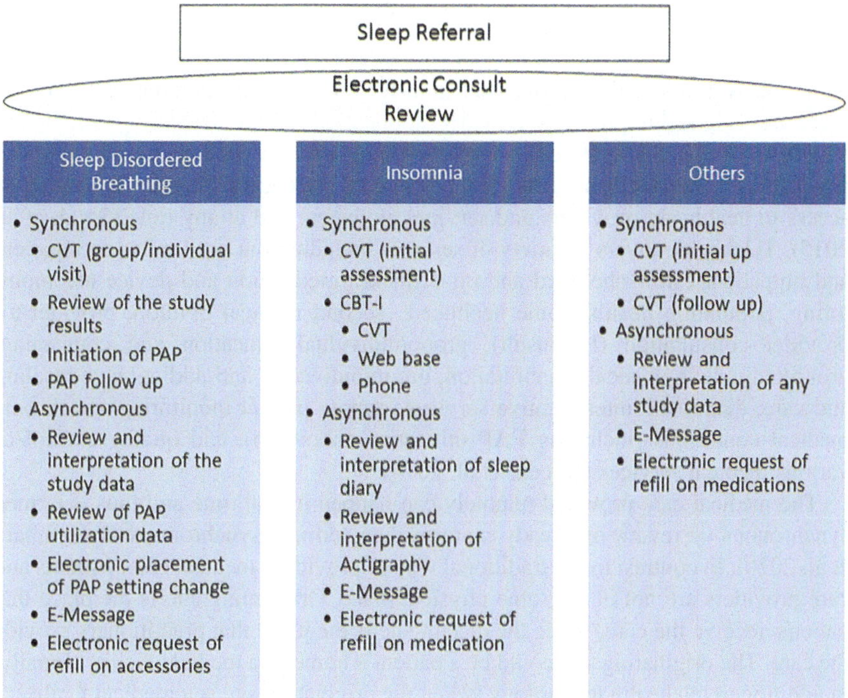

Fig. 22.1 Applications of telehealth to deliver sleep medicine services

particularly in hospital settings. Staffing of sleep centers in healthcare facilities strained by the pandemic resulted in almost complete closure of sleep laboratories and clinics during quarantine in the USA, Europe, and in other parts of the world (Grote et al. 2020; Wilson et al. 2020; American Academy of Sleep Medicine 2022).

Reopening of outpatient medical services encountered significant obstacles. Among these have been (a) scarcity of workforce, (b) lack of clear guidelines backed by data in how to open the sleep-related medical services while protecting the patients and healthcare providers at the same time, and (c) scarcity of medical resources due to diversion of the biotechnology industry in combating COVID and disruption of supply chain. Patient and provider safety significantly influenced traditional workflows across the healthcare industry, and telehealth took a center stage. Although a certain number of patients may require face-to-face contact, the shift to a new care paradigm, especially in the time of COVID-19, has been very critical.

22.3 Telemedicine: Concept and Definitions

Although both telemedicine and telehealth have been used interchangeably in the literature, telehealth provides a wider application of the technology including administrative and educational services of healthcare in contrast to telemedicine, which mostly relates to medical care of patients. Telehealth broadly is defined as access to healthcare providers and services anywhere and at any time (Singh et al. 2015). Telehealth covers a variety of services including but not limited to inpatient and outpatient care (scheduled and on-demand), medication and device use monitoring, population health, home healthcare, second medical opinion, provider-to-provider consultation (E-consult), group/individual education and counseling, group/individual physical rehabilitation, group/individual anti-addiction counseling and care, diagnostic interpretative services, remote patient monitoring for chronic medical conditions (including PAP utilization follow-up), and quality control of various medical services (Kvedar et al. 2014).

The medical care provided remotely can happen in real time and thus is termed synchronous vs. review of already captured data, termed asynchronous (Bahammam et al. 2020). In contrast to the traditional way of providing medical care, patients and care providers are not in the same physical place. Originating site is the place that patients receive the care, while the distant site is the place that practitioners provide the care. The originating site could be a patient's home or a medical facility. Finally, in addition to telehealth technology, when the originating site is a medical facility, a third person called presenter in the originating site is involved that will have healthcare background with training on how to operate the telehealth technology.

22.4 Requirements of Telehealth

The US healthcare market most likely is one with the most advanced telehealth worldwide. Although this market grew tremendously over the last few decades, its growth significantly accelerated with the COVID-19 pandemic. Use of telehealth in the time of pandemic seems intuitive; however, the implementation of telehealth needs a completely different mindset. Telehealth is not equal to information technology (IT). Presence of physician champions, trained staff in the use of telehealth, presence of required technology including hardware and software, and proper workflow are some of the very important components that are needed for implementation of telehealth.

One of the major issues with telehealth is the patients' and providers' acceptance. Studies show that patients view telehealth for sleep services as a positive experience (Parikh et al. 2011). Donovan and colleagues conducted a survey about patients' experiences with telehealth delivery of sleep services. COVID-19 pandemic started halfway during this study. Overall, the study clearly showed the acceptance of telehealth as an acceptable alternative. The patients reported that the telehealth is keeping them safe. One important concern was potential loss of privacy (Donovan et al. 2021).

Although telehealth use has been growing tremendously in part due to the pandemic, many aspects of telehealth are still evolving. Practice and business of healthcare vary significantly around the world, and each locality needs to address many issues related to telehealth before successful and sustainable implementation of virtual care. Important issues related to telehealth include technology-related issues (hardware and software), regulations and legal implications, payer-related issues, and licensing of practitioners (Sharafkhaneh and Kuna 2021; Khosla 2020; Cervenka and Iber 2020; Fields 2020; Abbasi-Feinberg 2020; O'Donnell et al. 2020). Figure 22.2 depicts the issues and stakeholders involved for a successful telehealth program (Sharafkhaneh et al. 2021).

22.5 Applications of Telemedicine in Sleep

There are multiple areas in which telehealth can be easily adopted in the evaluation of sleep disorders. Telehealth holds the promise not only of assessment and treatment but also of monitoring to ensure continued treatment and troubleshooting of the myriads of factors that can negatively impact numerous sleep disorders. Figure 22.1 shows potential applications of telehealth to sleep medicine services. The most common uses to date include orientation and demonstration for diagnostic testing or sleep-disordered breathing (SDB), individual sleep evaluations, and follow-up of disorders such as SDB, REM behavior disorder (RBD), narcolepsy, insomnia, and circadian rhythm disorder (Sharafkhaneh and Kuna 2021; Singh et al. 2015; Sharafkhaneh et al. 2022; Hirshkowitz and Sharafkhaneh 2014). The complications of telemedicine are relatively well understood; technology may impede the representation of medical conditions related to skin discoloration (leading to missed conditions such as jaundice, iron deficiency), olfaction which is sometimes an indicator of conditions such as infection, or other conditions. Identification of conditions which a practitioner might otherwise become aware of visually serendipitously may also be missed. However, assessments of *psychological* factors which influence the course of a sleep disorder are less vulnerable to these shortcomings. Telehealth for psychologically impacted sleep disorders may target a range of sleep complaints and broaden access to care (Sharafkhaneh et al. 2021).

Figure 22.3 depicts a simplified telehealth workflow for sleep services developed in our center. Initial evaluation of a patient with suspected sleep disorders entails evaluation by a sleep expert. Sleep medicine consultation relies mostly on obtaining medical history and limited physical exam; thus, sleep experts using a synchronous video visit can provide much-needed sleep consultation services with no need for patient and provider exposure to the virus and risk of infection with COVID-19. Available telehealth technologies including various secure applications for video visits make video consultation feasible. Diagnosis of sleep disorders heavily relies on the history obtained from patients, and telehealth is easily amenable to the

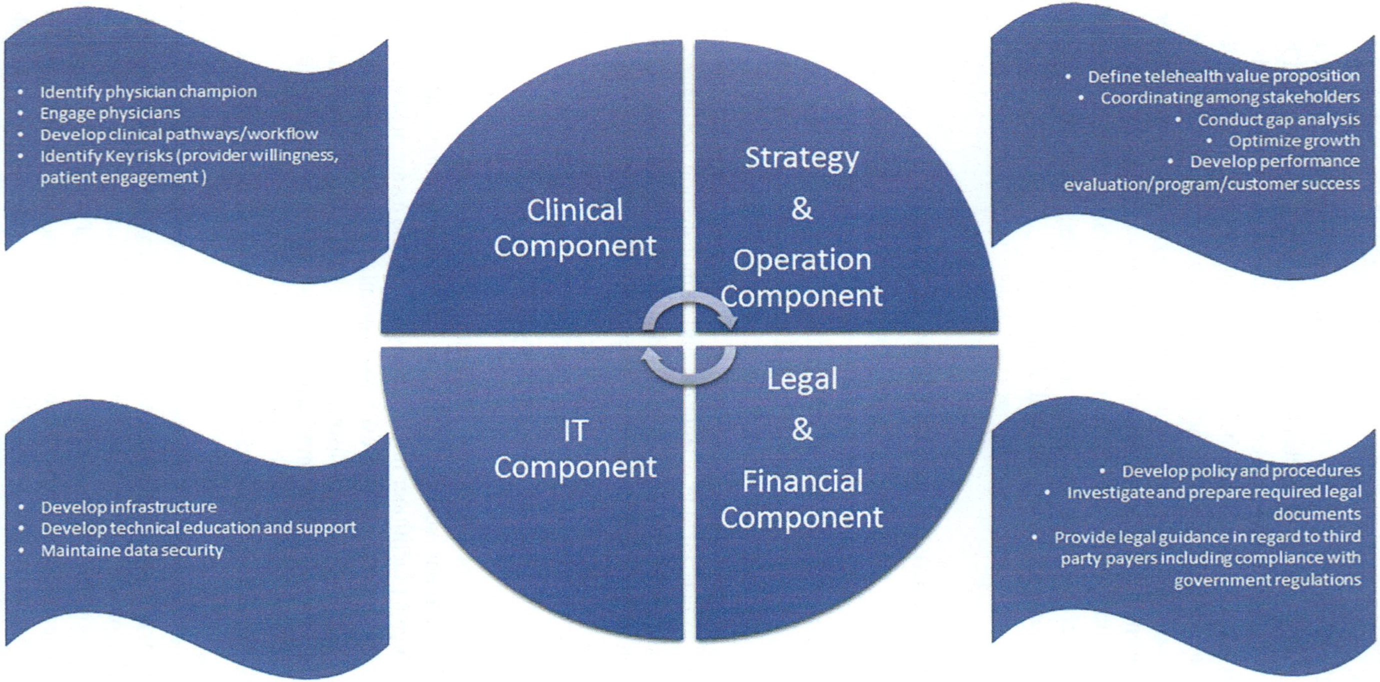

Fig. 22.2 Components required for a successful telehealth program

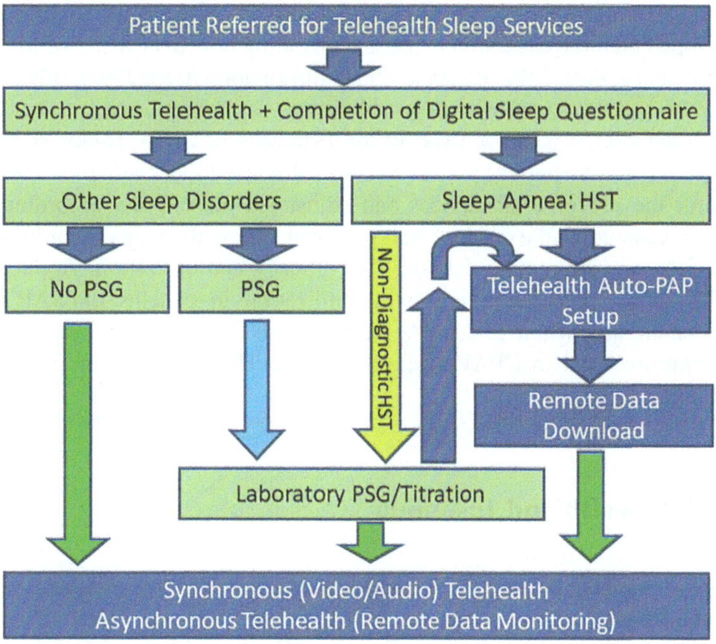

Fig. 22.3 A simplified telehealth sleep workflow

collection of the medical history information through the use of questionnaires in digital format. Further, confirmation of diagnosis of sleep disorders relies on objective diagnostic tools like polysomnography and home sleep testing. Scoring and interpretation of sleep studies are clear examples of store and forward or asynchronous service used in telehealth. Patients' follow-up after sleep studies by sleep experts and initiation of therapy can happen over video visits.

22.5.1 Telehealth and Sleep-Disordered Breathing

After diagnosis of sleep-disordered breathing, almost all aspects of sleep care can be delivered using telehealth technology (Sharafkhaneh and Kuna 2021; Hirshkowitz and Sharafkhaneh 2014; Verbraecken 2021). Implementation of telehealth for SDB care depends very much on the local healthcare organization and the already established workflows for the management of SDB. Discussion of the results of sleep studies with the patients can easily be done using synchronous telehealth including phone or video visits. Discussions about the type of treatments available and indicated and patient education are easily achievable through the use of telehealth. In our center, telehealth is used for initial setup and follow-up of PAP therapy of patients with sleep-disordered breathing (SDB). During COVID-19 pandemic, even mask fittings were conducted through video visits. In this program, we developed pictorial educational material including how to measure various points on human face to enable the patient to obtain information needed to decide about the mask size.

PAP utilization is a major issue in the treatment of SDB (Taylor et al. 2006; Yetkin et al. 2008; Stepnowsky Jr. et al. 2002). Telehealth technology has been used to monitor the use of PAP devices and adjust the prescription according to the efficacy data including the residual AHI provided. Currently, major manufacturers of PAP devices provide remote data on PAP efficacy and utilization. Application of telehealth shows promise in improving utilization in patients on PAP therapy (Sharafkhaneh and Kuna 2021). A meta-analysis by Aardoom and colleagues showed improvement in CPAP usage by use of eHealth interventions (Aardoom et al. 2020).

22.6 Telehealth and Insomnia

In all areas of sleep, behavioral medicine is a necessary tool in comprehensively understanding any sleep disorder, as well as issues that may block adherence to treatment(s). Behavioral therapy workflow is easily adaptable to telehealth as it does not require any physical exam. For example, behavioral therapy to improve PAP utilization can be effectively conducted using telehealth.

CBT-I is the first line of therapy for insomnia; however, access to CBT-I is significantly limited. Thus, telehealth may play a significant role in improving access to care. Telehealth can provide both self-driven (computer based) and clinician-driven CBT-I. Computerized cognitive behavioral therapy (CCBT) has rapidly expanded to offer structured CBT programs via the internet or mHealth (mobile health apps). These are generally grounded in the basic principles of CBT but are self-driven (Chan et al. 2017; Thorndike et al. 2013). In a recent meta-analysis, CCBT programs appear to have mild-to-moderate benefit on insomnia (Cheng and Dizon 2012) and to be overall moderately useable (Yu et al. 2019). Studies involving trained providers, however, indicate that telehealth delivery of CBT is effective for both depression and insomnia (Lichstein et al. 2013; Scogin et al. 2018). Contrary to internet programs in which no provider is involved, telehealth is a clinician-driven CBT-I. Telehealth-based CBT-I is as effective as face-to-face standard of care in multiple studies. A recent meta-analysis of telehealth in insomnia management showed significant improvement in access to services and therapeutic effects (Sharafkhaneh et al. 2022).

22.7 Summary

COVID-19 pandemic resulted in suspension of many non-emergent medical services to stop the spread of the virus, but also to divert the medical resources to much-needed COVID-19 care. At the same time, various forms of telehealth became the center stage for delivering medical care. Many governmental agencies around the world loosened the regulations related to telehealth to enable the providers to use communication technology to deliver medical care with minimal risk of spreading COVID-19 infection. Sleep services, being considered nonurgent in overwhelming majority of cases, encountered significant setback. Interestingly, as sleep services did not relay as much on detailed physical exam, many of the services related to sleep care did transfer to telehealth workflow. Many centers including ours expanded or adopted telehealth to provide sleep care. These services included initial and follow-up patient evaluation through telephone or video visits, initiation of SDB therapy using remote interface fitting, patient education, PAP setup (using auto-PAP when indicated), remote patient usage monitoring, remote patient counseling when usage is suboptimal, and remote CBT-I. With the widespread use of telehealth, obstacles to use also emerged. The obstacles include but not limited to availability of technology, design of effective workflow fitting any specific medical practice, legal issues including licensing of the remote providers, patient privacy, data security, connectivity, and financial aspects. Before the pandemics, many patients and providers did not accept that the telehealth can provide similar medical services as face-to-face visits will do. The pandemic significantly overcame the obstacles of patient and provider acceptance of telehealth services. Telehealth is only a tool to facilitate access to care including sleep services. However, telehealth should be customized to each locality with consideration given to major issues including the payers, patients' culture and acceptance, licensing requirements, workflow, existence of the technology, and cost. A telehealth model developed in one country may not fit in other places and thus needs to be adapted. Further research in legal, technological, and economic aspects of telehealth for sleep medicine is needed to optimize and expand its use.

References

Aardoom JJ, Loheide-Niesmann L, Ossebaard HC, Riper H (2020) Effectiveness of eHealth interventions in improving treatment adherence for adults with obstructive sleep apnea: meta-analytic review. J Med Internet Res 22(2):e16972

Abbasi-Feinberg F (2020) Telemedicine coding and reimbursement–current and future trends. Sleep Med Clin 15(3):417–429

American Academy of Sleep Medicine (2022) COVID-19 Mitigation strategies for sleep clinics and sleep centers-REOPENING. https://aasm.org/covid-19-resources/covid-19-mitigation-strategies-sleep-clinics-labs. Accessed 20 Sept 2022

Bahammam S, Asiri S, Al-Jahdali Y et al (2020) Attitudes and behaviours of physicians towards the relationship with the pharmaceutical industry in Saudi Arabia. East Mediterr Health J 26(3): 323–330

Bastier PL, Aisenberg N, Durand F et al (2020) Treatment of sleep apnea by ENT specialists during the COVID-19 pandemic. Eur Ann Otorhinolaryngol Head Neck Dis 137(4):319–321

Bruyneel M (2016) Technical developments and clinical use of telemedicine in sleep medicine. J Clin Med 5(12):116

Cervenka T, Iber C (2020) EHR integration of PAP devices in sleep medicine implementation in the clinical setting. Sleep Med Clin 15(3):377–382

Chan C, West S, Glozier N (2017) Commencing and persisting with a web-based cognitive behavioral intervention for insomnia: a qualitative study of treatment completers. J Med Internet Res 19(2):e37

Cheng SK, Dizon J (2012) Computerised cognitive behavioural therapy for insomnia: a systematic review and meta-analysis. Psychother Psychosom 81(4):206–216

Donovan LM, Mog AC, Blanchard KN et al (2021) Patient experiences with telehealth in sleep medicine: a qualitative evaluation. J Clin Sleep Med 17(8):1645–1651

Drummond M (2020) Sleep labs, lung function tests and COVID-19 pandemic–only emergencies allowed! Pulmonology 26(4):244–245

Fields BG (2020) Regulatory, legal, and ethical considerations of telemedicine. Sleep Med Clin 15(3):409–416

Grote L, McNicholas WT, Hedner J, ESADA collaborators (2020) Sleep apnoea management in Europe during the COVID-19 pandemic: data from the European Sleep Apnoea Database (ESADA). Eur Respir J 55(6):2001323

Grote L, Theorell-Haglow J, Ulander M, Hedner J (2021) Prolonged effects of the COVID-19 pandemic on sleep medicine services-longitudinal data from the Swedish sleep apnea registry. Sleep Med Clin 16(3):409–416

Hirshkowitz M, Sharafkhaneh A (2014) A telemedicine program for diagnosis and management of sleep-disordered breathing: the fast-track for sleep apnea tele-sleep program. Semin Respir Crit Care Med 35(5):560–570

Hwang D, Pépin JL (2020) Sleep telemedicine: a template for other specialties? Sleep Med Clin 15(3):xi

Johnson KG, Sullivan SS, Nti A, Rastegar V, Gurubhagavatula I (2021) The impact of the COVID-19 pandemic on sleep medicine practices. J Clin Sleep Med 17(1):79–87

Kanchan S, Saini LK, Daga R, Arora P, Gupta R (2021) Status of the practice of sleep medicine in India during the COVID-19 pandemic. J Clin Sleep Med 17(6):1229–1235

Khosla S (2020) Implementation of synchronous telemedicine into clinical practice. Sleep Med Clin 15(3):347–358

Kole A (2020) Home sleep apnea testing in the era of COVID-19: a community perspective. J Clin Sleep Med 16(9):1633

Kumar N, Goyal A, Hussain A et al (2022) SARS-CoV-2 infection is associated with increased odds of insomnia, RLS and dream enactment behavior. Indian J Psychiatry 64(4):354–363

Kvedar J, Coye MJ, Everett W (2014) Connected health: a review of technologies and strategies to improve patient care with telemedicine and telehealth. Health Aff (Millwood) 33(2):194–199

Lichstein KL, Scogin F, Thomas SJ, DiNapoli EA, Dillon HR, McFadden A (2013) Telehealth cognitive behavior therapy for co-occurring insomnia and depression symptoms in older adults. J Clin Psychol 69(10):1056–1065

O'Donnell C, Ryan S, McNicholas WT (2020) The impact of telehealth on the organization of the health system and integrated care. Sleep Med Clin 15(3):431–440

Parikh R, Touvelle MN, Wang H, Zallek SN (2011) Sleep telemedicine: patient satisfaction and treatment adherence. Telemed J E Health 17(8):609–614

Scogin F, Lichstein K, DiNapoli EA et al (2018) Effects of integrated telehealth-delivered cognitive-behavioral therapy for depression and insomnia in rural older adults. J Psychother Integr 28(3):292–309

Shamim-Uzzaman QA, Bae CJ, Ehsan Z et al (2021) The use of telemedicine for the diagnosis and treatment of sleep disorders: an American Academy of sleep medicine update. J Clin Sleep Med 17(5):1103–1107
Sharafkhaneh A, Kuna S (2021) Sleep telemedicine and remote PAP adherence monitoring. In: Kryger MH, Roth T, Goldstein CA (eds) Principles and practice of sleep medicine, 7th edn. Elsevier, Amsterdam
Sharafkhaneh A, Moghtader S, Farhadpour B, Singh S, Rose M (2021) Telesleep medicine. In: Khawaja IS, Hurwitz TD (eds) Sleep disorders in selected psychiatric settings: a clinical casebook. Springer, Cham, pp 165–174
Sharafkhaneh A, Salari N, Khazaie S et al (2022) Telemedicine and insomnia: a comprehensive systematic review and meta-analysis. Sleep Med 90:117–130
Singh J, Badr MS, Diebert W et al (2015) American Academy of sleep medicine (AASM) position paper for the use of telemedicine for the diagnosis and treatment of sleep disorders. J Clin Sleep Med 11(10):1187–1198
Stepnowsky CJ Jr, Marler MR, Ancoli-Israel S (2002) Determinants of nasal CPAP compliance. Sleep Med 3(3):239–247
Taylor Y, Eliasson A, Andrada T, Kristo D, Howard R (2006) The role of telemedicine in CPAP compliance for patients with obstructive sleep apnea syndrome. Sleep Breath 10(3):132–138
Thorndike FP, Ritterband LM, Gonder-Frederick LA, Lord HR, Ingersoll KS, Morin CM (2013) A randomized controlled trial of an internet intervention for adults with insomnia: effects on comorbid psychological and fatigue symptoms. J Clin Psychol 69(10):1078–1093
Verbraecken J (2016) Telemedicine applications in sleep disordered breathing: thinking out of the box. Sleep Med Clin 11(4):445–459
Verbraecken J (2021) Telemedicine in sleep-disordered breathing: expanding the horizons. Sleep Med Clin 16(3):417–445
Voulgaris A, Ferini-Strambi L, Steiropoulos P (2020) Sleep medicine and COVID-19. Has a new era begun? Sleep Med 73:170–176
Wilson KC, Kaminsky DA, Michaud G et al (2020) Restoring pulmonary and sleep services as the COVID-19 pandemic lessens. From an Association of Pulmonary, critical care, and sleep division directors and American Thoracic Society-coordinated task force. Ann Am Thorac Soc 17(11):1343–1351
Yetkin O, Kunter E, Gunen H (2008) CPAP compliance in patients with obstructive sleep apnea syndrome. Sleep Breath 12(4):365–367
Yu JS, Kuhn E, Miller KE, Taylor K (2019) Smartphone apps for insomnia: examining existing apps' usability and adherence to evidence-based principles for insomnia management. Transl Behav Med 9(1):110–119

Chapter 23
Changes in Positive Airway Pressure Application and Practices in Individuals with Sleep-Related Breathing Disorder During the COVID-19 Pandemic

Dennys Franco-Avecilla, Mouaz Saymeh, and Tetyana Kendzerska

Abstract The positive airway pressure (PAP) therapy prescription and use for individuals with sleep-related breathing disorder (SRBD) have been affected by the COVID-19 pandemic, given concerns for aerosolizing risk associated with PAP therapy use. Guidelines have recommended balancing the risks and benefits of continuing PAP therapy during the pandemic by assessing cohabitation status, severity of underlying SRBD, comorbidities, and occupation safety. At the same time, untreated SRBD has been shown to be associated with poor outcomes in patients contracting COVID-19 infection. Although PAP prescriptions have decreased during the pandemic and PAP initiation methods have shifted from face-to-face to remote setups, PAP use has not changed and potentially increased since the beginning of the COVID-19 pandemic. It has been found that individuals with the lowest PAP use before the pandemic demonstrated the most marked improvement in PAP use. Male gender and age 65 years and older were associated with increased PAP use compared to their female and younger counterparts. Conversely, higher stress levels, depression, poor sleep quality, remote PAP setup, and living with someone who experienced symptoms that could be attributable to COVID-19 have been suggested to be associated with lower PAP use. Critically, lower PAP usage was found in individuals newly started on PAP after the first wave of the pandemic. Further research is needed to identify populations at risk of reducing PAP use and target education strategies.

D. Franco-Avecilla · M. Saymeh
The Ottawa Hospital, Ottawa, ON, Canada
e-mail: dfranco@toh.ca; mosaymeh@ohri.ca

T. Kendzerska (✉)
Clinical Epidemiology Program, The Ottawa Hospital Research Institute, Ottawa, ON, Canada

Division of Respirology, Department of Medicine, Faculty of Medicine, The University of Ottawa, Ottawa, ON, Canada

The Ottawa Hospital Sleep Center, The Ottawa Hospital, Ottawa, ON, Canada
e-mail: tkendzerska@toh.ca

A. S. BaHammam et al. (eds.), *COVID-19 and Sleep: A Global Outlook*, Progress in Sleep Research, https://doi.org/10.1007/978-981-99-0240-8_23

Keywords Sleep-related breathing disorder · Positive airway pressure use · COVID-19 pandemic · Populations at risk · Positive airway pressure initiation

23.1 COVID-19 Transmission Risk Associated with Positive Airway Pressure Use

A severe respiratory infection started in Wuhan, China, in December 2019 caused by a novel coronavirus called SARS-CoV-2, which led to high morbidity and mortality. On March 11, the World Health Organization (WHO) declared a worldwide pandemic (Sencer 2020). As a result, many countries declared a nationwide state of emergency, forcing individuals to their homes and allowing only essential travel.

The COVID-19 pandemic has changed the operation of the healthcare system in an effort to decrease the risk of viral transmission. The spread of the virus is known predominantly through droplets; however, it is known to transmit also through airborne particles in high concentrations within enclosed environments, especially seen during aerosol-generating procedures (AGPs). AGPs include noninvasive ventilation (NIV) and positive airway pressure (PAP) devices, such as bilevel positive airway pressure (BPAP), continuous positive airway pressure (CPAP), and high-flow nasal prone (HFNP). The latter two showed limited dispersion and provided a good mask fitting (Hui et al. 2019).

23.2 Importance of Obstructive Sleep Apnea Treatment

Obstructive sleep apnea (OSA) is a prevalent chronic sleep-related disorder estimated to affect 3–7% of adult males and 2–5% of adult females in Western countries (Lindberg 2010). Worldwide, OSA prevalence has been estimated to be 936 million with mild-to-moderate disease and 425 million with moderate-to-severe OSA (Benjafield et al. 2019). The affected population is often comorbid with diseases such as hypertension, diabetes mellitus, heart failure, and atrial fibrillation, which similarly causes worse COVID-19 infection-related outcomes (Chen et al. 2020); moreover, even treated OSA has shown to be associated with a risk of mortality on day 7 of admission to hospital due to COVID-19 (Cariou et al. 2020). This risk demarcates the importance of ensuring proper OSA management.

The management of OSA has been affected by the COVID-19 pandemic, given concerns for aerosolizing risks among individuals using PAP at home. The WHO and the Centers for Disease Control (CDC) both acknowledged the risk of COVID-19 transmission with PAP use early at the beginning of the current pandemic (CDC 2022; WHO 2022). In May 2020, the Canadian Thoracic Society recommended balancing the risks and benefits of continuing PAP therapy in individuals with suspected or confirmed COVID-19 infection taking into account cohabitation status and OSA severity.

23.3 Impact of the COVID-19 Pandemic on the Diagnosis and Management of Sleep-Related Breathing Disorder

From the beginning of the COVID-19 pandemic in March 2020, several countries have observed the impact of the pandemic in the modes of health care in order to shift resources to care for critically ill individuals (ACS 2020) and to mitigate the risk of COVID-19 infection spread. Worldwide, primary care centers have transitioned to a remote care model with only selected in-person medical visits to optimize medical staff and protective equipment (PPE) distribution while protecting high-risk individuals with medical comorbidities from infection (Kendzerska et al. 2021).

The diagnosis and management of SRBD and PAP therapy use and prescription have been altered since the pandemic. Regarding the diagnosis of new SRBD, the use of sleep study methods has varied, and the availability has possibly decreased since the beginning of the pandemic. For example, in sleep centers across Europe, a questionnaire study showed a sharp decrease in in-lab diagnostic polysomnography (PSG) from 92.5% to 20.0% ($p < 0.001$) as well as a decrease in at-home polygraphy for the diagnosis of SRBD from 87.5% to 32.5%. Telemedicine-based SRBD diagnosis decreased from 30.0% to 27.5%; this shift in diagnostic approach happened during the first 1–2 months of the COVID-19 pandemic and may have been explained by the closure of centers that provided diagnostic services (Grote et al. 2020).

Conversely, in the UK, a study comparing diagnostic approaches for SRBD in sleep centers showed no significant change in their use. The study took place comparing 5 months after the beginning of the pandemic (July–August 2020) to the same time a year prior and showed that at-home PSG decreased marginally from 61.0% to 57.8%; laboratory PSG/polygraphy increased from 13.7% to 16.2%, and pulse oximetry increased from 22.9% to 23.1%; however, this increase was not significant ($p = 0.59$) (Turnbull et al. 2022).

The prescription of PAP therapy has been shown to decrease during the pandemic. For example, the study performed in Italy reported different aspects of care in individuals admitted to a rehabilitation facility with several respiratory conditions, including SRBD. A reduction was seen in the prescriptions for PAP for all individuals with respiratory conditions from 25.4% in the pre-pandemic era to 16.2% during the pandemic; this was felt secondary to a decrease in indication for PAP during the first year after the beginning of the pandemic (March 2020–March 2021) compared to the same time a year before (Vitacca et al. 2021). Earlier in the pandemic, across various sleep centers in Europe, in-lab titrations decreased significantly for CPAP and BiPAP during the first 1–2 months of the COVID-19 pandemic (from 90.0% to 17.5% for CPAP titrations and from 87.5% to 17.5% for BiPAP, $p < 0.001$).

Lastly, in the UK, Turnbull et al. described three different PAP therapy initiation methods after the COVID-19 first wave (July–August 2020) in eight sleep centers; these included (a) face-to-face (F2F) with a PAP machine turned on, (b) F2F with a PAP machine turned off, and (c) entirely remote setup. Expectedly, it was found that face-to-face PAP titration with the machine turned on decreased by 97.80%, from 98.9% to 1.1%. The alternative methods were F2F titration with a PAP machine turned off, which was used in 71.1% of the individuals, and the entirely remote setup used in 27.9% (Turnbull et al. 2022).

In-person sleep clinic follow-up appointments were also limited in Europe during the first year of the pandemic. Specifically, in-person follow-up appointments decreased by 75% because they were mainly performed via phone calls, and ambulatory titration decreased by the same percentage because of telemonitoring (Grote and McNicholas 2020). In the UK, telephone follow-up visits increased from 21% to 88% 4 months after the beginning of the pandemic ($p < 0.0001$) (Turnbull et al. 2022).

23.4 Positive Airway Pressure Use During the COVID-19 Pandemic

The use of PAP therapy has been studied since the pandemic using subjective (self-reported) (Thorpy et al. 2020; Batool-Anwar et al. 2020; Kendzerska et al. 2022) or objective (electronic monitoring) (del Campo et al. 2020; Fidan et al. 2021; Pepin et al. 2021) measures.

PAP therapy use has varied during the COVID pandemic across different countries (Table 23.1). Initially, during the first month since the beginning of the pandemic (March–April 2020), studies conducted in France and Spain showed an increase in PAP usage (del Campo et al. 2020; Pepin et al. 2021). A study conducted in the United States during the same time period demonstrated no change in PAP use (Batool-Anwar et al. 2020). Specifically, Pepin et al., who measured PAP use objectively and longitudinally in a large study (March–April 2020), found that the PAP use per night increased from 394 min (interquartile range [IQR], 305–458) to 414 min (IQR, 324–479). Notably, it was found that the subgroup with the lowest PAP use before the pandemic demonstrated the most marked improvement. In the same study, an increase in PAP use was seen more in men as compared to women (absolute difference, 7.6 min (95% CI, 3.3–11.6 min), $p < 0.01$) (Pepin et al. 2021) (Table 23.1). Interestingly, a mixed regression model demonstrated a 20.2-min increase in PAP use after the lockdown among individuals 65–75 years old compared to those under 65 years, and a 10.3-min increase among individuals over 75 years compared to those under 65. Overall, the interaction between age and assessment period was significant ($p < 0.01$) (Pepin et al. 2021). Furthermore, during the first month in France, areas with a higher COVID-19 infection burden were not significantly associated with changes in PAP use (Pepin et al. 2021). A study conducted in Spain demonstrated that CPAP use significantly increased during the first month (March–April 2020) since the pandemic from 75.1% to 79.5% ($p < 0.0001$); moreover, 27% of patients who were noncompliant prior became compliant during this period (del Campo et al. 2020).

Table 23.1 Population characteristics of studies focusing on the management and diagnosis of sleep-disordered breathing since the COVID-19 pandemic

Study	Population	Demographics	Changes in PAP use and characteristics since the pandemic	Variables associated with reduced PAP use
Fidan et al. (Turkey) Prospective observational study/objective data on PAP use	N = 34 individuals with prior auto-CPAP use (19 with COVID-19 infection history)	Age: 48.4 ± 11.1 Males: 64.5% BMI: 33.5 ± 9.4 Baseline AHI: 55.8 ± 14.1 Lowest SPO_2: 74.8 ± 11.9	**Changes in PAP use:** Not reported **Changes in PAP characteristics and OSA control before vs. 1 month after COVID-19 infection** – Average 95% APAP pressure: 8.56 ± 0.17 vs. 9.78 ± 0.21 cm H_2O ($p < 0.01$) – Median pressure: 7.49 ± 0.16 vs. 8.15 ± 0.19 cm H_2O ($p < 0.01$) – Residual AHI as recorded by the device: 3.61 ± 0.09 vs. 3.78 ± 0.09 ($p = 0.27$)	N/A
Thorpy et al. (USA) Telephone questionnaire survey study/subjective data on PAP use	N = 112 (10 reported a positive test for COVID-19)	Age: 60.6 (43–77) Males: 41% BMI: 36.2 ± 7.3 COVID-19 infection: 9% Asthma or COPD: 31% Hypertension: 59% Arrhythmias: 8% CAD: 11% Diabetes mellitus: 24%	**PAP use before vs. 1–2 months since the pandemic** – Continued: 85 (88%) From the continued group 21 (20%) reported increased use – Stopped: 11 (11%) **Among individuals with COVID-19** (N = 10) – Stopped: 6 (60%) – Increased: 2 (20%) – Continued: Not reported	N/A

(continued)

Table 23.1 (continued)

Study	Population	Demographics	Changes in PAP use and characteristics since the pandemic	Variables associated with reduced PAP use
Kendzerska et al. (USA and Canada) Cross-sectional survey study/ subjective data on PAP use	N = 570	**Continued PAP vs. stopped PAP** Age: 62 vs. 59 Males: 54.6% vs. 39.0% Chronic illness: 84.1% vs. 80.5% Anxiety: 4% vs. 7% Depression: 8% vs. 9% COVID-like symptoms: 32.7% vs. 43.9%	**PAP use before vs. up to 10 months since the pandemic** – Continued: 529 (93%) – Stopped: 41 (7%) **Getting tested for COVID-19**: – Among those who continued: 32/529 (6%) – Among those who stopped: 6/41 (15%) ($p = 0.034$)	**Continued PAP vs. stopped PAP**[a] – ↑ Stress level: OR 1.13 (1.02–1.25), $p = 0.02$ – Someone else in a household had – COVID-19: OR 3.05 (1.00–9.31), $p = 0.049$ – ↓ sleep time, sleep efficiency, and sleep quality (p values <0.001)
Tumbull et al. (UK) Multicenter retrospective observational study/objective data on PAP use	**Before vs. after COVID-19** N = 620 vs. 567	**Before vs. after COVID-19:** Age: 53.4 ± 13.6 vs. 53.4 ± 13.3 Males: 66.8% vs. 67% BMI: 34.9 vs. 34.8 ESS: 11.2 ± 5.5 vs. 11.8 ± 5.2 ODI (median): 23.0 vs. 21.5/h AHI (median): 25.5 vs. 25.7/h	**CPAP use before vs. 5 months since the pandemic in patients newly started on CPAP after the first wave of the pandemic** – Median CPAP use: 5.4 (2.7–6.9) vs. 4.2 (1.4–6.1) hours/night Unadjusted reduction: 0.9 h/night (95% CI: 0.5–1.2) ($p < 0.0001$) Adjusted reduction: 0.6 h/night (95% CI: 0.3–1.2) ($p < 0.0006$) – Adherent with CPAP (>4 h/night on ≥70% of nights): 53.3% (95% CI: 48.7–57.8%) vs. 41.2%	**5 months since the pandemic**: Remote setup vs. F2F: ↓ in CPAP use of 0.6 h/night ($p = 0.03$)

			(95% CI: 36.9–45.6%) ($p = 0.0002$) **CPAP setup before vs. 5 months since the pandemic** – F2F with CPAP machine turned on: 98.9% vs. 1.1% – F2F with CPAP machine turned off: 0% vs. 71.1% – Entirely remote setup: 0% vs. 27.9% **Sleep study methods before vs. 5 months since the pandemic ($p > 0.05$)** – Home polygraphy: 378 (61.0%) vs. 328 (57.8%) – Laboratory polygraphy/ polysomnography: 85 (13.7%) vs. 92 (16.2%) – Pulse oximetry 142 (22.9%) vs. 131 (23.1%) **Telephone follow-up before vs. 5 months since the pandemic:** 21% vs. 88%	
Vitacca et al. (Italy) Retrospective observational database study/objective data on PAP use	**Before vs. after COVID-19** N total = 6001: 3912 vs. 2089 OSA: 5.16% vs. 2.15%	**Before vs. after COVID-19** Age: 70.44 ± 10.95 vs. 69.82 ± 11.74 Males: 63.40% vs. 65.21% COVID-19: 12.7%	**CPAP prescription 1 year before vs. 1 year since the pandemic:** 25.4–16.2%	N/A

(continued)

Table 23.1 (continued)

Study	Population	Demographics	Changes in PAP use and characteristics since the pandemic	Variables associated with reduced PAP use
		COPD: 25.8% vs. 14.6%		
Grote et al. (across Europe) Prospective questionnaire-based study/subjective data on PAP use	N = 40 sleep centers No individual-level data	N/A	**CPAP treatment start procedures before vs. 1–2 months since the pandemic** – In-lab titration: 90% vs. 17.5% ($p < 0.001$) – Ambulatory titration: 55% vs. 22.5% – Telemedicine-based titration: 32.5% vs. 32.5% – Regularly use telemedicine, n: 6 vs. 4 **Bilevel PAP treatment start procedures before vs. 1–2 months since the pandemic** – In-lab titration: 87.5% vs. 17.5% ($p < 0.001$) – Ambulatory titration: 40% vs. 17.5% – Telemedicine-based titration: 20% vs. 12.5% – Regularly use telemedicine, n: 4 vs. 3	N/A
Del Campo et al. (Spain) Prospective observational study/objective data on PAP use	N = 2956	Age: 63 (IQR 54–71) Males: 77.7% CPAP baseline compliance: 75.1%	**CPAP use 3 months before vs. 1 month since the pandemic** ***PAP usage, hour/night*** ***(p values < 0.01):***	Depression was associated with worse CPAP use ($p = 0.017$) (details were not reported)

		CPAP follow-up compliance: 79.5% AHI residual baseline: 1.59/h	– In total sample: 6.4 (4.8–7.5) vs. 6.9 (5.4–7.9) – In compliers at baseline (>4 h/night on ≥70% of nights): 6.9 (6.1–7.8) vs. 7.3 (6.4–8.2) – In non-compliers at baseline: 3.0 (1.2–4.1) vs. 3.5 (1.3–5.0) ***% of nights using CPAP for at least 4 h (p values < 0.01):*** – In total sample: 93 (70–99) vs. 96 (85–100) – In compliers: 97 (90–100) vs. 100 (96–100) – In non-compliers: 39 (13–57) vs. 44 (8–73) – Noncompliant → compliant: 27.3% – Compliant → noncompliant: 3.7% **Residual AHI as recorded by the device, events/hour:** – In total sample: 1.59 (0.7–3.6) vs. 1.63 (0.7–3.9), $p < 0.01$ – In compliers: 1.57 (0.8–3.5) vs. 1.63 (0.8–3.8), $p < 0.01$ – In non-compliers: 1.71 (0.7–4.2) vs. 1.63 (0.6–4.2),	

(continued)

Table 23.1 (continued)

Study	Population	Demographics	Changes in PAP use and characteristics since the pandemic	Variables associated with reduced PAP use
			$p = 0.41$ **Mask leak, p95, l/min:** – In total sample: 11.7 (5.2–21.0) vs. 11.7 (4.5–21.7), $p = 0.35$ – In compliers: 11.9 (5.3–20.9) vs. 12.2 (4.9–21.8), $p = 0.07$ – In non-compliers: 11.2 (4.3–21.8) vs. 10.7 (3.5–21.1), $p = 0.213$	
Pepine et al. (France) Longitudinal retrospective observational database study/ objective data on PAP use	N = 14,543	Age: 66 (54–71) Males: 68.3%	**CPAP use before vs. 1 month since the pandemic**[a] – ↑PAP use: 394 (IQR: 305–458) vs. 414 min. (IQR: 324–479), median change of 3.7% (IQR: −4.9 to 14.7) ($p < 0.01$) – Men vs. women: ↑ in 48.7% (absolute difference of 7.5 min., 95% CI: 3.3–11.6, $p < 0.01$) – Among individuals with poor adherence: ↑ > 30 min **1 month since the pandemic (p-value for interaction with age**	High levels of COVID-19 infection parameters were not significantly associated with changes in CPAP use (details were not reported)

			>0.01) – 65–75 vs. <65 years: ↑ CPAP use by 20.2 min – > 75 vs. <65 years: ↑ CPAP use by 10.3 min	
Batool-Anwar et al. (USA) Retrospective review of medical records/objective data on PAP use	N = 123	Age: 63.5 ± 13.9 Males: 55% BMI: 31.8 ± 7.9 OSA severity: – Mild: 11.2% – Moderate: 29.3% – Severe: 59.5%	**PAP use before vs. 1 month since the pandemic** No change in PAP use: 5.97 ± 1.9 vs. 5.87 ± 2.1 h/night	Stratification by OSA severity did not change the results (data not shown)

APAP Auto-titrating positive airway pressure, *AHI* Apnea-Hypopnea Index, *BMI* body mass index, *CAD* coronary artery disease, *COPD* chronic obstructive pulmonary disease, *CPAP* continuous positive airway pressure, *ESS* Epworth Sleepiness score, *F2F* face-to-face, *NIV* noninvasive ventilation, *OSA* obstructive sleep apnea, *ODI* oxygen desaturation index [≥4%], *PAP* positive airway pressure, *PSG* polysomnography, *SaO2* oxygen saturation levels, *SD* standard deviation

[a] Adjusted for confounders in analysis

A study conducted in the United States demonstrated that PAP use, measured subjectively, did not significantly change during the first month of the pandemic (5.97 ± 1.9 vs. 5.87 ± 2.1 h/night) (Batool-Anwar et al. 2020). Interestingly, in this survey study, although the number of hours of PAP use did not change, individuals reported an increased incidence of insomnia, which may have been associated with anxiety associated with the pandemic (Batool-Anwar et al. 2020). Similarly, in the US telephone survey study conducted during the first 2 months since the pandemic (March–April 2020) by Thorpy et al., there was no significant difference in PAP use: only 11% of individuals stopped using PAP, with the larger proportion of individuals with the COVID-19-positive test. The survey demonstrated concerns around COVID-19 and OSA therapy; 38% of individuals reported being concerned about getting COVID-19 infection while having SRBD, 29% reported thinking that they were at a higher risk of acquiring COVID-19 infection due to their SRBD, and 63% were concerned about COVID-19-related severe complications (Thorpy et al. 2020).

Lastly, a survey study conducted in North America (April 2020–January 2021) showed that the majority of individuals continued PAP therapy during the pandemic, and only 7.2% of all responders stopped PAP (Kendzerska et al. 2022).

Since the beginning of the pandemic, PAP prescription and initiation may have been affected by limited access to PAP titration studies and the introduction of remote setup. Turnbull et al. evaluated PAP use in the 30 nights preceding the first follow-up visit within 5 months since the start of the pandemic (July–August 2020). Remote PAP setup showed a significantly lower PAP use compared to F2F setup (-0.6 h/night, 95% CI -1.1 to -0.1, $p = 0.03$) (Turnbull et al. 2022). This suggests that F2F setup may be essential to promote PAP initiation and use.

23.5 Positive Airway Pressure Use During Active COVID-19 Infection

In the United States, a survey study from April to May 2020 found that 10 out of 112 individuals with OSA contracted COVID-19 infection (Table 23.1). Two of ten individuals who tested positive for COVID-19 increased their PAP use during the early stages of their illness; however, six of ten stopped PAP use due to shortness of breath; the individuals with OSA affected by the infection had several comorbidities including obesity (100%), diabetes mellitus (30%), cardiovascular disease (80%), and asthma (20% of the cases) (Thorpy et al. 2020). A survey study performed in North America (April 2020–January 2021) reported that 32% of the individuals who continued PAP therapy had >3 symptoms associated with COVID-19 compared with 43.9% of the patients who stopped PAP therapy. Further, among individuals who continued PAP therapy, 6% reported getting tested for COVID-19 vs. 15% among those who stopped PAP therapy. However, there was no statistical significance between contracting COVID infection and PAP usage (OR = 0.96; 95% CI 0.37–2.49; $p = 0.94$) (Kendzerska et al. 2022).

Finally, a small study was performed in January 2021 in Turkey, where cases with COVID-19 infection were included, and the alteration of auto-PAP levels was assessed objectively 1 month post-illness showing an increase in average auto-CPAP 95thpp (from 8.6 to 9.8 cm H_2O, $p < 0.01$) and median CPAP pressure (from 7.5 to 8.2 cm H_2O, $p < 0.01$). There was no change in the residual apnea-hypopnea index (AHI) on the auto-titrating CPAP (3.6 vs. 3.8 events per hour, $p = 0.27$) (Fidan et al. 2021). In contrast, Del Campo et al. demonstrated a modest but significant increase in residual AHI since the pandemic in individuals who were compliant with CPAP (1.57 vs. 1.63, $p < 0.01$) (del Campo et al. 2020).

23.6 Risk Factors Associated with Reduced Use of Positive Airway Pressure During the Pandemic

Subjective measured PAP use reduction has been reported during active COVID-19 infection; a small study observed that 60% of the individuals reported stopping PAP use during their illness due to respiratory symptoms and difficulty breathing (Thorpy et al. 2020). Depression was associated with low PAP use during the pandemic ($p = 0.02$) (del Campo et al. 2020), and similarly, a high level of stress was significantly associated with stopping PAP since the pandemic (OR = 1.13, 95% CI: 1.02–1.25, $p = 0.016$) (Kendzerska et al. 2022) (Table 23.1). Cohabitation with a person tested positively for COVID-19 infection was associated with a decrease in PAP use (OR = 3.05, 95% CI: 1.00–9.31, $p = 0.049$) (Kendzerska et al. 2022), which speaks about the lack of information available for patients on this topic.

Finally, healthcare delivery setup plays an important role in PAP use. A study conducted in July–August 2020 has shown a significant decrease in PAP use during the pandemic associated with entirely remote PAP setup with telemonitoring follow-up (Turnbull et al. 2022).

23.7 Conclusion

The use of PAP therapy in SRBD has been affected during the COVID-19 pandemic. This could be explained by the uncertainty at the beginning of the pandemic and the precautious measures that took place after. Despite concerns about PAP use related to an increasing spread of COVID-19 infection, PAP use has not changed and has even potentially increased during the pandemic. We attribute this increase to fear of contracting the COVID-19 infection or having greater medical complications if they were to get COVID-19 infection due to the presence of SRBD, as untreated SRBD has been associated with poor outcomes in patients contracting the infection. In addition, individuals with SRBD expressed concerns about COVID-19 because of their associated comorbidities. However, even a small proportion of individuals who stopped using PAP since the pandemic cannot be ignored. It is important to note that individuals at risk of reducing PAP therapy use have a higher burden of mental comorbidities, such as higher stress levels and depression; moreover, remote PAP initial setup and living with someone with COVID-19-related symptoms placed individuals at risk of reducing their PAP use. These risk factors also place individuals with untreated SRBD at a higher risk of poor outcomes in COVID-19 infection. Critically, lower CPAP usage was found in individuals newly started on CPAP after the first wave of the pandemic. Conversely, the male gender and age older than 65 years old were associated with increased PAP use compared to their female and younger counterparts. Further research is needed to identify populations at risk of reducing PAP use and target education strategies, as well as to investigate the effectiveness of PAP therapy in preventing complications of COVID-19.

References

ACS (2020) COVID-19: recommendations for management of elective surgical procedures. In: ACS. https://www.facs.org/for-medical-professionals/covid-19/clinical-guidance/elective-surgery/. Accessed 28 Jun 2022

Batool-Anwar S, Omobomi OS, Quan SF (2020) Impact of the novel coronavirus disease on treatment adherence and sleep duration in patients with obstructive sleep apnea treated with positive airway pressure. J Clin Sleep Med 16:1917–1920. https://doi.org/10.5664/JCSM.8746

Benjafield AV, Ayas NT, Eastwood PR et al (2019) Estimation of the global prevalence and burden of obstructive sleep apnoea: a literature-based analysis. Lancet Respir Med 7:687–698. https://doi.org/10.1016/S2213-2600(19)30198-5

Cariou B, Hadjadj S, Wargny M et al (2020) Phenotypic characteristics and prognosis of inpatients with COVID-19 and diabetes: the CORONADO study. Diabetologia 63(8):1500–1515. https://doi.org/10.1007/s00125-020-05180-x

CDC (2022) Infection control: severe acute respiratory syndrome coronavirus 2 (SARS-CoV-2). CDC. https://www.cdc.gov/coronavirus/2019-ncov/hcp/infection-control-recommendations.html. Accessed 7 Apr 2022

Chen T, Wu D, Chen H et al (2020) Clinical characteristics of 113 deceased patients with coronavirus disease 2019: retrospective study. BMJ 368:m1091. https://doi.org/10.1136/BMJ.M1091

del Campo F, López G, Arroyo CA et al (2020) Study of the adherence to continuous positive airway pressure treatment in patients with obstructive sleep apnea syndrome in the confinement during the COVID-19 pandemic. Arch Bronconeumol 56:818–819. https://doi.org/10.1016/J.ARBRES.2020.06.019

Fidan V, Koyuncu H, Akin O (2021) Alteration of auto-CPAP requirements in obstructive sleep apnea patients with COVID-19 history. Am J Otolaryngol 42:102919. https://doi.org/10.1016/J.AMJOTO.2021.102919

Grote L, McNicholas WT (2020) Sleep apnoea management in Europe during the COVID-19 pandemic: data from the European Sleep Apnoea Database (ESADA). Eur Respir J 55(6): 2001323. https://doi.org/10.1183/13993003.01323-2020

Grote L, McNicholas WT, Hedner J, On behalf of the ESADA collaborators (2020) Sleep apnoea management in Europe during the COVID-19 pandemic: data from the European Sleep Apnoea

Database (ESADA). Eur Respir J 55(6):2001323. https://doi.org/10.1183/13993003.01323-2020

Hui DS, Chow BK, Lo T et al (2019) Exhaled air dispersion during high-flow nasal cannula therapy versus CPAP via different masks. Eur Respir J 53:1802339. https://doi.org/10.1183/13993003.02339-2018

Kendzerska T, Zhu DT, Gershon AS et al (2021) The effects of the health system response to the COVID-19 pandemic on chronic disease management: a narrative review. Risk Manag Healthc Policy 14:575–584. https://doi.org/10.2147/RMHP.S293471

Kendzerska T, Saad M, Ayas N et al (2022) Changes in positive airway pressure use in adults with sleep-related breathing disorder during the COVID-19 pandemic: a cross-sectional National Community-Based Survey. Sleep Vigil 6(2):281–286. https://doi.org/10.21203/rs.3.rs-547671/v1

Lindberg E (2010) Epidemiology of OSA. Eur Respir Monogr 50:51–68

Pepin JL, Daabek N, Bailly S et al (2021) Adherence to continuous positive airway pressure hugely improved during COVID-19 lockdown in France. Am J Respir Crit Care Med 204:1103–1106. https://doi.org/10.1164/RCCM.202103-0803LE/SUPPL_FILE/DISCLOSURES.PDF

Sencer DJ (2020) CDC Museum COVID-19 Timeline|David J. Sencer CDC Museum|CDC. https://www.cdc.gov/museum/timeline/covid 19.html. Accessed 23 Jun 2022

Thorpy M, Figuera-Losada M, Ahmed I et al (2020) Management of sleep apnea in New York city during the COVID-19 pandemic. Sleep Med 74:86. https://doi.org/10.1016/J.SLEEP.2020.07.013

Turnbull CD, Allen M, Appleby J et al (2022) COVID-19-related changes in outpatient CPAP setup pathways for OSA are linked with decreased 30-day CPAP usage. Thorax 77(8):839–841. https://doi.org/10.1136/thoraxjnl-2021-218635

Vitacca M, Ceriana P, Balbi B et al (2021) The respiratory rehabilitation Maugeri network service reconfiguration after 1 year of COVID-19. Monaldi Arch Chest Dis 91. https://doi.org/10.4081/MONALDI.2021.1843

WHO (2022) Infection prevention and control during health care when novel coronavirus (nCoV) infection is suspected. https://wwwwhoint/publications/i/item/10665-331495. Accessed 7 Apr 2022

Chapter 24
Melatonin's Benefits as a Treatment for COVID-19 Pandemic and Long COVID

Daniel P. Cardinali, Gregory M. Brown, and S. R. Pandi-Perumal

Abstract Melatonin has been recognized for its therapeutic potential as a chronobiotic cytoprotective drug to combat the effects of COVID-19 infection. Melatonin may be unique in mitigating the symptoms of SARS-CoV-2 infection due to its wide-ranging actions as an antioxidant, anti-inflammatory, and immunomodulatory chemical. Furthermore, melatonin is an efficient chronobiotic drug in treating delirium and reversing the circadian disturbance caused by social isolation. Melatonin is a cytoprotector that helps to treat various comorbidities, including diabetes, metabolic syndrome, and ischemic and nonischemic cardiovascular disease, all of which increase COVID-19 illness. As the COVID-19 pandemic continues, it has become known that clinical sequelae and symptoms for a considerable number of patients may linger for weeks to months beyond the acute stage of SARS-CoV-2 infection (long COVID). Based on indications of neurological sequelae in COVID-19-infected individuals, there is another possible use of melatonin based on its documented neuroprotective properties. Melatonin is an excellent agent for controlling cognitive deterioration (brain fog) and pain in myalgic encephalomyelitis (i.e., chronic fatigue syndrome); therefore, its therapeutic importance for the neurological consequences of SARS-CoV-2 infection should be investigated.

Keywords Cytokine storm · COVID-19 · Long COVID · Viral infection · Coronavirus · Sepsis

---

D. P. Cardinali (✉)
Faculty of Medical Sciences, Pontificia Universidad Católica Argentina, Buenos Aires, Argentina
e-mail: daniel_cardinali@uca.edu.ar

G. M. Brown
Centre for Addiction and Mental Health, Department of Psychiatry, University of Toronto, Toronto, ON, Canada

S. R. Pandi-Perumal
Somnogen Canada Inc, Toronto, ON, Canada

Saveetha Institute of Medical and Technical Sciences, Saveetha University, Chennai, Tamil Nadu, India

A. S. BaHammam et al. (eds.), *COVID-19 and Sleep: A Global Outlook*, Progress in Sleep Research, https://doi.org/10.1007/978-981-99-0240-8_24

24.1 Introduction

The pathological responses to acute respiratory syndrome coronavirus type 2 (SARS-CoV-2) infection range from asymptomatic to severe respiratory and multiple-organ failure. Clinical studies in highly infected patients show that increased inflammation, a weakened immune system, and an active cytokine storm with positive feedback loops controlling cytokine production overpower counter-regulatory systems (Aimrane et al. 2022; Merad et al. 2022; Lyra e Silva et al. 2022).

The COVID-19 pandemic's repercussions are catastrophic. Over 532 million persons worldwide have had proven infection with (SARS-CoV-2) virus as of June 6, 2022, and over 6.2 million have died (Center for Systems Science and Engineering (CSSE) 2022). Although the SARS-CoV-2 virus has a higher attraction for lung tissue, clinical and experimental research has revealed that the coronavirus family has also a strong affinity for the central nervous system (Ahmad et al. 2022). Neurological symptoms such as headache, dizziness, memory loss, lack of concentration, myalgia, and anosmia have been reported in COVID-19 patients, as have cases of encephalopathy, encephalitis, hemorrhagic necrotizing encephalopathy, stroke, epileptic seizures, and rhabdomyolysis associated with SARS-CoV-2 infection. Long-term consequences include weariness, cognitive issues (especially memory loss), persistent fever, myalgia, muscular weakness, and shortness of breath (particularly on exertion) (Hope and Evering 2022).

Increasingly, vaccinated individuals are being diagnosed with COVID-19 because of breakthrough SARS-CoV-2 infection. In a study of 33,940 individuals with breakthrough SARS-CoV-2 infection and several controls of people without evidence of SARS-CoV-2 infection at 6 months after infection, people with breakthrough SARS-CoV-2 infection exhibited a higher risk of death and incident post-acute sequelae, including cardiovascular, coagulation and hematologic, gastrointestinal, kidney, mental health, metabolic, musculoskeletal, and neurologic disorders (Al-Aly et al. 2022). Persons with breakthrough SARS-CoV-2 infection had reduced odds of mortality and incident post-acute sequelae than people with SARS-CoV-2 infection who had not previously been vaccinated. The authors of the study concluded that immunization before infection provides only limited protection in the post-acute phase of COVID-19 illness and that a single mitigation technique may not diminish the long-term health implications of SARS-CoV-2 infection optimally (Al-Aly et al. 2022).

These findings highlight the need of continuing to optimize options other than vaccinations for the primary prevention of SARS-CoV-2 infection. Melatonin, an effective chronobiotic/cytoprotective chemical, has been recommended as a therapy from the onset of the COVID-19 epidemic (Acuña-Castroviejo et al. 2020; Simko et al. 2020; Reiter et al. 2020; Cardinali 2020; Artigas et al. 2020). The justification of its use derived not only from its high-security profile, but also from its multiple beneficial effects in experimental and clinical studies related to pandemic. Melatonin (a) prevents SARS-CoV-2 infection; (b) is suitable as an effective anti-inflammatory/immunoregulatory/antioxidant agent; (c) counteracts chronodisruption; (d) combats several COVID-19 comorbidities such as diabetes, metabolic syndrome, and ischemic and nonischemic cardiovascular diseases; (e) exerts a neuroprotective effect in acutely and chronically affected SARS-CoV-2 patients; and (f) can be an adjuvant to potentiate anti-SARS-CoV-2 vaccines (Cardinali et al. 2020). Melatonin has a unique multifaceted therapeutic potential that exists for no other therapeutic medication candidate for COVID-19 pandemic.

However, neither the government nor the pharmaceutical industry has expressed an interest in its usage. Considering the quantity of scientific/medical studies that have suggested its usage, the inability of melatonin to garner attention as a viable therapy is disheartening. More than 180 papers on pubmed.gov have examined the use of melatonin as a safe and potentially effective therapy for the COVID-19 pandemic since its inception (PubMed 2022). This might be due to several factors, including the fact that no influential organization has promoted its therapeutic use for this condition. Melatonin is non-patentable and cheap; therefore, the pharmaceutical business has little motive to encourage its usage. Meanwhile, several potentially harmful and costly medications have been repackaged as therapies for this disease (Castle et al. 2021).

In this review, we discuss this subject paying particular attention to the potential use of melatonin in long COVID. As such, we further extend the hypothesis (Jarrott et al. 2022) that melatonin, by activating the expression of enzymes that synthesize intracellular glutathione, will quench toxic free radicals in long COVID.

24.2 Mechanism of Action of Melatonin in Acute SARS-CoV-2 Infection

24.2.1 Virus Entrance and Replication

Melatonin inhibits viral entry and replication of the SARS-CoV-2 virus (Reiter et al. 2022b; Tan and Reiter 2022). SARS-CoV-2 entry into cells is mediated by angiotensin-converting enzyme 2 (ACE2), transmembrane protease serine 2 (TMPRSS2), and metalloproteinase 17 (ADAM17) (Zlacká et al. 2021). SARS-CoV-2 entry into cells is mediated by angiotensin-converting enzyme 2 (ACE2), transmembrane protease serine 2 (TMPRSS2), and metalloproteinase 17 (ADAM17) (Shukla et al. 2015) and allosterically binds to human ACE2 of endothelial cells causing it to change shape (Cecon et al. 2022). Melatonin also reduces the increased activation of CD147 (another SARS-CoV-2 entry receptor) (Sehirli et al. 2020).

Another possible way melatonin may regulate the entrance of virus to the cell is via binding and inhibition of calmodulin (CaM) (Benítez-King et al. 1993, 1996). CaM modulates the surface expression and retention of ACE2 in the plasma membrane, and CaM inhibitors increase ACE2 ectodomain release by diminishing the interaction between CaM and ACE2 (Lambert et al. 2008). Melatonin may thus be considered an indirect inhibitor of ACE2-CoV-2 interaction during viral particle fusion.

In terms of SARS-CoV-2 virus replication, cell machinery initiates the protein translation mechanism at the ribosome level, and a viral polyprotein including all structural and enzymatic proteins required to make new viruses is produced (Zhang and Yang 2022). The viral polyprotein is selectively split into smaller fragments by specific proteases (Mpro), with these fragments being used for virus replication. SARS-CoV-2 Mpro is an enzyme conserved among the coronavirus species. By using silico tools to identify new SARS-CoV-2 Mpro inhibitors, molecular docking studies described the binding sites and the interaction energies of 74 ligand complexes (Feitosa et al. 2020). Inhibition of Mpro by nirmatrelvir (Paxlovid™, Pfizer) results in the termination of virus replication. Paxlovid reduces the hospitalization and death by 89% in the mild to moderately severe COVID-19 patients (Hammond et al. 2022). Melatonin, like Paxlovid, is found in the catalytic region of Mpro and binds to the catalytic amino acid residues C145 and H41 (Feitosa et al. 2020). Thus, melatonin may function as a Mpro inhibitor with high efficiency.

The indirect evidence of possible antiviral action of melatonin helps to explain the results obtained in a study monitoring 11,672 patients with a statistical model that predicted infection by COVID-19 (Jehi et al. 2020). Males, African American, older patients, and those with known COVID-19 exposure were at higher risk of being positive for COVID-19, while the risk was reduced in those who were on melatonin, paroxetine, or carvedilol treatment. However, a low daily dose of melatonin (2 mg) did not reduce mortality in adult hospitalized patients with COVID-19 (Sánchez-Rico et al. 2022) nor was effective as a prophylactic treatment for prevention of SARS-CoV-2 infection among healthcare workers (García-García et al. 2022).

Possibly, higher doses of melatonin are needed. For example, in a retrospective cross-sectional study of data from a closed population of 110 old adult patients treated with melatonin (mean daily dose: 46 mg) for at least 12 months prior to the availability of COVID-19 vaccination, there was no death in face of a lethality rate of 10.5% in the local population of elders suffering from acute COVID-19 (Valiensi et al. 2022). Indeed, animal studies support the use of high doses of melatonin to prevent infection in murine COVID-19 models (Cecon et al. 2022). From several animal studies, the potential human equivalent dose (HED) of melatonin for a 75 kg adult was calculated by allometry (Cardinali 2019a). Allometry is commonly used for determining the doses for Phase I human clinical drug trials (Reagan-Shaw et al. 2008). Noteworthy, theoretical HED doses (50–100 mg) calculated from animal studies for melatonin ranged from 2 to 3 orders of magnitude greater than those commonly employed in humans.

24.2.2 Cytokine Storm

Although SARS-CoV-2 infection is commonly thought to be a respiratory disease, the repercussions of this virus extend beyond the respiratory system (Bahmani et al.

2022). This disease eventually becomes systemic, resulting in severe sepsis or septic shock and multiple-organ failure, which can lead to death in individuals infected with the SARS-CoV-2 virus (Eskandarian Boroujeni et al. 2022). Sepsis can be caused by a viral, bacterial, or fungal infection, and from a pathophysiological standpoint, the damage to various organs and the patient's systemic collapse have the same source, a cytokine storm and hyperinflammation with enhanced oxidative damage (Montazersaheb et al. 2022).

The use of melatonin has been proposed to mitigate cytokine storm in SARS-CoV-2 infection (Zhang et al. 2020b; Kleszczyński et al. 2020; Reiter et al. 2022b; Tan and Reiter 2022). Melatonin, a methoxyindole present in all forms of life with aerobic respiration and whose primary function is cytoprotection, has anti-inflammatory, antioxidant, and immunostimulant effects (Maestroni 2000; Anderson et al. 2015) and is an extremely potent neuroprotective agent (Cardinali 2019b). Melatonin has anti-inflammatory properties via a variety of mechanisms (Reynolds and Dubocovich 2021). Sirtuin-1, for example, decreases macrophage polarization toward the proinflammatory type (Hardeland 2018). The anti-inflammatory effect of melatonin also includes the suppression of nuclear factor (NF)-κB activation (Shang et al. 2009; da C Pedrosa et al. 2010) and the stimulation of nuclear erythroid 2-related factor 2 (Nrf2) (Ahmadi and Ashrafizadeh 2020). Inflammation is commonly associated with elevated production of cytokines and chemokines. Melatonin reduces proinflammatory cytokines [tumor necrosis (TN)F-α, interleukin (IL)-1β, IL-6, and IL-8] while increasing anti-inflammatory cytokines such as IL-10 (Habtemariam et al. 2017; Hardeland 2018).

Melatonin has significant antioxidant and scavenging effects on free radicals in both the cytoplasm and the cell nucleus that are mainly independent of receptors (Manchester et al. 2015). These effects are exerted in three ways: melatonin is a free radical scavenger; it is converted to molecules with high antioxidant activity, and it is an indirect antioxidant, stimulating the production of antioxidant enzymes while inhibiting the synthesis of prooxidant enzymes (Galano et al. 2011). In addition, several antiapoptotic and cytoprotective effects of melatonin are exerted under conditions of ischemia (unrelated to free radicals) and can be attributed to the stabilizing activity of the mitochondrial membrane (Reiter et al. 2018).

The shift of the metabolic profile of activated immune cells from mitochondrial oxidative phosphorylation to cytosolic glycolysis (Warburg effect) is a distinguishing hallmark of septicemia, independent of whether caused by a virus or fungus (Bar-Or et al. 2018). The main exponent of the change in the oxidation of mitochondrial glucose (positive regulation of pyruvate metabolism in the cytosol) is often accompanied by the stimulation of the hypoxia-inducible factor-1α (HIF-1α) and the activation of NF-κB and other transcription factors that promote inflammation (Tian et al. 2021). M2 anti-inflammatory macrophages in COVID-19 patients are changed into M1 proinflammatory cells, which, together with other immune cells that undergo a similar transformation, contribute to the cytokine storm. In this context, melatonin can reduce the damage resulting from sepsis mediated by COVID-19 through different mechanisms, i.e., by mitigating the production of HIF-1α (Owczarek et al. 2020), by suppressing NF-κB (Moniruzzaman et al.

2018), by inhibiting NLRP3 inflammasome (Saha et al. 2022), and by transforming proinflammatory macrophages M1 into anti-inflammatory macrophages M2 while reversing the Warburg-type metabolism (Reiter et al. 2021). The discovery of a correlation between circulating secreted phospholipase-A2 (Group IIA) and severity of COVID-19 disease (Snider et al. 2021) prompted the consideration of cyclooxygenase inhibition by melatonin (Cardinali et al. 1980; Deng et al. 2006) as another potential mechanism by which the methoxyindole inhibits viral infection.

Melatonin, as a result, possesses a set of unusually diverse abilities and is endowed with unique properties to suppress events associated with hyperinflammation, exacerbated oxidative stress, mitochondrial dysfunction, and/or metabolic reprogramming, all of which are common pathophysiological mechanisms in people acutely infected with SARS-CoV-2. Melatonin medication might reduce the severity of SARS-CoV-2 infections by reducing the intensity of symptoms, reducing the requirement for hospitalization, minimizing the hospital admission period, removing the need for mechanical intubation, and decreasing the rate of fatality.

24.3 Melatonin in the ICU

In diseases showing an important level of inflammation, the application of melatonin showed promising results with strong attenuation of circulating cytokine levels. This was documented in patients with diabetes mellitus and periodontitis (Bazyar et al. 2019) and severe multiple sclerosis (Sánchez-López et al. 2018). In the acute phase of inflammation, during surgical stress (Kücükakin et al. 2008), cerebral reperfusion (Zhao et al. 2018), or reperfusion of the coronary artery (Shafiei et al. 2018), treatment with melatonin reduced the level of proinflammatory cytokines.

According to the COVID-19 clinical reports, patients with severe infection have an increased risk of sepsis and cardiac arrest (Chen et al. 2020; Huang et al. 2020). The available information indicates that the application of melatonin can improve septic shock through inhibition of the NLRP3 inflammasome pathway (Volt et al. 2016). Specifically, melatonin has a preventive effect against sepsis-induced kidney damage, septic cardiomyopathy, and liver damage (Dai et al. 2019; Chen et al. 2019; Zhang et al. 2020a). Melatonin has also been reported as beneficial in patients with myocardial infarction, cardiomyopathy, hypertensive heart disease, and pulmonary hypertension. In the ICU, deep sedation is associated with increased long-term mortality, and the application of melatonin reduces the use of sedation and the frequency of pain, agitation, and anxiety and improves the quality of sleep (Lewandowska et al. 2020). Therefore, the rationale for the use of melatonin in COVID-19 focuses not only on the attenuation of infection-induced respiratory disorders, but also on general improvement and prevention of possible complications, like the cardiac ones.

Circadian disruption by sleep loss, like that observed in COVID-19 patients admitted to ICU, affects every major system in the human body (Golombek et al.

2022). Several epidemiologic studies have reported associations between sleep/wake cycle disruption and cardiometabolic disease (Chellappa et al. 2019; Grandner 2020). Sleep deprivation and poor sleep quality have also been recognized as risk factors for cognitive decline, neurodegenerative illness, mood disorders, depression, and other neuropsychiatric diseases (Ferini-Strambi et al. 2020; Brownlow et al. 2020).

Several meta-analyses support the view that the chronobiotic/hypnotic properties of melatonin are useful in patients with primary sleep disorders to decrease sleep-onset latency and to increase total sleep time (Ferracioli-Oda et al. 2013; Auld et al. 2017; Li et al. 2019). Expert consensus reports also support such a role of melatonin in adult insomnia (Wilson et al. 2010; Geoffroy et al. 2019; Palagini et al. 2020; Vecchierini et al. 2020). Melatonin reduces the need for sedation in ICU patients (Ibrahim et al. 2006; Bourne et al. 2008; Bellapart and Boots 2012; Mistraletti et al. 2015; Foreman et al. 2015; Soltani et al. 2020).

Brusco et al. (2021) employed a 9 mg melatonin dose to improve clinical conditions and hastened recovery in a group of 37 hospitalized patients with COVID-19 pneumonia. All patients were on corticosteroid treatment. As a comparison, data from a laboratory-confirmed COVID-19 pneumonia study performed in the New York metropolitan region were employed, considering only the subgroup of patients under corticosteroid treatment (Majmundar et al. 2020). Patients who received melatonin exhibited reduced length of stay (from 10.7 to 4.9 days) and death rate (from 14.6 to 2.7%).

24.4 Clinical Trials Using Melatonin in Acute COVID-19 Disease

Three randomized controlled trials (RCTs; published before September 11, 2021) satisfied the inclusion criteria in a meta-analysis of RCTs in which melatonin was used as a therapy for COVID-19 (Farnoosh et al. 2021; Mousavi et al. 2022; Alizadeh et al. 2022). The duration of the trials was either 7 or 14 days with the total dose of melatonin being either 6 or 9 mg; 86 patients were treated with melatonin, and 85 received placebo. The results documented that melatonin treatment significantly improved the recovery rate, as well as provided evidence of a lowered requirement for intensive care, reduced mortality, and depressed inflammation (C-reactive protein levels) compared to those given placebo (Lan et al. 2022).

These findings were consistent with the outcomes of observational studies. Castillo and colleagues performed a case series study that included ten severely ill COVID-19 patients given 36–72 mg melatonin daily (Castillo et al. 2020); the melatonin-treated patients experienced a more rapid clinical improvement, reduced requirement for mechanical ventilation, shorter duration of hospitalization, and evidence of a lower mortality. Ramlall et al. (2020) examined the outcome of 112 COVID-19 patients who required mechanical intubation as respiratory support and observed that those who received melatonin had a better clinical outcome compared to those who did not receive melatonin. Sánchez-González et al. (2022) found that melatonin treatment of seriously ill COVID-19 patients led to a reduced mortality rate.

A retrospective clinical trial investigated the impact of melatonin on the mortality of COVID-diseased patients and found a significantly reduced death rate (Hasan et al. 2022). However, a retrospective cohort study on the effects of melatonin in COVID-19 ICU showed only marginal effect on disease-specific biochemical parameters and a non-statistically significant reduction in mortality in melatonin-receiving group (41%) vs. non-melatonin-receiving group (72%) (Karimpour-razkenari et al. 2022).

In a real-time meta-analysis of 16 studies (including RTC and observational studies) on melatonin for COVID-19 (CovidAnalysis 2022), statistically significant improvements were seen for mortality, ventilation, and recovery. Eight studies showed statistically significant improvements with melatonin (six for the most serious outcome). Meta-analysis using the most serious outcome reported showed 49% [33–62%] improvement with melatonin. Results were similar for RCT and slightly worse for peer-reviewed studies (Covid Analysis 2022). It must be noted that in an analysis like this including controlled and observational trials, the strength of the statistical evidence provided is weak.

The reduction in the mortality of COVID-19 patients who are treated with melatonin is a theme seen in most of the studies that have utilized melatonin (Pilia et al. 2022). Obviously, there is no more important endpoint than survival.

24.5 Melatonin and Long COVID

As the COVID-19 pandemic continues, it has become increasingly clear that clinical sequelae and symptoms may persist for weeks to months beyond the acute stage of SARS-CoV-2 infection for a significant proportion of patients. This chronic illness has been termed post-acute sequelae of SARS-CoV-2 (PASC) infection, post-COVID syndrome, or long COVID (Hope and Evering 2022). Patients with these syndromes may suffer from cardiovascular, gastrointestinal, neurologic, dermatologic, and mucocutaneous manifestations that result from a hyperinflammatory syndrome with persistent immune activation and dysregulation of the immune response (Peluso et al. 2021).

Accepted guidelines define acute infection as lasting up to 4 weeks, ongoing symptomatic COVID-19 infection as lasting up to 12 weeks, and post-COVID syndrome as a cluster of overlapping signs and symptoms across multiple systems that develop during or after the initial infection and last more than 12 weeks after SARS-CoV-2 infection (Shah et al. 2021). In general, the reported incidence of protracted COVID varies from 10% to 35% among individuals assessed or treated for acute COVID-19 in the outpatient environment (Greenhalgh et al. 2020).

According to research conducted by the Northwestern University's Neuro-COVID-19 Clinic, brain fog was the most often reported chronic (>6 weeks) neurologic symptom among survivors (81%) who did not have a history of acute sickness, hypoxia, or respiratory compromise (Nuzzo et al. 2021). A systematic review assessing the prevalence of symptoms of long COVID affecting all body systems found fatigue to be the most reported complaint (58%) (Diem et al. 2022). For chronic trouble concentrating and attention problems, the effects on cognition were found to be 31% and 27%, respectively. Short-term memory (32%) and attention (27%) abnormalities were also seen in patients with brain fog. Neuroinflammation resulting in hypometabolic lesions has been hypothesized as one cause for chronic dysfunction following moderate COVID disease (Hugon et al. 2022).

In a study including 84,285 individuals receiving a Great British Intelligence Test with biologically confirmed COVID-19 infection, people who had recovered, including those no longer reporting symptoms, showed significant cognitive deficits (Hampshire et al. 2021). Within the same dataset, the magnitude of the detected deficiencies was similar to the average 10-year loss in worldwide performance between the ages of 20 and 70. "Brain fog," or disorientation, forgetfulness, inability to focus, exhaustion, and poor mental energy, is, therefore, a developing important consequence of COVID-19 infection (Raj et al. 2018; Wells et al. 2020). Melatonin's neuroprotective qualities should be considered in this perspective (Cardinali 2019b). Cell line studies regarding Alzheimer's disease (AD) and melatonin have delineated important melatonin-mediated mechanisms in AD prevention. For comprehensive reviews on melatonin activity to reverse disrupted signaling mechanisms in neurodegeneration, including proteostasis dysfunction, disruption of autophagic integrity, and anomalies in the insulin, Notch, and Wnt/β-catenin signaling pathways, see reviews by Jeong (2017), Corpas et al. (2018), Shukla et al. (2019), Rehman et al. (2019), and Abramov et al. (2020).

In transgenic models of AD, results are compatible with the view that melatonin regulates amyloid-β (Aß) metabolism mainly at the initial phases of the pathological process (see for ref. Cardinali 2019c). From the doses of melatonin used in these different transgenic models, the HED of melatonin for a 75 kg adult could be calculated as ranging from 2 to 3 orders of magnitude greater than those usually employed in humans (Cardinali 2019c).

The exact mechanism by which melatonin inhibits the production of Aβ is unknown. Melatonin interacts with $Aß_{40}$ and $Aß_{42}$, inhibiting progressive sheet and/or amyloid fibrils, an interaction that appears to be dependent on structural melatonin features rather than antioxidant capabilities (Pappolla et al. 1998). Via blockage of formation of secondary sheets, melatonin may facilitate peptide clearance by increasing proteolytic degradation.

Aß-induced neurotoxicity and cell death involve oxidative stress, and melatonin effectively protects cells against it in vitro (Feng and Zhang 2004) and in vivo (Furio et al. 2002; Shen et al. 2002). Melatonin was found to protect against Aβ toxicity, particularly at the mitochondrial level. Melatonin effectively inhibits tau hyperphosphorylation in N2a and SH-SY5Y neuroblastoma cells through influencing protein kinases and phosphatases (Deng et al. 2005; Xiong et al. 2011).

Melatonin treatment to AD transgenic mice increases Aβ glymphatic clearance (Pappolla et al. 2018; Reiter et al. 2022a). Relevant to this, melatonin is known to preserve slow-wave sleep in patients (Duffy et al. 2022). During sleep, the elimination of Aβ peptides increases considerably (Boespflug and Iliff 2018). Thus, the sleep disturbance found as a comorbidity in AD may contribute to the development and progression of the disease via a failure of Aβ clearance.

Another factor in the pathogenesis of AD is the activation of microglia with the consequent increased expression of proinflammatory cytokines, epidemiological studies suggesting that the use of anti-inflammatory drugs may decrease the incidence of AD (Tan et al. 2019). Melatonin attenuated the microglial production of proinflammatory cytokines induced by Aß, NF-kB, and nitric oxide (Rosales-Corral et al. 2003).

Two meta-analyses endorsed the view that melatonin therapy is effective in improving sleep in patients with dementia (Zhang et al. 2016; McCleery and Sharpley 2020). An analysis of published data of the use of melatonin in the early stages of cognitive decline consistently showed that administration of melatonin, every night before retiring, improves the quality of sleep and cognitive performance in this phase of the disease (see for ref. Cardinali 2019b). In one of our laboratories, a retrospective analysis of 25 patients with minimal cognitive impairment (MCI), who in the past 3 years had received a daily dose of 3–9 mg of melatonin along with their usual medication, was performed. Compared to an untreated group, melatonin-treated patients significantly improved cognitive and emotional performance and quality of sleep/wake rhythm (Furio et al. 2007). Another series of 96 outpatients with a diagnosis of MCI, 61 of whom had received 3–24 mg of melatonin daily for 15–60 months, was reported (Cardinali et al. 2012). Patients treated with melatonin showed a significantly better performance in various neuropsychological tests. They also had lower scores in the Beck Depression Inventory concomitantly with improvement in the quality of sleep and wakefulness. Therefore, melatonin treatment can be effective in the early stages of neurodegenerative disease, as brain fog in long COVID patients is.

Another prominent neurologic consequence reported in long COVID patients is myalgic encephalomyelitis/chronic fatigue syndrome (ME/CFS). ME/CFS is a multisystemic condition with devastating and often lifelong symptoms (Renz-Polster et al. 2022). It is a common and disabling disorder primarily characterized by persistent fatigue and exercise intolerance, with associated sleep disturbances, autonomic dysfunction, and cognitive problems. Most commonly, the disorder develops in the aftermath of acute infections, predominantly from viruses, e.g., Epstein-Barr virus, SARS coronavirus, influenza virus, Ebola virus, enteroviruses, and others. A yet undefined proportion of persons with long COVID is predicted to also meet the criteria of ME/CFS, which may significantly add to the global disease burden (van Campen et al. 2021; Komaroff and Lipkin 2021).

Given the prominence of activity and sleep problems in ME/CFS, circadian rhythm disruption has been examined as a contributing factor. It has been proposed that disrupted transforming growth factor-β (TGF-β) signaling in ME/CFS plays a role in disrupting physiological rhythms in sleep, activity, and cognition, leading to the insomnia, energy disturbances, cognition problems, depression, and autonomic dysfunction associated with ME/CFS (McCarthy 2022). As a result, treatment with a chronobiotic/cytoprotective drug such as melatonin is particularly suited to the pathophysiology of ME/CFS (Cardinali 2019c).

A proinflammatory state, increased oxidative and nitrosative stress, disruption of gut mucosal barriers, and mitochondrial dysfunction together with dysregulated bioenergetics have been identified in ME/CFS, and melatonin has been proposed as an effective therapy to control this pathological process (Morris et al. 2019). The beneficial effects of melatonin on fibromyalgia (associated commonly with ME/CSF) have been documented (Hemati et al. 2020).

24.6 Conclusions

In an analysis of 27 publications related to the ability of drugs to successfully treat COVID-19, the authors concluded that melatonin is at least twice as effective as remdesivir or tocilizumab in reducing the inflammatory markers of a coronavirus 2019 infection (Castle et al. 2021). Both remdesivir (Veklury®) and tocilizumab (Actemra®) are FDA-authorized for use to treat select COVID patients suffering with a severe infection; both drugs have notable side effects and are given intravenously (Takahashi et al. 2020). In contrast, melatonin has a high-safety profile and can be taken orally or administered by any other route.

Additionally, although this chapter considers melatonin as a sole treatment for SARS-CoV-2 infections, it has also been suggested as a co-treatment with vaccines to improve their efficacy (Cardinali et al. 2021; Lee and Glickman 2021) and in combination with other drugs (Castle et al. 2021); this latter suggestion would be especially applicable when the medications have different but complementary mechanisms of action to those of melatonin.

In critical situations, such as an Ebola outbreak or a COVID-19 pandemic, it is ethical to use all accessible and safe medicines, even if their usefulness has not been fully demonstrated, especially if the therapy has no major adverse effects. Indeed, given the huge number of deaths caused by SARS-CoV-2 infections throughout the world, it may be immoral not to take advantage of any possibly safe therapies, especially if the vaccinations become less effective as the virus continues to evolve. People who are vulnerable and may become infected with such diseases should not have to wait for the development of a new vaccine, which might take months or years, during which many people die. Currently, there are ten clinical trial studies reported in Clinicaltrials.gov database proposing melatonin (nine studies) or a melatonin agonist (one study) for COVID-19 treatment in mild to moderate (four studies), or severe hospitalized (four studies), or as a prophylactic indication (one study). Except for two randomized, double-blind, placebo-controlled studies using melatonin in COVID-19 outpatients, all the clinical trials listed are testing in combination drug therapy and in low doses. Well-controlled and powered clinical trials are therefore needed to test the hypothesis that melatonin is safe and efficacious to treat COVID-19 and its sequelae.

Acknowledgements Our research did not receive any specific grant from funding agencies in the public, commercial, or not-for-profit sectors. D.P.C. is an Emeritus Superior Investigator from Argentine National Research Council and Emeritus Professor, University of Buenos Aires. G.M.B. is an Emeritus Professor, University of Toronto.

Conflict of Interest Declaration S.R.P. is a stockholder and the President and Chief Executive Officer of Somnogen Canada Inc., a Canadian Corporation. He declares that he has no competing interests that might be perceived to influence the content of this chapter. This does not alter the authors' adherence to all the journal policies. All remaining authors declare that they have no proprietary, financial, professional, nor any other personal interest of any nature or kind in any product or services and/or company that could be construed or considered to be a potential conflict of interest that might have influenced the views expressed in this manuscript.

References

Abramov AY, Potapova EV, Dremin VV, Dunaev AV (2020) Interaction of oxidative stress and misfolded proteins in the mechanism of neurodegeneration. Life (Basel) 10:1–14. https://doi.org/10.3390/life10070101

Acuña-Castroviejo D, Escames G, Figueira JC et al (2020) Clinical trial to test the efficacy of melatonin in COVID-19. J Pineal Res 69:e12683. https://doi.org/10.1111/jpi.12683

Ahmad SJ, Feigen CM, Vazquez JP et al (2022) Neurological Sequelae of COVID-19. J Integr Neurosci 21(3):77. https://doi.org/10.31083/J.JIN2103077

Ahmadi Z, Ashrafizadeh M (2020) Melatonin as a potential modulator of Nrf2. Fundam Clin Pharmacol 34:11–19. https://doi.org/10.1111/fcp.12498

Aimrane A, Laaradia MA, Sereno D et al (2022) Insight into COVID-19's epidemiology, pathology, and treatment. Heliyon 8:e08799. https://doi.org/10.1016/j.heliyon.2022.e08799

Al-Aly Z, Bowe B, Xie Y (2022) Long COVID after breakthrough SARS-CoV-2 infection. Nat Med 28:1461. https://doi.org/10.1038/S41591-022-01840-0

Alizadeh N, Dianatkhah M, Alimohamadi Y et al (2022) High dose melatonin as an adjuvant therapy in intubated patients with COVID-19: a randomized clinical trial. J Taibah Univ Med Sci 17:454. https://doi.org/10.1016/J.JTUMED.2022.04.012

Anderson G, Maes M, Markus RP, Rodriguez M (2015) Ebola virus: melatonin as a readily available treatment option. J Med Virol 87:537–543. https://doi.org/10.1002/jmv.24130

Artigas L, Coma M, Matos-Filipe P et al (2020) In-silico drug repurposing study predicts the combination of pirfenidone and melatonin as a promising candidate therapy to reduce SARS-CoV-2 infection progression and respiratory distress caused by cytokine storm. PLoS One 15: e0240149. https://doi.org/10.1371/journal.pone.0240149

Auld F, Maschauer EL, Morrison I et al (2017) Evidence for the efficacy of melatonin in the treatment of primary adult sleep disorders. Sleep Med Rev 34:10–22. https://doi.org/10.1016/j.smrv.2016.06.005

Bahmani M, Chegini R, Ghanbari E et al (2022) Severe acute respiratory syndrome coronavirus 2 infection: role of interleukin-6 and the inflammatory cascade. World J Virol 11:113–128. https://doi.org/10.5501/WJV.V11.I3.113

Bar-Or D, Carrick M, Tanner A et al (2018) Overcoming the Warburg effect: is it the key to survival in sepsis? J Crit Care 43:197–201. https://doi.org/10.1016/J.JCRC.2017.09.012

Bazyar H, Gholinezhad H, Moradi L et al (2019) The effects of melatonin supplementation in adjunct with non-surgical periodontal therapy on periodontal status, serum melatonin and inflammatory markers in type 2 diabetes mellitus patients with chronic periodontitis: a double-blind, placebo-controlled trial. Inflammopharmacology 27:67–76. https://doi.org/10.1007/s10787-018-0539-0

Bellapart J, Boots R (2012) Potential use of melatonin in sleep and delirium in the critically ill. Br J Anaesth 108:572–580. https://doi.org/10.1093/bja/aes035

Benítez-King G, Huerto-Delgadillo L, Antón-Tay F (1993) Binding of 3H-melatonin to calmodulin. Life Sci 53:201–207. https://doi.org/10.1016/0024-3205(93)90670-X

Benítez-King G, Ríos A, Martínez A, Antón-Tay F (1996) In vitro inhibition of Ca2+/calmodulin-dependent kinase II activity by melatonin. Biochim Biophys Acta Gen Subj 1290:191–196. https://doi.org/10.1016/0304-4165(96)00025-6

Boespflug EL, Iliff JJ (2018) The emerging relationship between interstitial fluid-cerebrospinal fluid exchange, amyloid-β, and sleep. Biol Psychiatry 83:328–336. https://doi.org/10.1016/J.BIOPSYCH.2017.11.031

Bourne RS, Mills GH, Minelli C (2008) Melatonin therapy to improve nocturnal sleep in critically ill patients: encouraging results from a small randomised controlled trial. Crit Care 12:R52. https://doi.org/10.1186/cc6871

Brownlow JA, Miller KE, Gehrman PR (2020) Insomnia and cognitive performance. Sleep Med Clin 15:71–76. https://doi.org/10.1016/j.jsmc.2019.10.002

Brusco LI, Cruz P, Cangas AV et al (2021) Efficacy of melatonin in non-intensive care unit patients with COVID-19 pneumonia and sleep dysregulation. Melatonin Res 4:173–188. https://doi.org/10.32794/mr11250089

Cardinali DP (2019a) Are melatonin doses employed clinically adequate for melatonin-induced cytoprotection? Melatonin Res 2:106–132. https://doi.org/10.32794/mr11250025

Cardinali DP (2019b) Melatonin: clinical perspectives in neurodegeneration. Front Endocrinol 10:480. https://doi.org/10.3389/fendo.2019.00480

Cardinali DP (2019c) Melatonin as a chronobiotic/cytoprotector: its role in healthy aging. Biol Rhythm Res 50:28. https://doi.org/10.1080/09291016.2018.1491200

Cardinali DP (2020) High doses of melatonin as a potential therapeutic tool for the neurologic sequels of covid-19 infection. Melatonin Res 3:311–317. https://doi.org/10.32794/mr11250064

Cardinali DP, Ritta MN, Fuentes AM et al (1980) Prostaglandin E release by rat medial basal hypothalamus in vitro. Inhibition by melatonin at submicromolar concentrations. Eur J Pharmacol 67:151. https://doi.org/10.1016/0014-2999(80)90025-4

Cardinali DP, Vigo DE, Olivar N et al (2012) Therapeutic application of melatonin in mild cognitive impairment. Am J Neurodegener Dis 1:280–291

Cardinali DP, Brown GM, Pandi-Perumal SR (2020) Can melatonin be a potential "silver bullet" in treating COVID-19 patients? Diseases 8:44. https://doi.org/10.3390/DISEASES8040044

Cardinali DP, Brown GM, Pandi-Perumal SR (2021) An urgent proposal for the immediate use of melatonin as an adjuvant to anti- SARS-CoV-2 vaccination. Melatonin Res 4:206–212. https://doi.org/10.32794/mr11250091

Castillo RR, Quizon GRA, Juco MJM et al (2020) Melatonin as adjuvant treatment for coronavirus disease 2019 pneumonia patients requiring hospitalization (MAC-19 PRO): a case series. Melatonin Res 3:297–310. https://doi.org/10.32794/mr11250063

Castle RD, Williams MA, Bushell WC et al (2021) Implications for systemic approaches to COVID-19: effect sizes of Remdesivir, tocilizumab, melatonin, vitamin D3, and meditation. J Inflamm Res 14:4859–4876. https://doi.org/10.2147/JIR.S323356

Cecon E, Izabelle C, le Poder S et al (2022) Therapeutic potential of melatonin and melatonergic drugs on K18-hACE2 mice infected with SARS-CoV-2. J Pineal Res 72:e12772. https://doi.org/10.1111/JPI.12772

Center for Systems Science and Engineering (CSSE) (2022) COVID-19 map–Johns Hopkins coronavirus resource center. https://coronavirus.jhu.edu/map.html. Accessed 5 Jun 2022

Chellappa SL, Vujovic N, Williams JS, Scheer FAJL (2019) Impact of circadian disruption on cardiovascular function and disease. Trends Endocrinol Metab 30:767–779. https://doi.org/10.1016/j.tem.2019.07.008

Chen J, Xia H, Zhang L et al (2019) Protective effects of melatonin on sepsis-induced liver injury and dysregulation of gluconeogenesis in rats through activating SIRT1/STAT3 pathway. Biomed Pharmacother 117:109150. https://doi.org/10.1016/j.biopha.2019.109150

Chen N, Zhou M, Dong X et al (2020) Epidemiological and clinical characteristics of 99 cases of 2019 novel coronavirus pneumonia in Wuhan, China: a descriptive study. Lancet 395:507–513. https://doi.org/10.1016/S0140-6736(20)30211-7

Corpas R, Griñán-Ferré C, Palomera-Ávalos V et al (2018) Melatonin induces mechanisms of brain resilience against neurodegeneration. J Pineal Res 65:e12551. https://doi.org/10.1111/JPI.12515

CovidAnalysis (2022) Melatonin for COVID-19: real-time meta-analysis of 16 studies. https://c19melatonin.com/meta.html. Accessed 10 Jun 2022

da C Pedrosa AM, Weinlich R, Mognol GP et al (2010) Melatonin protects CD4+ T cells from activation-induced cell death by blocking NFAT-mediated CD95 ligand upregulation. J Immunol 184:3487–3494. https://doi.org/10.4049/JIMMUNOL.0902961

Dai W, Huang H, Si L et al (2019) Melatonin prevents sepsis-induced renal injury via the PINK1/Parkin1 signaling pathway. Int J Mol Med 44:1197–1204. https://doi.org/10.3892/ijmm.2019.4306

Deng YQ, Xu GG, Duan P et al (2005) Effects of melatonin on wortmannin-induced tau hyperphosphorylation. Acta Pharmacol Sin 26:519–526. https://doi.org/10.1111/J.1745-7254.2005.00102.X

Deng WG, Tang ST, Tseng HP, Wu KK (2006) Melatonin suppresses macrophage cyclooxygenase-2 and inducible nitric oxide synthase expression by inhibiting p52 acetylation and binding. Blood 108:518–524. https://doi.org/10.1182/blood-2005-09-3691

Diem L, Fregolente-Gomes L, Warncke JD et al (2022) Fatigue in post-COVID-19 syndrome: clinical phenomenology, comorbidities and association with initial course of COVID-19. J Cent Nerv Syst Dis 14:117957352211027. https://doi.org/10.1177/11795735221102727

Duffy JF, Wang W, Ronda JM, Czeisler CA (2022) High dose melatonin increases sleep duration during nighttime and daytime sleep episodes in older adults. J Pineal Res 73(1):e12801. https://doi.org/10.1111/jpi.12801

Eskandarian Boroujeni M, Sekrecka A, Antonczyk A et al (2022) Dysregulated interferon response and immune Hyperactivation in severe COVID-19: targeting STATs as a novel therapeutic strategy. Front Immunol 13:888897. https://doi.org/10.3389/FIMMU.2022.888897

Farnoosh G, Akbariqomi M, Badri T et al (2021) Efficacy of a low dose of melatonin as an adjunctive therapy in hospitalized patients with COVID-19: a randomized, double-blind clinical trial. Arch Med Res 53(1):79–85. https://doi.org/10.1016/j.arcmed.2021.06.006

Feitosa EL, Júnior FTDSS, Neto JADON et al (2020) Covid-19: rational discovery of the therapeutic potential of melatonin as a SARS-CoV-2 main protease inhibitor. Int J Med Sci 17:2133–2146. https://doi.org/10.7150/ijms.48053

Feng Z, Zhang JT (2004) Protective effect of melatonin on beta-amyloid-induced apoptosis in rat astroglioma C6 cells and its mechanism. Free Radic Biol Med 37:1790–1801. https://doi.org/10.1016/J.FREERADBIOMED.2004.08.023

Ferini-Strambi L, Galbiati A, Casoni F, Salsone M (2020) Therapy for insomnia and circadian rhythm disorder in Alzheimer disease. Curr Treat Options Neurol 22:4. https://doi.org/10.1007/s11940-020-0612-z

Ferracioli-Oda E, Qawasmi A, Bloch MH (2013) Meta-analysis: melatonin for the treatment of primary sleep disorders. PLoS One 8:e63773. https://doi.org/10.1371/journal.pone.0063773

Foreman B, Westwood AJ, Claassen J, Bazil CW (2015) Sleep in the neurological intensive care unit: feasibility of quantifying sleep after melatonin supplementation with environmental light and noise reduction. J Clin Neurophysiol 32:66–74. https://doi.org/10.1097/WNP.0000000000000110

Furio AM, Cutrera RA, Thea VC et al (2002) Effect of melatonin on changes in locomotor activity rhythm of Syrian hamsters injected with beta amyloid peptide 25-35 in the suprachiasmatic nuclei. Cell Mol Neurobiol 22:699. https://doi.org/10.1023/A:1021805023906

Furio AM, Brusco LI, Cardinali DP (2007) Possible therapeutic value of melatonin in mild cognitive impairment: a retrospective study. J Pineal Res 43:404. https://doi.org/10.1111/j.1600-079X.2007.00491.x

Galano A, Tan DX, Reiter RJ (2011) Melatonin as a natural ally against oxidative stress: a physicochemical examination. J Pineal Res 51:1–16. https://doi.org/10.1111/j.1600-079X.2011.00916.x

García-García I, Seco-Meseguer E, Ruiz-Seco P et al (2022) Melatonin in the prophylaxis of SARS-CoV-2 infection in healthcare workers (MeCOVID): a randomised clinical trial. J Clin Med 11:1139. https://doi.org/10.3390/JCM11041139

Geoffroy PA, Micoulaud Franchi JA, Lopez R, Schroder CM (2019) The use of melatonin in adult psychiatric disorders: expert recommendations by the French institute of medical research on sleep (SFRMS). L'Encéphale 45:413–423. https://doi.org/10.1016/j.encep.2019.04.068

Golombek D, Pandi-Perumal S, Rosenstein RE et al (2022) Dysregulated light/dark cycle impairs sleep and delays the recovery of patients in intensive care units: a call for action for COVID-19 treatment. Chronobiol Int 39:903–906. https://doi.org/10.1080/07420528.2022.2056477

Grandner MA (2020) Sleep, health, and society. Sleep Med Clin 15:319–340. https://doi.org/10.1016/j.jsmc.2020.02.017

Greenhalgh T, Knight M, A'Court C et al (2020) Management of post-acute covid-19 in primary care. BMJ 370:m3026. https://doi.org/10.1136/BMJ.M3026

Habtemariam S, Daglia M, Sureda A et al (2017) Melatonin and respiratory diseases: a review. Curr Top Med Chem 17:467–488. https://doi.org/10.2174/1568026616666160824120338

Hammond J, Leister-Tebbe H, Gardner A et al (2022) Oral Nirmatrelvir for high-risk, nonhospitalized adults with Covid-19. N Engl J Med 386:1397–1408. https://doi.org/10.1056/NEJMOA2118542

Hampshire A, Trender W, Chamberlain SR et al (2021) Cognitive deficits in people who have recovered from COVID-19. EClinicalMedicine 39:101044. https://doi.org/10.1016/J.ECLINM.2021.101044

Hardeland R (2018) Melatonin and inflammation—story of a double-edged blade. J Pineal Res 65:e12525. https://doi.org/10.1111/jpi.12525

Hasan ZT, Atrakji DMQYMAA, Mehuaiden DAK (2022) The effect of melatonin on thrombosis, sepsis and mortality rate in COVID-19 patients. Int J Infect Dis 114:79–84. https://doi.org/10.1016/J.IJID.2021.10.012

Hemati K, Amini Kadijani A, Sayehmiri F et al (2020) Melatonin in the treatment of fibromyalgia symptoms: a systematic review. Complement Ther Clin Pract 38:101072. https://doi.org/10.1016/J.CTCP.2019.101072

Hope AA, Evering TH (2022) Postacute sequelae of severe acute respiratory syndrome coronavirus 2 infection. Infect Dis Clin N Am 36:379–395. https://doi.org/10.1016/J.IDC.2022.02.004

Huang C, Wang Y, Li X et al (2020) Clinical features of patients infected with 2019 novel coronavirus in Wuhan, China. Lancet 395:497–506. https://doi.org/10.1016/S0140-6736(20)30183-5

Hugon J, Msika EF, Queneau M et al (2022) Long COVID: cognitive complaints (brain fog) and dysfunction of the cingulate cortex. J Neurol 269:44–46. https://doi.org/10.1007/S00415-021-10655-X

Ibrahim MG, Bellomo R, Hart GK et al (2006) A double-blind placebo-controlled randomised pilot study of nocturnal melatonin in tracheostomised patients. Crit Care Resusc 8:187–191

Jarrott B, Head R, Pringle KG et al (2022) "LONG COVID"—a hypothesis for understanding the biological basis and pharmacological treatment strategy. Pharmacol Res Perspect 10:e00911. https://doi.org/10.1002/PRP2.911

Jehi L, Ji X, Milinovich A et al (2020) Individualizing risk prediction for positive COVID-19 testing: results from 11,672 patients. Chest 158:1364. https://doi.org/10.1016/j.chest.2020.05.580

Jeong S (2017) Molecular and cellular basis of neurodegeneration in Alzheimer's disease. Mol Cells 40:613–620

Karimpour-razkenari E, Naderi-Behdani F, Salahshoor A et al (2022) Melatonin as adjunctive therapy in patients admitted to the Covid-19. Ann Med Surg (Lond) 76:103492. https://doi.org/10.1016/J.AMSU.2022.103492

Kleszczyński K, Slominski AT, Steinbrink K, Reiter RJ (2020) Clinical trials for use of melatonin to fight against COVID-19 are urgently needed. Nutrients 12:1–12. https://doi.org/10.3390/nu12092561

Komaroff AL, Lipkin WI (2021) Insights from myalgic encephalomyelitis/chronic fatigue syndrome may help unravel the pathogenesis of postacute COVID-19 syndrome. Trends Mol Med 27:895–906. https://doi.org/10.1016/J.MOLMED.2021.06.002

Kücükakin B, Lykkesfeldt J, Nielsen HJ et al (2008) Utility of melatonin to treat surgical stress after major vascular surgery - a safety study. J Pineal Res 44:426–431. https://doi.org/10.1111/j.1600-079X.2007.00545.x

Lambert DW, Clarke NE, Hooper NM, Turner AJ (2008) Calmodulin interacts with angiotensin-converting enzyme-2 (ACE2) and inhibits shedding of its ectodomain. FEBS Lett 582:385–390. https://doi.org/10.1016/j.febslet.2007.11.085

Lan SH, Lee HZ, Chao CM et al (2022) Efficacy of melatonin in the treatment of patients with COVID-19: a systematic review and meta-analysis of randomized controlled trials. J Med Virol 94:2102–2107. https://doi.org/10.1002/JMV.27595

Lee RU, Glickman GL (2021) Sleep, circadian health and melatonin for mitigating COVID-19 and optimizing vaccine efficacy. Front Neurosci 15:711605. https://doi.org/10.3389/FNINS.2021.711605

Lewandowska K, Małkiewicz MA, Siemiński M et al (2020) The role of melatonin and melatonin receptor agonist in the prevention of sleep disturbances and delirium in intensive care unit–a clinical review. Sleep Med 69:127–134. https://doi.org/10.1016/j.sleep.2020.01.019

Li T, Jiang S, Han M et al (2019) Exogenous melatonin as a treatment for secondary sleep disorders: a systematic review and meta-analysis. Front Neuroendocrinol 52:22–28. https://doi.org/10.1016/j.yfrne.2018.06.004

Lyra E Silva NM, Barros-Aragão FGQ, de Felice FG, Ferreira ST (2022) Inflammation at the crossroads of COVID-19, cognitive deficits and depression. Neuropharmacology 209:109023. https://doi.org/10.1016/j.neuropharm.2022.109023

Maestroni GJM (2000) Therapeutic potential of melatonin in immunodeficiency states, viral diseases, and cancer. Adv Exp Med Biol 467:217–226. https://doi.org/10.1007/978-1-4615-4709-9_28

Majmundar M, Kansara T, Lenik JM et al (2020) Efficacy of corticosteroids in non-intensive care unit patients with COVID-19 pneumonia from the New York metropolitan region. PLoS One 15:e0238827. https://doi.org/10.1371/journal.pone.0238827

Manchester LC, Coto-Montes A, Boga JA et al (2015) Melatonin: an ancient molecule that makes oxygen metabolically tolerable. J Pineal Res 59:403–419. https://doi.org/10.1111/jpi.12267

McCarthy MJ (2022) Circadian rhythm disruption in Myalgic encephalomyelitis/chronic fatigue syndrome: implications for the post-acute sequelae of COVID-19. Brain Behav Immun Health 20:100412. https://doi.org/10.1016/J.BBIH.2022.100412

McCleery J, Sharpley AL (2020) Pharmacotherapies for sleep disturbances in dementia. Cochrane Database Syst Rev 11:CD009178. https://doi.org/10.1002/14651858.CD009178.PUB4

Merad M, Blish CA, Sallusto F, Iwasaki A (2022) the immunology and immunopathology of COVID-19. Science 375:1122–1127. https://doi.org/10.1126/science.abm8108

Mistraletti G, Umbrello M, Miori S et al (2015) Melatonin reduces the need for sedation in ICU patients: a randomized controlled trial. Minerva Anestesiol 81:1298–1310

Moniruzzaman M, Ghosal I, Das D, Chakraborty SB (2018) Melatonin ameliorates H_2O_2-induced oxidative stress through modulation of Erk/Akt/NFkB pathway. Biol Res 51:17. https://doi.org/10.1186/S40659-018-0168-5

Montazersaheb S, Hosseiniyan Khatibi SM, Hejazi MS et al (2022) COVID-19 infection: an overview on cytokine storm and related interventions. Virol J 19:92. https://doi.org/10.1186/S12985-022-01814-1

Morris G, Puri BK, Walker AJ et al (2019) Myalgic encephalomyelitis/chronic fatigue syndrome: from pathophysiological insights to novel therapeutic opportunities. Pharmacol Res 148: 104450. https://doi.org/10.1016/J.PHRS.2019.104450

Mousavi SA, Heydari K, Mehravaran H et al (2022) Melatonin effects on sleep quality and outcomes of COVID-19 patients: an open-label, randomized, controlled trial. J Med Virol 94: 263–271. https://doi.org/10.1002/JMV.27312

Nuzzo D, Vasto S, Scalisi L et al (2021) Post-acute COVID-19 neurological syndrome: a new medical challenge. J Clin Med 10:1947. https://doi.org/10.3390/JCM10091947

Owczarek A, Gieczewska KB, Polanska M et al (2020) Melatonin lowers HIF-1α content in human proximal tubular cells (HK-2) due to preventing its deacetylation by Sirtuin 1. Front Physiol 11: 572911. https://doi.org/10.3389/fphys.2020.572911

Palagini L, Manni R, Aguglia E et al (2020) Expert opinions and consensus recommendations for the evaluation and management of insomnia in clinical practice: joint statements of five Italian scientific societies. Front Psychol 11:558. https://doi.org/10.3389/fpsyt.2020.00558

Pappolla M, Bozner P, Soto C et al (1998) Inhibition of Alzheimer beta-fibrillogenesis by melatonin. J Biol Chem 273:7185–7188. https://doi.org/10.1074/JBC.273.13.7185

Pappolla MA, Matsubara E, Vidal R et al (2018) Melatonin treatment enhances Aβ lymphatic clearance in a transgenic mouse model of amyloidosis. Curr Alzheimer Res 15:637–642. https://doi.org/10.2174/1567205015666180411092551

Peluso MJ, Lu S, Tang AF et al (2021) Markers of immune activation and inflammation in individuals with Postacute sequelae of severe acute respiratory syndrome coronavirus 2 infection. J Infect Dis 224:1839–1848. https://doi.org/10.1093/INFDIS/JIAB490

Pilia E, Alborino E, Covello RD (2022) Does melatonin reduce mortality in COVID-19? Ann Med Surg 78:103817. https://doi.org/10.1016/J.AMSU.2022.103817

PubMed (2022) Melatonin covid–Search Results–PubMed. https://pubmed.ncbi.nlm.nih.gov/?term=melatonin+covid&sort=pubdate&size=50. Accessed 7 Jun 2022

Raj V, Opie M, Arnold AC (2018) Cognitive and psychological issues in postural tachycardia syndrome. Auton Neurosci 215:46–55. https://doi.org/10.1016/j.autneu.2018.03.004

Ramlall V, Zucker J, Tatonetti N (2020) Melatonin is significantly associated with survival of intubated COVID-19 patients. medRxiv. https://doi.org/10.1101/2020.10.15.20213546

Reagan-Shaw S, Nihal M, Ahmad N (2008) Dose translation from animal to human studies revisited. FASEB J 22:659–661. https://doi.org/10.1096/FJ.07-9574LSF

Rehman S, Ikram M, Ullah N et al (2019) Neurological enhancement effects of melatonin against brain injury-induced oxidative stress, Neuroinflammation, and neurodegeneration via AMPK/CREB signaling. Cell 8:760. https://doi.org/10.3390/cells8070760

Reiter RJ, Tan DX, Rosales-Corral S et al (2018) Melatonin mitigates mitochondrial meltdown: interactions with SIRT3. Int J Mol Sci 19:2439. https://doi.org/10.3390/ijms19082439

Reiter RJ, Sharma R, Ma Q et al (2020) Melatonin inhibits COVID-19-induced cytokine storm by reversing aerobic glycolysis in immune cells: a mechanistic analysis. Med Drug Discov 6: 100044. https://doi.org/10.1016/j.medidd.2020.100044

Reiter RJ, Sharma R, Rosales-Corral S (2021) Anti-Warburg effect of melatonin: a proposed mechanism to explain its inhibition of multiple diseases. Int J Mol Sci 22:1–24. https://doi.org/10.3390/ijms22020764

Reiter RJ, Sharma R, Rosales-Corral S et al (2022a) Melatonin in ventricular and subarachnoid cerebrospinal fluid: its function in the neural glymphatic network and biological significance for neurocognitive health. Biochem Biophys Res Commun 605:70–81. https://doi.org/10.1016/J.BBRC.2022.03.025

Reiter RJ, Sharma R, Simko F et al (2022b) Melatonin: highlighting its use as a potential treatment for SARS-CoV-2 infection. Cell Mol Life Sci 79:143. https://doi.org/10.1007/S00018-021-04102-3

Renz-Polster H, Tremblay ME, Bienzle D, Fischer JE (2022) The pathobiology of myalgic encephalomyelitis/chronic fatigue syndrome: the case for neuroglial failure. Front Cell Neurosci 16:888232. https://doi.org/10.3389/FNCEL.2022.888232

Reynolds JL, Dubocovich ML (2021) Melatonin multifaceted pharmacological actions on melatonin receptors converging to abrogate COVID-19. J Pineal Res 71:e12732. https://doi.org/10.1111/JPI.12732

Rosales-Corral S, Tan DX, Reiter RJ et al (2003) Orally administered melatonin reduces oxidative stress and proinflammatory cytokines induced by amyloid-beta peptide in rat brain: a comparative, in vivo study versus vitamin C and E. J Pineal Res 35:80–84. https://doi.org/10.1034/J.1600-079X.2003.00057.X

Saha M, Manna K, das Saha K (2022) Melatonin suppresses NLRP3 Inflammasome activation via TLR4/NF-κB and P2X7R signaling in high-fat diet-induced murine NASH model. J Inflamm Res 15:3235–3258. https://doi.org/10.2147/JIR.S343236

Sánchez-González MÁ, Mahíllo-Fernández I, Villar-Álvarez F, Llanos L (2022) What if melatonin could help patients with severe COVID-19? J Clin Sleep Med 18:335–336. https://doi.org/10.5664/JCSM.9554

Sánchez-López AL, Ortiz GG, Pacheco-Moises FP et al (2018) Efficacy of melatonin on serum pro-inflammatory cytokines and oxidative stress markers in relapsing remitting multiple sclerosis. Arch Med Res 49:391–398. https://doi.org/10.1016/j.arcmed.2018.12.004

Sánchez-Rico M, de la Muela P, Herrera-Morueco JJ et al (2022) Melatonin does not reduce mortality in adult hospitalized patients with COVID-19: a multicenter retrospective observational study. J Travel Med 29:taab195. https://doi.org/10.1093/JTM/TAAB195

Sehirli AO, Sayiner S, Serakinci N (2020) Role of melatonin in the treatment of COVID-19; as an adjuvant through cluster differentiation 147 (CD147). Mol Biol Rep 47:8229. https://doi.org/10.1007/S11033-020-05830-8

Shafiei E, Bahtoei M, Raj P et al (2018) Effects of N-acetyl cysteine and melatonin on early reperfusion injury in patients undergoing coronary artery bypass grafting: a randomized, open-labeled, placebo-controlled trial. Medicine (Baltimore) 97:383. https://doi.org/10.1097/MD.0000000000011383

Shah W, Hillman T, Playford ED, Hishmeh L (2021) Managing the long term effects of covid-19: summary of NICE, SIGN, and RCGP rapid guideline. BMJ 372:n136. https://doi.org/10.1136/BMJ.N136

Shang Y, Xu SP, Wu Y et al (2009) Melatonin reduces acute lung injury in endotoxemic rats. Chin Med J 122:1388–1393. https://doi.org/10.3760/cma.j.issn.0366-6999.2009.12.006

Shen YX, Xu SY, Wei W et al (2002) Melatonin blocks rat hippocampal neuronal apoptosis induced by amyloid beta-peptide 25-35. J Pineal Res 32:163–167. https://doi.org/10.1034/J.1600-079X.2002.1O839.X

Shukla M, Htoo HH, Wintachai P et al (2015) Melatonin stimulates the nonamyloidogenic processing of βAPP through the positive transcriptional regulation of ADAM10 and ADAM17. J Pineal Res 58:151–165. https://doi.org/10.1111/JPI.12200

Shukla M, Chinchalongporn V, Govitrapong P, Reiter RJ (2019) The role of melatonin in targeting cell signaling pathways in neurodegeneration. Ann N Y Acad Sci 1443:75–96. https://doi.org/10.1111/NYAS.14005

Simko F, Hrenak J, Dominguez-Rodriguez A, Reiter RJ (2020) Melatonin as a putative protection against myocardial injury in COVID-19 infection. Expert Rev Clin Pharmacol 13:921. https://doi.org/10.1080/17512433.2020.1814141

Snider JM, You JK, Wang X et al (2021) Group IIA secreted phospholipase A2 is associated with the pathobiology leading to COVID-19 mortality. J Clin Invest 131:e149236. https://doi.org/10.1172/JCI149236

Soltani F, Salari A, Javaherforooshzadeh F et al (2020) The effect of melatonin on reduction in the need for sedative agents and duration of mechanical ventilation in traumatic intracranial hemorrhage patients: a randomized controlled trial. Eur J Trauma Emerg Surg 48:545. https://doi.org/10.1007/s00068-020-01449-3

Takahashi T, Luzum JA, Nicol MR, Jacobson PA (2020) Pharmacogenomics of COVID-19 therapies. NPJ Genom Med 5:35. https://doi.org/10.1038/S41525-020-00143-Y

Tan DX, Reiter RJ (2022) Mechanisms and clinical evidence to support melatonin's use in severe COVID-19 patients to lower mortality. Life Sci 294:120368. https://doi.org/10.1016/J.LFS.2022.120368

Tan SH, Karri V, Tay NWR et al (2019) Emerging pathways to neurodegeneration: dissecting the critical molecular mechanisms in Alzheimer's disease, Parkinson's disease. Biomed Pharmacother 111:765–777. https://doi.org/10.1016/J.BIOPHA.2018.12.101

Tian M, Liu W, Li X et al (2021) HIF-1α promotes SARS-CoV-2 infection and aggravates inflammatory responses to COVID-19. Signal Transduct Target Ther 6:308. https://doi.org/10.1038/S41392-021-00726-W

Valiensi SM, Folgueira A, Vera VA et al (2022) Pre-pandemic melatonin treatment for sleep disorders and COVID-19 infection. A retrospective cross-sectional study. Vertex XXXIII:13–24. https://doi.org/10.53680/VERTEX.V33I155.132

van Campen CMC, Rowe PC, Visser FC (2021) Orthostatic symptoms and reductions in cerebral blood flow in long-haul COVID-19 patients: similarities with myalgic encephalomyelitis/chronic fatigue syndrome. Medicina (Kaunas) 58:28. https://doi.org/10.3390/MEDICINA58010028

Vecchierini MF, Kilic-Huck U, Quera-Salva MA (2020) Melatonin (MEL) and its use in neurological diseases and insomnia: recommendations of the French medical and research sleep society (SFRMS). Rev Neurol 177:245. https://doi.org/10.1016/j.neurol.2020.06.009

Volt H, García JA, Doerrier C et al (2016) Same molecule but different expression: aging and sepsis trigger NLRP3 inflammasome activation, a target of melatonin. J Pineal Res 60:193–205. https://doi.org/10.1111/jpi.12303

Wells R, Paterson F, Bacchi S et al (2020) Brain fog in postural tachycardia syndrome: an objective cerebral blood flow and neurocognitive analysis. J Arrhythm 36:549–552. https://doi.org/10.1002/joa3.12325

Wilson SJ, Nutt DJ, Alford C et al (2010) British Association for Psychopharmacology consensus statement on evidence-based treatment of insomnia, parasomnias and circadian rhythm disorders. J Psychopharmacol 24:1577–1600. https://doi.org/10.1177/0269881110379307

Xiong YF, Chen Q, Chen J et al (2011) Melatonin reduces the impairment of axonal transport and axonopathy induced by calyculin A. J Pineal Res 50:319–327. https://doi.org/10.1111/J.1600-079X.2010.00846.X

Zhang C, Yang M (2022) Newly emerged antiviral strategies for SARS-CoV-2: from deciphering viral protein structural function to the development of vaccines, antibodies, and small molecules. Int J Mol Sci 23:6083. https://doi.org/10.3390/IJMS23116083

Zhang W, Chen XY, Su SW et al (2016) Exogenous melatonin for sleep disorders in neurodegenerative diseases: a meta-analysis of randomized clinical trials. Neurol Sci 37:57–65. https://doi.org/10.1007/S10072-015-2357-0

Zhang J, Wang L, Xie W et al (2020a) Melatonin attenuates ER stress and mitochondrial damage in septic cardiomyopathy: a new mechanism involving BAP31 upregulation and MAPK-ERK pathway. J Cell Physiol 235:2847–2856. https://doi.org/10.1002/jcp.29190

Zhang R, Wang X, Ni L et al (2020b) COVID-19: melatonin as a potential adjuvant treatment. Life Sci 250:117583. https://doi.org/10.1016/j.lfs.2020.117583

Zhao Z, Lu C, Li T et al (2018) The protective effect of melatonin on brain ischemia and reperfusion in rats and humans: in vivo assessment and a randomized controlled trial. J Pineal Res 65: e12521. https://doi.org/10.1111/jpi.12521

Zlacká J, Stebelová K, Zeman M, Herichová I (2021) Interactions of renin-angiotensin system and COVID-19: the importance of daily rhythms in ACE2, ADAM17 and TMPRSS2 expression. Physiol Res 70:S177–S194. https://doi.org/10.33549/PHYSIOLRES.934754

Chapter 25
Overview and Trends on Sleep and COVID-19 Publication Output

Gabriel Natan Pires and Sergio Tufik

Abstract The emergence of the COVID-19 pandemic led researchers and scientists from many different fields to generate a surmount amount of data, which eventually was published, becoming available, to both the health professionals and to the society. However, the body of knowledge continued growing in an unprecedented way. A condition of "infodemic" was characterized, referring to the growth in the available information following epidemic patterns. Likewise, the publication output of sleep medicine was subjected to important adaptations, which include the number of articles published, the research methods employed, and the motivations of researchers to publish. This chapter describes the publication output of sleep-related research during the pandemic.

Keywords COVID-19 · Bibliometrics · Infodemic · Meta-research · Sleep

25.1 Introduction

The emergence of the COVID-19 pandemic brought uncertainties to many areas of the society, especially into medicine and health sciences. At a fast pace, a disease until unknown spread rapidly through people and countries. Very limited information was available regarding the etiological factor, its infectious potential, epidemiology, prognosis, outcomes, risk factors, prevention, and treatment.

In face of such insecurity, researchers and scientists all over the world have had to devote enormous efforts to understand the virus, the disease, and the pandemic as fast as possible, in a way never seen before. That happened in many fields of research: virologists and molecular biologists were focused on understanding better the virus and its structures, immunologists devoted their efforts to develop vaccines, clinical researchers observed and described the disease progression and its

G. N. Pires (✉) · S. Tufik
Departamento de Psicobiologia, Universidade Federal de São Paulo (UNIFESP), São Paulo, Brasil
e-mail: Gabriel.Pires@sono.org.br

A. S. BaHammam et al. (eds.), *COVID-19 and Sleep: A Global Outlook*, Progress in Sleep Research, https://doi.org/10.1007/978-981-99-0240-8_25

prognosis, epidemiologists monitored the spread of the disease and their populational parameters, and psychologists and psychiatrists described the behavioral and psychopathological consequences of social isolation policies. This just to name a few, among several other classes of professionals involved in the response to the pandemic. All of them left their homes, exposed themselves to the unknown, and took the risks, all for the sake of science, knowledge, and public good.

That massive global scientific effort obviously generated a surmount amount of data, which eventually was published, becoming available, to both the health professionals and to the society. Eventually, the process of generating information grew faster than the spread of the disease, being able to foresee it, improving prediction and prevention. However, the body of knowledge continued growing in an unprecedented way (Lou et al. 2020; Škorić et al. 2020; Kambhampati et al. 2020; Gale 2020). In most areas of research, the number of published articles about COVID-19 was so high, which was impossible to keep updated. This condition has been named as "infodemic," referring to the growth in the available information following epidemic patterns.

These issues apply equally to all research areas within the biomedical sciences. Likewise, the publication output of sleep medicine was subjected to important adaptations, which include the number of articles published, the research methods employed, and the motivations of researchers to publish. This chapter describes the publication output of sleep-related research during the pandemic.

An important disclaimer: this chapter brings criticism to some research practices that took place during the pandemic and approaches some sensible topics regarding scientific conduct. We expect that readers understand it as call for reflection, rather than as a disapproval to any specific individual researcher and/or their practices. Conclusions should not be individualized but put in context and applied to the whole body of sleep researchers. We have gone through unprecedented times, and in face of such uncertainties, it is just natural that we succeeded in some and failed in some other aspects of life, of which scientific publication is just a part. Finally, every possible conclusion and criticism in this chapter apply to its authors as well.

25.2 Sleep Medicine Publication Through the Pandemic

Lee et al. (2022) categorized the evolution of knowledge synthesis and clinical standards during the pandemic into three "eras" (Fig. 25.1). The first era was characterized by the need for an immediate reaction even in the face of limited or inexistant medical evidence. In such case, clinical practice and public health measures were very much depending on theoretical reasoning, anecdotal observations, and hypothesis, borrowing knowledge from other respiratory viral diseases and previous pandemics. The dissemination of knowledge was also nonstandard, as social media was much faster in widespread information (including misinformation) than the regular scientific publication framework (Lacsa 2022; Gisondi et al. 2022). Another hallmark of the first era was the publication of some position statements

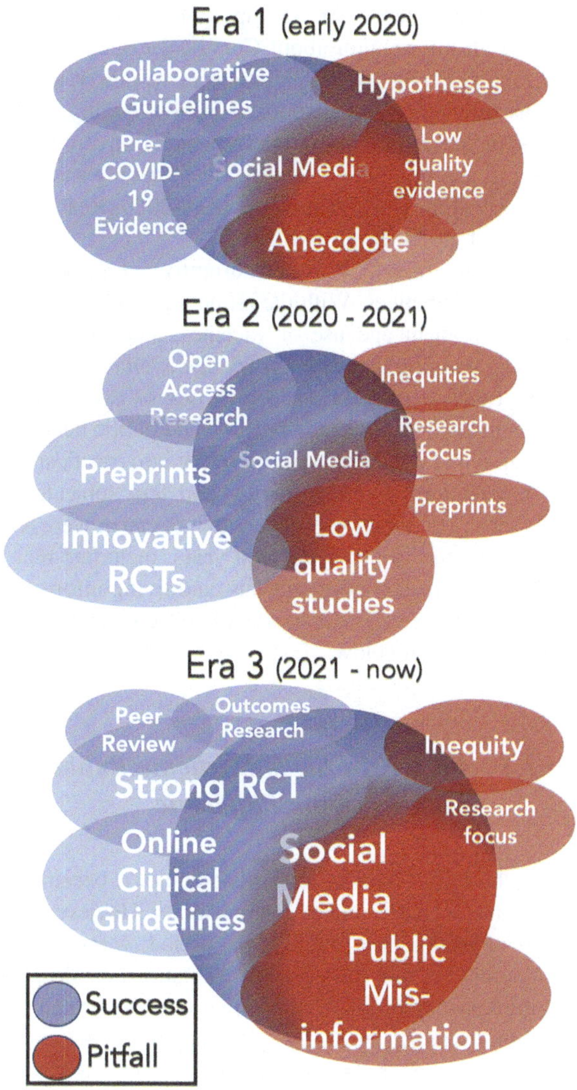

Fig. 25.1 The eras of evidence synthesis and research during the COVID-19 pandemic. Each era had its own successes (blue) and pitfalls (red). Reprinted with authorization from Lee et al. (2022)

from medical associations, which although not properly evidence based were needed to provide early directions to health professionals and the society.

The second era was characterized by a rapid increase in the available evidence and the implementation of innovative approaches to the traditional process of generating and synthesizing evidence. This led to an unprecedented increase in the total number of articles published, which was then characterized as an infodemic. The publication of preprints became much more common than before, playing an important role in the initial dissemination of information (Fraser et al. 2021; Gianola et al. 2020; Gordon et al. 2022; Kodvanj et al. 2022; Nelson et al. 2022). Another positive consequence was an important move from publishers and journals towards making all information about COVID-19 openly available (Besançon et al. 2021; Arrizabalaga et al. 2020). On the negative side, it was noticed that an important part of this infodemic was composed by low-quality studies (caused by either honest errors or scientific misconduct), which impaired a proper process of evidence synthesis and brought noise into policy making and implementation of public health and medical measures (Ramanan et al. 2020; Reardon 2021; Alexander et al. 2020; Nieto et al. 2020; Bramstedt 2020).

The third era, which we are currently living, show a slight return into traditional publication practices. With the reduction in number of cases and deaths, success on the development and use of vaccines, and loosening of social distancing norms, urgently applying any new knowledge available became less needed. Unlike during the first and second eras, in which one or two studies with positives results were sufficient to change bedside practice and public health measures, it now takes longer from an initial description or discovery into practical implementation. This turns the implementation of research safer and less subjected to biases and frauds. Other research and publication practices that began or were reinforced during the pandemic might continue valid for the next years, including real-word evidence studies, fast-track randomized controlled trials, and living information sources.

All these three eras can be identified for sleep medicine. Each of them is discussed in the subsection below.

25.2.1 First Era: Sleep Medicine Publication in the First Months of the Pandemic

Since the initial months of the pandemic, it became clear that COVID-19 had no direct specific sleep-related manifestation. However, many concerns related to sleep medicine were raised in this initial period. A few distinguishable concern areas can be identified in this early phase: adaptations on sleep medicine practices (including the implementation of telemedicine) (Hollander and Carr 2020; Penzel et al. 2021; Shamim-Uzzaman et al. 2021), adaptations and consequences of COVID-19 on sleep among healthcare professionals, treatment to sleep disorders during the pandemic (especially the safety of CPAP), relationship between obstructive sleep apnea and COVID-19, and relationship of insomnia with the context of social isolation and home confinement.

However, as very few studies were available during the initial months of the pandemic, practical guidance came from either indirect evidence (from other viral diseases or other pandemics) or theoretical reasoning. In parallel, several speculative studies started appearing, suggesting possible relationships between sleep and COVID-19, which could not be tested or proved at the time due to the absence of proper data.

Among these initial and indirect evidence were some position statements from sleep-related medical societies and other important stakeholders in sleep medicine. These were not available through peer-reviewed papers but produced by these societies and distributed directly to their affiliated members, or available in their websites. As an example, societies such as the World Sleep Society (WSS), the American Academy of Sleep Medicine (AASM), the European Sleep Research Society (ESRS), and the Federation of Latin American Sleep Societies (FLASS) provided recommendation and practical advice to their members. Similarly, many national societies published similar guidance on local languages, which is reasonable considering that frontline health practitioners and sleep medicine professionals in many countries may not follow international literature or are unable to read in English.

Among the topics on these first wave of position statement papers were discussions regarding the safety of performing in-lab polysomnographies, possibility of CPAP devices producing aerosol and spreading the virus, advice about the implementation of telemedicine into sleep medicine, and discussions about the effects of social isolation on sleep.

25.2.2 *Second Era: Infodemic and Preprints in Sleep Medicine*

Era 2 is characterized by the publication of the first wave of studies about sleep and COVID-19. In a matter of a couple of months, a rapid and extraordinary increase in the number of sleep studies was observed. This increase was so evident that a condition of infodemic was characterized (more information on *Publication Output Analysis*).

In this era, the first official statements from sleep-related societies became available at PubMed and other databases, complementing the information that was restricted to societies' websites so far. The first official society publication to become available on PubMed was probably the recommendations to deal with and treat insomnia and other sleep problems during the pandemic, by the European CBTi academy (Altena et al. 2020). Following that, an overall guidance to the professional practice of sleep medicine during the pandemic was published, by the Indian Society for Sleep Research (Gupta et al. 2020), and guides for the diagnosis and treatment of sleep-disordered breathing during the pandemic, by the German Society for Sleep Medicine (Büchner et al. 2020) and by the Italian Thoracic Society (Insalaco et al. 2020). All these practice standard papers were published up to August 2020 and were followed by many other ones from other societies.

Most of the initial studies in this area were not properly focusing on sleep medicine, but rather evaluating insomnia and other self-reported sleep complaints among broader behavioral evaluations, most of them among healthcare professionals. These were among the first populations to be studied during the pandemic, as due to the changes in work characteristics and due to the psychological burden they were subjected to. Following this, studies with the general population and then with clinical populations were available. The methods employed in these studies were also different than usual in the second era. The use of online surveys became common for the evaluation of insomnia and overall self-reported sleep characteristics. For more objective sleep assessments, database studies (mostly cross-sectional and retrospective cohorts) were more common than actual patient-recruiting studies. Systematic reviews and meta-analyses also became increasingly common (Abbott et al. 2022; Pires et al. 2021a) (more information on *Research Methods* subsection).

One important characteristic of this era was the rise of highly speculative papers. Those articles proposed interventions and pathophysiological relations that might seem logical and reliable, but that at the time were not properly supported by evidence (more information on *Motivations to Publish* subsection). Among the main topics that were the focus of speculative research are the relationship between obstructive sleep apnea and severe COVID-19 (Tufik et al. 2020; Tufik 2020), the potential of melatonin as an adjuvant treatment (Cardinali et al. 2020; Hardeland and Tan 2020; Öztürk et al. 2020), and the effect of sleep disorders on the efficacy of vaccination against COVID-19 (Kow and Hasan 2021; Salles et al. 2021; Benedict and Cedernaes 2021).

Another important characteristic of this era was the publication of sleep studies in preprint servers. Just as for other research areas, preprints were a good alternative for disseminating research results immediately, bypassing the inherent delays posed by the standard peer-review system. Although that was considered an important measure to expedite the availability of data in the early phase of the pandemic, preprints have been questioned due to the possibility of being a means through which misleading and even fraudulent studies gain visibility (Gianola et al. 2020; Kupferschmidt 2020; Majumder and Mandl 2020) (more information on *Journals* subsection).

A final characteristic of this era is the average low quality of research. Several studies have disclosed that flawed and misleading studies were more common during the pandemic than before for most areas. Some studies about the quality of the research during the pandemic have been published, demonstrating that the average quality is lower, and the proportion of methodologically flawed studies is higher than before in general COVID-19 research (Bramstedt 2020; Abbott et al. 2022; Khatter et al. 2021; Jung et al. 2021), and probably also in sleep medicine.

25.2.3 Third Era: Consolidation of the Relations Between Sleep and COVID-19

After almost 3 years of pandemic, the society is slowly getting back to normal (or at least getting used to the social changes imposed by the pandemic). This was the time needed for knowledge to be settled, analyzed, and understood by the health sciences community. Also, there is no longer an urgency to provide new information from one day to the other, so researchers now have the time to analyze their data accordingly.

In such a stage, it can be said that the research between sleep and COVID-19 has reached a more mature level. Some research characteristics commonly observed during the initial eras of the pandemic, such as speculative articles and online surveys, are not as important as they were before (although they still are being published). Now, proper controlled trials, large cohorts, and robust meta-analyses are being conducted, considering the time required to properly perform these studies. Also, many questions and doubts that were important during the beginning of the pandemic are deeply understood now (such as the impact of social isolation on the prevalence of insomnia (AlRasheed et al. 2022) and the relationship between obstructive sleep apnea and severe COVID-19 (Strausz et al. 2021)). On the other hand, there are fields for which a high-level certainty of evidence could not be reached so far (such as for the use of melatonin as an adjuvant treatment for COVID-19 and for the relationship between sleep disorders and vaccination).

Also, important practice standards are properly described and are now being implemented with little margin for doubts. That includes the implementation of telemedicine in sleep medicine (which was expedited due to the pandemic) and the safety of CPAP devices. These practical issues are properly covered by position statements, consensus, and guidelines provided by international sleep medicine societies. These publications were essential for the standardization and consolidation of knowledge and practice of sleep medicine in relation to COVID-19 and the pandemic.

Research about sleep and COVID-19 will obviously continue on the forthcoming years, but it tends to get into a more stable pattern than the frantic growth observed so far. We should expect better evidence to be reached based on more robust study designs, even if the number of articles stalls or even decreases (both in absolute numbers and proportionally).

25.3 How Authors Dealt with Research During the Pandemic

Due to the COVID-19 pandemic, most sleep research centers (from clinical sleep labs to preclinical facilities) remained closed in 2020. Researchers had to reinvent themselves, to sustain their academic production and maintain sleep science performing at a high level, even far from their labs and with limited resources.

Even with such difficulties, good-quality research and robust evidence were achieved as fastest as never before. This is true for most research fields, and in sleep medicine as well. However, it cannot be said that all research during the pandemic was performed with such auspicious intents. In many cases, the publication output and the way articles were put forward were more as a consequence of the "*publish or perish*" paradigm, than a real need for more research. This section discusses the motivation for research during the pandemic, the rise of speculative research, and the research methods that became more common during the pandemic.

25.3.1 Motivations to Publish

The reasons for research to be published vary in many different ways, but one interesting way to categorize it as extrinsic and intrinsic research motivations (Table 25.1). Extrinsic research motivation drives the researcher based on the benefit the results of that research might offer to people other than the researchers itself. It might be to the society or to patients, or simply for the sake of science. On the other hand, intrinsic research motivation is the one performed to provide return and benefits to the researcher himself or herself, either directly or indirectly.

Any researcher's publication record is always a consequence of a balance between altruistic and egocentric drives, and researchers would hardly be driven by only one of these two sides. The society and the scientific community in general want researchers to publish based on altruistic values, and their results to serve in some way for the good of all. However, researchers benefit personally from having a new article published and from the repercussion of it. Our feeling though is that the balance between intrinsic and extrinsic research motivations has changed, and the weight of intrinsic motivations has become heavier. This seems like a natural shift due to the struggles researchers have been through during these last 3 years, and it can be understood differently, based on two scenarios: for researchers on the

Table 25.1 Extrinsic and intrinsic research motivations

Extrinsic research motivation	Intrinsic research motivation
For the common good	For individual professional growth
To develop new technologies	To get promoted or for career progression
To increase the body of knowledge	To get respect by peers and superiors
To understand and describe scientific processes	To get attention
To promote better healthcare and improve clinical practices	To increase the likelihood of being funded by research-supporting agencies
To promote social development	To comply with institutional publication goals
To generate evidence	To be "on the spotlight"
To truly contribute with and generate science	For the sake of one's ego

Extrinsic research motivation regards research performed towards or for the benefit of others. It might also be called eccentric or altruistic research motivation. Intrinsic research motivation is the one performed for the benefit of the researcher itself and might be also called egocentric research motivation. Most research comes from the balance between altruistic and egocentric research motivations relationship between sleep and COVID-19, and for researchers about sleep in general, during the pandemic.

In the first case, for researchers studying the relationship between sleep and COVID-19, the intrinsic research motivation is evident. There were important research questions about it to be addressed since the beginning of the pandemic. Understanding how COVID-19 affected sleep or how sleep disorders could predispose to more severe COVID-19 cases was indeed very relevant. However, soon a huge wave of research appeared, proposing and supposing pathophysiological, epidemiological, behavioral, and therapeutic relationships between sleep and COVID-19. Most of these studies, which were more common during the first and early second era of the pandemic publication history, were purely theoretical or based on very preliminary data. In such cases, the intrinsic benefit comes from being the first to describe a certain phenomenon, even if that eventually implies in publishing excessively speculative and low-quality studies (more information on *Speculative Research* subsection).

In the second case, for sleep researchers dealing with topics other than COVID-19, the struggles were different. As aforementioned, most researchers were kept away for their regular sources of data for research, either it being sleep medicine clinics or preclinical sleep laboratories. Most of them remain closed during the pandemic due to safety measures. Therefore, many research protocols were either interrupted or postponed (Ramanan et al. 2020; Dobler 2020). However, the pressure to publish remained on the back of the researcher regardless of it all, and the "publish or perish" paradigm seems to have stroked harder than ever before.

Very negative effects were observed from the hazardous combination of sustained publication output goals and increased pressure for productivity on one side, with limited funding and restricted career opportunities on the other side (Suart et al. 2022). Important consequences were observed in research integrity (including reduced overall quality and cases of research misconduct) (Bramstedt 2020; Anderson et al. 2021) and in researchers' mental health (including depression and burnout among researchers) (Suart et al. 2022).

25.3.2 Speculative Research

Considering the struggles sleep researchers have gone through during the pandemic, including the "publish or perish" duality most of us have been subjected to, some alternatives had to be found. One of them was the publication of theoretical papers, including narrative reviews, hypothesis, and letters to the editor.

Theoretical studies are indeed very important when we have no information at all. That was exactly what happened during the beginning of the pandemic, especially during the first and the early second eras. Considering the severity and spread of the disease, researchers of any area had to consider what would be the possible implications of COVID-19 to their practices and to their patients. In such a case,

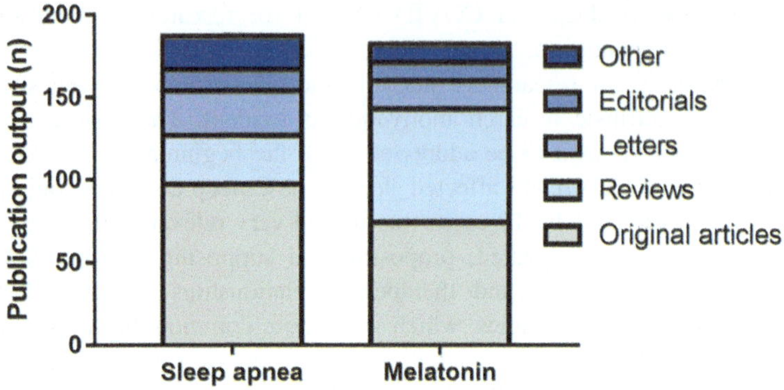

Fig. 25.2 Publication output relating to COVID-19 and either OSA or melatonin. Data are split by publication type and include only studies published up to 2021, in order to include only years for which publication output has been completed. Data acquired through Scopus

theoretical reasoning, hypothetical considerations, and indirect evidence brought from other respiratory viral diseases and other pandemics became important tools.

However, the publication of speculative studies and hypothesis implies in some way to the willingness of testing it and subjecting it to scientifically rigorous methods. So, when the pandemic proceeded to its third era, with relaxation of social isolation norms and decrease in the number of new cases and deaths, one would expect that the theoretical studies would be slowly replaced by the actual primary research, so scientific speculation would give room to the actual evidence-based research. That did not happen, and it was no different in sleep medicine.

One illustrative case is the suggestion of melatonin as an adjuvant treatment to COVID-19. Such assumption indeed makes sense from a pathophysiological standpoint, and important articles have been published suggesting it in the early phase of the pandemic (Cardinali et al. 2020; Hardeland and Tan 2020; Öztürk et al. 2020; Shneider et al. 2020). However, the amount of theoretical and speculative articles on this matter was outnumbering the number of actual primary research. By the end of 2021, 60% of all publication output about melatonin and COVID-19 was composed by theoretical articles (including reviews, letters to editor, editorials, and other types—Fig. 25.2).

A similar case happened with the relationship between obstructive sleep apnea and COVID-19. This was an early concern, considering that the phenotype of patients with either severe OSA or severe COVID-19 shared very similar characteristics at that time (men, overweight, older adults, with cardiometabolic comorbidities) (Tufik et al. 2020; Tufik 2020). Important studies at this early stage suggested four different possible relationships (De Mello et al. 2020): (1) CPAP therapy might increase the spread of the virus as it can release aerosols. (2) OSA increases the risk of COVID-19 infection due to immunological impairment. (3) OSA is a risk factor for negative COVID-19 outcomes (including ICU admission, mechanical ventilation, and death).

The level of evidence today is satisfactory. It allows concluding that CPAP is safe to use and does not increase the spread of the virus, that OSA might increase the chances of getting infected (although there are divergent studies), and that OSA is indeed a risk factor for a negative prognosis of COVID-19. However, regardless of the good amount of primary data, theoretical research still strives. By the end of 2021, slightly less than a half of the publication output about COVID-19 and OSA was composed by theoretical papers (Fig. 25.2).

More recently, especially since vaccines against COVID-19 became available, many theoretical studies ventilated the possibility that sleep disorders and sleep deprivation could decrease vaccination efficacy (Kow and Hasan 2021; Salles et al. 2021; Benedict and Cedernaes 2021; Zhu et al. 2021). That assumption was completely valid and relevant, since it was based on indirect evidence about the decreased antibody titers after vaccination for other viral diseases among sleep-deprived individuals. However, it took a while since the initial narrative reviews and comments on the topic and the first primary studies to address it. Also, when the few studies on this matter became available, it demonstrated that the relationship is not that clear, that a reduced vaccination efficacy against COVID-19 is not a robust and evident consequence of sleep disorders, and that there might be many confounding factors involved in this relationship (Tufik et al. 2022; Jolliffe et al. 2022; Yamanaka et al. 2022). As of today, some evidence is being built regarding the impact of the time of the day in which vaccination took place (Jolliffe et al. 2022; Yamanaka et al. 2022; Matryba et al. 2022; Wang et al. 2022; Zhang et al. 2021; Filippatos et al. 2022), but results are conflicting and inconsistent so far. In any case, regardless of the actual results, theoretical speculation still outnumbers the number of data-driven studies.

The reason for this surge in speculative research is multifactorial. Some researchers might not have access to patients to test their hypothesis, although still considering that their ideas are worth sharing. Some others have been impacted by lockdowns and other enforced social distancing rules, so theoretical articles were all that was available. In other cases, the proposed relationship was indeed new and has not been properly evaluated, so theoretical reasoning is indeed a good starting point. In all these cases and in others, one might think, the drive of a researcher to write a theoretical essay is usually legitimate, not likely to be a case of research misconduct or something to be criticized at an individual level. However, the group behavior of the sleep research community seems to have prioritized theoretical papers over data-driven ones.

25.3.3 Research Methods

Many interesting and innovative research tools have been implemented during the pandemic. This might be another consequence of the "publish or perish" paradigm, but it might be more auspicious to think that researchers were able to reinvent themselves and their research methods in order to collect data and generate evidence, even when social distancing was in place and sleep laboratories were mostly closed. A few of them were remarkably used and deserve special attention: research on medical personnel, online surveys, database research, real-world evidence, and meta-analyses.

25.3.3.1 Research on Medical Personnel

Evaluating sleep among medical and healthcare professionals was probably the first widely used research approach in sleep medicine during the pandemic. In a time in which most of the population was locked up in their homes, these professionals were among the few who were in the field. Most of these studies were not focusing only on sleep, but rather having sleep-related complaints and symptoms being assessed as a part of broader psychological and behavioral evaluations. In this sense, two clear relations were described: First, insomnia complaints were common among healthcare professionals working on COVID-19 wards, which is most likely a result of the psychological burden they have been subjected to. Second, sleep deprivation was a common feature among these professionals, being a direct consequence of either insomnia or an increased workload and decreased workforce. Research on medical professionals remains being performed and will likely remain being relevant. However, their novelty potential decreased once the phenomenon was properly characterized, and other study populations became available.

25.3.3.2 Online Surveys

Right by the beginning of the pandemic, in-person interaction had to rapidly adapt to virtual and online modalities. This was true for telemedicine, education, and business, among others. Likewise, researchers found virtual online surveys to be a good tool to continue collecting data even at a distance and considering the social isolation norms.

Sleep-related online surveys were already a relevant research tool even before the pandemic, but they became much more popular during it. Sleep studies based on online surveys started being published right at the early second era of the pandemic, providing some important benefits. First, it allowed collecting data from large sample sizes at a reduced timeframe. Second, by being purely online, it circumvented all the drawbacks and limitations imposed by social isolation. Third, it was easy to adapt the most used sleep questionnaires to an online and self-applicable mode. Google Forms, SurveyMonkey, REDCap, and other similar platforms were widely used for these purposes. Also, social media and messaging apps were widely used to promote these surveys.

A good number of interesting papers using such tools were published, most of them describing the prevalence of insomnia symptoms and other subjective sleep-related complaints during the pandemic. Indeed, most meta-analyses evaluating insomnia prevalence during the pandemic are based on studies employing online surveys. Research on chronobiology and circadian disruption was also performed consistently with online methods.

However, as the pandemic evolved, the relevance of such studies was reduced and their average novelty potential decreased. With that, at least two important research biases inherent to online surveys (and to open call studies in general) became evident: interest bias and recall bias. Interest bias refers to an increased likelihood of attracting people who have a certain condition of interest. As an example, if someone promotes an online survey about poor sleep habits in children during the pandemic through social media, it is possible that parents whose children have been sleeping poorly feel more motivated to answer it. It creates a biased and unreliable measure of prevalence, not being able to assure the level of representativeness of a sample in comparison to the whole population. Recall bias refers to the likelihood of recalling information in an unprecise way. This seems to be a bigger problem for studies trying to compare some sleep-related feature in the pre-pandemic times and during the pandemic. In several cases, participants were asked to fill up questionnaires about their average sleep habits before the pandemic. Asking someone how they used to sleep 2 years ago (i.e., before the pandemic) subjects to a big risk of recall bias, in a way that the collected information is not the actual sleep pattern before the pandemic, but rather one's perceptions of its own sleep, deturbed by time.

Studies based on online surveys continue to be published. They might be relevant and will certainly continue being used as a relevant research tool after the pandemic is over. However, researchers employing these methods should be aware of its limitations.

25.3.3.3 Database Research

Due to the impossibility of recruiting new research participants and performing in-lab polysomnography, secondary analysis of available databases became an important research alternative.

Most large sleep study databases were widely used for research purposes during the pandemic. Also, open databases containing sleep information were employed as a source of data and publications, based on either polysomnographies and other objective sleep assessment tools (such as the National Sleep Research Resource—NSRR) or self-reported and subjective sleep data (such as the National Health and Nutrition Examination Survey—NHANES). Clinical databases of large sleep clinics were also considered for research purposes in this period, but in these cases, more serious concerns about data protection, data management, and other research ethics issues took place, since these data were collected for clinical, rather than research, purposes.

25.3.3.4 Real-World Evidence

Real-word evidence is not a new term but has been refashioned and widely used in many different areas. It traditionally refers to the use of real-world data to reach clinical evidence about the benefits and possible harms of medical devices, products, and treatments. Real-world data refers to data that are collected as part of regularly running medical services or devices, reflecting a cohort of real customers, users or patients (therefore, not research participants). Real-world evidence is especially relevant for postmarked surveillance, but has been commonly used for exploratory and descriptive studies, as well as to refine the methods of future clinical trials.

However, during the pandemic, the use of real-world data and the publication of studies using real-world evidence became increasingly common and their usage was widened. In many cases, it was used almost as a replacement for clinical trials. Although they are not intended for this purpose, it provides some level of evidence, especially when randomized controlled trials were not possible to perform.

One important source of data for the use of real-world evidence is sleep technology companies, ranging from more traditional ones (such as CPAP manufacturers) to innovative ones (such as sleep trackers and digital cognitive behavioral therapy apps). All these companies have databases including massive amounts of sleep-related data, which are indeed a great source of information for sleep research. Such real-world evidence studies have already been performed before the pandemic, with some very interesting studies performed. But the pandemic served as a way to accelerate the development of portable treatments, wearable devices, and digital therapy in sleep medicine. Therefore, the number of sleep studies based on real-world data seems to have increased considerably in this period.

Real-world evidence should not be considered as a definitive source of evidence though. In most cases, it is basically a large convenience sample, and as such, there are some important biases. First, it does not provide a proper control of confounders and biases as randomized controlled trials do, therefore always being subjected to selection bias. Another concern is interest bias, as for example, people who monitor their sleep through sleep trackers or who undergo digital CBTi are more likely to being concerned either about its own sleep patterns or to have insomnia. Therefore, estimating prevalence might be problematic based on real-world data. Finally, industry sponsorship bias is an important concern, since most of these studies are either directly or indirectly supported by commercial companies with interests in the sleep market, which can be directly impacted by the results of a given article.

25.3.3.5 Systematic Reviews and Meta-analyses

We were already living in a period of massive production of systematic reviews and meta-analyses even before the pandemic. That was associated with an overall low quality of these studies, including misleading, redundant, unnecessary, and flawed meta-analyses (Ioannidis 2016). A similar pattern is observed in sleep medicine,

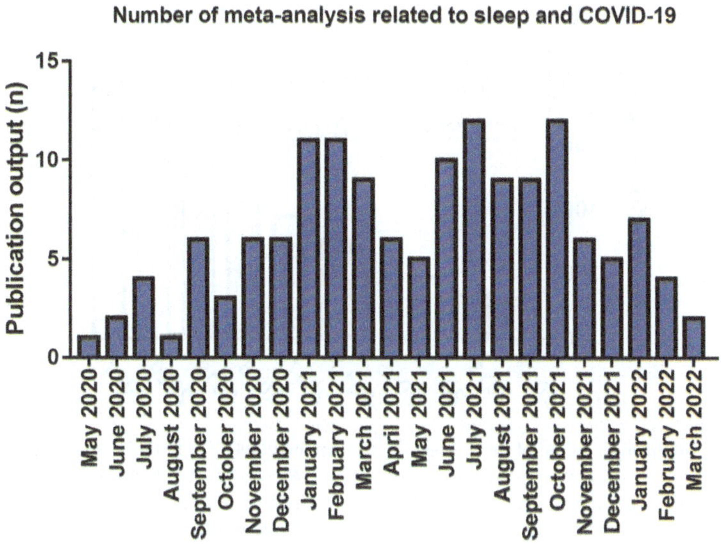

Fig. 25.3 Number of meta-analyses about sleep and COVID-19, by month. The first meta-analyses were published in May 2020. Since then, the number of meta-analyses has been growing, reaching its peak in October 2021. The drop in the number of meta-analyses in 2022 is likely to be an artifact, resulting from the delays between availability and actual publication of these studies. Data acquired through Scopus

with an average growth rate of 30% in the number of published meta-analyses per year (Pires et al. 2021a).

During the pandemic, the number of systematic reviews and meta-analyses in sleep medicine increased considerably. This is directly related to the "publish and perish" paradigm, associated with the ease of performing it. In other words, sleep researchers were locked up in homes during the pandemic, away from their usual research places, but still subjected to a pressure to continue publishing. In such case, systematic reviews seemed to be a good alternative, as the only requirements to perform it are a computer and some colleagues to be involved in eligibility analysis. Even specialized software are not necessary, as many open-source and free options are widely available.

The first meta-analysis about sleep and COVID-19 was published in May 2022, being followed by an increased amount of other similar studies in the following months (Fig. 25.3). It calls for attention that meta-analyses were being published so early. As being a study design directly dependent on the availability of previous studies, one might question whether there were sufficient data to be meta-analyzed in the initial months.

The number of meta-analyses about sleep and COVID-19 indeed increased rapidly, and a new meta-analysis in this area was published every 6 days (Fig. 25.3). Such a great number of published meta-analyses probably reflects the growing pattern of publication of this type of studies in sleep medicine as well, but it

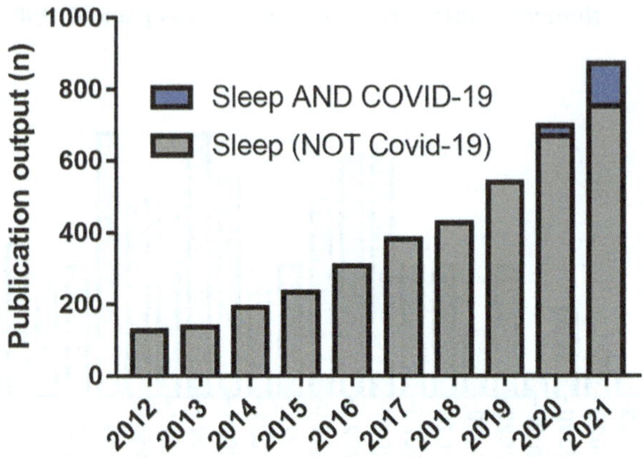

Fig. 25.4 Publication output of meta-analyses in sleep medicine. The number of meta-analyses has been growing consistently over the last couple of years. Meta-analyses about COVID-19 and sleep appeared already in the first year of the pandemic, although representing a small proportion of the total publication output of that year. In 2021, 871 sleep-related meta-analyses were published, and meta-analyses about COVID-19 accounted for about 10% of it. Only studies published up to 2021 were considered, in order to include only years for which publication output has been completed. Data acquired through Scopus

is undoubtable that meta-analyses about sleep and COVID-19 took an important share of it. Only in 2021, 871 sleep-related meta-analyses were published, and meta-analyses about sleep and COVID-19 accounted for about 10% of it (Fig. 25.4). It really calls for attention that only 2 years after the description of the disease, meta-analyses about it took such share among the whole publication output.

Some concerns regarding the quality of these meta-analyses should be highlighted. Systematic reviews and meta-analyses are a very pragmatic research method, guided by well-described guidelines, and it requires some time and efforts to master it. The pandemic brought the demand for more systematic reviews to be performed and published, to deal with the sustained pressure to publish, but it cannot be assured that individuals performing systematic reviews under these conditions were aware of or applying the best methods. As a result, the quality of these systematic reviews was very low.

Actually, the average methodological quality of systematic reviews and meta-analyses was already considered poor even before the pandemic (Xu et al. 2021). During the pandemic, the average methodological quality level of meta-analyses on all areas was remarkably low (Abbott et al. 2022; Pires et al. 2021b), and less than 10% of them could be considered as of high confidence level (Pires et al. 2021b). There are no studies specifically about the methodological quality of meta-analyses about sleep and COVID-19, but it is supposed that they would present the same pattern of low quality observed in other fields and in sleep medicine in general.

25.4 Journals

Just like the researchers, most stakeholders involved in biomedical research adapted procedures to assure that COVID-19-related research was performed rapidly (Song and Karako 2020). This was indeed critical, especially during the beginning of the pandemic, as a matter of responding to the urgent need of information and assuring that the time from initial research to the availability of the final results was as short as possible. For example, governments and research-supporting agencies were providing special research support grants for research about COVID-19, while regulatory agencies and research ethics committees were prioritizing projects related to COVID-19 (Zhang et al. 2020). Publishers and journals have also adapted their regular procedures to implement an efficient publication outflow during the pandemic.

During the beginning of the pandemic, all the most renowned international publishers have applied open-access policies to articles related to COVID-19 (Besançon et al. 2021; Arrizabalaga et al. 2020; Song and Karako 2020). This was done as a matter of social responsibility, to assure that no knowledge was blocked by paywalls during the pandemic. This was easily implemented, as it required very few technical resources and deviations from the standard publication procedures but has had an almost immediate effect on assuring that data were made available as soon as possible. The decision to implement open-access practices is something taken primarily at a publisher level but has certain effect on the journals individually. Specifically to the field of sleep medicine, the journals that were published on a subscription-based or hybrid format made their articles available as open access.

Another important measure was the implementation of fast-track peer-review processes (Shah et al. 2020; Treweek et al. 2020). This was a very important aspect of the publication process during the pandemic, as it was related to the critical balance between the need for making studies available as fast as possible and the need of sticking with the peer-review process. Although there are several different formats of peer review (each of them with its advantages and drawbacks), it seems consensual that having manuscripts undergoing appraisal by peers is a matter of responsibility, which increases the likelihood of only publishing studies that are relevant and reliable, as well as decreasing the chances of scientific fraud. The exact way by which fast-track peer review was implemented varied across journals, but it usually involved higher level editorial board members in the peer-review process (such as the editor-in-chief itself, or deputy, section, or associate editor) assuring high commitment with the journal's scope and short turnaround time.

Important discussions took place on the quality of peer review on these fast-track reviewing policies (Shah et al. 2020; Smith et al. 2020; Bagdasarian et al. 2020; Peyrin-Biroulet 2020), and it has been argued that it might increase the odds of overlooking important methodological errors and research frauds, or allowing low-priority manuscripts to undergo fast-track peer review.

Another publication modality that gained importance during the pandemic was the preprint servers. Depositing research manuscripts and related materials at preprint sources (such as MedRXiv or BioRXiv) was a growing open research practice before the pandemic, but it became considerably more common during it (Gianola et al. 2020; Kupferschmidt 2020; Majumder and Mandl 2020). It is estimated that during the first months of the pandemic, about a third of the information was available in preprint sources (Fraser et al. 2021). The number of studies available as preprints only is so significant that including preprint sources in systematic review search strategies became common during the pandemic (AlRasheed et al. 2022; Pires et al. 2021b). This has also been observed in sleep medicine, as at least 35 sleep-related studies are available on MedRXiv via PubMed.

The main benefit of preprints relies on the immediate availability of data, bypassing the delays that the peer-review system might have. On the other hand, important criticism and concerns have been raised (van Schalkwyk et al. 2020; Sheldon 2018), and it is not consensual whether its benefits outweigh its potential risks to research integrity (van Schalkwyk et al. 2020). The most important of them is the lack of a proper peer review (or the availability of the manuscript before peer review has been completed), which might be a way of disseminating wrong, poor-quality false information (King 2020).

The overall quality of COVID-19 studies available in preprint servers seems to be adequate (Kodvanj et al. 2022; Nelson et al. 2022), and the likelihood of these manuscripts being published in regular peer-reviewed studies is high (Gordon et al. 2022). Therefore, it can be concluded that preprints are a safe and reliable source of information in most cases. However, even being the minority, important problems have happened indeed, including the publication of flawed or fraudulent studies, which were made publicly available without proper peer appraisal. The most renowned of these cases are those involving a supposed structural relationship between SARS-CoV-2 and HIV, questioning the severity of SARS-CoV-2 infections and promoting the use of inefficient pharmacological treatments (Reardon 2021; Watson 2022). This became even more problematic when these non-peer-reviewed materials reached the general population, the press, and the media, which might not have all the attributes necessary to judge the quality and reliability of a scientific study.

Another consequence of the fast-track peer-review policies during the pandemic was the increased number of retractions (Bramstedt 2020; Anderson et al. 2021; Abritis et al. 2021; Piller and Travis 2020). This was observed since the first months of the pandemic, and the total number of retractions related to COVID-19 was tremendously higher than for other viral diseases (Peterson et al. 2022). This points out to this excessive number of retractions being a consequence of the combination of high need for new information and flexibilization of peer-review policies (Bramstedt 2020). Fortunately, sleep medicine does not seem to have been affected, as only a single study about COVID-19 and sleep has been retracted so far (Liu et al. 2020), according to the Retraction Watch Database (http://retractiondatabase.org/), which has been accounted as an error of the publishers, rather than of the authors.

Finally, most sleep medicine and chronobiology journals published a significant number of COVID-19 studies. Some of them had open calls for papers or created special collections related to sleep and COVID-19. An example is the Journal of

Clinical Sleep Medicine, which was waiving their submission fee for articles in this topic. These special collections and call for papers were primarily driven by a sense of social responsibility, as journals were concerned on generating evidence and sharing it to the public as soon as possible. But it is undeniable that they also benefit from publishing these articles, as they are more likely to be cited in the short term.

25.5 Publication Output Analysis

As previously discussed, as soon as the pandemic was declared, researchers tried their best to generate as much evidence and new information as possible, at a fast pace. This was done reasonably well, and most aspects of the virus and the disease were properly described in a couple of months. However, the amount of information generated was so massive that no one could follow it all. This led to the characterization of a condition of infodemic, which refers to the accumulation of information following epidemic growth patterns.

The first article about sleep and COVID-19 to be available on PubMed was probably a letter to the editor commenting on the possible use of melatonin to reduce the severity of COVID-19 (Shneider et al. 2020). It was published in the International Reviews of Immunology and made available online on April 29, 2020. Of note, this information just reflects what is currently available on PubMed based on a simplified search strategy ((sleep[tiab] OR sleep[mh]) AND COVID-19). It is possible that other articles might have been published before, mainly on preprint servers and on non-indexed sources. Another possibility is that the primary studies were available in Chinese mandarin only and published in local Chinese journals, reflecting the fact that the disease began there. For this reason, most of the initial meta-analyses about COVID-19 usually included Chinese databases, in order to assure including the first articles that were published in local language and not available for the international audience.

As expected, this first article was a theoretical reflection on the possible relationship between sleep and COVID-19, which is justifiable in the first era of the pandemic and considering that there was no actual data available. In such case, theoretical and speculative essays were indeed necessary, to being the discussion about the relevance of sleep in the scenario of the pandemic. The following papers were also theoretical and speculative essays, including topics such as the overall importance of sleep during the pandemic (Gulia and Kumar 2020), sleep health among frontline healthcare professionals (Singh et al. 2020; Ferini-Strambi et al. 2020), use of CBTi for essential workers (Benhamou and Piedra 2020), and management of sleep-disordered breathing during the pandemic (Pirzada et al. 2020). All of them were published in the early phase of the pandemic, only 3 months after it has been declared.

Shortly after, some studies started to be published, initially using online survey tools to evaluate sleep-related psychological consequences of social isolation or evaluating sleep in medical personnel. That evolved quickly into more refined research approaches, as discussed in the previous sections. As of today, living in the third era of the pandemic, the availability of sleep research tools seems to have been back to normal, as sleep laboratories, clinics, and research facilities seem to have returned to their normal functioning. Still, those research approaches that became common during the pandemic continue to be used.

25.5.1 Number of Articles Published

The infodemic is characterized by a surge in the amount of available information, and there are no doubts it happens in the publication output between sleep and COVID-19. It shall be mentioned that the overall publication record of sleep medicine has been increasing consistently since a long time, but this growth increased considerably more during the pandemic (Fig. 25.5). The percentual growth rate of the publication record from 1 year to the following was about 5% in the years before the pandemic. But it increased to an average growth of 20% during the first two pandemic years. Importantly, before the pandemic, the articles were distributed among a plethora of sleep-related themes, while most of those accounting for this huge increase in the publication record during the pandemic focused on a single one. This reinforces the infodemic nature of the publication profile of sleep and COVID-19.

As of today, a simple PubMed ((sleep[tiab] OR sleep[mh]) AND COVID-19) retrieves 3769 records (search updated on November 8, 2022). It results in an

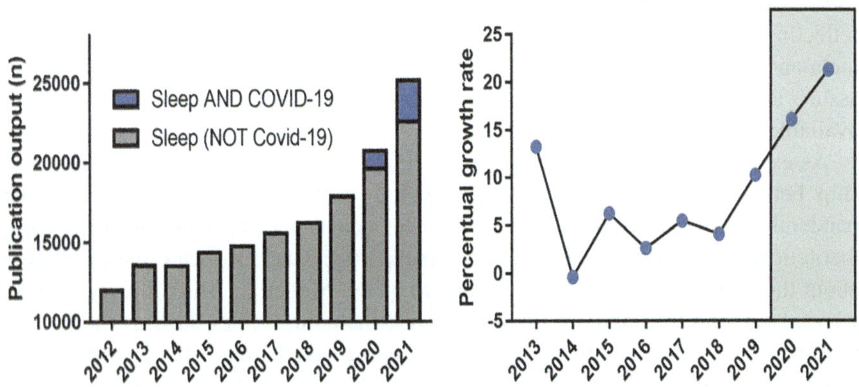

Fig. 25.5 Sleep-related publication record, highlighting the pandemic period. A: Total number of articles published. It can be seen that the publication record was increasing consistently in the last years. But from 2020 to 2021, an important growth could be attributed to the studies about sleep and COVID-19. B: Percentual growth rate of the publication output from 1 year to the other. A consistent growth was always observed, being around 5% per year before the pandemic. However, during the pandemic (gray shaded area), it was considerably higher, being around 20%. Only studies published up to 2021 were considered, in order to include only years for which publication output has been completed. Data acquired through Scopus

average of 4.1 published articles per day. It might be argued that not all of these studies are directly related to sleep, approaching it indirectly or figuratively, rather than properly focusing on sleep medicine and sleep research. Previous studies calculated that about 60% of the records retrieved from a simple sleep research are actually related to sleep medicine. Therefore, in a corrected and conservative estimation, it can be said that there are about 2261 articles about COVID-19 and sleep, resulting in an average of 2.4 articles published per day.

Sleep medicine corresponds to a small proportion of the total publication record related to COVID-19. A simple search strategy for COVID-19 retrieves 311,120 articles since 2019. Among these, 93,673 were published in 2020, 139,324 in 2021, and 109,994 in 2022. Therefore, sleep medicine corresponds to about 1.21% of the total COVID-19 publication output (or to 0.72%, considering the correction mentioned above).

These publication records were not constant during the pandemic. It seems that the first year (2020) resulted in a small number of articles available on PubMed (721 articles), but an important increase was observed in 2021 (1769 articles—145% increased from 1 year to the other). In 2022, it seems that the number of articles remained stable in comparison with the previous year (1651, as of November 8).

25.5.2 Research Quality

In the expense of a rapid publication output to serve to the need of immediate information during the pandemic, the quality of research seems to have been compromised. Since the beginning of the pandemic, important concerns regarding the quality of research were raised. There are numerous studies describing how low-quality studies became a problem during the pandemic, for reasons that range from honest errors to methodological flaws and misconduct (Ramanan et al. 2020; Nieto et al. 2020; Bramstedt 2020; Dobler 2020; Peyrin-Biroulet 2020; Jung et al. 2022). Some remarkable cases are the ones involving chloroquine, hydroxychloroquine, and ivermectin as potential treatments for COVID-19 (Reardon 2021; Alexander et al. 2020), which were evidently flawed, leading to misleading conclusion and generating chaos, especially when those results reached general media.

Among common quality errors were studies with small sample sizes, lack of sample size calculations, wrong biostatistical analysis, lack of control groups, non-randomized trials, excessive use of convenience samples, and uncertainties about the protocol of these studies (Nieto et al. 2020; Dobler 2020; Ordak 2022).

This low-quality profile was strongly associated with both the fast-track peer-review policies and the need to generate evidences when nothing was available (Ramanan et al. 2020; Peyrin-Biroulet 2020). Once both conditions are no longer needed, one might expect that the average quality of research will be back to normal eventually.

25.6 Conclusion

The COVID-19 pandemic brought the challenge of generating evidence about a new disease as fast as possible. Scientists were successful in this task, but it led to an infodemic as a consequence. We have never generated so much information in a short period of time. The whole process by which we generate medical evidence was changed in order to adapt to emergent need. Likewise, the way we publish and disseminate scientific results changed as well. That is true for most research areas, including sleep medicine.

Now, the society is slowly returning to its normal functioning, as social distancing norms are no longer in place and the number of new cases and deaths has been reduced. However, some acquired behaviors and practices from the pandemic will linger on. The way we deal with scientific information (either in general or specifically in sleep medicine) follows the same rule.

The annual growth rate of the total publication output might be lowered in the next years, as the number of articles about COVID-19 and sleep might stall or even reduce. Also, the urgency for making research results available is no longer a reality. Considering that this was one of the reasons for the average low quality of research during the pandemic, this should be better in the forthcoming years. Finally, many of the research methods that became popular during the pandemic will probably remain being used, including online surveys, meta-research, and real-world evidence.

In conclusion, the pandemic was a good opportunity to understand how scientific publication works. It included some positive aspects (such as the immediate reaction and implementation of evidence-based practices) but some negative ones (including low-quality studies and scientific misconduct). We hope that we as researchers can learn from that, taking the good sides of it for the future, assuring that the development of the publication output in sleep medicine grows in a responsible and reliable way.

Acknowledgments GNP is a shareholder at SleepUp© and founder of P&P Metanálises. ST has no conflicts of interest to disclose. This work was supported by grants from the Associação Fundo de Incentivo à Pesquisa (AFIP), the Coordenação de Aperfeiçoamento de Pessoal de Nível Superior (CAPES), and the Conselho Nacional de Desenvolvimento Científico e Tecnológico (CNPq).

References

Abbott R, Bethel A, Rogers M, Whear R, Orr N, Shaw L et al (2022) Characteristics, quality and volume of the first 5 months of the COVID-19 evidence synthesis infodemic: a meta-research study. BMJ Evid Based Med 27(3):169–177

Abritis A, Marcus A, Oransky I (2021) An "alarming" and "exceptionally high" rate of COVID-19 retractions? Account Res 28(1):58–59

Alexander PE, Debono VB, Mammen MJ, Iorio A, Aryal K, Deng D et al (2020) COVID-19 coronavirus research has overall low methodological quality thus far: case in point for chloroquine/hydroxychloroquine. J Clin Epidemiol 123:120

AlRasheed MM, Fekih-Romdhane F, Jahrami H, Pires GN, Saif Z, Alenezi AF et al (2022) The prevalence and severity of insomnia symptoms during COVID-19: a global systematic review and individual participant data meta-analysis. Sleep Med 100:7–23
Altena E, Baglioni C, Espie CA, Ellis J, Gavriloff D, Holzinger B et al (2020) Dealing with sleep problems during home confinement due to the COVID-19 outbreak: practical recommendations from a task force of the European CBT-I academy. J Sleep Res 29:e13052
Anderson C, Nugent K, Peterson C (2021) Academic journal retractions and the COVID-19 pandemic. J Prim Care Community Health 12:21501327211015592
Arrizabalaga O, Otaegui D, Vergara I, Arrizabalaga J, Méndez E (2020) Open access of COVID-19-related publications in the first quarter of 2020: a preliminary study based in PubMed. F1000Res 9:649
Bagdasarian N, Cross GB, Fisher D (2020) Rapid publications risk the integrity of science in the era of COVID-19. BMC Med 18(1):192
Benedict C, Cedernaes J (2021) Could a good night's sleep improve COVID-19 vaccine efficacy? Lancet Respir Med 9(5):447–448
Benhamou K, Piedra A (2020) CBT-informed interventions for essential workers during the COVID-19 pandemic. J Contemp Psychother 50(4):275–283
Besançon L, Peiffer-Smadja N, Segalas C, Jiang H, Masuzzo P, Smout C et al (2021) Open science saves lives: lessons from the COVID-19 pandemic. BMC Med Res Methodol 21(1):117
Bramstedt KA (2020) The carnage of substandard research during the COVID-19 pandemic: a call for quality. J Med Ethics 46(12):803–807
Büchner N, Woehrle H, Dellweg D, Wiater A, Young P, Hein H et al (2020) Management of diagnostic procedures and treatment of sleep related breathing disorders in the context of the coronavirus pandemic. Somnologie (Berl) 24(4):274–284
Cardinali DP, Brown GM, Pandi-Perumal SR (2020) Can melatonin be a potential "silver bullet" in treating COVID-19 patients? Diseases 8(4):44
De Mello MT, Silva A, Guerreiro RC, da-Silva FR, Esteves AM, Poyares D et al (2020) Sleep and COVID-19: considerations about immunity, pathophysiology and treatment. Sleep Sci 13(3): 199–209
Dobler CC (2020) Poor quality research and clinical practice during COVID-19. Breathe (Sheff) 16(2):200112
Ferini-Strambi L, Zucconi M, Casoni F, Salsone M (2020) COVID-19 and sleep in medical staff: reflections, clinical evidences, and perspectives. Curr Treat Options Neurol 22(10):29
Filippatos F, Tatsi EB, Efthymiou V, Syriopoulou V, Michos A (2022) Time of day of BNT162b2 COVID-19 immunization affects Total SARS-CoV-2 antibody levels but not neutralizing activity. J Biol Rhythm 37(5):562–566
Fraser N, Brierley L, Dey G, Polka JK, Pálfy M, Nanni F et al (2021) The evolving role of preprints in the dissemination of COVID-19 research and their impact on the science communication landscape. PLoS Biol 19(4):e3000959
Gale RP (2020) Conquest of COVID-19. Publish it to death? Br J Haematol 190:358
Gianola S, Jesus TS, Bargeri S, Castellini G (2020) Characteristics of academic publications, preprints, and registered clinical trials on the COVID-19 pandemic. PLoS One 15(10):e0240123
Gisondi MA, Barber R, Faust JS, Raja A, Strehlow MC, Westafer LM et al (2022) A deadly infodemic: social media and the power of COVID-19 misinformation. J Med Internet Res 24(2): e35552
Gordon M, Bishop M, Chen Y, Dreber A, Goldfedder B, Holzmeister F et al (2022) Forecasting the publication and citation outcomes of COVID-19 preprints. R Soc Open Sci 9(9):220440
Gulia KK, Kumar VM (2020) Importance of sleep for health and wellbeing amidst COVID-19 pandemic. Sleep Vigil 4(1):49–50
Gupta R, Kumar VM, Tripathi M, Datta K, Narayana M, Ranjan Sarmah K et al (2020) Guidelines of the Indian Society for Sleep Research (ISSR) for practice of sleep medicine during COVID-19. Sleep Vigil 4(2):61–72

Hardeland R, Tan DX (2020) Protection by melatonin in respiratory diseases: valuable information for the treatment of COVID-19. Melatonin Res 3(3):12

Hollander JE, Carr BG (2020) Virtually perfect? Telemedicine for Covid-19. N Engl J Med 382(18):1679–1681

Insalaco G, Dal Farra F, Braghiroli A, Salvaggio A, Sacco, On behalf of the Italian Thoracic Society (ITS-AIPO) (2020) Sleep breathing disorders in the COVID-19 era: Italian Thoracic Society organizational models for a correct approach to diagnosis and treatment. Respiration 99(8): 690–694

Ioannidis JP (2016) The mass production of redundant, misleading, and conflicted systematic reviews and meta-analyses. Milbank Q 94(3):485–514

Jolliffe DA, Faustini SE, Holt H, Perdek N, Maltby S, Talaei M et al (2022) Determinants of antibody responses to SARS-CoV-2 vaccines: population-based longitudinal study (COVIDENCE UK). Vaccines (Basel) 10(10):1601

Jung RG, Di Santo P, Clifford C, Prosperi-Porta G, Skanes S, Hung A et al (2021) Methodological quality of COVID-19 clinical research. Nat Commun 12(1):943

Jung H, Kim D, Lee W, Seo H, Seo J, Choi J et al (2022) Performance evaluation of a wrist-worn reflectance pulse oximeter during sleep. Sleep Health 8(5):420–428

Kambhampati SBS, Vaishya R, Vaish A (2020) Unprecedented surge in publications related to COVID-19 in the first three months of pandemic: a bibliometric analytic report. J Clin Orthop Trauma 11:S304

Khatter A, Naughton M, Dambha-Miller H, Redmond P (2021) Is rapid scientific publication also high quality? Bibliometric analysis of highly disseminated COVID-19 research papers. Learn Publ 34(4):568–577

King A (2020) Fast news or fake news?: the advantages and the pitfalls of rapid publication through pre-print servers during a pandemic. EMBO Rep 21(6):e50817

Kodvanj I, Homolak J, Virag D, Trkulja V (2022) Publishing of COVID-19 preprints in peer-reviewed journals, preprinting trends, public discussion and quality issues. Scientometrics 127(3):1339–1352

Kow CS, Hasan SS (2021) Do sleep quality and sleep duration before or after COVID-19 vaccination affect antibody response? Chronobiol Int 38(7):941–943

Kupferschmidt K (2020) Preprints bring 'firehose' of outbreak data. Science 367(6481):963–964

Lacsa JEM (2022) COVID-19 infodemic: the role of social media and other digital platforms. J Public Health (Oxf) 44(2):e341

Lee CK, Merriam LT, Pearson JC, Lipnick MS, McKleroy W, Kim EY (2022) Treating COVID-19: evolving approaches to evidence in a pandemic. Cell Rep Med 3(3):100533

Liu Y, Jiang TT, Shi TY, Liu YN, Liu XM, Xu GJ et al (2020) Erratum to "Withdrawal notice to the effectiveness of diaphragmatic breathing relaxation training for improving sleep quality among nursing staff during the COVID-19 outbreak: a before and after study" [Sleep Medicine X (2) (2020) 100026]. Sleep Med X 2:100030

Lou J, Tian SJ, Niu SM, Kang XQ, Lian HX, Zhang LX et al (2020) Coronavirus disease 2019: a bibliometric analysis and review. Eur Rev Med Pharmacol Sci 24(6):3411–3421

Majumder MS, Mandl KD (2020) Early in the epidemic: impact of preprints on global discourse about COVID-19 transmissibility. Lancet Glob Health 8(5):e627–ee30

Matryba P, Gawalski K, Ciesielska I, Horvath A, Bartoszewicz Z, Sienko J et al (2022) The influence of time of day of vaccination with BNT162b2 on the adverse drug reactions and efficacy of humoral response against SARS-CoV-2 in an observational study of young adults. Vaccines (Basel) 10(3):443

Nelson L, Ye H, Schwenn A, Lee S, Arabi S, Hutchins BI (2022) Robustness of evidence reported in preprints during peer review. Lancet Glob Health 10(11):e1684–e16e7

Nieto I, Navas JF, Vázquez C (2020) The quality of research on mental health related to the COVID-19 pandemic: a note of caution after a systematic review. Brain Behav Immun Health 7: 100123

Ordak M (2022) COVID-19 research: quality of biostatistics. Arch Med Sci 18(1):257–259

Öztürk G, Akbulut KG, Güney Ş (2020) Melatonin, aging, and COVID-19: could melatonin be beneficial for COVID-19 treatment in the elderly? Turk J Med Sci 50(6):1504–1512
Penzel T, Dietz-Terjung S, Woehrle H, Schöbel C (2021) New paths in respiratory sleep medicine: consumer devices, e-health, and digital health measurements. Sleep Med Clin 16(4):619–634
Peterson CJ, Alexander R, Nugent K (2022) COVID-19 article retractions in journals indexed in PubMed. Am J Med Sci 364(1):127–128
Peyrin-Biroulet L (2020) Will the quality of research remain the same during the COVID-19 pandemic? Clin Gastroenterol Hepatol 18:2142
Piller C, Travis J (2020) Authors, elite journals under fire after major retractions. Science 368(6496):1167–1168
Pires GN, Niyama A, Andersen ML, Tufik S (2021a) Publication of meta-analyses in sleep medicine: a scoping review. J Clin Sleep Med 17(4):811–817
Pires GN, Bezerra AG, Oliveira TB, Chen SFI, Malfatti VDA, Mello VFF et al (2021b) COVID-19 meta-analyses: a scoping review and quality assessment. Einstein (Sao Paulo) 19:eAO6002
Pirzada AR, Aleissi SA, Almeneessier AS, BaHammam AS (2020) Management of aerosol during noninvasive ventilation for patients with sleep-disordered breathing: important messages during the COVID-19 pandemic. Sleep Vigil 4(2):89–94
Ramanan M, Stolz A, Rooplalsingh R, Billot L, Myburgh J, Venkatesh B (2020) An evaluation of the quality and impact of the global research response to the COVID-19 pandemic. Med J Aust 213(8):380–380.e1
Reardon S (2021) Flawed ivermectin preprint highlights challenges of COVID drug studies. Nature 596(7871):173–174
Salles C, Freitas MCS, Meira E, Cruz M (2021) Comorbid insomnia and sleep apnea (COMISA) as an additional risk factor for reduced response to the COVID-19 vaccination? Sleep Med 83:159
Shah SS, Kulkarni N, Mahant S (2020) Rapid publication, knowledge sharing, and our responsibility during the COVID-19 pandemic. J Hosp Med 15(5):261
Shamim-Uzzaman QA, Bae CJ, Ehsan Z, Setty AR, Devine M, Dhankikar S et al (2021) The use of telemedicine for the diagnosis and treatment of sleep disorders: an American Academy of Sleep Medicine update. J Clin Sleep Med 17(5):1103–1107
Sheldon T (2018) Preprints could promote confusion and distortion. Nature 559(7715):445
Shneider A, Kudriavtsev A, Vakhrusheva A (2020) Can melatonin reduce the severity of COVID-19 pandemic? Int Rev Immunol 39(4):153–162
Singh M, Sharda S, Gautam M, Hawa R (2020) Optimal sleep health among frontline healthcare workers during the COVID-19 pandemic. Can J Anaesth 67(10):1471–1474
Škorić L, Glasnović A, Petrak J (2020) A publishing pandemic during the COVID-19 pandemic: how challenging can it become? Croat Med J 61(2):79–81
Smith MJ, Upshur REG, Emanuel EJ (2020) Publication ethics during public health emergencies such as the COVID-19 pandemic. Am J Public Health 110:e1–e2
Song P, Karako T (2020) Scientific solidarity in the face of the COVID-19 pandemic: researchers, publishers, and medical associations. Glob Health Med 2(2):56–59
Strausz S, Kiiskinen T, Broberg M, Ruotsalainen S, Koskela J, Bachour A et al (2021) Sleep apnoea is a risk factor for severe COVID-19. BMJ Open Respir Res 8(1):e000845
Suart C, Neuman K, Truant R (2022) The impact of the COVID-19 pandemic on perceived publication pressure among academic researchers in Canada. PLoS One 17(6):e0269743
Treweek S, Jüni P, Li T, Collin J, Briel M, Chan AW et al (2020) COVID-19 randomised trial protocols: rapid publication without barriers. Trials 21(1):327
Tufik S (2020) Obstructive sleep apnea as a comorbidity to COVID-19. Sleep Sci 13(3):181
Tufik S, Gozal D, Ishikura IA, Pires GN, Andersen ML (2020) Does obstructive sleep apnea lead to increased risk of COVID-19 infection and severity? J Clin Sleep Med 16(8):1425–1426
Tufik S, Andersen ML, Rosa DS, Tufik SB, Pires GN (2022) Effects of obstructive sleep apnea on SARS-CoV-2 antibody response after vaccination against COVID-19 in older adults. Nat Sci Sleep 14:1203–1211

van Schalkwyk MCI, Hird TR, Maani N, Petticrew M, Gilmore AB (2020) The perils of preprints. BMJ 370:m3111

Wang W, Balfe P, Eyre DW, Lumley SF, O'Donnell D, Warren F et al (2022) Time of day of vaccination affects SARS-CoV-2 antibody responses in an observational study of health care workers. J Biol Rhythm 37(1):124–129

Watson C (2022) Rise of the preprint: how rapid data sharing during COVID-19 has changed science forever. Nat Med 28(1):2–5

Xu C, Furuya-Kanamori L, Kwong JSW, Li S, Liu Y, Doi SA (2021) Methodological issues of systematic reviews and meta-analyses in the field of sleep medicine: a meta-epidemiological study. Sleep Med Rev 57:101434

Yamanaka Y, Yokota I, Yasumoto A, Morishita E, Horiuchi H (2022) Time of day of vaccination does not associate with SARS-CoV-2 antibody titer following first dose of mRNA COVID-19 vaccine. J Biol Rhythm 37(6):700–706

Zhang H, Shao F, Gu J, Li L, Wang Y (2020) Ethics committee reviews of applications for research studies at 1 Hospital in China during the 2019 novel coronavirus epidemic. JAMA 323(18): 1844–1846

Zhang H, Liu Y, Liu D, Zeng Q, Li L, Zhou Q et al (2021) Time of day influences immune response to an inactivated vaccine against SARS-CoV-2. Cell Res 31(11):1215–1217

Zhu J, Zhang M, Sanford LD, Tang X (2021) Advice for COVID-19 vaccination: get some sleep. Sleep Breath 25:2287

The manufacturer's authorised representative in the EU is Springer Nature Customer Service Centre GmbH, Europaplatz 3, 69115 Heidelberg, Germany. If you have any concerns regarding our products, please contact ProductSafety@springernature.com

Printed and bound by CPI Group (UK) Ltd, Croydon, CR0 4YY
30/09/2026
02225303-0015